The Economics of
Climate Change in China

The Economics of Climate Change in China

Towards a Low-Carbon Economy

Edited by
Fan Gang, Nicholas Stern, Ottmar Edenhofer, Xu Shanda,
Klas Eklund, Frank Ackerman, Li Lailai and Karl Hallding

publishing for a sustainable future

London • Washington, DC

First published in 2011 by Earthscan

Copyright © Stockholm Environment Institute 2011

Managing editor: Tom Gill

Earthscan Ltd, Dunstan House, 14a St Cross Street, London EC1N 8XA, UK
Earthscan LLC, 1616 P Street, NW, Washington, DC 20036, USA

Earthscan publishes in association with the International Institute for Environment and Development

For more information on Earthscan publications, see www.earthscan.co.uk
or write to earthinfo@earthscan.co.uk

ISBN   978-1-84971-174-6 hardback

Typeset by MapSet Ltd, Gateshead, UK
Cover design by Andrew Corbett

A catalogue record for this book is available from the British Library

Library of Congress Cataloging-in-Publication Data
The economics of climate change in China : towards a low carbon economy / edited by Fan Gang ... [et al.].
     p. cm.
 Includes bibliographical references and index.
 ISBN 978-1-84971-174-6 (hardback)
 1. Environmental policy—Economic aspects—China. 2. Climatic changes—Economic aspects—China. 3. Carbon dioxide mitigation—Government policy—China. 4. Greenhouse gas mitigation—Government policy—China. I. Gang, Fan.
 HC430.E5E26 2010
 363.738'7460951—dc22
                                                                          2010036475

At Earthscan we strive to minimize our environmental impacts and carbon footprint through reducing waste, recycling and offsetting our CO_2 emissions, including those created through publication of this book. For more details of our environmental policy, see www.earthscan.co.uk.

Printed and bound in the UK by
CPI Antony Rowe.
The paper used is FSC certified.

Contents

PART I – THE ECONOMICS OF CLIMATE CHANGE IN CHINA: AN OVERVIEW OF THE POSSIBLE   1

Fan Gang, Nicholas Stern, Ottmar Edenhofer, Xu Shanda,
Klas Eklund, Frank Ackerman, Li Lailai and Karl Hallding

PART II – TOWARDS CLIMATE PROTECTION FOR DEVELOPMENT

PART III – GROWTH, OPPORTUNITY AND SUSTAINABILITY

PART IV – CLIMATE CHANGE MITIGATION: A FAIR, EFFECTIVE AND EFFICIENT GLOBAL DEAL

List of Figures,
Tables and Boxes

Figures

Tables

Boxes

List of Abbreviations

ACE	accumulated consumption emissions
AF	Adaptation Fund
AWG-LCA	Ad-hoc Working Group for Long-Term Cooperative Action
BAPS	Beyond Alternative Policy Scenario
BAT	best available technology
BAU	business as usual
BLS	baseline scenario
BOF	basic oxygen furnace
C&C	contraction and convergence
CAD	climate accession deal
CAIT	Climate Analysis Indicators Tool
CBA	cost–benefit analysis
CCAP	Center for Clean Air Policy
CCS	carbon capture and storage
CDM	Clean Development Mechanism
CE50	Chinese Economists 50 forum
CER	certified emission reduction
CGE	Computational General Equilibrium
CHUEE	China Utility-Based Energy Efficiency Finance Program
CIF	Climate Investment Fund
CMP	Parties to the Kyoto Protocol
CO_2	carbon dioxide
COD	chemical oxygen demand
COP	Conference of the Parties to the United Nations Framework Convention on Climate Change
CTF	Clean Technology Fund
DCRS	deep carbon reduction scenario
DNA	designated national authority
DOE	designated operational entity
EAF	electric arc furnace
EEA	European Environment Agency
EE/RE	energy efficiency and renewable energy
EIA	Energy Information Administration

EIB	European Investment Bank
EIT	economies in transition
EJ	exajoule
EKC	Environmental Kuznets Curve
eq	equivalent
ETP	Energy Technology Perspectives
ETS	emissions trading system
EUA	European Union Allowances
GCFM	Global Climate Financing Mechanism
GDP	gross domestic product
GDR	greenhouse development rights
GEF	Global Environment Facility
GHG	greenhouse gas
GJ	gigajoule
GRP	gross regional product
Gt	gigatonne
GVA	gross value added
GW	gigawatt
GWP	global warming potential
HFC	hydrofluorocarbon
HDI	Human Development Index
HVAC	heating, ventilation and air conditioning
IATAL	international air travel adaptation levy
ICAP	International Carbon Action Partnership
ICP	inter-country joint mitigation plan
IEA	International Energy Agency
IFC	International Finance Corporation
IMF	International Monetary Fund
IPCC	Intergovernmental Panel on Climate Change
IPO	initial public offering
IPR	intellectual property right
IRR	internal rates of return
K	degrees Kelvin
kg	kilogram
km	kilometre
kWh	kilowatt hour
LBNL	Lawrence Berkeley National Laboratory
LCA	long-term cooperative action
LDCF	Least Developed Country Fund
LEAP	Long-range Energy Alternatives Planning System
LULUCF	Land Use, Land Use Change and Forestry
MAC	marginal abatement costs
MDB	multilateral development bank
M-GDR	modified greenhouse development rights
MDG	Millennium Development Goal

MIGA	Multilateral Investment Guarantee Agency
MJ	megajoule
MRIO	Multi-Region Input–Output
MRV	monitoring, reporting and verification
MSW	municipal solid waste
Mt	million tonnes
MTAF	Multilateral Technology Acquisition Fund
NA1	non-Annex 1
NAMA	Nationally Appropriate Mitigation Action
NAP	National Allocation Plan
NDRC	National Development and Reform Commission
NGCC	natural gas combined-cycle
NGO	non-governmental organization
NPV	net present value
ODA	official development assistance
OECD	Organisation for Economic Co-operation and Development
OPEC	Organization of Petroleum Exporting Countries
PDD	project design document
PIK	Potsdam Institute for Climate Impact Research
PIN	Project Idea Note
p-km	passenger-kilometres
PoA	programme of activities
ppm	parts per million
PPP	purchasing power parity
PV	photovoltaics
R&D	research and development
RCA	revealed comparative advantage
R-CER	retirement certified emission reduction
RCI	Responsibility and Capacity Indicator
REDD	Reducing Emissions from Deforestation and Forest Degradation in Developing Countries
R-GDR	revised greenhouse development rights
RGGI	Regional Greenhouse Gas Initiative
SCCF	Special Climate Change Fund
SCE	standard coal equivalent
SCF	Strategic Climate Fund
SD-PAMs	sustainable development policies and measures
SDR	Special Drawing Right
SEFI	Sustainable Energy Finance Initiative
SEI	Stockholm Environment Institute
SIDS	Small Island Developing State
SNLT	sectoral no-lose target
SOE	state-owned enterprise
SOP	share of proceeds
t	tonne

tce	tonnes of coal equivalent
TFP	total factor productivity
TFR	total fertility rate
TJ	terajoule
t-km	tonne-kilometre
TVE	township and village enterprise
TWh	terawatt hour
UAE	United Arab Emirates
UNCTAD	United Nations Conference on Trade and Development
UNEP	United Nations Environment Programme
UNFCCC	United Nations Framework Convention on Climate Change
UNIDO	United Nations Industrial Development Organization
WCCF	World Climate Change Fund
WCI	Western Climate Initiative
WDI	World Development Indicators
WEO	*World Energy Outlook*
WTO	World Trade Organization

Foreword

The international community is increasingly recognizing the risks of global warming caused by human activities. The estimated potential costs of non-action on climate change, based on available scientific research, are so high that any action to reduce carbon emission may be regarded as non-regrettable. Therefore, climate change can be seen as the defining issue of our age, as it may, in the words of Ban Ki-moon, 'rewrite the global equation for development, peace and security'. Our generation is at a crossroads, facing the greatest challenge of the modern industrial era – to drastically reduce our reliance on fossil fuels while at the same time moving forward economic growth. China's participation, along with other rapidly emerging countries, and developed countries, is integral to any climate change solution.

To achieve this transformation to a low-carbon global economy, we need both better international cooperative mechanisms, and new growth approaches and policy strategies for individual countries. China has already recognized the vulnerabilities of its ongoing pattern of economic growth and the need for low-carbon development. Even in these difficult economic times, climate change action may present more opportunities than costs, for both developed and developing nations. China, like much of the world, would benefit from early mitigation efforts.

This book is unique in delivering a common research partnership between the Chinese and international economic and climate change community. It is the result of a two-year collaborative research project by Chinese, Swedish, German, British and American experts to address some of the most challenging economic issues underpinning China's low-carbon transformation. This book is based on a project established by the Stockholm Environment Institute (SEI) and the Chinese Economists 50 forum (CE50). It presents academic analyses and policy recommendations for the international community to improve cooperation on climate change and for the Chinese authorities to adapt economic policies and market instruments needed for the country to make an immediate, rapid and effective transition to a low-carbon economy.

This book demonstrates that it is feasible for China to drastically reduce its carbon emissions in key carbon-intensive industries, while still maintaining economic growth and its development aspirations. Moreover, these reductions can be achieved within the finite global carbon budget for greenhouse gas emissions, as determined by the hard constraints of climate science. *The Economics of Climate Change in China* details the key economic instruments, policies and institutions which would enable China to cost-effectively reduce its emissions. Market mechanisms such as price liberalization, carbon pricing and a global carbon market, and inter-country joint mitigation plans need to be supported by substantial investment in clean technologies, and international finance and technology transfer. Such a transformation, for China and the rest of the world, will not be easy. But it is possible, necessary and worthwhile to pursue.

Johan Rockström
Executive Director
Stockholm Environment Institute

Fan Gang
Director
National Economic Research Institute

Preface

Events in the global economy over the past two years have proved very challenging for the world. After the worst recession since the Second World War, the world is moving towards recovery, though growth in the developed world is fragile compared with emerging economies. There are tensions, too, over balance of payments and currency issues, in particular between China and the US. The policy challenges to maintaining fragile confidence and kick-starting the global economy are severe for both the financial sector and the real economy, requiring international coordination and cooperation, and a strong stand against protectionism.

At the same time, the challenge of global climate change remains real and urgent, and cannot be postponed while we 'fix' the economy. The notion of being able to postpone action on climate change for a year or two for the sake of the greater global economic challenge is false for five reasons. First, because waiting a few more years increases the risk of failure to hold concentrations below or in the range of 450ppm–500ppm CO_2e and makes mitigation more expensive in the future. If current emissions continue, the world will be at or above 450ppm CO_2e within five or ten years, which entails a greater than 50 per cent chance of an eventual temperature rise of above 2°C. The science tells us this is already a dangerous place for the world to be. Temperature rise on this scale could lead to water stress for billions of people, mass loss of species and flooding of some coastal, populous cities. If we fail to act we may well be on a pathway to 550ppm and above before too long, levels which most would deem unacceptable from the point of view of managing risk.

Second, economic policies decided upon and implemented today to support the global economy will affect the sustainability of growth tomorrow. Failure to acknowledge and learn from past mistakes will mean that we both miss important opportunities and lock in high-carbon technologies, thus sowing the seeds for a future crisis. Unless the global economy is placed on a pathway to a low-carbon future, with clear and mutually agreed stabilization goals, the potential impacts of climate change will ultimately be far more damaging to growth than the current economic crisis. Taking strong action now, on the other hand, will create exciting new opportunities for growth and jobs in future.

Third, the technology already exists today to begin strong and decisive action to tackle climate change. According to recent analysis by McKinsey's, we already have most of the technologies that we need in order to get started on a 450ppm path. The 75 per cent of abatement required for 2030 can be generated from technologies that are available today. But only by early implementation can we get the scale and experience to bring down costs in good time. Further, innovation remains crucial particularly for new technologies, such as carbon capture and storage (CCS), that require further development in order to test their potential and become marketable. Early and strong action will help to incentivize innovation and lower the costs of new technology. Any delay in the development of new technology raises the mitigation costs.

Fourth, the UN Conference of the Parties in Cancun in late 2010 will be crucial to agreeing the post-2012 frameworks and institutional arrangements. It is important to build positive international momentum towards these negotiations, which means making plans and agreeing the principles of a global deal as soon as possible. Delay in reaching an agreement will disrupt the expectations of investors who will have to make the crucial investments and commitments.

Fifth, this time of less pressure on energy resources is a time to act at lower cost. Improving insulation, for example, and bringing forward green projects will bring down the costs of acting in future.

As a global economic power, the domestic path that China's economy takes at this point continues to have strong implications for the rest of the world. As growth in advanced economies stalls, the world will increasingly look to China over the coming years as a key driver of global growth and demand. It is for this reason that the November 2008 fiscal stimulus package was very timely, demonstrating China would move quickly to sustain domestic demand. The RMB4 trillion package has had a positive impact on the global economy over the past two years, and supported domestic demand and employment.

However, the deeper economic challenge of restructuring and rebalancing growth to achieve a harmonious society remains vital. The goals of the 11th Five-Year Plan, to restructure industrial structures, promote regional balanced growth, deepen institutional reform and protect the environment, are ever integral to future growth and prosperity. The progress already achieved across all these goals is remarkable. However, they become ever more complex in the current global economic environment.

Looking ahead, there are four fundamental questions for consideration as part of an effective domestic economic response to climate change in China:

1   Is the aggregate and microeconomic understanding of the economic impacts and potential damages to China from climate change sufficient to enable effective planning for adaptation and mitigation responses?
2   What are the key policies to support domestic growth in China, which are both an effective response to the slowing global economy while at the same time placing China on the pathway to achieve a low-carbon economy?

3   How can the economic costs of mitigation during a transition to new economies be contained? How can the impact on employment be kept as low as possible, while creating new opportunities?
4   How can China best foster the new medium-term low-carbon investments that will be the economic drivers of growth over the next few decades? Indeed it must be recognized that the low-carbon path is the only possible medium and long-term growth path. High-carbon growth will eventually choke itself off through high hydro-carbon prices and, more fundamentally, a deeply hostile physical environment.

We focus here on the second, third and fourth issues, given that these are the focus of the SEI/CE50 study. However, it should be remembered that a complete story of the 'economics of climate change' for China needs to start with an understanding of the science as well as the impacts and likely damages of climate change at a range of possible stabilization and temperature scenarios. This is for two reasons. First, because adaptation and mitigation policies need to be grounded in a rigorous understanding of the science and potential impacts in order to be effective and efficient. And second, because any discussion of the costs of action is more compelling when put in the context of the costs of inaction, both to the world and to individual countries. This was the approach taken in the Stern Review (2006). Also the shorter term issue of questions 2 and 3 must be examined in the longer-term context of question 4.

Key issues in the pathway to a low-carbon, high-growth economy

The scale of emissions reductions required globally makes it necessary for developing countries, particularly fast-growing emerging economies, to play an active role in cutting emissions. The scale of the challenge is apparent in this simple arithmetic example. Assuming a global population of 9 billion by 2050, and a halving of global greenhouse gas (GHG) emissions by 2050 relative to 1990 from 40Gt CO_2e to 20Gt CO_2e, global per capita emissions by 2050 would need to fall to 2t per capita. This is not to advocate that all countries need to be at 2t per capita by 2050, but rather to illustrate what the average tells us needs to happen in order to hold concentrations below 500ppm. China is currently around 5t per capita. Given the inevitable pace of China's economic growth between now and 2050 (a possible factor increase of 10 to 16), the necessary reductions in emissions per unit of output will be multiplicative. This means decarbonizing the main sectors of the economy including power and surface transport.

China has already made significant strides towards achieving a low-carbon economy and is an example to the world in many respects. Very ambitious objectives in relation to energy intensity, renewable energy and forestation are already in place, and China is now a world leader in renewable energy production, renewable installed capacity and production of low-carbon technologies.

Moreover, the economic reform priorities of the Chinese government, drawn from the objectives of achieving a more harmonious society, are wholly consistent with the ambition of moving towards a low-carbon economy. Rebalancing the economy away from exports and investment in heavy industry towards a services-based economy and high-value added exports will lead to a less GHG-intensive growth path for the future.

The domestic policy framework to support these objectives – particularly fiscal, monetary and financial sector reforms – is a crucial consideration in support of these goals. Policies to increase the efficient use of energy are a priority in the context of reducing GHG intensity and future energy supply. In China, this needs to include a reduction of subsidies on fossil-fuel-based energy and a price being placed on carbon. Placing a price on carbon is at the heart of any strategy to cut emissions, but is inevitably a socio-political challenge. Nonetheless, it is made easier in the current global environment of falling energy prices. A price on carbon could be phased and take the form of a tax on coal, a tax on oil and gas, or carbon quotas that are traded (with the proportion of quotas which are auctioned rising quickly). Through applying a price, these measures would promote energy efficiency, while taking into account the damage to the environment caused by hydrocarbons (internalizing the market failure). Trading internally would also promote linkages with other quota trading systems around the world, ultimately resulting in increased financial flows to China. In some sectors price alone may not produce results sufficiently rapidly, and standards, regulations and explicit energy efficiency targets are likely to be required. This could include building and surface transport.

Policies on the expenditure side of the budget are particularly important and relevant in the context of China's fiscal stimulus. Expenditure programmes can focus on possible drivers of future growth, while supporting development of human capital, regionally balanced growth and net employment gains. For example, the drive towards energy efficiency could include a major programme for more efficient energy use in (and insulation of) buildings, use of technology for temperature control, new forms of renewable energy, combined heat and power generation, and modernization of coal-fired electricity to improve efficiency and to prepare for carbon capture and storage. Many forms of insulation are labour-intensive and can be implemented quickly, supporting the need to create and sustain high levels of employment as export industries shed labour. A large infrastructure programme, including mass transit (including high-speed rail to control the expansion of air traffic), could include closer linking of rural areas to urban markets, and facilitate access to power, water and sanitation for rural and peri-urban areas. This could be coupled with the development of mass transit urban infrastructure to reduce congestion and population in urban centres. A programme to decentralize the power supply, to include local wind, solar and small-scale hydroelectric power, would help to create new opportunities in low-carbon technologies. Re-tooling car and steel industries would get ahead of the game and allow continued access to world markets.

Beyond the direct focus on energy use, there are complementary measures in the financial sector to support China's transition away from a model of GHG-intensive industry and exports to an economy based on high-value added exports and services. Medium-term growth prospects depend on young and small industry becoming established in new growth areas of the economy. Continued financial sector reforms are important to support the efficiency of capital allocation across the economy, lower the savings rate and ensure availability of finance to smaller producers and consumers. This will also support the internal rebalancing process away from traditional heavy industry towards new growth areas. The expansion of secondary and university education is also important to provide sustained supply of skilled labour to sustain China's rise up the value chain and transition to a services-based economy.

Globally, future arrangements for financing mitigation and adaptation, and for the sharing and diffusing of low-carbon technology, will be of great importance to China. Both finance and technology will be key elements of a global deal on climate change, and it is important that the Chinese perspective on both these matters continues to be clearly articulated in the context of global negotiations. While significant strides have already been made domestically in energy-efficient and renewable technologies, much of this is for export rather than domestic absorption, and international cooperation remains important for the world to meet the scale of the challenge. For example, international cooperation on CCS use and retro-fitting is particularly important for China, given China's abundance in coal.

The challenge of containing costs

The Stern Review concluded that the cost of stabilization at around 500–550ppm would be around 1 per cent of global GDP in 2050. Delay and piecemeal action are among the biggest global risks to increasing these costs. If we wait even a few years to begin mitigation in earnest, we will incur greater costs for a given stabilization level. Uncoordinated action also holds potential to raise costs. If the world were to adopt a piecemeal approach to mitigation this would limit trading and potentially slow the pace of technology innovation. An international emissions trading scheme and a global price for carbon are the most efficient means of abatement, enabling mitigation to be done where it is cheapest, and thereby enabling large-scale financial flows to developing countries. Given changes in technology and rising prices of hydrocarbons, these costs may be on the high side. Further, they will provide short-term benefits in terms of pollution and energy security, as well as longer-term management of climate change.

At the national level, the concern that mitigation would 'cost too much' is widely heard, particularly in the current economic context. However, with good domestic policies, supported by a strong global framework, the costs can be contained and opportunities supported. Creating a price for carbon is central to this, as discussed above, enabling the damage caused by carbon to be priced in to the market and encouraging more efficient use of energy. However, a price signal alone is not enough, given that significant market failures exist in energy markets. Complementary measures such as energy efficiency standards and building regulation can complement market signals and support behavioural change. Urban planning can also make a difference, particularly to facilitate mass transport.

With any structural transformation, however, there will be winners and losers. Firms that are energy-intensive will incur relatively greater costs than those that are energy-efficient or use renewable energy, for a given price of carbon. However, as China increases its energy efficiency and moves away from heavy industry, consistent with its goals for harmonious and sustainable development, its economy should become less GHG-intensive. Placing a price on carbon will thus stimulate dynamic efficiency gains and encourage new and cleaner industry.

While the costs of strong action on climate change must be debated, we should not forget that there is opportunity and growth potential inherent in the transition to a low-carbon economy. Young and dynamic firms investing in future growth sectors and new technologies will create a sustainable job base for the future. Low-carbon investments in particular can be pro-growth and pro-job creation (particularly as new capital tends to be labour-intensive in the early stages). Energy savings can also create the space for financial savings, enabling consumers to afford other goods and services, boosting demand. These are all positive and dynamic gains that should be seen as possible outcomes from a strong policy framework to address climate change.

Concluding remarks

In many respects, the next few years are years of opportunity for China. First, China can go further in rebalancing the domestic economy and support the goals of harmonious development. The roots of sustainable recovery are in future growth drivers, including low-carbon, high-value-added industry and a rebalancing towards domestic consumption. Supporting this rebalancing through a well-targeted fiscal stimulus plan should help to revive growth in the near term. These goals are all wholly consistent with the objectives of the 11th Five-Year Plan.

Second, the coming years are an opportunity for China to give an example to the world in pathways to transform to a low-carbon economy. China is already a global leader on climate change in many respects. At this stage, the world needs a strong signal that China is ready to play its part at the appropriate time. The planning and articulation of the 12th Five-Year Plan, with low carbon growth as a central theme, will demonstrate to the world that China will play its part in the global challenge of climate change and achieving global stabilization targets, while putting in place policies to ensure it is reaping the gains of new opportunities and 'first mover advantage'.

Third, China can play a lead role in global discussions on the economics of climate change and reform of the international financial institutions. Global cooperation on both the economy and climate change is vital at this point. New and strengthened global institutions are undoubtedly required for the future. China should be central to world leaders' discussions on these issues, advancing its own view, as a major economic power, of the shape of future global institutions and processes.

Professor Lord Nicholas Stern
Chair of the Grantham Research Institute on Climate and the Environment

Melinda Bohannon
Research Fellow at Grantham Institute of Climate Change
and the Environment

A Note on Names

In this book we have tried to adopt a consistent approach to Chinese personal names of authors and others. We have followed the convention of listing the family name first, followed by the given name, and we have carried this through in listings (such as the references at end of chapters) where the convention for Western names is to list alphabetically by family name. Hence a Western name such as 'Frank Ackerman' would be listed under 'Ackerman, F.' in the references, while a Chinese name such as 'Fan Gang' would be listed under 'Fan, G.'.

Acknowledgements

This book, and the background papers which it builds on, were developed under the project Research and Forum on Economics of Climate Change: Towards a Low-Carbon Economy in China. The project was jointly led by the Stockholm Environment Institute (SEI) and the Chinese Economists 50 Forum (CE50).

The project originated from discussions between CE50 and SEI in the winter of 2006/2007 to establish an informal Chinese–international forum for dialogue and research between economists (including environmental specialists), as well as experts, scientists and policy-makers on climate change and China's development. Led by CE50 and SEI, an inception meeting to initiate this process was held in Stockholm in February 2008, which identified key research issues and formed a research agenda. The project carried out targeted research with scientific input from SEI, the Potsdam Institute for Climate Impact Research (PIK), the National Environmental Research Institute (NERI) and the London School of Economics Grantham Institute, culminating in 15 background papers. A mid-term meeting was organized in Beijing in December 2008 to review the drafts of the background papers and discuss cross-cutting issues, and a final Global Forum was convened in Beijing in September 2009 to present and discuss key findings.

The project has involved, at different stages, many leading Chinese and international economists working with climate change policies on a global level. The Chinese economists are: Liu He, Fan Gang, Wu Xiaoling, Xu Shanda, Lin Justin Yifu, Tang Min, Cai Fang, Hu Angang, Chen Dongqi and Xie Ping. The international economists are: Nicholas Stern, Ottmar Edenhofer, Frank Ackerman, Assar Lindbeck, Karl-Göran Mäler, Thomas Sterner, Kai Schlegelmilch, Laurence Tubiana, Tariq Banuri, Klas Eklund and Yuichi Moriguchi.

SEI and CE50 would like to express their appreciation to the various institutions that contributed to the project, and ultimately to this book: the National Economic Research Institute (NERI), China; the Potsdam Institute for Climate Impact (PIK), Germany; the Deutsche Gesellschaft für Technische Zusammenarbeit (GTZ), Germany; and the Grantham Research Institute on Climate Change and Environment, UK.

Naturally, this book and the project on which it is based would not have been possible without the generous support of funding agencies: the Swedish Ministry of Environment; the German Federal Ministry of Environment; Shell (China); the UK Department of Energy and Climate Change (DECC); and the Rockefeller Brothers Fund.

The project team is grateful to Göran Carstedt for moderating the project inception meeting, Måns Lönnroth for providing advice and guidance throughout the project process, Frédéric Cho for participation in the process and valuable input to the text, and Zhu Saini of NERI for assisting with project coordination. Many thanks also to Ying Li for translation services.

Work on synthesizing the research findings for the first part of this book was carried out by a team led by Karl Hallding. The team members were Klas Eklund (SEB); Frank Ackerman, Guoyi Han, Sivan Kartha, Marie Olsson and Helen Thai (SEI); Su Ming (Peking University); Cao Jing (Tsinghua University); and Gunnar Luderer (PIK).

Thanks are also due to the publisher Earthscan, in particular to Nick Bellorini and Hamish Ironside for their patience, hard work and support.

Particular credit is owed to the managing editor, Tom Gill, for his diligence and commitment in organizing the project material and coordinating the input of the many people involved in this book.

Part I

The Economics of Climate Change in China: An Overview of the Possible

Fan Gang, Nicholas Stern, Ottmar Edenhofer, Xu Shanda, Klas Eklund, Frank Ackerman, Li Lailai and Karl Hallding

Introduction

The science is unequivocal – climate change is progressing rapidly towards critical thresholds and shrinking the window for action. Climate change is the defining issue of our age, as it 'rewrites the global equation for development, peace and security' (UN News Centre, 2009). Our generation is at a crossroads, facing the greatest challenge of the modern, industrial era – to drastically reduce our reliance on fossil fuels while at the same time moving economic growth forward. China's participation, along with other rapidly emerging countries and developed countries, is integral to any climate change solution.

To achieve this transformation to a low-carbon global economy, new growth models and policy strategies will be required. China has already recognized the vulnerabilities of its own model of economic success and the need for low-carbon modernization. As Pan Yue, former Vice Minister of China's Ministry of Environmental Protection, memorably warned in 2005, 'the [economic] miracle will end soon because the environment can no longer keep pace' (Economy, 2007). Even in these difficult economic times, climate change action presents more opportunities than costs, for both developed and developing nations. China, like much of the world, would benefit from early mitigation efforts.

The research project on which this book is based is unique in delivering a common research partnership between the Chinese and international economic and climate change community. It is the result of a year-long collaborative effort by Chinese, Swedish, German, British and American experts to address some of the most challenging economic issues underpinning China's low-carbon transformation. The project, *Economics of Climate Change for China: Towards a Low Carbon Economy in China*, was established by the Stockholm Environment Institute (SEI) and Chinese Economists 50 forum (CE50) to bridge the policy and research gaps on the economic policies and market instruments needed for China to make an immediate, rapid and effective transition to a low-carbon economy.

The project demonstrates that it is feasible for China to drastically reduce its carbon emissions in key carbon-intensive industries, while still maintaining economic growth and development aspirations. Moreover, these reductions can be achieved within the finite global carbon budget for greenhouse gas (GHG) emissions, as determined by the hard constraints of climate science. This book details the key economic instruments, policies and institutions that would enable China to cost-effectively reduce its emissions. Market mechanisms such as price liberalization, carbon pricing (either through taxes or cap-and-trade regime) and a global carbon market need to be supported by substantial investment in clean technologies, and international finance and technology transfer. Such a transformation, for China and the rest of the world, will not easy. But it is possible and necessary.

The climate challenge

Climate change is advancing, and its effects and risks are increasingly clear. Doing nothing is not an option. Unless the global economy is placed on a pathway to a low-carbon future, the impacts of climate change will damage growth and living standards to an extent far beyond that of the current economic crisis.

In the preface to this book Nicholas Stern emphasizes the need for making climate change a priority. Postponing action against climate change, even for a few more years, increases the likelihood that we will be unable to hold GHG concentrations within the limits of acceptable risk, and makes mitigation more expensive in the future. This is largely due to the historical build-up of carbon-intensive energy infrastructure that, along with the long economic lifetime of investments, results in a lock-in to a global high-emission development trajectory. Even if we manage to limit the global temperature rise to 2°C, the socioeconomic impacts could be devastating: water stress for billions of people, mass loss of species and flooding of populous coastal cities.

It is within this context of overwhelming scientific evidence and the need for urgent action that this book makes its key point of departure. Combating climate change – and combating it now – is a shared global priority. The developed countries must lead, as they have committed to do under the Climate Convention.[1] But the participation of the developing countries – especially emerging economies such as China – is indispensable. Even if the world's developed countries were to make draconian cuts in their emissions, emissions in developing countries would need to be curbed significantly to keep atmospheric GHG emissions concentrations within acceptable levels. This book shows how it is possible for China, in cooperation with the world, to move to a low-carbon economy and put China on a 2°C pathway.

Science says: A finite global budget for greenhouse gas emissions

The hard constraints of climate science in determining a finite global budget for greenhouse gas emissions are clear. As recently presented in *Nature*,[2] to preserve a reasonable chance of keeping warming below 2°C requires limiting global carbon dioxide (CO_2) emissions to less than 1000Gt (gigatonne – 1 thousand million metric tonnes) CO_2 between 2000 and 2050, for both land-based and fossil fuel-based CO_2 emissions (Meinshausen et al, 2009). Since 2000 we have already emitted approximately 280Gt CO_2 from the use of fossil fuels, leaving a carbon budget of 720Gt CO_2 up to 2050. If heroic efforts are taken to bring deforestation and land degradation to a halt within one decade, then land-based emissions could be limited to approximately 60Gt CO_2. We

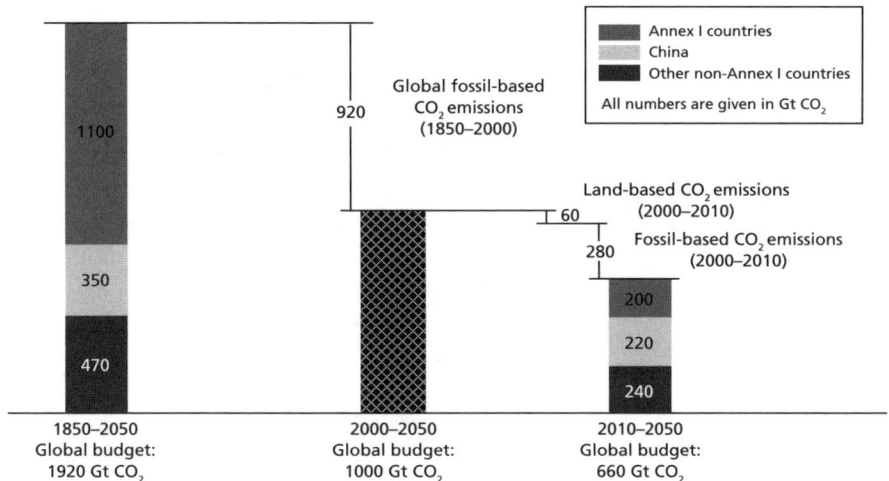

Source: SEI (based on Meinshausen et al, 2009)

Figure 0.1 *A global carbon budget*

are therefore left with a global carbon budget of just 660Gt CO_2 for fossil-fuels use over the next four decades (see Figure 0.1).

The global carbon budget provides boundary limits for discussions on individual countries' share of the emissions budget, and highlights the sensitivity of changes to individual emissions budgets. For instance, if the industrialized (Annex 1) countries were to commit to more ambitious targets of reducing their emissions[3] to 40 per cent below 1990 levels by 2020, and 95 per cent below 1990 levels by 2050, their future emissions would amount to 200Gt CO_2. This would leave 460Gt CO_2 for the non-Annex 1 countries. If we assume that China's part of this remaining budget is proportional to its share of current non-Annex 1 emissions,[4] its future budget would be 220Gt CO_2 (see Figure 0.1).

This combination of scientific facts and simple arithmetic provides a reasonable – and bracing – estimate of the emissions budget available to China. While there is a range of possible emissions paths that would keep China within this budget, all these paths imply bold and ambitious action.

For example, China's emissions could peak in 2015 and then decline at a rate of 5 per cent annually. Alternatively, China's peak in emissions could be delayed to 2020, but would then need to be followed by a much more rapid decline of 11 per cent annually. If China's peak in emissions did not occur until 2025, the decline would need to occur at a virtually unattainable rate of 35 per cent every year. And, if the annual 5 per cent rise in emissions continues beyond 2026, the full budget of 220Gt CO_2 will have been expended.[5] While each path requires unprecedented mobilization, it is clear that the longer that transformation to a low-carbon economy is delayed, the less feasible it becomes (see

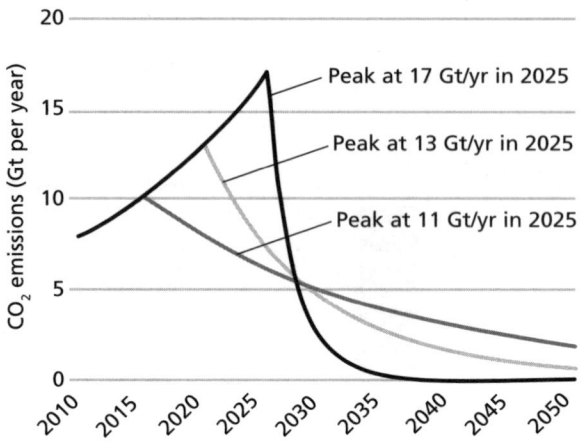

Source: Heaps (this volume)

Figure 0.2 *Carbon reduction scenarios for China with different peak years and a budget of 220Gt CO_2 from 2010 to 2050*

Figure 0.2). As shown in Chapter 3, the transformation is indeed feasible, but will have to be launched in the very near future.

There are two further points that must be emphasized about this arithmetic exercise. First, while it tells us that emissions from China must be rapidly curbed to address climate change, it does not tell us how the effort required to do so should be shared. Financial and technological resources will be critical for the transformation to occur, and China cannot be expected to bear the burden of generating and providing all those resources on its own. The same, of course, goes for other developing countries. The allocation of that effort among nations will need to be based on historical and political context, and grounded in the principles of equity that are clearly embodied in the United Nations Framework Convention on Climate Change (UNFCCC). In other words, the fact that China's *emissions* must be sharply curbed does not mean that China's *emission rights* have to be similarly curbed. The discrepancy between the two would then have to be covered by a commensurate amount of financial and technological support.[6] This support could be implemented through a global cap-and-trade system, where developed countries buy emissions rights from China and China receives financial resources in return. This would provide China with an economic incentive to sell a portion of its emissions rights and follow a more ambitious emissions trajectory domestically.[7]

Second, the above calculation is based on a target that still entails risks that warming will exceed 2°C. There is a scientific case to be made for adopting a more restrictive goal with lower risk for climate change impacts (Rockström et al, 2009). Delaying climate change mitigation also increases the need for more forceful action on climate change further down the track, and large-scale deployment of mitigation technologies to meet these targets. The key message here from a climate science and risk management perspective is that early mitigation not only reduces the costs but also the risks of delayed climate change action.

Rising to the global climate change challenge will require a concerted, coordinated and committed response from the entire international community. Governments will need to lead the charge in formulating a long-term strategy and create the institutions, mechanisms, technological alternatives and market environment to carry it out. It is vital that these conditions are established soon in order to guide the expectations of both investors, who will have to make critical long-term commitments, and decision-makers, who determine policies at sub-national levels. Citizens, too, must be prepared for the lifestyle changes that a transition to a low-carbon world will bring.

The message is loud and clear, the world needs to act now and act together, and China has a key role to play.

China's sustainability challenge in the climate change context

Since the start of economic reforms in the late 1970s, 30 years of spectacular economic growth in China have brought great prosperity and reduced poverty. China is today the second largest economy in the world, when measured in purchasing power parity terms, and its economic growth has followed a resource-intensive path similar to many developed countries. China managed to slow down emissions growth in relation to economic growth considerably between the 1980s and the end of the century. But despite the overall decline in emissions per unit of economic output, China is today one of the most energy- and carbon-intensive economies. At the same time, China faces many serious sustainability challenges that affect its ability to respond to the challenge of climate change.

China's economic growth has come at a high social and environmental cost, and there is a rapidly widening gap between the rich and poor, placing additional pressure on already scarce and unevenly distributed resources. China's macroeconomic structure, with its high dependency on the coal and industrial sectors, and its export strength in low-value goods, means that it is a big challenge to quickly transform its developing economy to a low-carbon, knowledge-based economy.

The speed of the modernization process adds stress to the task of simultaneously managing domestic, energy-intensive development and global climate change mitigation. China knows too well its own vulnerability to climate change impacts, from increased extreme events such as flood and drought, to melting glaciers and water shortages, to reduction of key grain production. Climate change impacts will reverse years of development gains and act as environmental and economic constraints on China's ability to modernize. Thus, seeking a low-carbon development path is not just *an* option for China but *the only* option.

China is attempting to modernize at a time when climate change mitigation is moving up the international agenda, and it will have to find fuels other than coal, gas and oil for its economic growth. This will require massive structural transformation for China, not only of its energy mix, but also of its energy-intensive transport and building sectors. No country has previously taken on a modernization process as extensive as the one China is currently struggling to balance.

A global low-carbon transformation will be practically viable and politically acceptable only if it does not compromise development and growth in developing countries. China is clearly still a developing country, with per capita emissions less than half of the Annex 1 average, and only one-quarter of US levels. Viewed in terms of cumulative contributions to climate change, this contrast with industrialized countries is even more stark, with China's per capita contribution less than one-fifteenth that of the US.

However, China's contribution to the climate problem is undeniable. The country as a whole currently accounts for almost a quarter of global CO_2 emissions and makes up for almost 60 per cent of the global increase in carbon emissions within this decade. China's per capita emissions have now risen above the world average. In Chapter 1, Fan et al show that China's historical contribution to CO_2 in the atmosphere may still be comparatively small but this is changing rapidly as its emissions grow. China would exceed its share of global cumulative emissions at some point in the next two or three decades, in the absence of mitigation measures beyond 'no-regrets' policies (Chapter 1).[8]

China realizes that addressing climate change brings profound economic and developmental opportunities. Still in the process of modernization, China can absorb lessons learned from developed countries. Increasing energy efficiency and reducing its reliance on exported fossil fuels will increase China's energy security and international economic competitiveness. However, avoiding 'lock-in' to carbon-intensive infrastructure and energy sources over the coming years will depend on the choices China makes, as well as its determination and capacity to travel the new path.

China's present climate change policies are heading in the right direction, and it realizes that addressing climate change in the framework of sustainable development is in line with its national strategic interest and would bring tremendous benefits for the environment, health, long-term energy security and for many other areas. From the guiding principle of the Scientific Outlook on Development (Hallding et al, 2009b) to an extensive set of laws and regulations, to major national programmes and policies, China has started on the road towards a low-carbon future. It continues to invest a significant amount of money in climate change mitigation and adaptation, and has enhanced its efforts in the current fragile economic environment.

However, more needs to be done by all. China in partnership with the rest of the world will need to embark on a new era of greatly deepened financial, technological and institutional cooperation if global emissions are to be reduced sufficiently to hold the increase in global temperatures below 2°C.

This book

This book, and the background papers that it builds on, has been developed under the project *Research and Forum on Economics of Climate Change – Towards a Low-Carbon Economy in China* (see Acknowledgements). The project addressed the issue of how China could 'bend the curve' from that of the trajectory followed by most countries in their development process towards that of a low-carbon development path (see Box 0.1). In doing so, we address some of the most challenging macroeconomic issues concerning climate change adaptation and mitigation for China, such as: what are the key policies that can support domestic growth while at the same time placing China on a low-carbon pathway? How can the economic and social costs of mitigation during a transition to a new economy be contained? How can China best foster the opportunities of low-carbon growth?

This book presents the project's research results in a broader policy context. It does not aim to provide a prescriptive blueprint for climate policy but instead to contextualize China's climate challenges and to provide food for thought on the policy options for China and their feasibility. Our research findings include:

- comparative cross-country analysis of the relationship between emissions and living standards;
- development of alternative burden-sharing frameworks for determining China's share of emissions reductions, including a Chinese proposal for a consumption-based burden-sharing concept;
- use of a deep carbon reduction scenario to demonstrate the feasibility of, and conditions for, reducing carbon emissions in China;
- analysis of two major mitigation mechanisms, carbon tax and carbon trade, and their implications for China;
- economic measures to mitigate negative effects of China's shift to a low-carbon economy;
- policies and mechanisms to increase innovation, technology transfer and investment in a low-carbon China, including a proposal for a new regime for international cooperation on technology transfer.

This book also attempts to bridge gaps in economic research on climate change policies for China. Still, more work needs to be done and throughout we have identified areas for further research. Key areas include:

- *Capturing economic opportunities of a low-carbon economy*: Specific policy and reform measures to enable China to simultaneously best capture opportunities for future comparative advantage, scientific and economic development, dynamic growth and technology innovation. Additional areas for study include China's building, transport, industry and energy sectors – all identified areas in the deep carbon reduction scenario that hold

BOX 0.1 BENDING THE CURVE – CHINA'S CARBON EMISSION CHALLENGE

Figure 0.3 illustrates a range of emission projections for China that have been assessed by different international and Chinese institutions and research groups. These projections can be categorized into three groups (Hallding et al, 2009a, updated with new data):

- At the top end of the range there are *business-as-usual (BAU) scenarios* that presuppose that China's development will continue with no or only small technology gains and at near constant energy intensity.
- In the middle range there are a number of *baseline* or *reference scenarios* that fall near to the trajectory that would be the result of successful implementation of policies to reduce China's energy intensity by 20 per cent per five-year period – i.e. the energy efficiency scheme set up by the present Five-Year Plan.
- Finally, there is a group of *deep reduction scenarios* that are based either on backcasting how much reduction is needed in China to keep global temperature within the 2°C target, or assuming the maximum technically realistic abatement opportunity.

The range of projections shows both the opportunity to make considerable progress towards climate security in China and the importance of coming to an international agreement that will enable China to harness the significant opportunities for continued development while slowing down emission growth.

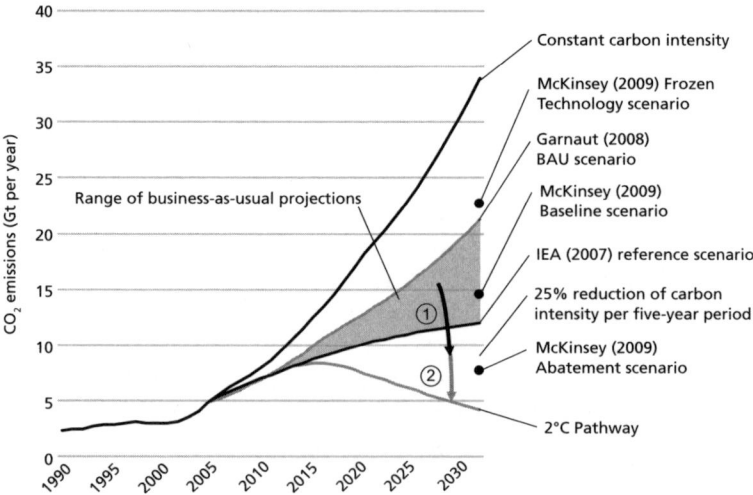

Source: International Energy Agency (2007); Garnaut et al (2008); Jiang (2008); McKinsey & Company (2009a)

Figure 0.3 *Schematic overview of possible future emission pathways for China*[9]

The first arrow ➡ illustrates the considerable curb of carbon emissions that China's current ambition to reduce energy intensity and switch to non-fossil fuel would imply if it were extended to 2030.[10] The second arrow ➡ shows how much more would be needed for China to reach a low-carbon emission trajectory in line with what would be needed for the world to meet the 2°C target. This report describes how this move to low-carbon development might be made possible.

clear mitigation potential, but also industries where transformative change (in private investment, business practices, skills upgrading, rapid diffusion of new technologies, practices and standards or shifts in social behaviour) could be problematic. Further research on regulatory measures to remove market barriers that prevent consumers from responding efficiently to carbon price signals.

- *Social resilience and adaptability to, and support for, a low-carbon transformation*: Social and labour market policies, and public participation policies to facilitate low-carbon transition in China. Policies to avoid or offset regressive distributive impacts of a low-carbon transition, including through transportation, housing, land use and revenue recycling policies, which encourage public support for a low-carbon economy. Research into China's social resilience and adaptability, and the conditions or requirements to improve social resilience and adaptability, to improve the effectiveness of domestic low-carbon policies.

- *Carbon market instruments*: Research gaps include China's institutional and administrative capacities to affect this macroeconomic structural change, and priority capacity building and policy measures to enable a successful low-carbon shift. More detailed analysis could be conducted on implementation and expansion of carbon market instruments, their integration with other policies, and international cooperation in these areas. Such research could consider the social and economic implications of various policy measures and appropriate compensation mechanisms to mitigate negative effects.

- *Low-carbon pathways*: Notwithstanding the deep carbon reduction scenario developed in this book (see Chapter 3), there could be further development of alternative pathways for China to achieve environmental, developmental and economic growth aspirations. Future research could include cost estimates for each scenario and distinct narratives leading to alternative policy directions.

Although the climate change policies explored here are grounded in an economic context, it is important to remember the limitations of cost–benefit principles. For instance, cost–benefit calculations may not properly account for the risk of large or catastrophic climate change damage, or the vital non-market impacts of climate change that a price cannot be placed on, such as health impacts, stress of physical relocation or secure access to clean water (Ackerman, 2009).

The remainder of Part I is structured as follows: the next section explores the meaning of a fair global deal and reviews different perspectives in terms of their implications for China's share of emissions rights. It lays the groundwork for examining the nature of the international financial and technological cooperation that will be needed to enable China to undertake a low-carbon transformation. Then we illustrate the nature of that transformation from a technological perspective, showing that it is feasible if it is launched promptly.

We look at the implications of a low-carbon strategy in various sectors, including building, transportation and electricity generation.

Next, Part I focuses on some of the policy options that China and the international community face as we look forward to a low-carbon future. First we explore some of the key market mechanisms and institutions for pricing, regulating and trading carbon, while measures to foster domestic and international innovation and investment are subsequently discussed. The final section examines the daunting challenge of modernization facing China.

The need for a fair deal

Meeting the global climate challenge must start from the increasingly alarming scientific evidence of climate change, as well as an understanding that development for the disadvantaged must not be compromised. In this section we search for an understanding of how China and the global community could act together to put China on an emissions pathway that is consistent with scientific advice on climate change, while allowing for continued poverty alleviation and increased living standards.

Emissions, living standards, and consumption

Chapter 3 shows that the relationship between living standards and emissions is far from straightforward. Climate change impacts are becoming increasingly serious as a result of rising GHG emissions. But these same emissions have been essential to maintaining high-consumption lifestyles in rich countries and progressing development for poorer and middle-income countries.

The case of China illuminates a difficult issue of international equity: how can the international community balance each individual's right to an adequate standard of living with the imperative of reducing global GHG emissions? Bluntly speaking, only the poorest countries have per capita emissions that are currently on a level that is commensurate with a long-term global 2°C target. Countries with higher standards of living, not just measured in terms of private consumption but also by life expectancy and literacy, show much higher emissions per capita.

While there is great diversity in development levels within China, average per capita emissions are today already higher than the world average.[11] For countries below China's emissions per capita, the trend is clear: GHG emissions are highly correlated with the level of development. Above China's level of per capita emissions, improvements in life expectancy and literacy are slow, but gains in income per capita remain strong.

Although there have been remarkable gains in energy efficiency and carbon intensity throughout the reform period, China still belongs to the ten most carbon intensive economies in the world.[12] In Chapter 3, Figures 3.1 and 3.3 show that few countries share the Chinese provinces' high emissions intensity with a similar range of income per capita.[13]

What can China learn from its own development?

The three richest provinces in China – Tianjin, Beijing and Shanghai – all have considerably higher per capita emissions than China as a whole. Being the most developed regions in China, these provinces should be those most able to transform to a low-carbon economy. Yet their relatively high emission levels foreshadow a considerable rise in per capita GHG emissions as China continues to urbanize – far above levels compatible with the global 2°C target. At the same time, it is apparent from comparing Beijing with its surrounding Hebei province that it should be possible for China to raise per capita income levels considerably with only small emission increases. Yet the Beijing/Hebei example indicates a future for China of roughly 7t (tonnes) per capita, which is far above the global levels compatible with the global 2°C target.

A more positive example is provided by China's two Special Administrative Regions, Hong Kong and Macau. Both have shown consistent income development with only limited growth in per capita emissions – despite the fact that energy supply relies almost entirely on fossil fuels. Part of the explanation for this low-energy intensity lies in high population density, abundant public transport and a service-based economy. The remarkable difference between Hong Kong and Macau and other prosperous urban areas in China (for example Shanghai, Beijing and Tianjin) could also be partly explained by the fact that a considerable part of the wealth of the former (particularly Hong Kong) is built on manufacturing that takes place in Guangdong province. Still, the combination of Hong Kong, Macau and Guangdong province would add up to an economically more diverse entity with per capita emissions at around 5t per capita. This is better than the examples provided by the Beijing/Hebei example above.

What can China learn from other countries?

In order to stand a reasonable chance of keeping within a 2°C target, global per capita emissions must decrease steadily throughout this century from the current roughly 4.5t to well below 4t by 2020, less than 2t by 2050 and roughly 0.5t by 2100. If global emissions are to decrease over time two things will be necessary: first, rich countries must cut their per capita emissions drastically; and second, low-income and middle-income countries must find ways to increase their per capita incomes while maintaining or reducing emissions intensity.

One conclusion is clear from this analysis: there are no examples of high-income countries with per capita emissions low enough to serve as a blueprint for a low-carbon development path. The high-income countries with the lowest emissions, such as France, Sweden and Switzerland, are still at about 6t per capita. These countries also share the trait of producing a higher share of energy from renewables and nuclear. China's ambitious plans to expand the share of non-fossil fuel energy sources are important steps in this direction.

Nevertheless, historical growth trajectories for individual countries, based on Gapminder analysis, show a wide range of different development pathways.

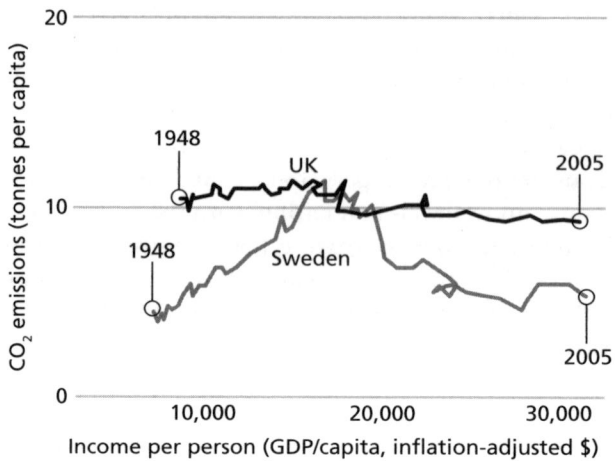

Source: Gapminder (2008)

Figure 0.4 CO_2 *emissions and income per capita for Sweden and the UK, 1948–2005*

Several countries – albeit developed ones – have bent their emissions curve downwards or contained emissions levels while still increasing per capita income levels (see Figure 0.4). For example the UK and Sweden have shown positive emissions trends in important sectors and policy has played a crucial role in reducing emissions while at the same time creating space for growth. The stories of how this came about differ between Sweden, where a nuclear programme and carbon tax have played important roles (see Box 0.4), and the UK, where fuel switching from coal to natural gas was a key factor. However, although this indicates promising opportunities, it needs to be emphasized that for cases such as Sweden, where absolute emissions have decreased since the 1980s while incomes increased, the emission cuts have to some extent been offset by increasing imports of embedded carbon (see Box 0.4).

But more importantly, there are an increasing number of positive stories on carbon reduction from sectors, cities or regions where considerable progress towards low-carbon transformation is being demonstrated. We provide several references to such examples in Chapter 5 where we lay out the deep carbon reduction scenario for China.

Consumption, carbon emissions, and the new global order of production

As noted above, Sweden and the UK, along with many other countries around the world, have used the increased income generated by growth to consume more and more imported goods. And here is where China comes into the picture. Over the last decade or two, China has dominated the global increase in production capacity, particularly for consumer products. The main reasons for this have been comparative advantages in factors such as labour costs and productivity, as well as China's accession to the World Trade Organization (WTO) in the early 2000s. During the same period, however, China acquired a huge net surplus in embedded carbon exports, which has led to suggestions that China benefits from its dirty production. In Chapter 4, Ackerman shows that China does not have comparative advantages in carbon-intensive exports. Instead the monumental amount of embedded carbon in China's exports is simply a consequence of how the high carbon intensity rubs off on its huge trade surplus.

The result of these circumstances is that the combination of low-cost and high-energy intensity production in China (and other countries) has become a global driver of carbon emissions.

Questions, reflections and the consumption dilemma

The relationship between consumption, emissions, living standards and how globalization has reorganized production contains crucial knowledge for our understanding of how the world might be improved for all, and for solving the climate dilemma. Further investigation in this area is needed, but based on Chapter 4 and other work done by SEI outside of this project (Wiedmann et al, 2008; Barrett et al, 2009; Scott, 2009) we can identify two key issues.

First, to the extent that China continues to supply the world with consumer products, consumption-based per capita emissions in importing countries will only decrease if China can reduce its overall carbon intensity. Ackerman (see Chapter 4) concludes that the incentive for high-income countries to support China in lowering its carbon intensity would be stronger if consumers were responsible for the carbon emissions embedded in import products. This would help to promote policies leading to joint goals of global low-carbon development. The most effective mechanism to ensure that costs are transferred to consumers is through the market, which supports the argument that both China and high-income countries should increase efforts to establish carbon-market instruments, and to help build China's capacity to work effectively with such instruments. Chapter 4 analyses the pros and cons of market mechanisms to price carbon.

Our second conclusion is far reaching and potentially more ambiguous. As noted above, the limited emission reductions shown in some high-income countries, such as Sweden and the UK, are offset by increasing private material consumption. Therefore the question must be asked: can the target of restraining a rise in global temperatures to below 2°C be attained if there are no effective policies to radically reduce material consumption? And conversely, if such policies were in place, what would that mean for poor countries in terms of their development opportunities?

This book does not examine this dilemma in depth, and there is certainly a need for further research. However, a few reflections can be offered. Once a basic level of living standards has been reached there is no clear relationship between material consumption and quality of life. Still, consumption plays a central role for maintaining and fuelling economic growth. Policy measures to take the world out of the predicament of consumptive growth, driven by fossil fuels, will have to establish strong incentives for growth through non-material consumption while at the same time providing opportunities for poorer countries to increase consumption in order to meet basic needs. It will be particularly important to make existing technologies available and to research, develop and implement cutting-edge low-carbon technologies that are appropriate for use in developing countries.

In terms of development, China stands in the middle. It has both a vast number of the population living far below a reasonable development threshold and a growing, increasingly high-consuming, middle class. In its ambition to move from a model of export-led growth to one that is increasingly dependent on domestic consumption, the question of material consumption is a very challenging one. The deep carbon reduction scenario presented in Chapter 5 argues that it would be possible to find a low-carbon path for China. Chapter 6 outlines a policy portfolio of cost-effective and ecologically effective mechanisms. These include a price on carbon and promotion of global cap-and-trade, combined with complementary regulation and joint technology development and diffusion, as well as measures to compensate for regressive impacts on groups disadvantaged by climate policy.

Frameworks for burden sharing

Climate change raises serious questions about global equity in reducing emissions. Chapter 11 shows that most of the carbon emissions to date have occurred in the industrialized countries, and there is a strong link between historical emissions and accumulation of wealth (see Figure 11.1). At the same time, developing countries are particularly vulnerable to the impacts of climate change.

Any meaningful solution to the climate crisis must induce an urgent and sweeping transformation of the global emission trajectory. This will be practically viable and politically acceptable only if it does not compromise development and growth in developing countries. Well-off people, wherever in the world they live, will have to accept taking on a larger responsibility for the global climate transition so as to safeguard the right to development for the most disadvantaged.

An institutional framework with the power to bring about the urgently needed transformation of the global emission trajectory (i.e. as discussed in Chapter 1) has to reflect both equity (taking into account different countries responsibilities for the climate crisis and capabilities to mitigate emissions and respond to climate impacts) and cost efficiency (reducing emissions and responding to impacts at the least total cost for the global community). A number of frameworks for burden sharing have been developed that propose how countries might agree on allocating obligations within a global climate regime. Each framework applies different distributional principles that affect the emission rights of different countries and financial flows between countries. Different principles often put forward in the discussion of allocation rules include:

- egalitarian (same rights to every person);
- ability to pay (capacity to contribute to the mitigation effort);
- polluter pays principle (accounting for historic responsibility).

Each of these principles provides a basis for allocating access to a global commons resource. As such, they naturally come into competition with a fourth principle – the principle of sovereignty. Sovereign countries tend to privilege their own short-term national interest and to resist joint management of commons resources. This is often expressed (often implicitly) as an appeal to the principle of a 'customary right to emit'. While this is a strong force within the climate negotiations, it provides little basis for a long-term solution to any problem of the global commons.

Specific proposals for allocation can be characterized by the relative importance they assign to distinct principles and none will satisfy them all. Several approaches have emerged, each based on its own set of principles, each with different implications for different countries. For this research effort, we have reviewed a set of key proposals with respect to their implications for China's emissions allocation. The proposals we have reviewed are as follows:

- *Equal per capita emission rights.* A straightforward approach premised on the equal rights to the atmospheric commons, with all countries being awarded emission permits in proportion to their population, which they would be free to trade (Agarwal and Narain, 1991).
- *Grandfathering.* The direct expression of the 'customary right to emit' whereby countries' allocations are determined by granting all countries permits in proportion to their prior emissions, constrained by an overall decline in global emissions consistent with the temperature target.
- *Contraction and convergence* (C&C). A hybrid framework combining grandfathered emission rights with per capita emission rights, with a gradual transition from the former to the latter over a specified number of decades (GCI, 2008).
- *Equal cumulative per capita emission rights.* Extends the concept of equal per capita rights to cover the entire carbon budget, rather than just the portion of the budget remaining for the future.
- *The Indian proposal.* A proposal whereby India's per capita emissions would not exceed developed country emissions. Up to that point, its emissions allocation is equal to its unmitigated requirements.
- *Greenhouse development rights* (GDR). This proposal shares burdens among countries according to capacity and responsibility, with each of these defined with respect to a development threshold so as to explicitly safeguard a right to development (Baer et al, 2008). We have also modelled

variants of the GDR approach, as developed by Fan et al in Chapter 1) (see Box 0.2 for details of the GDR approach.)

Among these proposals, a distinction can be made between *resource sharing* and *effort sharing* approaches. *Resource sharing* approaches (for example equal per capita emission rights, grandfathering and C&C) assume a limited global greenhouse gas budget and define rules for allocating this resource. Some resource sharing approaches, however, do not account for historical responsibility and ability to pay.

Effort sharing approaches (for example GDR) seek to equitably share the effort required to reach the climate target. They can directly take into account historic responsibility and the ability to pay, but have to rely on the definition of a baseline against which the reduction effort can be measured. As a consequence, effort-sharing approaches tend to favour carbon-intensive countries.

What is China's fair share of global emission reductions?

Figure 0.5 shows the emissions rights that would be conferred to China under different burden-sharing frameworks. The most notable feature of this analysis is that the three well-known resource-sharing frameworks (the Indian proposal, C&C and grandfathered emissions, all shown in light grey) would all allocate *fewer* emissions rights to China than would be sufficient to cover even the very demanding emission pathway needed to keep China within the 2°C budget (outlined in Chapter 1). In other words, expressed in terms of a global cap-and-trade system, China would under these frameworks be a net purchaser (rather than seller) of emission permits. It would need to invest its own resources in reaching the 2°C-consistent pathway, as well as purchase permits from other countries that have excess permits.

Also shown (in dark grey) are a set of variants of the GDR framework (see Box 0.2). Compared to most other frameworks the GDR approach puts a greater weight on a country's own capacity to combat climate change. As Figure 0.5 shows, the GDR-based scenarios all allocate fewer emission rights than the International Energy Agency (IEA) reference scenario for China (black line), but more than would be needed to meet the 2°C-consistent trajectory. In other words, China would be expected to invest its own resources in a certain amount of mitigation, but would be able to rely on international resources for the remainder.

Also shown (in dotted black) are two different versions of equal cumulative per capita allocations, one of which equalizes per capita emissions since 1850, the other since 1950. Both of these, like the GDR approaches, allocate fewer emission rights than the IEA reference scenario, but more than is required to reach the 2°C pathway. The choice of year from which cumulative per capita emissions are equalized is an important parameter, and, as would be expected, the later the year the lower China's allocation.

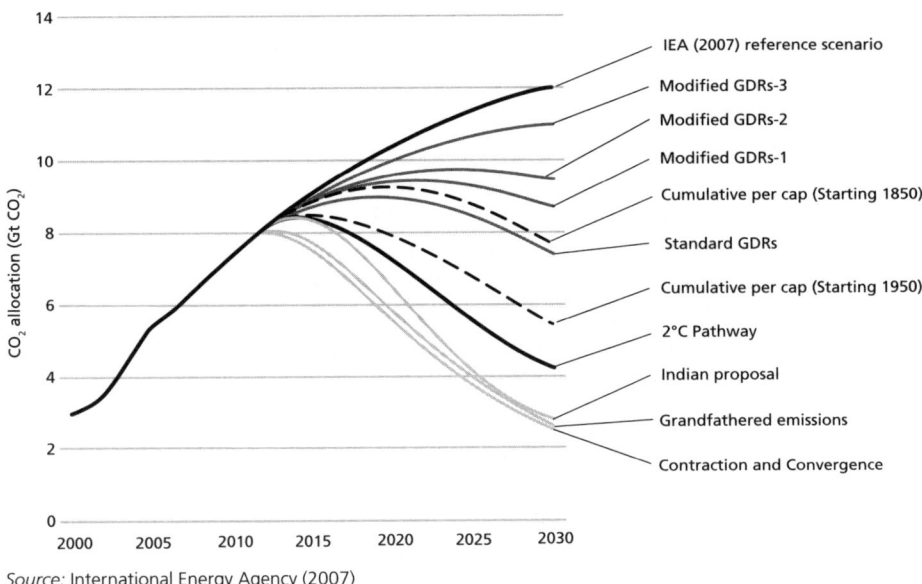

Source: International Energy Agency (2007)

Figure 0.5 *Emission allocations for China under various frameworks*

Figure 0.5 shows four trajectories based on different variants of the GDR framework, each of which defines differently the capacity and responsibility of countries. The most stringent of the trajectories (the bottommost dark grey line) reflects the standard GDR analysis as presented by Kartha in Chapter 2. Three main modifications that soften the demands on China have been proposed by Fan in Chapter 1. The first of these moves the historical responsibility back to 1850, rather than 1990, as in the standard GDR analysis. The second modification introduces consumption-based accounting, adjusting for

---

BOX 0.2 THE GDR FRAMEWORK

The GDR framework aims at ensuring on the one hand that global emissions are capped and cut to meet the 2°C target, while on the other that developing countries' right to development is safeguarded. It defines burden sharing in a manner that shields from all climate costs those individuals who fall below a specified 'development threshold'. People below this threshold are taken as having development as their overriding priority, and thus have no carbon reduction obligations. People above the threshold are taken as having realized their right to development and obligated to help preserve that right for others.

The GDR calculates the burden share of countries by quantifying *responsibility* and *capacity*. Responsibility is interpreted as the contribution to the climate problem and capacity as financial wherewithal to invest in climate solutions. The indicators for responsibility and capacity are GHG emissions and income levels, respectively. The GDR differs from conventional approaches in that it interprets both of these indicators with respect to the development threshold. More specifically, it defines capacity as *income above the development threshold*, and responsibility as *emissions corresponding to consumption above the development threshold*.

emissions embedded in internationally traded goods. The third corrects the carbon intensity of the domestic energy mix by using an average global carbon intensity. These modifications, taken together, reduce China's share of the global obligation for reducing global emissions by roughly a factor of three.

Although the differences between Fan et al's and Kartha's results are not trivial, the range of GDR variants all yield emission rights for China that significantly exceed what China's emissions would be under the 2°C pathway. This contrasts sharply with the three resource-sharing regimes (shown by the light grey lines) and means that under the GDR variants a large portion, if not the overwhelming majority, of emission reductions in China would be eligible for support from international sources. This underscores that a burden-sharing regime that is focused on explicitly safeguarding a right to development ultimately places significantly less of the burden on the developing countries, including China, and correspondingly more on the industrialized countries.

The equity implications of a burden-sharing regime rely not just on its underlying principles but also on how it might be implemented. In particular, there is a difference between an approach that pays for the incremental cost of reductions (such as a managed climate fund might do) and an approach in which permits are purchased at a common global price (such as a market-based system including cap-and-trade). In the latter type of approach, actual emissions will deviate from allocated emissions rights because countries can raise money (in this context called rent or producer surplus) by selling mitigation opportunities at a market price that exceeds their production costs.

It is important to note that under a global cap-and-trade scheme the economic costs of climate policy for a particular country not only depend on the allocation of emission rights but also on domestic circumstances, such as the energy system structure and the availability of low-cost climate-friendly alternatives (for example renewables).

In the long run, a global carbon market will be a key ingredient for achieving climate stabilization cost effectively. In that respect, burden-sharing frameworks can be applied to allocate permits between countries in an international cap-and-trade regime, but how those permits are allocated will affect the distribution of mitigation costs between countries. Some burden-sharing frameworks would mean that, because China has a low emissions allowance, it would be a net buyer of emission permits. Other frameworks provide China with a higher emissions allowance and thus position it as a net seller. Generally speaking, effort-sharing approaches, because they take into account historical responsibility and capacity to pay, provide China with higher emissions allowances. For China, a minimum requirement of any approach will be that it can act as a net seller of emission permits, thus generating some of the revenues required to finance its low-carbon transformation from the global market. However, Chapter 11 shows that a higher emissions allowance to China would shift the burden of mitigation costs to other countries. Any international burden-sharing approach will need to be perceived as fair and acceptable for all countries.

A 'graduation threshold'

As part of their modifications of the GDR framework, Fan et al in Chapter 1 have developed a specific recommendation regarding when the onset of binding commitments might be justified for China. Their proposal suggests that countries above a 'graduation threshold' for emissions should automatically face compulsory mitigation. This threshold is defined as the level of accumulative consumption emissions per capita of the Annex 1 country with the lowest level of such emissions (currently, this happens to be Romania). Countries below the graduation threshold could pursue voluntary mitigation activities but would not be compelled to mitigate.

Such voluntary mitigation would delay the point when the threshold is crossed and mitigation becomes compulsory. If a country accepts international technology and financial transfers, its graduation threshold would be reduced. Under this proposal, it would be possible for China to delay by a few years mitigation through legally binding means, provided that China devotes itself to voluntarily mitigating its energy intensity by 20 per cent under its 11th five-year programme, and pursuing this target into its 12th and 13th five-year programmes (2011–2020).

In this context, however, it should be noted that if a functioning global carbon market is established, most countries – including China – would actually benefit from early mitigation. This is because they can participate in the market for emission rights (i.e. to gain revenue from selling such rights) and that the risks decrease of being locked-in to expensive emission-intensive investment.

In any event, any meaningful effort to protect the climate will require ambitious mitigation action to be undertaken in China even *before* the onset of binding reduction commitments. If this is to occur, that action will have to be supported and enabled by technology and financing, based on the commitments of industrialized countries. A fair and equitable burden-sharing framework can help to clarify not only what levels of domestic mitigation countries must undertake, but just as importantly what levels of financial and technological support they must provide.

The art of the possible – a deep carbon reduction scenario

A development pathway to a low-carbon economy, although challenging, would be technically feasible for China to achieve. The scenario presented here is one of several possible scenarios to radically reduce China's current emissions. What makes this deep carbon reduction scenario (DCRS) so ambitious is that it is an attempt to drastically cut emissions while at the same time allowing China to grow rapidly and expand the material welfare of its population.

The DCRS demonstrates the *technical* feasibility of reducing China's emissions to a level that is compatible with protecting the planet. At the same time it gives enough 'emissions space' for China to continue to develop. In this scenario, emissions are decoupled from GDP growth; while production and income continue to grow, emissions are reduced.

The deep carbon reduction scenario

Chapter 5 shows how China can meet development and economic growth goals over the next four decades, while at the same time keeping GHG emissions within the very tight budgets of a global 2°C target (see Box 0.3).

Given the overall momentum for growth and development in China, and the lack of availability of technologies that can be deployed immediately on a large scale, it is simply inevitable that China's emissions will continue to climb in the next decade, even under the most ambitious of mitigation scenarios. This makes the requirements for reducing emissions later particularly challenging.

Our analysis shows that a DCRS is technically feasible with strong mitigation potential in the buildings, industry, transport and electric generation sectors. The result of successful capture of mitigation potential in these sectors is shown in Figure 0.6.

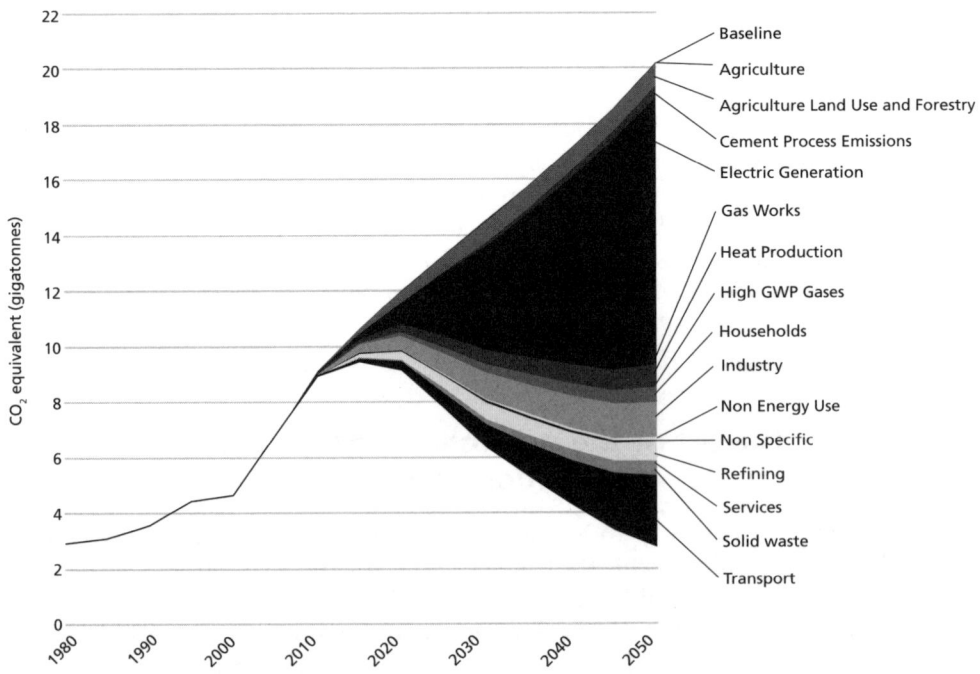

Source: Heaps (this volume)

Figure 0.6 CO_2 *emissions in baseline and deep carbon reduction scenarios*

BOX 0.3 WHAT IS THE DEEP CARBON REDUCTION SCENARIO?

The DCRS examines the feasibility of massively reducing China's CO_2 emissions in 2050 to only 10 per cent of the 2005 levels projected in a baseline scenario. Expressed in terms of Figure 0.5, the DCRS is the combined result of low-carbon technology options across a range of sectors. This pathway for China is designed to match China's 2°C pathway outlined in Figure 0.3 and to stay within an overall emissions budget for energy sector emissions of about 230Gt CO_2 between 2005 and 2050. This budget for China's emissions is consistent with the global emission pathway presented in Chapter 1.

The scenario focuses on mitigation technologies that are either already commercialized or are expected to become commercialized in the next decade, in time for their deployment to have a significant impact in reducing China's emissions. However, given the long time horizon, this development necessarily includes far-reaching assumptions about the degree and direction of technological change. For instance, we assume that energy intensity will continue to decline rapidly, particularly for the period after 2030, which inevitably implies a massive and sustained level of research and development (R&D).

The DCRS and the baseline scenario assume a general continuation of current policies that include some significant efforts to address sustainability and the climate challenge, but it does not foresee any fundamental shifts in energy policy. Both scenarios rely on a similar set of assumptions about structural changes in China's economy: for instance, that the share of GDP coming from services is projected to increase from 40 per cent in 2005 to 52 per cent in 2050; that the share from agriculture decreases markedly from around 12 per cent in 2005 to only 2 per cent in 2050; and that the share from industry stays almost constant going from around 47 per cent in 2005 to around 46 per cent in 2050.

The baseline scenario relies upon a sector by sector review of historical trends, as well as an examination of how future energy and CO_2 emissions patterns can be expected to evolve as China develops and average income levels increase. Up until 2030 the scenario closely matches trends in energy use and CO_2 emissions foreseen in the IEA's *World Energy Outlook 2008* for China (International Energy Agency, 2008). Additional analysis has been done to extrapolate energy and CO_2 emissions patterns to 2050.

It is important to understand that the DCRS is not a proposal about the level of obligation that China should take on in any international climate negotiations. Nor is it a proposal about what financial burdens different parties should carry, or a prescriptive technology or policy roadmap. The DCRS is not intended to represent a least-cost development path. Other positive economic effects, such as innovation, comparative advantage, technological leadership and export gains are discussed throughout the book.

A range of alternative pathways could potentially yield the same total emissions. But these alternative pathways would need to be much less materials intensive. This can be achieved either by lowering the economic growth ambitions or by drastically changing the composition of growth, emphasizing the provision of welfare more through the delivery of services than through the consumption of goods. For example, they might include less consumption of meat, more consumption of vegetables, better urban planning to reduce the need for transport and more emphasis on health care and environmental protection. The net result of these measures would be a smaller industrial sector but a larger service sector with a consequent lowering of energy and GHG emissions. China's choice in the development pathway it takes will partly depend on the priorities it places on social needs, consumption of goods, equity of welfare and income distribution, and other policy objectives.

The net result of implementing the four groups of measures in the buildings, industry, transport and electricity generation sectors is that total energy sector emissions are lowered dramatically compared to the baseline scenario.

The deep carbon reductions result in energy sector CO_2 emissions peaking in about 2017 at 7.4Gt CO_2/year, before falling to 1.9Gt CO_2/year in 2050. Cumulative emissions between 2005 and 2050 are 229Gt CO_2, equal to the budget set for China by the GDR framework described above. Annual CO_2 emissions intensities show a similar shaped curve, growing from 3.6t/capita in 2005, peaking in 2016 at 5.3t/capita and then falling to only 1.3t/capita in 2050 – clearly in line with what is necessary on a global scale.

Construction

The construction sector has huge potential for mitigation of CO_2 as rapid urbanization and development in China continues to maintain high demand for building construction. Policy action is critical in this sector to avoid lock-in effects for future generations, particularly as rates of construction can be expected to decline as China's population peaks, its workforce ages and its economy matures.

Energy intensities in the construction sector can be reduced dramatically. The sector can also be largely decarbonized by replacing direct use of fuels (for example driving a car on gasoline), with greater use of electricity, excess heat and solar energy. Energy intensity reductions can be achieved through measures including:

- better design of new buildings to incorporate passive heating and cooling principals;
- retrofitting of existing building shells to reduce heating and cooling loads;
- more efficient heating, ventilating and air conditioning systems;
- efficient lighting;
- introduction and enforcement of stringent appliance efficiency standards for such as refrigerators, washing machines, dryers and TVs;
- improvements in the construction of buildings (for example to use more sustainable and less energy-intensive building materials).

Of these measures, improved building design is likely to have the largest long-term potential and the best economic benefit–cost ratio.

Implementing efficient new building designs will call for wholesale retraining of architects and engineers, and policy changes at the highest level to require that these new types of buildings become the norm. Retraining and education will take time. Failure to act quickly will mean that greater numbers of buildings will eventually have to be retrofitted for energy efficiency; an approach that is more costly and provides less emissions reduction.

The mitigation benefits of energy intensity improvements in the household and services sectors would be further magnified through changes in the fuels used in the buildings sector, with a nearly complete shift away from coal, oil and natural gas in favour of electricity, district heating and solar energy (the last of these primarily for hot water). As explained below, this shift away from

---

BOX 0.4 SWEDEN – THINKING OUTSIDE THE CARBON BOX

Sweden shows it is possible to combine economic growth with reduced emissions, although as noted earlier, part of its emission cuts are offset by increased imports of embedded carbon. Nevertheless, in the 1980s emissions decreased largely due to a government-led expansion of nuclear power and a carbon tax. Furthermore, since 1990, GHG emissions in Sweden have fallen a further 10 per cent, while GDP has increased by nearly 50 per cent. Switching energy systems and energy efficiency improvements are key drivers behind this trend.

For instance, energy for heating residential and service sector buildings is largely sourced from renewable biomass and waste, combined with efficient combustion and gas heat recovery. By exploiting surplus heat from industry, Swedish companies achieve substantial energy and cost cuts. Industrial waste heat is supplied by forest industries, refineries, steelworks and chemical and food-processing industries.

Source: Ministry of the Environment and Ministry of Enterprise, Energy and Communications (2009)

---

small-scale combustion of fuels can be coupled with a dramatic decarbonization of electricity and heat production. In the DCRS, a significant level of biomass fuel use is retained, reflecting the development of cleaner, more efficient and more convenient ways of using biomass fuels in rural households. In the service sector there is also potential for the near-complete phase-out of fossil fuels, to be replaced by electricity, heat and (to a smaller extent than in the residential sector) solar energy.

Transport

Emissions from China's transport sector are rising rapidly. Increasing income leads ever more households to purchase cars and travel more, while efficiency improvements in business practices and the industrial supply system (for example through just-in-time deliveries) are increasing total emissions output. If continued population growth and increased production result in a similar trend in emissions from transport, China's emissions budget will not hold.

The DCRS assumes three potential major shifts in transportation policy. First, we expect policies to slow the rapid overall growth in passenger transportation. The DCRS assumes a range of actions, such as higher fuel prices, improved urban planning to reduce the need for commuting, congestion charging, a revitalization of cycling, parking restrictions and restricted airport developments.

Second, the DCRS assumes that, contrary to baseline trends, rail and road transport will retain their current shares of the transport sector. This implies increased public awareness and use of public transport to constrain rising car use. Road transport as a share of total passenger kilometres (km) is assumed to stay roughly constant even as the total passenger-km value grows enormously – by a factor of 3.75. The share of road transport made up by buses is also assumed to stay roughly constant. Similarly, the rail share of passenger-km is assumed to decrease only slightly from 35 per cent in 2005 to 30 per cent in 2050. Again, because of population growth, this represents a huge increase in absolute terms from about 606 billion passenger-km in 2005 to about 1900 billion passenger-km in 2050 in rail transport. This would be a high value for a developed nation but not unprecedented. The share for air travel, both domestic and international increases from 12 per cent in 2005 to 15 per cent in 2050, below the 18 per cent reached in 2050 in the baseline scenario.

The third shift, and perhaps the most challenging of all, is a massive move away from dependence on oil-based internal combustion engines towards electrification of passenger road transport. Such a transition could happen in a number of ways, but is likely to follow a gradual path through hybrids and plug-in hybrids to fully electric vehicles. In the DCRS, this is modelled as a gradual transition that would begin slowly in about 2015, ramp up after 2030 and culminate in 2050 with 90 per cent of all road passenger-km being delivered by electrical vehicles. The transition from oil to electricity not only yields a significant decarbonization of transport but also yields important efficiency benefits.

Similar but much less dramatic transitions are assumed for freight transport. In addition to these fundamental shifts, the DCRS also assumes modest penetration of biofuels in the air travel and maritime sectors. Finally, in terms of rail transport, we assume the complete phase-out of coal-fired railways and a complete transition by 2030 to electric-powered trains. Furthermore, gradual efficiency improvements are predicted for rail travel due to the introduction of new technologies such as regenerative braking.

Industry

In China, industry is the largest consumer of energy, accounting for 55 per cent of final energy consumption. Compared to the baseline scenario, the DCRS assumes more concerted efforts to improve energy efficiency and switch to lower-carbon fuels wherever possible.

Iron and steel

In the iron and steel sector, large emissions reductions can be achieved with the use of electric arc furnaces for steel production. Notwithstanding efforts to switch to such modern techniques, traditional coal-based basic oxygen furnace production will remain important in China given the country's reliance on coal. Here, reductions in emissions will have to rely on energy efficiency improvements as well as carbon capture and storage. We assume energy intensity improvements of 30 per cent per tonne of steel by 2050 combined with carbon capture and storage for the CO_2 generated from 50 per cent of oxygen furnace-produced steel by 2050.

Cement

In the coming decades, a number of options are available to reduce emissions from the cement industry. These include a more rapid and complete switch to the advanced and less energy-intensive dry-process pre-calciner kilns; further reductions in the clinker-to-cement ratio (to reduce the overall need for cement in concrete); the use of carbon capture and storage for up to 50 per cent of cement kilns by 2050; and the limited use of low-carbon agricultural residues or biofuels for co-firing of kilns. In the longer run a number of new technologies are being researched for the production of new 'eco-cements' that could potentially reduce emissions more dramatically. However, as these processes are only design concepts at present and face significant hurdles before they can be commercialized, they have not been included in our scenario. Nevertheless, there is reason to think that cement production could be significantly decarbonized beyond 2050 and perhaps even before.

Other manufacturing

In other manufacturing sectors – chemicals, non-ferrous metals, transport equipment, machinery, food and tobacco, paper and pulp, wood and wood products, textiles and leather – our DCRS projects energy consumption using fairly simple assumptions. Industrial energy intensity in 2050 in this scenario is only 47 per cent of its starting value in 2005. This assumption of a 53 per cent reduction is clearly optimistic but it is not without precedent, especially in China where IEA statistics suggest that, in the 25 years since 1980, energy intensity in the major industrial sectors other than cement, and iron and steel, have been reduced by between 50 per cent and 90 per cent. Of course these reductions may reflect the gathering of 'low-hanging fruit' and the benefits of economies of scale that cannot easily be repeated. But it is at least possible that the reductions assumed in our DCRS can be engineered in the coming 40 years if the international political will emerges to do so. Coupled with these assumptions on future intensity improvements, the scenario examines the potential within each sector for switching to lower-carbon fuels, based largely on an acceleration of past trends towards greater use of electricity and central production of heat. In two sectors – paper and pulp, and wood and wood products – the scenario also assumes the use of biomass and agricultural residues. Carbon capture and storage is not assumed to be viable in sectors other than the largest and most carbon-intensive sectors: cement, and iron and steel.

Dematerialization

In addition to improvements to energy efficiency, there is a variety of opportunities for China to begin to pursue less material-intensive forms of development. Dematerialization implies that a larger fraction of GDP will come from services and light industry and a smaller fraction from the more energy- and carbon-intensive heavy industrial sectors. However, despite the mitigation potential of dematerialization in China, this has not been assumed to occur as part of the DCRS. This is because such shifts might not necessarily imply genuine dematerialization – they might simply imply that production shifts away from China to other perhaps equally carbon-intensive nations while material consumption in China continues to increase. Such a transition would in fact be replicating how most nations of the Organisation for

---

BOX 0.5 RIZHAO AND BAODING – CHINESE CITIES EMBRACING RENEWABLE ENERGY

Rizhao

In Rizhao, a city of 3 million people in Shandong province, northern China, a shift to renewable energy has resulted in a halving of CO_2 emissions. As of 2007, 99 per cent of Rizhao's households are using solar-power heaters, most street lights are powered by photovoltaic cells and solar water heaters must be integrated into the construction of all new buildings. This shift has been made possible due to government policies that encourage solar energy use and provide financial support for R&D. Local solar panel industries have also capitalized on the opportunity to improve and popularize their products.

These environmental improvements have been achieved alongside a growing economy. Between 2001 and 2005, GDP growth rate for Rizhao averaged 15 per cent per year – implying almost a doubling of income level in only five years. By comparison, the average growth was 13 per cent in the broader Shandong province over the same period. The improved environmental profile of Rizhao may also have helped attract high-profile universities to the city, rapidly increase foreign direct investment and boost the travel industry.

Baoding

Baoding, a city of 10 million people in Hebei province, south of Beijing, has transformed itself from a city reliant on the automobile and textile industries into the fastest growing hub of solar, wind and biomass energy-equipment makers in China.

Since 2002, nearly 200 renewable energy companies have emerged in Baoding, some of which are now among China's biggest manufacturers in their fields. Baoding now has the highest growth rate of any city in Hebei Province: in 2007, Baoding's GDP grew by 17 per cent, with the renewable energy sector contributing 12 per cent of this rise. Within the next two years, the renewable energy industry is forecast to overtake the auto and textile sectors as the city's most important engine of growth, and by 2050 is expected to contribute 40 per cent of the city's GDP.

Like Rizhao, Baoding's economic shift was enabled by a combination of factors. Its growth as a hub for the renewable energy sector was supported by government policies that included targeted tax benefits for companies and investors. The local government also provided subsidies for residential areas that adopt solar power.

Source: ChinaGate (2009); Ford (2009); HSBC (2009); World Bank (2009)

Economic Co-operation and Development (OECD) have grown in recent decades – apparently reducing their energy intensities, even while substantial amounts of production have been shifted oversees (much of it to China).

Electric generation

The fourth and most difficult area for policy intervention is China's energy supply – specifically its electricity generation sector – that is heavily coal dependent. Here, the DCRS includes a massive undertaking to decarbonize the supply of electricity and heat to the fullest extent possible given China's resource base. This pathway is designed to work in tandem with efforts on the demand side to eliminate the localized combustion of fossil fuels and replace them with electricity and centrally supplied heat.

Early retirement of existing inefficient coal-fired electricity generation

The DCRS assumes the accelerated retirement by 2045 of all existing coal-, oil- and gas-fired power plants. Since a large proportion of these plants were built in the current decade during China's most rapid phase of economic expansion, this represents the early retirement of a significant level of electricity-generating capacity. It is an expensive proposition but essential for keeping China's CO_2 emissions within the overall budget specified for the scenario.

Large-scale deployment of efficient coal-fired power with carbon capture and storage

The DCRS assumes that all new coal plants built from 2010 onwards will be much more efficient than the current stock of coal-fired power plants. The average efficiency of new coal plants is assumed to be 43 per cent, versus the roughly 29 per cent average efficiency of the current stock. The DCRS assumes that carbon capture and storage (CCS) will not be available until 2020, but after that date all new power plants are assumed to use a CCS system that captures 90 per cent of the CO_2 emitted by the plants. These plants are expected to operate at a lower efficiency (39 per cent) due to the energy needs of the CCS process. In addition, by 2050, 90 per cent of the existing power plants built between 2010 and 2020 are assumed to be retrofit- ted for CCS capture (which again entails a loss of efficiency). Under the DCRS it is also predicted that fewer natural gas plants are constructed than in a baseline scenario, and that these plants are subject to the same assumptions about CCS.

Large-scale development of non-fossil energy

The DCRS assumes that large amounts of new power will be generated from wind (offshore as well as onshore), solar energy (concentrating solar panels and solar photovoltaic), municipal solid waste and biomass and small hydropower plants. Given the growing resistance to large-scale hydropower schemes, the capacity for these plants is assumed to be the same as in the baseline scenario. Finally, the DCRS also takes into account the potential for significant increases in nuclear power. While nuclear power is still unpopular in parts of the world, mainly due to concerns over nuclear proliferation, safety, waste management and storage, it is included in the DCRS because of its potential for CO_2 mitigation and the preference shown for this technology by policy-makers in China.

Combined heat and power generation

The DCRS also assumes that all new thermal power plants will be designed for the combined production of both heat and power, as well as operating at greater efficiencies. By 2050, 25 per cent of all energy inputs to thermal and nuclear plants are assumed to be captured as usable process or district heat. This is likely to be hard to achieve, particularly since many power plants will

BOX 0.6 CITIES AROUND THE WORLD AIMING FOR ZERO EMISSIONS

Munich, Germany

A study of Munich, a city of 1.3 million inhabitants, has found that it can use cost-effective measures to transform itself into an almost carbon-free city over the next five decades. Moreover, such a shift is possible without any fundamental change in the standard of living of its residents. Ways of reducing carbon emissions include better insulation in buildings, more efficient heating and power generation systems, energy-efficient appliances and lighting systems, and a changeover to renewable and low-carbon power plants. Government policies can encourage consumers to invest more consistently in environmentally friendly, cost-effective technology and to increase their use of environmentally friendly transportation.

Tangshan, China

Tangshan municipality in north eastern China is traditionally a centre of heavy industry, based on steel production and coal resources. At the start of its efforts to achieve a circular economy (one which balances economic development with environmental and resources protection), the city used Caofeidian island, 80km from the Tangshan city centre, as a proving ground for realizing new development visions. Construction is now under way of a new eco city for 1–1.5 million people, in conjunction with a new industrial zone that will host heavy industries previously located in Tangshan. The eco-city's buildings will be climate neutral and self-sustained with renewable energy resources. The Swedish company Sweco was commissioned to undertake the project, and the work will be underpinned by advanced Swedish environmental technology solutions. An ecological park will also be constructed to allow public access to Caofeidian's unique ecosystem.

Masdar, United Arab Emirates

By the end of a ten-year project initiated in 2006, it is envisaged that Masdar city, in the middle of the desert in the United Arab Emirates (UAE), will be entirely carbon free. The city aspires to create a silicon valley for green technologies that will host 1500 companies, 40,000 residents and 50,000 commuters. The city also aims to use 100 per cent renewable energy, implement building design that incorporates passive heating and cooling, and roll out transport powered by renewable energy. Cars will be banned in favour of a city design that encourages walking, and the introduction of high-technology solutions such as personal rapid transit and freight rapid transit. Furthermore, it is hoped that awareness raising and behavioural change will reduce energy consumption by 25 per cent relative to comparable cities in the UAE. Investments in zero-waste efforts, geothermal cooling and wastewater recycling will also contribute to the city's zero-emissions target.

Source: Lans (2008); Masdar City (2009)

be built away from the potential consumers of heat. However, this assumption is key to lowering emissions since it allows an essentially zero-carbon resource (waste heat) to be used to meet the increasing need for energy in China's industrial sector. Such a scenario is only likely to be plausible if future industries and power plants are developed in a much more integrated fashion.

Managing the challenges and disruptive effects

Two key points are clear from our research: first, if the DCRS is to be achieved, the expected challenges are massive. All of the elements identified in our scenario must happen: electrification of vehicles, massive deployment of renewables, a complete switch to CCS-based coal-fired generation, huge improvements in energy efficiency, significant changes to passenger transportation modes and so on. Although extremely ambitious, the DCRS would only barely remain within an emissions budget that matches China's 2°C pathway. Consequently, a loss of any single major element would make it impossible to reach these mitigation goals.

Second, time is short. Not only do all of the options need to be implemented, but they also need to happen quickly. Any delay in implementing options will make it almost impossible to meet the overall target budget of 230Gt CO_2. Furthermore, the later the peak in carbon emissions, the more difficult it will be for China to comply with the emissions budget.

A drastic reduction of carbon emissions, such as proposed in our DCRS, is a strategy of forced retirement of existing capital and replacing it with new forms of capital that are technologically, socially and institutionally advanced. In the spirit of Joseph Schumpeter, this process could be described as 'creative destruction' or 'forced disruption'. This process, especially on the scale required for ambitious climate action, is likely also to have disruptive effects. Central to these concerns are the anticipated job losses in existing heavy industry, such as would occur from the early and large-scale retirement of much of China's existing power sector, and the costs of mitigation and adaptation.

Historically, countries have gone about this 'creative destruction' in different ways. Europe in general, and Scandinavia in particular, has put strong emphasis on social and labour market policies, underpinned by the idea that skills and people – not specific jobs in existing industries – should be protected. When jobs are made redundant, policies should be implemented to retrain people in new sectors. It is the quality of social and labour market policies that will determine the extent to which a particular rate of structural change, such as that required for an ambitious climate change policy, will be politically feasible and socially acceptable for China.

Cai et al in Chapter 9 reaffirm the importance of implementing active labour policies to soften the negative employment impacts caused by a transition to a low-carbon economy. Cai et al note that job losses are likely to occur in certain sectors and regions if the labour market is not responsive enough. Some sectors will be hit especially hard by a shift to a low-carbon economy – coal mining and some heavy industries in particular. However, it is also worth noting that although heavy industry may be the number one energy consumer and polluter in China, it does not do well at creating or absorbing employment. In fact, the present dependence on capital-intensive heavy industry may make it more difficult for China to absorb surplus rural labour.

In a low-carbon transition, sectors such as light industry and services will grow more rapidly, which could make it easier to absorb rural labour. China has good opportunities in the global race for green jobs, supported by its recent economic stimulus packages. For instance, China surpassed the US in the number of new wind turbines built in the first half of 2009. In wind power, local demand often means local jobs, particularly in China, where domestic regulations require that companies must recruit 70 per cent of its employees locally (The Climate Group, 2009). Green job opportunities in China are likely to increase and support an overall shift to a knowledge-based and services-oriented economy. Thus, in the long run and with sound government policies in place, employment should not be negatively affected by the transition to a low-carbon economy.

Moreover, such a transition presents a multitude of other opportunities for China that sit well with its development aspirations. By developing service sectors, scientific and technological R&D and a knowledge-based economy, China may move up the value chain in production of goods and services and capture future comparative advantage. This will bring better-paid jobs, and diversification of its economic strengths will provide for more dynamic and healthy growth. And the earlier that China takes strong action towards a low-carbon economy the greater these opportunities will be. Thus, a future of deep carbon reductions is also a future of significant opportunity for China.

The labour market will, however, need time to adapt, and the extent of the transformation that China will need to undergo in pursuit of low-carbon development must not be underestimated. This underlines the need for careful implementation of a broad emissions reductions strategy – one that would help to stabilize the economy, enable economic restructuring and also facilitate structural shifts in the labour market. Our research suggests that, complementing pricing and market policies, a labour market strategy for a smoother adjustment path should include the following elements:

- Active policies to improve the functioning of labour markets and move labour to expanding light industries and services.
- Increased labour mobility across regions within China, where integration of the welfare system across regions may encourage mobility.
- Promotion of education, training and employment opportunities in service sectors and other new sectors created by a carbon-constrained global economy. This must include the expansion of secondary and university education to provide a sustained supply of skilled labour to enable China's transition to a services-based economy.
- Provision of training services to enterprises to facilitate dissemination of energy saving technology and growth of these sectors. Given that skilled human resources are a bottleneck limiting the development of low-carbon industries, the Chinese government could play a more active role in basic research, training and other public services to support these enterprises. This could help achieve the necessary balance between emission reductions

---

Box 0.7 China goes green

China has already made considerable progress in developing low-carbon technologies in a wide range of sectors. Policies and innovation strategies that encourage green innovation and investments are being introduced. Improvement targets are set for many sectors and strata of society, as well as compulsory laws and better city planning.

There are currently over 50 million electric bikes and motorcycles in China, and the array of available low-carbon transportation is being expanded further. China is now leading the mass production of electric hybrid cars. Incentives for energy-saving markets are being created through a number of government policies. Energy efficiency improvements in industries such as metals and cement have helped China to save more than 900 billion kilowatt hours (kWh) over the last four years. In 2008 China generated 12 million kWh of wind power, a figure set to double every year in the upcoming years. Furthermore, the government's energy conservation targets promote the use of low-carbon building materials, and several 'eco-cities' are in advanced stages of planning

Source: The Climate Group (2009)

---

and reducing unemployment costs. This is also an important area for international cooperation.

The next two sections of Part I explore key macroeconomic policies for enabling deep carbon reductions and offsetting some of the consequential social costs. Nevertheless, further research is required to better grasp the implications of such a full-scale and complex transformation of China's economy. This research should focus on: detailed sectoral-level policies in areas with the highest mitigation potential; regulatory measures to ensure consumers can overcome market barriers and respond efficiently to price signals; and the measures necessary to avoid or offset potential negative social and economic distributive impacts.

Market mechanisms to price carbon

Climate change is frequently described as the greatest market failure ever. Current market mechanisms and institutions do not adequately internalize – that is, reflect – and pass on the actual costs of using fossil fuels to producers or consumers. Thus, a critical first step to correct this market failure must be to internalize the costs associated with GHG emissions.

Placing a higher price on carbon is at the heart of any strategy to cut emissions and increase energy efficiency. It would create incentives both for companies and individuals to produce and consume less carbon-intensive goods and services, and to undertake abatement opportunities to reduce their overall carbon footprint. Furthermore, the higher the price of carbon, the higher would be the potential gains from the sale of excess emissions rights in an international emissions trading system. Also, anticipation of higher future carbon prices sets an incentive to develop low-carbon technology and products, and can thus steer investments in this direction. Economists tend to prefer a global price, which would create a level playing field for all producers.

In practice, however, a price on carbon means more expensive goods and services, which means that implementing it is a major political challenge, even in the current global environment of relatively low energy prices. Levels of income vary between different countries, as do local prices for labour, land and other inputs. Therefore, a global price would cause sharply divergent effects on both consumption levels and the distribution of income in different countries. In particular, a high carbon price could also have a huge impact on the distribution of wealth between different countries (this would also depend on the allocation scheme adopted under a global climate regime). A measure such as a uniform, across-the-board tax could have unacceptable equity implications. An urban slum dweller struggling to pay for fuel for daily cooking would be hit substantially harder than the owner of an oversized personal passenger vehicle. The same kinds of problem (though on a smaller scale) could also occur on a national level, should countries take autonomous decisions to hike the carbon price. Any price – global or national – on carbon or carbon emissions therefore must be accompanied by actions to soften these negative consequences.

This section explores the suitability, effectiveness and implications for China of various carbon pricing instruments. It then explores policy issues that China and the international community should consider when establishing and participating in international emissions trading systems. It also offers suggestions for managing the political reality and economic consequences of carbon pricing and cap-and-trade regimes emerging unevenly throughout the world.

Phasing out subsidies

To place an accurate price on carbon, an essential first step is to continue to phase out the subsidies on fossil fuels that are still common in many countries, including China. Such subsidies have many forms, including direct subsidies through low, regulated prices or tax levels (or depreciation rules) that are lower than on comparable products. These subsidies worsen the existing market failure because consumers are given the incentive to use even more fossil fuels. They also counteract efforts to improve energy efficiency and energy security as producers are given incentives to waste fuel and reduced incentives to invest in clean energy.

The use of subsidies is common in both developing and developed countries. In 2007, the total amount of subsidies to fossil fuels in the 20 largest developing economies amounted to more than $300 billion. Eliminating fossil fuel subsidies worldwide would reduce global GHG emissions by 10 per cent in 2050, according to the OECD and IEA (*The Economist*, 2009a). In addition, removing subsidies frees up much-needed resources that can then be directed to climate policies, such as funding adaptation measures and low-carbon technologies. For instance, the UNFCCC estimates that slightly over $200 billion per year in incremental investment and financial flows would be required to return global emissions to 25 per cent below 2000 levels by 2030 (UNFCCC, 2007; McKinsey & Company, 2009a, 2009b).

This issue has been recognized by China and the G20 in a recent communiqué at the Pittsburgh Summit, where China and G20 leaders backed a proposal to phase out and rationalize inefficient fossil fuel subsidies over the medium term, while encouraging targeted support for the poorest (Reuters, 2009; US Department of State, 2009). In China's case (although energy subsidies have gradually decreased, allowing prices to converge towards international market prices) further lowering subsidies and eventually getting rid of them is an integrated part of the overall price reform. Furthermore, although China is making some efforts to rebalance the allocation of capital and lending away from heavy industry, this process of macroeconomic restructuring and financial reform will need to be continued.

Carbon tax

Liberalization of prices will not alone be sufficient to reach climate targets. Prices on fossil fuels must be raised further, either through taxes or a tradable permit system. Both mechanisms work to reduce emissions in different ways. Carbon taxes, as an environmental tax on CO_2, act indirectly to reduce demand and thereby emissions, and emissions trading systems limit emissions directly by defining and restricting the available amount of emissions permits.

In Chapter 7, Cao indicates that even a fairly small carbon tax would bring substantial emission reductions in China at only modest cost. If carbon tax revenues are recycled, such a tax reform would actually bring benefits for consumption and economic growth, since the tax system as a whole would be less distortive.

A carbon tax regime would bring substantial co-benefits to the macroeconomy by changing industrial structures and energy sources, improving economic efficiency and reducing local damage to health and the environment. Revenue from carbon taxes can be recycled to lessen the current distortions within China's fiscal structure, for instance through improving the efficiency of capital and value added taxes, removing highly distortive subsidies on energy prices and further reforming electricity regulation.

Ideally, the carbon tax should cover all the significant GHGs. Practical difficulties usually arise, however, in levying a charge on methane produced by agriculture and measurements of non-CO_2 GHGs. Therefore, in the initial stages only fossil fuels should be taxed, based on their carbon content. A carbon tax also presupposes a number of administrative and institutional reforms, such as a robust monitoring mechanism for measuring emissions output.

Another challenge lies in finding the right level of tax. An initial national carbon tax should be high enough to significantly affect behaviour, but not so high that it would lead to unacceptable effects on employment. Chapter 7 suggests an initial tax level that will increase the price on coal by 10–20 per cent. However, if a tax is accompanied by other measures to stimulate employment and a shift of labour to other sectors, the carbon tax level may turn out to be higher. In the longer term, the price on carbon must rise much higher,[14] if the DCRS presented in Chapter 5 is to be implemented.

Higher prices on carbon will have distributional effects. China's main political concern over any energy tax reform is that it will be regressive, so that increases in energy prices may stimulate structural inflation and hurt lower income groups. In fact, even if the government waived the carbon tax for poorer sections of the population, empirical studies suggest that most energy taxes or carbon taxes are likely to be regressive, so the poor will be hit even with carbon tax exemption.

Cao (2008) explores the consequences of implementing an environmental tax, in particular its potential effects in exacerbating pre-existing tax distortions and thus driving up welfare costs associated with environmental tax reform. In her analysis, she compares the effects of two environmental tax regimes: a direct fuel tax (input tax) on primary fuels and an indirect tax (output tax) on output products. Cao concludes that an output tax is more likely to exacerbate rural–urban migration distortions than an input tax. Thus, an input tax on fuel might be preferable over an output environmental tax. The government should consider the option of combining a carbon tax with other measures aiming at balancing these effects, for example by redistributing some of the carbon tax revenues to mitigate the negative social impacts. This may require different types of compensation regimes. The topic merits more detailed study.

In Chapter 6, Edenhofer et al demonstrate that, to achieve ambitious climate mitigation targets, environmental taxes should be applied to all sectors, or at least strongly limit tax exemptions, as each tax exemption increases the cost of achieving abatement targets. This is further emphasized by experiences

---

BOX 0.8 ENVIRONMENTAL TAX EXPERIENCES IN NORWAY

Beginning in 1991, Norway introduced a hybrid between an energy tax and a resource tax. A carbon tax was put on fossil fuels, while energy taxes were levied on electricity as well as oil and natural gas used for heating. However, a large share of emissions-intensive industries and other sectors are exempt (coal, pulp and paper, fisheries, cement, metal processing and air transport) and there are differentiations in the tax. This means that many resource-intensive industries have to pay only a small share of the normal tax rate so that, in effect, the tax burden is mostly borne by the consumers. Norway's environmental tax is indexed to inflation.

Since its introduction, Norway's energy and CO_2 taxes have led to a decrease in CO_2 emissions, although the exact amount of this reduction that can be attributed to tax effects is uncertain. The most visible effect of the Norwegian environmental tax was the early deployment of CCS technology at the Sleipner gas field in the North Sea by StatoilHydro – the oldest CCS project in operation. To avoid paying carbon taxes levied on venting excess CO_2 into the air, Statoil decided to dispose of CO_2 in a deep saline aquifer. Since 1996, Sleipner has been storing about one million tonnes of CO_2 a year (see Chapter 6 for more details).

BOX 0.9 ENVIRONMENTAL TAX EXPERIENCES IN GERMANY

In 1999, Germany introduced an end-user eco-tax on electricity, transport fuels, natural gas and light and heavy oils, repeatedly raising the rates until 2003. The tax generated a total revenue of just under 1 per cent of national gross domestic product (GDP). The revenues from the eco-tax were in the first years exclusively used to reduce employers' and employees' social security contributions. In later years, roughly 10 per cent of the revenues were transferred to the general national budget, while 1 per cent was used to fund the promotion of renewable energies.

Germany's eco-tax has delivered both economic and environmental benefits. Social security contributions are estimated to be about 10 per cent lower than without the energy tax, and fuel consumption has decreased. Fears about reducing companies' competitiveness have not materialized; rather, most companies actually gained from the introduction of the eco-tax.

Criticism of Germany eco-tax generally focuses on three issues: the tax is too limited to have a large impact on the ratio of labour versus growth rates in energy productivity; the tax is inefficient since it sets a price on energy instead of directly on fossil-fuel carbon; and major tax exemptions (for example for manufacturing industry, agriculture and forestry) strongly reduce the tax's cost effectiveness (see Chapter 6 for more details).

---

from Germany and Norway (see Boxes 0.8 and 0.9). When the main goal is to reduce CO_2 emissions, it is also important to tax the carbon content of fossil fuel input rather than energy output in order to exploit all cost-efficient abatement options along the supply chain. The efficiency of environmental taxes is weakened by factors such as infrastructure lock-in, which reduce the flexibility to substitute polluting activities/fuel sources with cleaner alternatives. This underscores the importance of supplementing any price mechanism with other policy instruments, such as subsidies for R&D or efficiency standards, to increase the possibilities for substitution.

Environmental taxes can and should be designed, marketed and implemented to address public and business concerns without jeopardizing environmental objectives. For instance, indexing taxes to inflation may ensure their influence over the long term and at the same time increase their political acceptability. Repeated tax increases may face major political opposition and therefore create political risk that has to be carried by firms planning long-term investments. However, taxes that are automatically adjusted to inflation create more stable planning conditions. Also, most energy taxes recycle the revenues to enable reductions of social security contributions or income taxes – the so-called 'double dividend' – thereby leading to a net reduction of tax at the firm level.

Additionally, Ackerman in Chapter 13 indicates that the fear of losing international competitiveness through implementation of an environmental tax is largely misplaced (see also above). Other factors such as infrastructure, wage levels, education and proximity to growing markets have a higher influence on competitiveness and choice of location than environmental regulation. Furthermore, the ongoing shift towards energy taxes in many countries will lead towards a more level playing field, reducing the competitive pressure on energy-intensive firms as competitors face similar increases in energy prices.

Cap-and-trade system

Chapter 6 shows that emissions trading holds some key advantages to tackling climate change, and building a functioning global cap-and-trade system should be the key mid-term goal of global climate policy. By making pollution rights explicit and transferable, the market is able to value and trade these rights. As a result, emission reductions can occur wherever they are cheapest. National or regional emission trading systems can also be linked into a global system or regionally interlinked systems of emissions trading, with appropriate burden sharing and harmonized carbon prices. In addition, a global cap-and-trade system also allows for integration into a global carbon market, as discussed further above. A national emissions trading system in China would facilitate the long-term goal of international harmonization of carbon prices.

To set up a cap-and-trade system requires effective administration as well as a mature legal system. Emissions must be measured and monitored, individual emitters identified and charged correctly, and trading of the permits requires a well-functioning market with a high degree of sophistication. Consequently, such a system cannot be built rapidly. But China is now creating pilot projects to start trading in small volumes. We expect that, within a number of years, this will have the potential to develop into a fully fledged system for emissions rights trading. This is an area where international support and experience, particularly from the European trading system, could help China move more quickly towards a domestic carbon market and eventually to link that up with international carbon trading systems.

As China builds a future trading system, a significant concern over its design and implementation centres on how to distribute permits. This step is

---

BOX 0.10 THE 'GREEN PARADOX' DOES NOT HOLD FOR CHINA

One potential problem of imposing a carbon tax that increases over time is that owners of natural resources have an incentive to extract resources at a faster rate, resulting in short-term increases in resource use and emissions. This is called the 'green paradox'.

This paradox makes it difficult to estimate how patterns of resource extraction and resource prices will change in response to an increasing carbon tax. This leads to strategic uncertainties and complicates the process of achieving optimal negotiated emission trajectories. This is particularly important with respect to oil extraction. Because of the dynamic strategic behaviour of fossil-resource owners, it will hardly be possible to implement taxes that fully internalize the social costs of burning fossil fuels. However, because a quantity cap on emissions imposes a binding upper limit for resource extraction, resource owners cannot extract more than allowed under the cap. This is an additional advantage of cap-and-trade systems in achieving high environmental effectiveness.

However, the 'green paradox' argument may not hold for China, since the national government controls all fossil resources. Thus, unlike the global oil market where there are numerous competing producers, there should be no conflict between a higher price on carbon on the one hand, and the short-term interests of competing oil producers to maximize profits on the other.

contentious because it critically determines the policy's distributional impact. Permits can either be sold, usually through auction, or allocated for free. The key question is who receives the revenue created by capping emissions. Chapter 8 suggests that free allocation of permits may distort incentives for reducing emissions. There are three main objections to free allocation:

- First, the expectation that the baseline year for allocations will be updated may encourage polluters to invest in dirty technology, or refrain from investing in clean technology, in order to increase or maintain emissions levels and thus receive more free permits in future. Thus, the system may cause unwanted lock-in effects into old technologies.
- Second, free allocation may present a barrier to market entry. If existing polluters receive permits for free but new plants must pay for them, free allocation reduces competition. At the same time, it may present a barrier to market exit. The requirement that an installation must be kept open in order to receive free permits may prevent the closure of inefficient plants, freezing emissions at higher levels than would otherwise be the case.
- Third, free allocation leads to increased lobbying because emission permits, which have a monetary value, are given out for free. Lobbying of powerful producer groups may put governments under considerable pressure.

Auctioning offers several advantages over free allocation:

- First, it places upfront costs on polluters. This will tend to enhance management's awareness of carbon cost, leading to more efficient decisions.
- Second, auctioning raises revenues that can be used to address equity issues through reductions in taxes or other distributions to low-income groups. Governments can also use these revenues to invest in the development and deployment of cleaner technologies, or provide finance for other countries' efforts for climate change mitigation and adaptation. This is analogous to the 'double dividend' feature of carbon taxes. Recycled revenues from auctioned permits can have efficiency benefits if they are used to reduce distortive taxes, whereas free allocation of tradable permits may be regressive because it can transfer income towards higher income groups (i.e. shareholders) at the expense of poorer households (i.e. consumers with high income shares of energy expenditure).
- Third, auctioning provides stronger incentives for technological innovation. Under free allocation, some sources are buyers and others sellers. Sellers have an incentive to behave strategically and keep prices high by avoiding technological innovation. In auction markets, all emitters are buyers. Hence, all sources benefit from low-carbon technologies because of the decreased cost of abatement and lower permit prices.

Overall, auctioning can avoid many of the problems associated with free allocation and offers distinct advantages, although even with the auctioning system, good design is necessary to avoid inefficiencies. Small, frequent auctions may be effective in limiting the market power of large bidders. Such auctions may also encourage learning processes, help players to adjust bids and promote price stability. In contrast, one large auction at the beginning of each trading period may minimize administrative cost. However, it may also enable large polluters to buy the bulk of permits and use the permits to extract higher payments on the secondary market.

China's choices for a carbon pricing mechanism

So should China choose taxes or a cap-and-trade system? Economic theory gives no universal conclusion whether a tax or a cap-and-trade system is better suited for putting a social price on carbon. The uncertainties around the cost of mitigation and the cost of climate damages are significant and difficult to quantify. The previous sections have shown that both systems have pros and cons, and a number of economists recommend a hybrid of the two instruments in the form of trading system with a price floor and ceiling.

Our conclusion is that a domestic carbon tax is probably the more robust instrument for cutting carbon emissions at this stage of China's development, and it is also the favoured option of China's policy-makers. For instance, auctioning permits (as under a cap-and-trade system) is a politically sensitive issue in China. It is expected to use free allocation rather than auctioning of permits in the sulphur trading regime for the electricity sector. China has little experience with trading permits and it will take time to cut down the transaction costs of a trading regime.

A carbon tax in China is also the more practical option from a policy perspective. Existing institutions will find it easier to implement them, they may involve smaller transaction costs, and their use would also align with the broader reform of the resources tax base that is currently under way. A trade mechanism requires more sophisticated systems for monitoring and implementation. In addition, in the event that a carbon tariff is introduced, China might face tremendous pressure to put forward a domestic carbon tax, so that exported commodities can be waived under the current WTO rules.

China could choose to adopt a carbon taxation system in the early stages of its transition to a low-carbon economy, complemented by a cap-and-trade system that could become the more dominant carbon pricing instrument further down the track. Because China would have excess emissions rights to sell to OECD countries, such a strategy could generate export revenues. China could also extend its offsets programme to gradually access the global carbon market. This would eventually drive down the transaction costs while building monitoring and reporting capacities for such a trading regime in the future.

Thus, in practice, both carbon pricing systems can coexist in the immediate future, with a domestic carbon tax existing in parallel with cap-and-trade until such time as China chooses to join an international emissions trading regime. Both instruments, however, need to set a credible long-term carbon price for investors, sensibly reallocate revenue from auctioning permits, and ensure that there is broad coverage of all GHG-emitting sectors. Furthermore, both pricing instruments have to be extended by technology support mechanisms to account for market failures in clean technology markets, a subject further addressed in Chapter 5.

A global carbon market

A global carbon market can be established in many ways, either top-down through governments establishing global institutions to cap, trade and monitor emissions, or bottom-up by linking existing efforts to building national or regional emissions trading systems. Clearly, the top-down approach implies considerable challenges for the coordination of international economic policy, and it may be difficult to achieve global consensus on issues such as distribution of permits. Thus, a global carbon market is more likely to emerge step by step, for example, by continuing the current approach of the Kyoto Protocol beyond 2012, or by linking national and/or regional emissions trading systems. Similarly, the market should ideally stretch across all sectors and countries. In reality, emissions trading in some sectors and countries will not be feasible due to uncertainties in monitoring emissions, insufficient legal structures and weak property rights.

Chapter 11 shows that most countries would benefit from engaging in early mitigation and participation in international emissions trading, given that there will be a credible global carbon market in the future. China would benefit from early participation in a global carbon market along with Annex 1 countries, compared to a scenario in which only Annex 1 countries are the first movers in implementing deep carbon reductions. Countries adopting early mitigation targets would be accepting a higher overall reduction burden. However, in most cases, this is more than compensated for by the benefits of early action, namely, avoiding stranded investments and greater future mitigation costs, and collecting revenues from the sale of emission permits.

In a transition period, developing countries might adopt trading mechanisms, for instance on the sectoral level. This would have a number of advantages. Countries could begin decarbonizing their economic growth and build the infrastructure required for a more comprehensive future approach. Countries establishing carbon trading mechanisms would gain experience in market-based climate policy instruments. Finally, this may lessen the concerns of industrialized countries over carbon leakage (see above).

On the government level, in addition to carbon market instruments, instruments such as taxes, trading or standards need to be implemented, possibly utilizing funds fed by international trading to foster low-carbon development.

National or regional carbon markets can address equity issues through creating equitable access to lowest abatement opportunities. They can also

BOX 0.11 LESSONS LEARNT FROM THE EU CAP-AND-TRADE SYSTEM

The European Union has led the world in regional cap-and-trade systems since introducing its EU emissions trading system (ETS) in 2005. Despite being considered broadly successful, the European ETS has suffered from several problems. Robust emissions data were absent at the beginning of its operation and the cap was originally set too high, resulting in very low prices. Also, not enough sectors are in the system, and emission rights were not sold but handed out for free. The review by Brunner et al in Chapter 8 of the EU ETS experience highlights characteristics that should be included in the design and implementation of a comprehensive and robust future trading system for China:

- The trading system coverage, both in terms of sectors and GHGs, should be as broad as possible in order to maximize market liquidity and the range of potential abatement options.
- Sectors with highly dispersed emission sources (for example transport) should be covered upstream in order to keep transaction cost low. Sources which are difficult to measure and monitor (for example agriculture) should be excluded from coverage to safeguard the cap's environmental integrity.
- Setting and communicating long-term trajectories for cap and coverage is of paramount importance for enhancing predictability and investor confidence.
- Free allocation of permits can significantly distort incentives but may ease the introduction of the system in its early stages. Increasing the use of auctioning is likely to generate benefits in terms of cost effectiveness, distribution and public finances.
- Monitoring, reporting and verification of emissions are critical for ensuring a trading system's transparency, integrity and credibility. Reliable historic data are essential for determining caps at appropriate levels. Stringent verification is essential to correct market distortions and build trust among investors. Institutions need to be put in place to oversee the carbon market and ensure its effectiveness.

BOX 0.12 A CARBON BANK

Future permit markets can be distorted by insecure property rights, imperfect information, limited access to markets in the future or uncertainty about the regulator's future policies. Hence, governments have a key role in managing expectations and supporting the provision of information about long-term abatement options and their costs and risks to all market participants. A carbon bank (see Chapter 6) can assist by stabilizing the expectations of firms and regulating the permit market, while guaranteeing the long-term credibility of the carbon budget.

A carbon bank – which could be set up on both the global level and nationally – can manage permits, define trading ratios, monitor transactions, provide banking and borrowing, offer transparent information and create credibility regarding the fixed amount of permits over time, and hence provide planning security for economic actors. As an independent institution, a carbon bank can reduce regulatory uncertainty about future policies that might be exposed to political pressure (elections or public finance).

address international concerns over the cost of burden sharing and the costs of climate change and emissions abatement by adjusting regional caps and allowing for inter-regional carbon credit and permit trade. With national carbon taxes, this would be more difficult to achieve through redirection of international burden-sharing funds derived from tax revenues.

China in a global carbon market

With international emissions trading regimes emerging worldwide, how can China participate in a future global carbon market? In this section, we introduce a new Chinese proposal called 'inter-country joint mitigation plans' (ICPs). ICPs would act as an intermediate framework for collaborative mitigation action for the period before the adoption of binding mitigation targets. ICPs aim to cost-effectively realize potential for emission reductions in developing countries, while ensuring the required financial flows and technology transfer from industrialized countries.

Additionally, Chapter 11 suggests that developing countries such as China have a range of choices about how to participate in a future global carbon market. These options have varying benefits and shortfalls with respect to the policy's environmental effectiveness, distributional considerations, institutional feasibility and co-benefits.

Economy-wide cap

China could adopt economy-wide caps and have the government trade on behalf of the economy. China could still implement domestic policy instruments for reducing emissions, with options including emissions trading for companies, carbon taxes, combinations of these approaches, standards and R&D support schemes for low-carbon technologies (including subsidies for low-carbon technologies). Moreover, revenues from international emissions trading can be used to implement these domestic instruments, such as via tax cuts.

Adopting a national emissions cap does not necessarily mean that China will be required to reduce emissions. If accepted by industrialized countries, this cap may initially be implemented on a 'no-lose' basis, ensuring that developing countries do not need to buy permits if emissions rise beyond the cap. Nevertheless, there would be an incentive to reduce emissions below the cap via selling emissions. This implies that as long as China's emissions are below the cap, there is an opportunity cost if emissions increase because permits used for emitting cannot be sold on the international market. Thus, even no-lose trading can provide an incentive.

Absolute sectoral caps – government trading

Instead of setting a cap for the entire economy, China could set caps only for particularly emissions-intensive sectors, and the government could trade internationally on behalf of these sectors. Again, domestic policies are required to translate the signal from the international carbon price for domestic companies.

Setting a series of sectoral caps in major developing countries such as China is one way of implementing an international sectoral cap-and-trade regime that would mitigate concerns over carbon leakage. If a developing country cannot sell a permit due to the expansion of domestic emissions, for example, from the aluminium or steel sector, it suffers an opportunity cost. In fact, addressing carbon leakage in such a manner could enable more ambitious reduction targets in industrialized countries. If sectoral caps are implemented, it is important to ensure that the boundaries of the sector are clearly defined so that facilities cannot 'leak' out of a sector (for example through redefining affected facilities). This issue of carbon leakage is further explored in the next section.

Sectoral entity-level cap-and-trade with international linking

China may also implement cap-and-trade systems for certain sectors at the facility level (for example electricity, iron and steel) and enable covered entities to trade permits internationally. Such an approach may also be part of an international sectoral agreement. It would eliminate the need for government trading on behalf of companies. In fact, this approach is identical to the efforts by industrialized countries to link domestic cap-and-trade systems internationally. As with all other approaches discussed in this section, where the cap is set will crucially determine environmental and distributional effects. Also, the covered sectors will directly face the international permit price, and in a competitive market polluters will pass on the costs to consumers wherever possible.

Economy-wide intensity target

An economy-wide intensity-based trading system is characterized by an intensity benchmark for the entire economy, such as in the form of emissions per unit GDP. Credits are generated by taking into account actual GDP and emission levels in relation to the intensity benchmark. As the intensity target is applied to the whole economy, credits are issued to the government. Again, domestic policy instruments are required for this system to succeed.

The major argument in favour of intensity-based systems over absolute targets is that with uncertainty about economic development, there is less risk that a booming economy will exceed the cap (in the case of binding absolute targets, purchasing permits may be necessary). Vice versa, in the case of an economic recession, no excess permits become eligible for sale (as would be the case with absolute caps), which may be a concern for some industrialized countries. The major argument against intensity targets is that they do not provide certainty about emissions reductions, as expanding GDP (or another benchmark indicator) enables more emissions. However, intensity targets may be a useful instrument for transition periods where the major policy objective is to start decreasing emissions growth rates of developing countries.

Sectoral intensity targets

Analogous to absolute caps, intensity targets can also be applied for single sectors instead of the entire economy. The Center for Clean Air Policy (CCAP) (2008) has worked out a widely discussed proposal featuring no-lose intensity targets at the sectoral level. In this proposal, developing countries would implement intensity targets for emissions-intensive sectors such as electricity or iron and steel. When actual emission intensities are below the baseline, the corresponding emission reductions are certified and can be sold on the international carbon market. If emission intensities exceed the baseline, there is no need to purchase permits for developing countries.

For this proposal to be effective, benchmarks need to reflect the technological potential for reducing intensities to ensure that there is an incentive to actually overachieve on the target. Also, benchmarks may reflect the contribution of developing countries to emissions mitigation by setting them lower. As long as the intensity of sectoral emissions is below the benchmark, there is a disincentive to operate facilities with high emission intensities.

However, there are a number of additional issues to consider with such a proposal. It introduces an incentive for operating additional facilities with intensities below the baseline, possibly giving rise to concerns over subsidies for leakage. As with all other government-based trading mechanisms, this proposal leaves open the question of which domestic policy instruments to choose. Furthermore, this mechanism fails to incentivize the abatement option of reducing the demand for a sector's products, which may be used more efficiently or substituted by other materials given an appropriate price signal.

Sectoral projects

Sectoral projects operate by implementing some kind of programme or policy (for example a technological standard or a tax) and then calculating the amount of induced emissions reductions that determine the amount of credits for sale. In general, both governments and companies could initiate such projects. The major challenge is to calculate the emission reductions actually induced by any such programme. Also, it needs to be established whether a policy would not have been implemented in the absence of the emissions trading mechanism (additionality criterion).

Single projects, bundles of projects

Instead of designing policy instruments for the entire economy or sectors, GHG abatement may be rewarded on the project level, an approach employed by the current Clean Development Mechanism (CDM). Methodologies for setting baselines and determining additionality need to be developed for each type of emission reduction activity and transaction costs reduced.

It is also conceivable that several individual projects could be bundled together under one single baseline. This can broaden the scope of single projects, facilitate the process of baseline development, and reduce transaction costs.

International harmonization of carbon prices

From an economic perspective, if a global carbon pricing policy is to be effective, carbon pricing mechanisms should be implemented in most regions of the world and pricing needs to be relatively uniform. Price harmonization can enable deeper emissions reductions and ensure efficiency in the worldwide distribution of abatement effort: with appropriate market institutions, investment in emissions reduction should flow to the countries where the costs of reduction are lowest. Again, emission trading systems offer advantages over national carbon taxes in this respect, as international harmonization of carbon prices is easier with trading systems other than carbon taxes.

However, as noted in the opening of this chapter, a global carbon price can have disproportionately negative effects for developing countries, where incomes and welfare standards are lower. Depending on the level of carbon price and the allocation scheme, a high carbon price could also have a huge impact on the distribution of wealth *between* different countries. For developing countries, carbon emissions or the credits for avoiding them, will account for a much larger fraction of the value of production. The potential dissonance between expensive carbon and cheaper local inputs, as explored by Ackerman in Chapter 13, creates both an obstacle and an opportunity.

The obstacle is that development may be distorted in the direction of activities that yield marketable carbon reductions. Even undesirable activities may be promoted in order to generate carbon credits. In circumstances where governments intend to grandfather permits nationally, and/or need baselines to allocate permits internationally, safeguards are needed to prevent 'carbon-allowance-seeking' investments. Auctioning of permits and/or adopting a resource-sharing approach can assist in avoiding these difficulties.

The opportunity created by this same pattern of prices is that much deeper reductions in carbon emissions will be economical in developing countries. In the simplest terms, saving a tonne of carbon is 'worth' more hours of labour at a lower wage. So there may be a category of carbon-saving investments and technologies that are profitable only in developing countries, where the trade-off between carbon and other inputs is more favourable to emission reduction. With appropriate public initiatives and financing for these technologies, developing countries could 'leapfrog' beyond the patterns of energy use in higher income countries, establishing a new frontier for carbon reduction. Chapter 5 explores the need for R&D in clean energy technologies and details the ICP proposal to facilitate technology transfer to developing countries.

International competitiveness and carbon tariff proposals

One of the obstacles to international action on climate change is the concern that if only some countries introduce a price on carbon emissions they will place their industries at a competitive disadvantage. Other countries with lower carbon prices, or none at all, will have lower costs of production and could win an increased share of world markets. Additionally, some part of the expected reduction in emissions could be lost through 'leakage', as carbon-intensive industries migrate to carbon-tax-free locations.

Carbon tariffs (or border tax adjustments) have been proposed as a measure to eliminate any unfair advantage from low-carbon prices. They are essentially a tariff on the carbon embedded in imports, bringing the price of the embedded carbon up to the importing country's standard. Large-scale emissions leakage can only occur in industries that are both internationally competitive and highly carbon intensive (i.e. energy-intensive, primary materials industries). Thus, proposals have focused on targeting policies specifically at such industries, where international differences in carbon prices could conceivably cause leakage of carbon emissions to lower-priced regions.

Chapter 13 concludes that carbon tariffs targeted specifically at such industries would not be beneficial for developed countries. At the same time they would do little harm to China. Internationally competitive and highly carbon-intensive industries are few and account for only a very small fraction of US and European economies. And as China has a comparative advantage in only one such industry (see Box 0.13), carbon tariffs would also have little effect on China's trade. Thus, China does not have to fear a global environment where carbon prices are higher – but neither does it stand to gain significantly.

Consequently, as long as carbon tariffs stay low, they are not a climate tool that would bring a significant distortion to the overall trade flows between China and the rest of the world. Nor would they place internationally competitive, energy-intensive industries at a disadvantage. However, there may be a risk that, even if limited to affected industries, such tariffs can trigger retaliatory tariffs and an escalation of protectionist measures in other sectors.

Moreover, there are numerous practical problems with carbon tariffs. They would have to be differentiated by country of origin, since carbon prices could vary around the world. The taxes would also depend on elaborate calculations of embedded carbon: complex manufactured goods often contain components from more than one country, with differing carbon intensities and, perhaps, differing carbon prices. Carbon tariffs thus hardly seem feasible in practice, as they would require a complete life-cycle assessment for each single product that is imported. Neither are they a well-functioning instrument for combating climate change. Still, there is a risk that nationalistic, beggar-thy-neighbour politicians will be tempted to use them for domestic policy reasons.

Innovation and investment

Bending the global and Chinese emissions curves to stay below the 2°C target requires new solutions. The market mechanisms for pricing and trading carbon discussed above are essential to achieving deep carbon emissions reductions in an environmentally effective and cost-efficient manner. However, the weaknesses inherent in such market mechanisms mean that other policy and regulatory measures – such as innovation, technology, finance, and judicial and

---

BOX 0.13 EMBEDDED CARBON IN CHINA'S TRADE

There are some fears in China that a curb on carbon emissions would hurt China's trade since China's industrial production is relatively carbon intensive. Ackerman explores this topic in Chapter 4 and draws several important conclusions.

First, China's success in world trade is not closely tied to carbon intensity or cheap carbon, rather it is based on cheap labour. China's position as a net exporter of carbon does not result from exporting uniquely carbon-intensive products. China is of course remarkably successful in world trade. And China also has very carbon-intensive industries. However, these two facts have little to do with each other – China's most successful export sectors are not its most carbon-intensive ones. China's comparative advantage resides in a combination of advanced and traditional manufacturing, with only a minor role for natural resources. China is a net exporter of many manufactured goods, including both high-technology products such as electronics and machinery and traditional manufactures such as leather goods, apparel and textiles. None of the leading export industries are extraordinarily carbon intensive. Indeed, other economic sectors, where China does not have a comparative advantage, are on average more carbon intensive.

Second, as China develops, technological change may well eliminate the surplus of embedded carbon in trade, even if China retains a large trade surplus in monetary terms. China uses energy less efficiently and relies more heavily on coal than many of the developed countries it trades with. That is, it has higher carbon emissions per dollar of output. As a consequence, exports from China are more carbon intensive than almost all of the imports into the country. In addition, large fractions of China's carbon emissions occur as a result of demands from international trade: emissions that occur in China are, in many cases, incurred in order to satisfy final demand in other countries. Economic growth, however, is likely to bring more advanced, carbon-efficient technology into use, and may therefore narrow the carbon-intensity gap between exports and imports.

Thus, a vigorous climate policy does not threaten China's trade if the country modernizes and shifts towards new technologies. This does not mean that embedded carbon can be ignored in policy debate. Assigning responsibility for exported carbon to the consuming country will provide incentives for high-income importing countries to aid the development process. If the US 'owns' the share of China's carbon emissions embedded in its imports from China, then investing in modernizing China will be a much higher priority for the US – and likewise for other developed countries. This will help to promote policies that lead to the joint goals of climate and development.

Furthermore, as noted in the previous section, implementation of international sectoral cap-and-trade regimes is a more effective means of mitigating concerns over carbon leakage than carbon tariffs. In addition to the positive incentives of revenues gained through adopting an international cap-and-trade system, sectoral caps in China could also assist in reducing emissions of targeted industries, as those sectors with expanding emissions would lose out on the economic opportunity to sell excess emissions permits on the international emissions trading market.

---

administrative reform – must also be used to promote the required shifts in economies, businesses and consumer behaviour.

Innovation is needed in the energy sector, as well as in other climate-related areas such as transport, building, water management, urban design (World Bank, 2009), technology and technology transfer between countries, if we are to break the link between emissions and economic growth. International negotiations on technology transfer negotiations must reach a compromise between protecting intellectual property rights (mostly held in developed countries) and loosening protection of such rights to enable fast technology diffusion (mostly in developing countries). There needs to be a substantial, stable and predictable source of international finance, accompanied by market reform and regulatory mechanisms that can recognize, support and deepen domestic mitigation and adaptation efforts. Industry also needs to upgrade skills in a variety of areas, ranging from workers to (not least) management.

Technology and domestic innovation policy

Although most of the technologies for a deep carbon reduction path are technically available, they must also become commercially feasible. However, market failures are causing underinvestment and inertia in technological innovation. This is largely due to the prolonged R&D periods prior to commercialization of technology, and capital-intensive start-up phases. Moreover, if there is no early investment, cost reductions will not happen, making cheap, large-scale abatement in the future very unlikely. China therefore must encourage domestic innovation in, and widespread diffusion of, decarbonization technologies. At the same time, international efforts are needed to accelerate technology innovation and transfer, and share experiences in fostering governance capacity.

Fan et al in Chapter highlight the technological demands set by China to shift to a low-carbon economy. To cope with climate change, Premier Wen Jiabao has made five proposals, and among them he states that climate change must be tackled in part through technological progress. According to China's White Paper, *China's Policies and Actions for Addressing Climate Change* (State Council Information Office, 2008) the country will focus R&D on development of:

- technologies that save energy and enhance its efficiency;
- technologies for renewable energy and new energy;
- technologies that can control, dispose of or utilize GHGs, such as CO_2 and methane in industries;
- biological and engineering carbon fixation technology;
- technologies for the clean and efficient exploitation and utilization of coal, petroleum and natural gas;
- technologies for manufacturing advanced equipment for coal- and nuclear-generated power;
- technologies for capturing, utilizing and storing CO_2;
- technologies that control GHG emissions in agriculture and how land is used.

At the top of this list is energy efficiency in both energy supply and end-use, particularly in industries, transportation and building. Renewable energy technologies are prioritized, including lower-cost wind power and low wind-speed turbines, photovoltaic building materials and large-scale solar systems, advanced bio-refineries and cellulosic biofuels, water photolysis and energy storage options. Recovery and utilization of GHGs come as number three on the list.

However, despite China's advances in domestic technological innovation, research by Yang and Xing (2009) indicates that there is further scope to improve its framework conditions for innovation. Some of the measures include increased public investment in R&D of low-carbon technologies, support for demonstration projects, tax and other incentives for R&D partnerships, measures to promote competition and innovation, phasing out subsidies to established energy technologies, and establishing institutional mechanisms to facilitate and coordinate efficient public investment.

A fully fledged domestic innovation environment and policy for low-carbon technology would maximize opportunities available through international technology transfers and will also harbour other spin-off benefits. Some of these are opportunities to rapidly advance through or bypass stages of technological development, assist macroeconomic shifts to a knowledge-based economy, and create new industries and job opportunities in sectors such as renewable energy, environmental technology and energy efficiency. For instance, a recently completed study by the China Greentech Initiative estimates that China could build a green tech market worth $1 trillion per year (China Greentech Initiative, 2009).

Chapter 6 notes that the gains from innovation will not only be limited to improving energy and carbon intensity, but will also continue to drive improvements in labour productivity in different sectors. Hence, technological change influences both the speed of growth and the direction of growth.

There is still substantial uncertainty about the regional distribution of mitigation costs. Recent studies suggest that technological innovation is not only of key importance for reducing the overall costs of climate change mitigation, but also for making the regional distribution of mitigation costs less dependent on the allocation of emission permits. The larger the technological innovation, and the higher the institutional flexibilities for achieving the overall mitigation target, the lower is the global carbon price. Thus, international cooperation and innovation will reduce the financial flows associated with carbon trade and also help to take the pressure off the negotiations on the distribution of emission rights under a global cap-and-trade framework (Luderer et al, 2009; Lüken et al, 2009).

A new plan to boost technology transfer

The domestic innovation environment can be enhanced by international technology transfer. Today, the primary international institution for facilitating technology transfer is the CDM under the Kyoto Protocol. There are well-known problems with today's CDM system. It is cumbersome and slow, not least because it is difficult to ascertain that the projects really add 'additional' mitigation. Furthermore, a country receiving assistance under the CDM does not have a cap on emissions and so CDM projects may actually increase total emissions. Fan et al's assessment of the CDM experience in China in Chapter 12 reveals several other limitations of the current system:

- The Chinese CDM shows a weak link between technology transfer practice and the technology demands identified at the national level. Institutionally, the CDM is not linked to the country's national strategy or targets and therefore the CDM is of little benefit to China's national emission reduction actions.
- In the present CDM market, there is little incentive to transfer technology on the side of buyers of emissions rights and low interest in technology transfer from project owners.
- The insufficient technology transfer in CDM practice reflects major institutional gaps in promoting technology transfer. The CDM does not have a technology transfer mandate, nor is it a criterion for the Executive Board when approving or registering a CDM project. Furthermore, there are no specific proposals or mechanisms for promoting the transfer of environmentally sound technologies.

On top of these problems, the financial benefits of the CDM are minor compared to the investment needs in China. In other words, while China is significant for the CDM, the CDM is insignificant for China. With the inadequacy of CDM practice and the urgency and scale of action required to tackle climate change, new mechanisms will have to play a dominant role in bringing up the level and scale of coverage and operation. In the case of China, it would be beneficial to design a new mechanism that can better respond to its national development strategies and be better connected to the mainstream governance.

Inter-country joint mitigation plans

Chapter 12 proposes the establishment of ICPs. An ICP serves the same goal identified for the CDM but aims at large-scale emission reductions – not only project based or sector wide but economy wide. This puts the sustainable development interests of developing countries centre stage. The ICP is not a new mechanism per se, but rather brings together components from a range of existing proposals. The central idea underpinning the ICP is to broaden the channels and extent of international cooperation. Furthermore, ICPs could serve as an intermediate framework in the near term until binding targets are assumed. Their main strength is that they ensure technology transfer and finance from industrialized countries.

Such a joint plan is formed on the basis of cooperation between a host country (developing country) where the emission reduction takes place and one or more partner countries (developed countries) who share the necessary technology and finance, and also the reduction credits. An ICP begins when a host country prepares a national emission reduction plan with voluntary quantified targets, which could be aimed at specific sectors or projects, or even the construction of infrastructure. For example, in China this could mean building smarter electricity grids or developing infrastructure for electrical vehicles. Partner countries review the proposal for joint implementation of the ICP and an agreement is signed by the governments of all participating countries.

An ICP not only establishes emission targets but also specifies the technology and investment required to meet the targets. Therefore, the results of the ICP – emissions reduction, technology transfer and financing – must be measured, reported and verified. Furthermore, the technology transfers should also obey the international rules on intellectual property rights.

An ICP ensures that the partner countries gain the low-cost emission reduction in the host country – just as with the CDM but on a larger scale. ICPs also hold significant commercial opportunities for partner countries and companies because they access the market of mitigation technologies in developing countries. Partner countries can also gain long-term leverage on CO_2 emissions with joint research projects to build new technology demonstration programmes. An ICP with China would be attractive to countries or companies that hold technologies on China's mitigation technology list.

ICPs also provide strong incentives for the host country. The long-term benefits are defined and covered, because an ICP starts with a proposal made by the host country, putting its national interests in the centre and drawing the emission reduction targets from the national sustainable development targets. The host country of an ICP can also leverage foreign investments linked with technology transfers to meet the emission targets and upgrade targeted sectors. In this sense, the ICP has a long-term impact on the sustainable development of the host country, phasing out polluting technologies and phasing in the new ones, avoiding lock-in effects and supporting the structural change towards a low-carbon economy. Furthermore, as developing countries accumulate technological capacity and corresponding knowledge, it will help them to reduce future mitigation costs, so China and the other developing countries would participate in compulsory mitigation earlier. This effect would go some way to help achieve the global emission target.

With the existing interests of the host country and the partner countries as the starting point, and consensus reached through negotiations, ICPs are likely to become self-enforcing. However, managing an ICP is much more complex than running a CDM project. To operate an ICP, two other necessary conditions are needed: the first is the existence of adequate institutional capacity to manage an ICP (particularly at the national level in host countries and at the international level through an institution established to assess the ICP proposals, support and coordinate negotiation, and supervise and assess the ICP implementation). The second is the existence of an international fund, which needs to be set up and operated by a multilateral agency to support the ICP process.

While the ICP proposal addresses a number of the CDM's shortcomings (especially with regard to technology transfer and upscaling) other limitations remain. Similar to the CDM, it will be very difficult to quantify emission reductions against a counterfactual baseline and to ensure 'additionality' of emission reductions (i.e. that the emission reductions achieved by the project are genuine and that the project would not have occurred anyway). Further research is needed on detailed implementation of ICPs to address issues such as: who will receive the revenues for 'low-hanging fruit' – for example, abatement potential that is very cheap compared to the global price of CO_2? Will that rent be taxed, thus generating revenue for the government? Or will only the partner country pay the actual costs? In the latter case, how will projects be managed to avoid one industrialized country taking low-cost projects while others are left with more costly abatement projects?

Investment and financing

Domestic policies and measures in China have been, and are likely to continue to be, the dominant drivers of activity and finance for emission reductions – a point that is illustrated by China's renewable energy policies. For the world to achieve a low-carbon future, a significant fraction of mitigation investment will need to occur in China. According to Chapter 10, this will amount to 20–40 per cent of incremental investment by 2030.[15] In fact, the more ambitious the emission reduction goal, the greater the share of investment that China must receive. As such, mechanisms to recognize, support and deepen domestic policies – as implied by the Bali Action Plan – will be central to future international climate agreements.

Estimates of the scale of investment needs in China vary widely (see Table 0.1) and should be treated with caution due to uncertainty over future economic drivers, the pace of change, and technology costs and availability. At the global level, it is estimated that $200 billion to $1000 billion will be required over the next two decades, and for China, from $35 billion to $250 billion – the lower figures are UNFCCC (2007) estimates, the high ones McKinsey & Company's (2009a, 2009b). Much of this anticipated investment is for emerging technologies such as electric vehicles or CCS – for which there is limited domestic and international support today. Our DCRS outlined in Chapter 5 does not quantify investment costs, but it does seem clear from preliminary analysis that the high ambitions of our scenario will push costs closer to the McKinsey estimates than those of the UNFCCC. This is an important area of further research.

The numbers reported above may make the reader baulk: the size of the required investment is indeed tremendous. However, a significant amount of reductions can be achieved with positive economic returns or with only slight to moderate costs (although with high initial investment) once the savings from lower energy use and other efficiencies are taken into account. In China, energy efficiency in buildings and appliances, and recovery of industrial waste present such opportunities. Investment needs can also be partly financed from utilizing revenues gained from auctioning permits in a cap-and-trade system, levying carbon taxes or removing subsidies on fossil fuels.

Table 0.1 Annual emission reductions and investment needs: UNFCCC Mitigation Scenario (in 2030) and McKinsey Abatement Scenario (average 2026–2030 for global estimates, 2010–2030 for China estimates)

	UNFCCC					McKinsey			
	GHG emission reductions [Gt CO$_2$ equivalent]		Investment and financial flows [US$ billion]			GHG emission reductions [Gt CO$_2$ equivalent]		Investment and financial flows [US$ billion]	
Sectors	Global[c]	Non-Annex I	Global	Non-Annex I	China	Global	China	Global	China
Mitigation investment									
Power generation, of which	8.4	5.0	$148	$73	$36	14.4/10	2.8	$185	$60
Renewables, hydro, nuclear[a]		2	$85	$47	$19				
CCS			$64	$27	$17				
Industry, of which	3.8	2.3	$36	$19	$12	5.0/7.3	1.6	$140	$20
Energy-related			$20	$7	$3				
CCS			$14	$11	$9				
Other			$2	$1	$0				
Transport, of which	2.1	0.9	$88	$36	$11	3.2	0.6	$375	$80
Efficiency (incl hybrids)			$79	$32	$11				
Biofuels			$9	$4	$1				
Buildings (and appliances)	0.6	0.3	$51	$34	$4	1.3/3.5	1.1	$250	$60
Waste	0.7	0.5	$0.9	$0.6	$0.1	1.5		$10	
Agriculture	2.7	0.4	$35	$13	$4	4.6	0.6	$0	
Forestry	12.5	12.4	$21	$21		7.8		$55	
Technology R&D			$45						
Other									
Total Mitigation investment	317	21.7	$425	$176	$68				
Avoided fossil fuel and energy Infrastructure investment								Included in figures above	
Transmission & distribution[b]			-$101	-$41	-$18				
Fossil fuel generation			-$65	-$31	-$11				
Fossil fuel supply			-$58	-$32	-$3				
Investment savings			-$215	-$111	-$32				
Net mitigation			$210	$65	$36	37.8	6.7	$1000	$190–250

Notes: a Combined additional investment needed in renewables, nuclear and hydropower; b Does not consider increased transmission and distribution needed to provide electricity access to unserved populations in developing countries; c Where two numbers are shown for emission reductions, the first number assigns all electricity emission reduction to the power generation sector (similar to UNFCCC), while the second allocates emission reductions due to decreased electricity demand to industry and buildings sectors.

Source: UNFCCC (2007); McKinsey & Company (2009a, 2009b)

Moreover, these estimates do not include benefits to other public policy goals, such as energy security, air quality, cost savings and increased competitiveness. Climate change investments will yield social, environmental and financial returns – a cleaner, more modern, energy-efficient and healthier society, as well more rapid development of labour-intensive service sectors. Furthermore, they will reduce the risks of climate change-induced natural disasters, for which the socioeconomic costs are likely to exceed any mitigation efforts. In the longer run, the true costs of climate change are likely to be much higher than the amount of investment now required to tackle it.

Public domestic finance has been, and will continue to be, an important source in the early stages of transition to a low-carbon economy. In China for instance, state-owned utilities have directly financed much of the investment in renewable energy, providing equity and raising debt – often as high as 80 per cent of total finance – largely from state-owned banks. To the extent that domestic capital is abundant and domestic renewable energy policies themselves are the principal driver for investment, as has been the case in China, favouring domestic investment may not pose a constraint on overall renewable energy investment, at least for commercial renewable energy technologies.

Private financing and the carbon market will play an increasingly important role in meeting China's investment needs. For instance, the successful initial public offering (IPO) of Chinese solar company Suntech at the end of 2005 sparked investor interest. Since then, a steady stream of solar IPOs has followed, with many more in the pipeline. Overall, there is still relatively little private venture capital in China, though the China Environment Fund and others are actively seeking to scale up 'clean tech' investment.

However, to substantially upscale green technologies such as wind and biomass capacity may ultimately require a greater diversity in financing instruments and international support. International bilateral and multilateral funds have helped to increase the uptake of renewable energy and energy efficiency and build capacity in China and other developing countries. But to meet ambitious climate targets, far greater efforts will be needed in the future.

China's investment needs are predominantly in the power generation sector over the next decade, in CCS from 2020 onwards and in the transportation sector (for example electric vehicles) in the longer run. Our research suggests the types of financing mechanisms best-suited to China's needs are as follows:

- A combination of domestic policies and international funds providing debt finance, capacity-building grants and access to advanced technologies. These could be driven towards low-cost and cost-effective options in energy efficiency for vehicles, industry and construction as well as waste management and urban design. International funds can play an important role in guaranteeing private capital for small-scale activities and financing early-stage technologies. Funds (for example Nationally Appropriate Mitigation Action Plans) and technology agreements can support acceleration or deepening of efficiency targets and standards, and the commercialization of more costly and innovative energy efficiency technologies such as hybrid vehicles or smart grid devices (advanced meters).

- Mitigation opportunities where the abatement costs are expected to be in the range of anticipated carbon prices – such as many renewable electricity technologies, biofuels, some advanced efficiency technologies – represent the 'sweet spot' for carbon finance. Sectoral target approaches can support renewable energy standards or incentives, low-carbon fuel standards or product-based emissions intensity goals. For industrial sectors that are amenable to sectoral approaches due to high emissions intensity and homogeneous products (steel, cement, aluminium), sectoral 'no-lose' targets could support both low-cost abatement options (through policies and measures) and higher-cost technologies (through carbon finance). Project approaches, such as project-based CDM, can be enhanced through lists to target technologies and fuels that exceed common practice or high performance threshold.
- Where the marginal abatement costs are significantly higher than carbon prices in the near term, international grant-based funds could support R&D to bring down costs and to support demonstration and deployment, such as in CCS, electric and other vehicle technologies, and higher-cost renewable energy technologies.

There are a large number of proposals on international climate financing that attempt to bridge the shortfall between current investment and projected needs. Several of these reflect more innovative mechanisms to raise carbon finance and leverage private–public partnerships. They include recycling revenue from global carbon taxes; investing a portion of developing countries' foreign exchange reserves into mitigation and adaptation; modifying conditions for currency provision (for example through donating Special Drawing Rights, or debt-for-clean-energy programmes); and using public money to partially guarantee green investments in developing countries (*The Economist*, 2009b). Ultimately, an integrated approach or an international coordinating body will be essential to ensure that a mix of carbon market and fund-based mechanisms will function together in an efficient and adequate manner.

A low-carbon China is a modern China

The transition to a low-carbon economy will require huge investment but this investment will also provide massive benefits. To name only a few, China would be more energy efficient as well as more energy secure; the air would be breathable for millions of people with respiratory problems caused by carbon-fired power plants and heating; and transportation would be safer and cleaner.

In this sense, China's transition to a low-carbon economy is an integrated part of the modernization of China. In our view, the way towards what has been called a harmonious society goes hand in hand with the low-carbon strategy we have outlined above.

Box 0.14 Chinese capital seeking productive investment opportunities

China's relative position in the world is changing rapidly, and this will have implications for the issues analysed in this book. China continues to build its financial strength through the accumulation of the largest foreign exchange reserves in the world; massive savings from domestic banks; the profitability of the largest Chinese companies; the growing clout of Chinese institutional investors such as the sovereign wealth funds, banks, insurance companies, private equity funds; and through large state-owned companies such as China Energy Conservation Investment Corporation. Some of these funds could be used to nurture profitable green investment: there will be growing green business opportunities in the areas of renewable energy, environmental technology, energy efficiency, sustainable transport and vehicles, tourism and agriculture.

As noted by Nicholas Stern in the preface to this book, taking strong action now will create exciting new opportunities for growth and jobs. The next long upswing can be driven by the transformation to fossil-free technologies. Young and dynamic firms investing in future growth sectors and new technologies will create a sustainable job base for the future. Low-carbon investments in particular can be pro-growth and pro-job creation. Energy savings can also create the space for financial savings, enabling consumers to afford other goods and services, boosting demand. And China, by investing early in this new growth, could give itself a greater competitive edge internationally.

Implementing innovative policies for private–public partnerships could attract more foreign capital and technology into China to accelerate a transition to a low-carbon economy. For instance, China could give preferential treatment to foreign participants who provide either technology or capital. It could also allow first-mover foreign investors a guaranteed rate of return in order to give these investors comfort and predictability, and to encourage the transfer of technology, know-how and expertise.

China's rapid growth over the past three decades has been immensely successful in that it has lifted more people out of poverty than at any time in man's history. Hundreds of millions of Chinese have risen to a level of security and prosperity that nobody thought would be possible only a couple of decades ago. But this growth strategy now needs to be renewed for the following reasons:

- It has been based on cheap coal, creating environmental degradation on a grand scale. The health of the population and access to fresh water and unpolluted land now necessitate a move away from coal as the primary energy source of China.
- The rapid rise of manufacturing has made China the shop floor of the world. However, this development path is now closed. China cannot rely on rapidly rising exports as the main driver of growth, now that it is the world's largest exporter and the second largest economy in the world, in purchasing power parity terms.
- In China, the manufacturing sector has a much larger share of GDP – about 10 per cent higher than in comparable countries. This creates environmental problems and makes job creation more difficult. In the future, growth must be built to a much larger extent on consumption and an expanding service sector.

- As urbanization continues and millions of rural people move to the cities, a service-based economy will make it easier to create jobs. The service and consumption sectors are more labour-intensive, whereas the old heavy manufacturing sectors are not only bigger polluters but also more capital intensive.
- The global imbalances that contributed to the recent financial crisis – huge American deficits financed by borrowing from China and its surpluses – must be amended. This can only be done by the US increasing its savings and reducing its imports, while China shifts towards a less export-dependent growth strategy.
- In the longer term, the Chinese capital markets need to be modernized. To a large extent this means that old habits of furnishing capital-intensive manufacturing with cheap credit must be abandoned, and that the consumption and service sectors must be able to compete for capital on equal terms.

All of the above point in the same direction. China's modernization fits hand in glove with the 'greening' of China. The necessary investments in new low-carbon energy sources, R&D and education, services, a new transport infrastructure and forestation all contribute to creating a modern, harmonious China. Furthermore, such investment will make China more competitive at the global level as it takes the lead in new low-carbon products and sectors.

The innovative, entrepreneurial and pragmatic approach that has permeated China since the launch of the reform era in 1978 will serve the country well in her endeavours to become a low-carbon economy. Technology, pragmatism and financial resources of various forms will be the catalysts that catapult China into a leading position.

China's quest to become a strong and responsible member of the global economy through accelerated modernization will allow it to emerge as a global leader in the transformation towards a low-carbon economy. Looking at China's industrial evolution over the past 30 years, it has more often than not been the case that, once the direction has been set, China has managed to make rapid leaps in various sectors of the economy (for example the telecom/IT/internet and the automobile industries). In both these sectors China is now the world leader, both in terms of absolute size as well as in terms of technological development.

We hope and trust that China's next Five-Year Plan will be a 'green' plan, reflecting not only a more modern growth strategy to create jobs and increase prosperity, but also a plan for a low-carbon China, which takes a leading role in the global fight against climate change.

The world needs global leadership. China can contribute to this, acting as a role model for many developing countries. And by investing in China's climate action, the world can share in the dividends of a more climate-secure future.

Notes

1   Developed country commitments as Annex 1 signatories to the UNFCCC. See the appendix to Part I for a list of Annex 1 and non-Annex 1 countries
2   The relevant science and its implications have been presented quite clearly by Meinshausen et al (2009), who state that 'Limiting cumulative CO_2 emissions over 2000–2050 to 1,000 Gt CO_2 yields a 25 per cent probability of warming exceeding 2°C'. The Intergovernmental Panel on Climate Change (IPCC) would thus describe this scenario as being 'likely', but not 'very likely' to keep warming below 2°C (IPCC, 2007b). An assumption of Meinshausen et al (2009) is that comparably ambitious efforts to limit emissions from non-CO_2 GHGs such as methane, nitrous oxide and halocarbons are concurrently undertaken.
3   These reduction levels are at the stringent end of the ranges presented in the IPCC's Fourth Assessment Report for Annex 1 (25–40 per cent by 2020 and 80–95 per cent by 2050, relative to 1990 levels) for emission scenarios consistent with stabilization at 450 parts per million (ppm) CO_2 equivalent (IPCC, 2007a). If Annex 1 countries were less ambitious, and their reductions reached only the lower end of these ranges (20 per cent by 2020 and 80 per cent by 2050), they would occupy a significantly greater fraction of the available budget: roughly 305Gt CO_2 (between 2010 and 2050), leaving 355Gt CO_2 for non-Annex 1 countries to emit. Three independent analyses (AOSIS, 2009; UNFCCC Secretariat, 2009; Wagner and Amann, 2009) have examined the current Annex 1 countries' pledges and found that the aggregate implied Annex 1 reductions for 2020 fall shy of even the less ambitious end of the 25–40 per cent range.
4   If, alternatively, we assume that China's proportion is proportional to its share of non-Annex 1 population, its budget would be approximately 110Gt CO_2.
5   The three examples given here are simple indicative paths intended to clarify the size of the budget and the significance of timing of reductions. In each path, emissions continue along recent trends, growing at 5 per cent/year until the specified year (2015, 2020 or 2025), reaching a peak level (approximately 10Gt CO_2, 13Gt CO_2 or 17Gt CO_2, respectively), and then steadily declining at the stated annual rate (5 per cent, 11 per cent and 35 per cent, respectively). (Note that these are rates of decline in emissions, not in emissions intensity.) In each of the three paths, China's total emissions over the period 2010 to 2050 would amount to 220Gt CO_2.
6   In addition to providing the required financial and technological support, one could also imagine Annex 1 countries undertaking domestic emissions cuts greater than that assumed here (40 per cent by 2020 and 95 per cent by 2050), and thereby making more of the global emission budget available to the non-Annex 1 countries. If, hypothetically, Annex 1 entirely eliminated its emissions by 2025, it would consume 100Gt CO_2 of the remaining budget (instead of 200Gt CO_2), and leave 560Gt CO_2 to the non-Annex 1 countries, of which approximately 270Gt CO_2 might be used by China. In this case, China's emissions could peak in 2020 and then decline by 7 per cent/year (rather than 11 per cent/year). This is a significant quantitative difference but not sufficient to qualitatively change the picture.
7   In the absence of any other externality than the climate externality, China's emissions rights (in excess of their physical emissions) are compensated by a 'below physical emissions' allowance of the developed countries. In practice, this would amount to developed countries buying rights from China to equalize marginal

abatement costs. The allowance price would reflect the overall scarcity of the budget to reach the 2°C target. Thus, China would receive a money transfer that scales with the number of excess rights (indicating equity principles) and the allowance price (indicating the difficulty to meet the target globally). This money transfer is the economic incentive for China to sell a portion of its rights and follow a more ambitious emissions trajectory domestically. In the presence of additional externalities such as, for instance, a technological spill-over externality, additional measures directed to raise R&D investment and deployment levels and to foster technology diffusion will be necessary.

8 'No regrets' policies refer to policies that reduce GHG emissions with no negative costs because they generate direct or indirect benefits that are large enough to offset the costs of implementing these policies. In other words, there would be no regret in implementing these policies as there would be no costs involved and may even provide benefits, even when climate change is not taken into account. 'No regrets' policies are distinct from 'win–win' policies that provide climate change and non-climate change benefits (for example energy security) (IPCC, 2001).

9 The reference trajectory for constant energy intensity and 25 per cent carbon intensity reduction is based on economic growth assumptions from Jiang (2008) and the International Monetary Fund (2009). These do not account for improvements in fuel mix. The scenario with 25 per cent carbon intensity reduction per five-year period is based on the assumption that China meets its national targets of (1) 20 per cent reduction in energy intensity per five-year period, and (2) 20 per cent renewables in the primary energy mix by 2020. It is assumed that the increase of renewables is the same in 2010–2020 as in 2020–2030. In this scenario, no other national targets are taken into account (for example reforestation targets). Note, the McKinsey projections for 2030 are given in CO_2 equivalent.

10 The trajectory is drawn on the assumption of 25 per cent carbon intensity gains per five-year period, which would be the combination of (1) rolling 20 per cent energy intensity gains, and (2) a non-fossil share of the energy mix that reaches 20 per cent by 2020.

11 In the 2004 dataset used by Stanton in Chapter 3, China's emissions per capita place in the first decile above the median on a scale of greenhouse emissions in relation to living standards.

12 Chapter 3 shows that in a global dataset from 2004 China had the 8th highest emission intensity 0.9t per $1000 (purchasing power parity) per capita.

13 Out of 30 Chinese provinces, 23 are above the 90th percentile by emissions intensity in the international dataset used in Chapter 3.

14 According to the latest IEA *World Energy Outlook*, in industrialized countries the price of a permit to emit a tonne of CO_2 will need to reach $50 by 2020 and $110 by 2030. In developing countries the price would need to reach $30 a tonne by 2020 and $50 by 2030, as reported by *The Financial Times* (2009).

15 UNFCCC (2007) estimates suggest a value of 20 per cent. McKinsey & Company (2009a) provides an estimate for China of $400 billion for 2026–2030, which is roughly 40 per cent of their global estimate.

References

Ackerman, F. (2009) *Can We Afford the Future? The Economics of a Warming World*, London: Zed Books

Agarwal, A. and Narain, S. (1991) *Global Warming in an Unequal World: A Case of Environmental Colonialism*, New Delhi: Centre for Science and Environment

AOSIS (2009) 'Aggregate Annex-l reductions for 2020', in Compilation and Analysis presented by the Alliance of Small Island States to the Ad Hoc Working Group on Further Commitments for Annex 1 Parties under the Kyoto Protocol (AWG-KP), Bonn, 21 June

Baer, P., Athanasiou, T., Kartha, S. and Kemp-Benedict, E. (2008) *The Greenhouse Development Rights Framework: The Right to Development in a Climate Constrained World*, Publication Series on Ecology, Berlin: Heinrich Böll Foundation, Christian Aid, EcoEquity and the Stockholm Environment Institute

Barrett, J., Wiedmann, T., Paul, A., Owen, A. and Scott, K. (2009) *Evaluating Sweden's Emissions: At Home and Abroad*, Stockholm: Stockholm Environment Institute

Cao, J. (2008) *A Dynamic Computable General Equilibrium Analysis of Environmental Taxation and 'Rural–Urban' Migration Distortions in China*, Singapore: Economy and Environment Program for Southeast Asia (EEPSEA)

Center for Clean Air Policy-Europe, The Centre for European Policy Studies (CEPS), Climate Change Capital (CCC), The Centre for European Economic Research (ZEW) and The Institute for Sustainable Development and International Relations (IDDRI) (2008) 'Sectoral approaches: A pathway to nationally appropriate mitigation actions', in *Center for Clean Air Policy Interim Report*, Washington DC: Center for Clean Air Policy-Europe, The Centre for European Policy Studies (CEPS), Climate Change Capital (CCC), The Centre for European Economic Research (ZEW) and The Institute for Sustainable Development and International Relations (IDDRI)

ChinaGate (2009) 'Rizhao xunhuan jingji tuidong GDP "lüshe" zhenzhang' (Rizhao's circular economy promotes 'green' GDP growth), China Gate, 27 February, http://cn.chinagate.cn/chinese/jj/56465.htm

China Greentech Initiative (2009) *The China Greentech Report 2009*, Belmont, MA, Mango Strategy

The Climate Group (2009) *China's Clean Revolution II: Opportunities for a Low-Carbon Future*, Beijing: True North

The Economist (2009a) 'Fossilised policy', *The Economist*, 1 October

The Economist (2009b) 'Investors and climate change: Green backing', *The Economist*, 17 September

Economy, E. C. (2007) 'The great leap backward?', *Foreign Affairs*, www.foreignaffairs.com/articles/62827/elizabeth-c-economy/the-great-leap-backward

Fan, G., Han, G. and Li, L. (2009) *Meeting Global Targets through International Cooperation – the Inter-country Joint Mitigation Plan (ICP)*, Beijing: China Economics of Climate Change

The Financial Times (2009) 'IEA warns carbon price must double', *The Financial Times*, 10 November, www.ft.com/cms/s/0/ab9e8ffc-ce2f-11de-a1ea-00144feabdc0,dwp_uuid=d68cb1fc-a38d-11de-a435-00144feabdc0.html

Ford, P. (2009) 'The world's first carbon positive city will be … in China?' *ABC News International*, we page accessed 16 August

Gapminder.(2008) Gapminder World, podcast, http://graphs.gapminder.org/world

Garnaut, R., Jotzo, F. and Howes, S. (2008) 'China's rapid emissions growth and global climate change policy', in L. Song and W. T. Woo (eds) *China's Dilemma*, Washington DC: Brookings Institution Press

GCI (The Global Commons Institute) (2008) *The Carbon Countdown – The Campaign for Contraction & Convergence*, London: Global Commons Institute

Hallding, K., Han, G. and Olsson, M. (2009a) *A Balancing Act: China's Role in Climate Change*, Stockholm: Commission on Sustainable Development

Hallding, K., Han, G. and Olsson, M. (2009b) 'China's climate- and energy-securities dilemma: Shaping a new path of economic growth', *Journal of Current Chinese Affairs*, vol 38, no 3, pp119–134

HSBC (2009) 'Wind turbine blades: A sizeable production, Baoding, China', HSBC, www.hsbc.com/1/2/sustainability/case-studies/wind-turbine-china

International Energy Agency (2007) *World Energy Outlook 2007: China and India Insights*, Paris: OECD/IEA

International Energy Agency (2008) *World Energy Outlook 2008*, Paris: OECD/IEA

International Monetary Fund (2009) *World Economic Outlook Database*, Washington DC: IMF

IPCC (Intergovernmental Panel on Climate Change) (2001) 'Climate change 2001: Mitigation', in *A Report of Working Group III of the Intergovernmental Panel on Climate Change*, Washington DC: IPCC

IPCC (2007a) 'Policies, instruments and co-operative arrangements', in B. Metz, O. R. Davidson, P. R. Bosch, R. Dave and L. A. Meyer (eds) *Climate Change 2007: Mitigation. Contribution of Working Group III to the Fourth Assessment Report of the Intergovernmental Panel on Climate Change*, Cambridge, UK and New York: Cambridge University Press

IPCC (2007b) 'Technical summary', in S. Solomon, D. Qin, M. Manning, Z. Chen, M. Marquis, K. B. Averyt, M. Tignor and Miller, H. L. (eds) *Climate Change 2007: The Physical Science Basis. Contribution of Working Group I to the Fourth Assessment Report of the Intergovernmental Panel on Climate Change*, Cambridge, UK and New York: Cambridge University Press

Jiang, K. (2008) 'Energy and emission scenario up to 2050 for China', paper presented at the International Forum on Policy Options for Climate Change, 29 April 2009, Beijing

Lans, K. (2008) 'Abu Dhabi bygger koldioxidfri stad' (Abu Dhabi constructs carbon free city)', *DI*, web page accessed 12 February

Luderer, G., Bosetti, V., Steckel, J., Waisman, H., Bauer, N., Decian, E., Leimbach, M., Sassi, O. and Tavoni, M. (2009) *The Economics of Decarbonization – Results from the RECIPE Model Comparison*, RECIPE Working Paper, Potsdam Institute for Climate Impact Research, Potsdam

Lüken, M., Bauer, N., Knopf, B., Leimbach, M., Luderer, G. and Edenhofer, O. (2009) 'The role of technological flexibility for the distributive impacts of climate change mitigation policy', paper presented at the International Energy Workshop, 28 May, Venice, Italy

Masdar City (2009) 'Masdar City', www.masdarcity.ae/en/index.aspx

McKinsey & Company (2009a) *China's Green Revolution: Prioritizing Technologies to Achieve Energy and Environmental Sustainability*, Beijing: McKinsey & Company

McKinsey & Company (2009b) *Pathways to a Low-Carbon Economy: Version 2 of the Global Greenhouse Gas Abatement Cost Curve*, Global: McKinsey & Company

Meinshausen, M., Meinshausen, N., Hare, W., Raper, S., Frieler, K., Knutti, R., Frame, D. and Allen, N. (2009) 'Greenhouse-gas emission targets for limiting global warming to 2°C', *Nature*, no 458, pp1158–1162

Ministry of the Environment and Ministry of Enterprise, Energy and Communications (2009) *Towards an Eco-efficient Economy – 12 Swedish Examples*, Stockholm: Ministry of the Environment and Ministry of Enterprise, Energy and Communications

Reuters (2009) 'G20 agrees to phase out fuel subsidies', 25 September, www.reuters.com/article/GCA-GreenBusiness/idUSTRE58O18U20090926

Rockström, J., Steffen, W., Noone, K., Persson, A., Chapin, F. S., Lambin, E. F., Lenton, T. M., Scheffer, M., Folke, C., Schellnhuber, H. J., Nykvist, B., de Wit, C. A., Hughes, T., van der Leeuw, S., Rodhe, H., Sörlin, S., Snyder, P. K., Costanza, R., Svedin, U., Falkenmark, M., Karlberg, L., Corell, R. W., Fabry, V. J., Hansen, J., Walker, B., Liverman, D., Richardson, K., Crutzen, P. and Foley, J. A. (2009) 'Planetary boundaries: Exploring the safe operating space for humanity', *Ecology and Society*, vol 14, no 2, article 32

Scott, K. (2009) *A Literature Review on Sustainable Lifestyles and Recommendations for Further Research*, Stockholm: Stockholm Environment Institute

State Council Information Office (2008) 'White paper: China's policies and actions on climate change', 29 October, www.china.org.cn/government/news/2008-10/29/content_16681689.htm

UN News Centre (2009) 'Secretary-General Ban Ki-moon General Assembly', 22 September, www.un.org/apps/news/infocus/sgspeeches/statments_full.asp?statID=582

UNFCCC (United Nations Framework Convention on Climate Change) (2007) *Investment and Financial Flows to Address Climate Change*, Bonn: UNFCCC

UNFCCC Secretariat (2009) 'Compilation of information relating to possible quantified emission limitation and reduction objectives as submitted by Parties', 12 June, Bonn: UNFCCC Secretariat

US Department of State (2009) 'The Pittsburgh Summit: Acting on our global energy and climate change challenges', www.pittsburghsummit.gov/resources/129661.htm

Wagner, F. and Amann, M. (2009) *Greenhouse Gas: Air Pollution Interactions and Synergies Gains: Analysis of the Proposals for GHG Reductions in 2020 Made by UNFCCC Annex I Countries*, Laxenburg: IIASA

Wiedmann, T., Wood, R., Lenzen, M., Minx, J., Guan, D. and Barrett, J. (2008) 'Development of an Embedded Carbon Emissions Indicator – Producing a Time Series of Input-Output Tables and Embedded Carbon Dioxide Emissions for the UK by Using a MRIO Data Optimisation System*, Report to the UK Department for Environment, Food and Rural Affairs by Stockholm Environment Institute at the University of York and Centre for Integrated Sustainability Analysis at the University of Sydney, London UK: Defra

World Bank (2009) 'Overview: Changing the climate for development', *World Development Report 2010*, Washington DC: World Bank

Yang, H. and Xing, W. (2009) *China's Mitigation Strategies, Policies and Institutions*, Tsinghua: Tsinghua University

Appendix: List of Annex 1 and non-Annex 1 Countries

The UNFCCC divides countries into different groups according to their differing commitments. The Annex 1 and the non-Annex 1 countries are listed in this appendix.

Annex 1 Parties include the developed countries that were members of the OECD in 1992, plus countries with economies in transition (the EIT Parties), including the Russian Federation, the Baltic States and several Central and Eastern European states:

Australia	Austria
Belarus*	Belgium
Bulgaria	Canada
Croatia*	Czech Republic*
Denmark	Estonia
European Community	Finland
France	Germany
Greece	Hungary
Iceland	Ireland
Italy*	Japan
Latvia	Liechtenstein*
Lithuania	Luxembourg
Monaco*	Netherlands
New Zealand	Norway
Poland	Portugal
Romania	Russian Federation*
Slovakia*	Slovenia*
Spain	Sweden
Switzerland	Turkey*
Ukraine*	United Kingdom of Great Britain and Northern Ireland
	United States of America

Non-Annex 1 Parties are mostly developing countries:

Afghanistan	Albania*
Algeria	Angola
Antigua and Barbuda	Argentina
Armenia*	Azerbaijan
Bahamas	Bahrain
Bangladesh	Barbados
Belize	Benin
Bhutan	Bolivia
Bosnia and Herzegovina	Botswana
Brazil	Brunei Darussalam
Burkina Faso	Burundi
Cambodia	Cameroon
Cape Verde	Central African Republic
Chad	Chile
China	Colombia
Comoros	Congo
Cook Islands	Costa Rica
Cuba	Cyprus
Côte d'Ivoire	Democratic People's Republic of Korea
Democratic Republic of the Congo	Djibouti
Dominica	Dominican Republic
Ecuador	Egypt
El Salvador	Equatorial Guinea

Eritrea
Fiji
Gabon
Georgia
Grenada
Guinea
Guyana
Honduras
Indonesia
Israel
Jordan
Kenya
Kuwait
Lao People's Democratic Republic
Lesotho
Libyan Arab Jamahiriya
Malawi
Maldives
Malta
Mauritania
Mexico
Mongolia
Morocco
Myanmar
Nauru
Nicaragua
Nigeria
Oman
Palau
Papua New Guinea
Peru
Qatar
Republic of Moldova*
Saint Kitts and Nevis
Saint Vincent and the Grenadines
San Marino
Saudi Arabia
Serbia
Sierra Leone
Solomon Islands
Sri Lanka
Suriname
Syrian Arab Republic
Thailand
Togo
Trinidad and Tobago
Turkmenistan*
Uganda
United Republic of Tanzania
Uzbekistan*
Venezuela (Bolivarian Republic of)
Yemen
Zimbabwe

Ethiopia
The former Yugoslav Republic of Macedonia
Gambia
Ghana
Guatemala
Guinea-Bissau
Haiti
India
Iran (Islamic Republic of)
Jamaica
Kazakhstan*
Kiribati
Kyrgyzstan
Lebanon
Liberia
Madagascar
Malaysia
Mali
Marshall Islands
Mauritius
Micronesia (Federated States of)
Montenegro
Mozambique
Namibia
Nepal
Niger
Niue
Pakistan
Panama
Paraguay
Philippines
Republic of Korea
Rwanda
Saint Lucia
Samoa
Sao Tome and Principe
Senegal
Seychelles
Singapore
South Africa
Sudan
Swaziland
Tajikistan
Timor-Leste
Tonga
Tunisia
Tuvalu
United Arab Emirates
Uruguay
Vanuatu
Viet Nam
Zambia

* Party for which there is a specific COP and/or CMP decision

Part II

Towards Climate Protection for Development

1

Fair Emissions: Rights, Responsibilities and Obligations

Fan Gang, Cao Jing and Su Ming

History, reality and future trends

Climate change results from the accumulation of historical greenhouse gas (GHG) emissions. Therefore, the accumulative amount of GHG historical emissions is one of the main tools for determining the climate change responsibilities of each country. Currently, the major cause of global warming is the large number of GHG emissions coming from the large-scale use of fossil fuels by developed countries through their industrial activities since the pre-industrial era.

Based on historical accumulative emission data from the Climate Analysis Indicators Tool (CAIT) version 5.0 for the period 1850–2004, we can see that the Annex 1 countries account for 75 per cent of the world's GHG emissions, in which the US and Europe (25) account for 29.4 per cent and 26.4 per cent, respectively. China alone counts for 8.1 per cent; however, China's per capita carbon emissions are ranked 92nd, about the world average of 5t CO_2 equivalent (eq) per capita (see Figure 1.1). Though developing countries emit less accumulative historical GHG emissions, if developing countries follow the same energy consumption pattern as developed countries, in the future developing countries might even exceed developed countries.

If we only count historical carbon emissions from 1990 rather than from 1850, the share of developing countries is higher. For example, China and India's burden is almost doubled, while the shares of developed countries decline from about 76 per cent to 61 per cent. Thus, the starting point used when counting

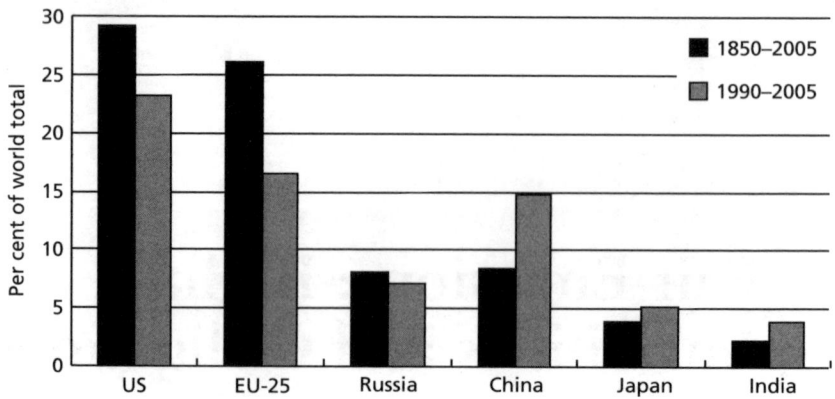

Note: 1850–2004 data are from CAIT and 2005 data are from the Energy Information Administration (EIA).
Source: World Resources Institute (2007) and EIA

Figure 1.1 *Accumulative historical carbon emissions over different timespans*

historical emissions is a major factor affecting how large a burden each country should take when combating the global climate change crisis. A fair burden-sharing rule should go back to the time when the industrial revolution started, when GHGs began to accumulate and affect the global climate system.

The core issue for the future international climate system is how to restrict GHG emissions. In the future, different countries and regions might differ greatly from each other in GHG emission trends for differential policies and actions. Regarding industrialized countries, the future emissions of the EU are flattening while the emissions of the US have been continually increasing. From 2005 to 2030, the global GHG emissions from fossil fuel combustion will increase by 114 per cent, while the proportion in global GHG emissions will increase from 40 per cent to 55 per cent. At that point, China and India will both have increased emissions and faster growth rates.

Consumption and emissions

In this chapter we emphasize the link between the consumption and the GHG emissions, and propose a new accounting system for calculating consumption-based emissions.

What are the relationships between consumption and GHGs? Why should we pay attention to them? Based on economic theory, the ultimate goal of production is final consumption. Therefore, total demand is equivalent to total supply. If people's consumption preferences stay the same as today and developing countries follow the same pattern of favouring high carbon-intensity commodities, there will be no way to reduce GHGs.

Currently, 'embedded carbon' (also called 'embodied carbon' or 'trans-ferred carbon') has become an important concept and is becoming increasingly

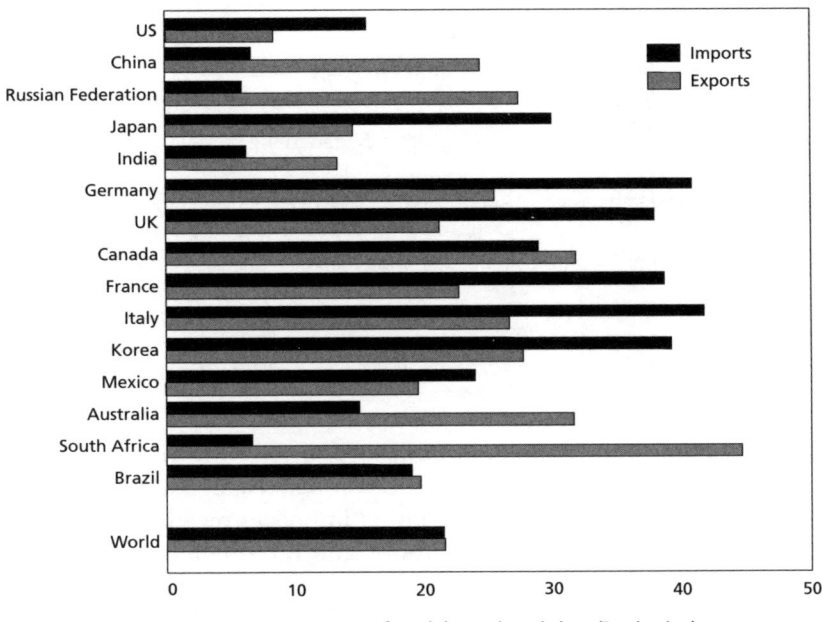

Note: CO_2 emissions embodied in exports (grey) and imports (black) are shown as a percentage of total domestic emissions.
Source: Peters and Hertwich (2008)

Figure 1.2 *Carbon emissions embodied in international trade*

prominent with the accelerating rate of globalization and specialization after World War II. This may be illustrated by a simple example: country A provides capital (K), country B provides labour (effective labour, defined as LE), country C provides energy inputs (E), while the firm itself is located in country D and GHGs are also produced in country D during the production process. In this situation, one can see that it is the consumption of several countries that causes the great emissions of country D. However, it could be possible that commodities are eventually consumed mostly in country E. Obviously, when determining the level of responsibility of each country, it is unfair to simply refer to domestic GHG emissions in one country regardless of the carbon emissions embedded in commodities and consumed by other countries.

Researchers have paid much attention to the unbalanced production and consumption locus; however, their research is only focused on international trade flows, or rather, the transferred emissions of one country that can be obtained by calculating the emissions from the goods of net exports. For example, Peters and Hertwich (2008) calculate the trade-related emissions of all countries in 2001, and find that these emissions are quite large and different among countries (see Figure 1.2). There is also similar research focusing especially on the embedded emissions of China, all of which highlight that the emissions embodied in net exports contribute much to the total emissions of China (see Table 1.1). However, there are limitations to all of this research.

Table 1.1 *A summary of embedded carbon emissions studies for China*

Researchers	Source	Major research results
Bin Shui and Robert Harriss (NARC)	*Energy Policy* (2006)	About 7–14 per cent of the CO_2 of China was embodied in the goods exported to the US from 1997 to 2003
Tyndall Center of Britain	Research report	About 1.1Gt CO_2 were embodied in the net exports of China in 2004, i.e. 23 per cent of total domestic emissions
Christopher Weber, Glen Peters, Dabo Guan and Klaus Hubacek	*Energy Policy* (2008)	In 2005, around a third of Chinese emissions was due to the production of exports, and this proportion rose from 12 per cent in 1987 to 21 per cent as recently as 2002. It is likely that consumption in the developed world is driving this trend

First, by mainly adopting input–output analysis, the research merely obtains a cross-section result; however, although each country's export-embedded emissions can be derived from its input–output data, there is no differentiation between consumption goods and intermediate goods. More importantly, fixed capital investment, together with consumption and exports, is treated as the final demand in the input–output analysis (Weber et al, 2008), which may underestimate the embodied emissions in trade in a dynamic economy. Kainuma et al (2000) compare general equilibrium analysis and input–output analysis when calculating embodied CO_2 emissions and find that the former is better, especially dealing with the changes in consumption structure and policy effects.

Second, by relating embedded emissions to international trade, the results are likely to result in trade protection effects or may be used by political trade conservatives with effects that contradict World Trade Organization (WTO) principles. Last but not least, the research does not directly point out how consumption causes such large levels of GHG emissions and does not reveal how the problem depends on the mode of consumption per se.

The model

Given the above, we propose a new accounting framework for calculating consumption-based accumulative GHG emissions. A one-sector economic-growth model is adopted based on the following assumptions:

- All countries participate in the global market economy; the distribution of factors of production, together with the demand structure, determine relative prices and trade patterns across countries.
- Non-labour factors of production – specifically, capital and energy – can be flexibly allocated to any country in the world. Multinational corporations choose where in the world to produce based on maximizing their profit. This makes the production function and the marginal product of non-labour factors[1] the same in different countries. Only the distribution of

factors across countries and the differences in labour productivity determine the countries in which different goods are produced.

- All investments are eventually turned into consumption. At the same time, it is assumed that during some periods large investment is required for urbanization and public goods in every country. In this way, consumed goods can eventually be used to determine responsibility for GHG emissions as embodied in capital.

The detailed theoretical model is described below.

Households

Households are assumed to be identical; we model the behaviour of one representative household. In each period, the endowments are: physical capital, effective labour and energy inputs. These endowments are given in the first period and are set grow at exogenous, constant growth rates. Each household also has an exogenous share of the stock of each firm. Thus a rational household will maximize its inter-temporal utility as given in equation 1.

$$\max \sum_{t=0}^{\infty} \beta^t U(C_{it}) \tag{1}$$

Here β is the discount factor, U is the utility function, and C_{it} is the consumption of household i in time period t. Given interest rate r_t, effective wage rate w_t and energy price p_t, the representative household needs to provide production inputs and receives income, and also receives dividends from the firms. Then it makes decisions regarding its consumption and investment. Thus the budget constraint can be represented as:

$$C_{it} + I_{it} = r_t K_{it} + w_t L_{it} + p_t E_{it} + \sum_j \theta_{ij} \pi_{jt} \tag{2}$$

Here j indexes the firms in the economy, and the new variables introduced in equation 2 are (omitting subscripts): I for the household's investment; K, L and E, respectively, for the capital, labour and energy the household supplies to the market; θ for the household's share of ownership of each firm; and π for the profits of each firm. If the depreciation rate of physical capital is δ, then the capital investment equation is as follows:

$$K_{it+1} = I_{it} + (1 - \delta) K_{it} \tag{3}$$

Production

Assuming that the production market is perfectly competitive, in each period given the interest rate r_t, wage rate w_t and energy price p_t (the product price is normalized as unity), the producer will maximize his/hers profit as equation 4 shows:

$$\max_{K_{jt},H_{jt}E_{jt}} \pi_{jt} = F(K_{jt},L_{jt},E_{jt}) - r_t K_{jt} - w_t L_{jt} - p_t E_{jt} \tag{4}$$

If we assume constant returns to scale, the profit of a perfectly competitive firm is zero. Thus the total income Y is equal to the total outputs:

$$Y_t = F(K_t,L_t,E_t) = F\left(\sum_j K_{jt}, \sum_j L_{jt}, \sum_j E_{jt} \right) \tag{5}$$

Equilibrium

For factor market equilibrium, supply is equal to the demand for all the inputs:

$$\sum_i K_{it}, = K_t, \sum_i L_{it} = L_t, \sum_i E_{it} = E_t \tag{6}$$

The commodity market is also in equilibrium, so we have:

$$\sum_i C_{it} + \sum_i I_{it} = F(K_t,L_t,E_t) \tag{7}$$

GHG emissions

If the ultimate goal of investment is consumption, then the emissions associated with physical capital should be attributed to the output produced by that capital. So we can assume that when a fraction δK_t of physical capital is consumed in the production process, there will be a release of the corresponding share of the carbon emissions required to produce the physical capital, some of which feeds into consumption and some of which feeds into new investment. If the GHG emissions deposited in physical capital are G_t^K in period t, and those allocated to investment and consumption are G_t^I and G_t^C respectively, then the stock and flow relationship can be denoted in equation 8:

$$G_{t+1}^K = G_t^I + (1 - \delta)G_t^K, G_0^K \tag{8}$$

In each period, consumption and investment will release GHG emissions of $G_t^E + \delta G_t^K$. GHG emissions from investment can be derived as:

$$G_t^I = \frac{\sum_i I_{it}}{Y_t}(G_t^E + \delta G_t^K) \tag{9}$$

Thus GHG emissions emitted directly from consumption can be derived as:

$$G_t C = \left(1 - \frac{\sum_i I_{it}}{Y_t}\right)(G_t^E + \delta G_t^K) = \frac{\sum_i C_{it}}{Y_t}(G_t^E + \delta G_t^K) \tag{10}$$

Therefore each household produces GHG emissions due to consumption of:

$$G_{it}^C = \frac{C_{it}}{\sum_i C_{it}} G_t^C \tag{11}$$

Finally, in the period of time between t_1 and t_2, all the GHG emissions from consumption satisfy the following equation:

$$G_{t_1}^K + \sum_{t=t_1}^{t_2} G_t^E = G_{t_2}^K + \sum_{t=t_1}^{t_2} \sum_i G_{it}^C \tag{12a}$$

Here the left side of the equation shows the GHG emissions deposited in physical capital at t_1 and the sum of all the GHGs combusted from energy inputs, and the right side of the equation denotes the emissions deposited in physical capital at t_2 as well as the sum of GHG emissions for household consumption.

When the time horizon is long enough, we can see that equation 12a can be simplified as:

$$\sum_{t=t_1}^{t_2} G_t^E = \sum_{t=t_1}^{t_2} \sum_i G_{it}^C \tag{12b}$$

where the GHG emissions deposited in the first and last periods can be ignored.

Preliminary results of consumption-based carbon accumulative emissions

Based on the above theoretical model, we calculate accumulative consumption-based carbon emissions from 1950 to 2005 for each country, as well as the related ratio of global consumption emissions. Currently about 24 per cent of global warming potential (GWP)-weighted GHGs in the year 2000 are comprised of the non-CO_2 GHGs, which include methane (CH_4), nitrous oxide (N_2O) and a number of high-GWP fluorinated gases (USEPA, 2006). For this study, due to data limitations, we only provide accumulative carbon emissions.

The database used has been assembled from a variety of publicly available sources. The emissions component is complied from World Resources Institute 1950–2004 data and the US Department of Energy's Energy Information Administration's 2005 data. Data on purchasing power parity and consumption are from Maddison's statistics and the IMF's International Financial

Table 1.2 *Global percentage shares of accumulative domestic emissions and accumulative consumption emissions 1950–2005 for selected countries and groups of countries*

Country/group	Domestic emissions	Consumption emissions	Country/group	Domestic emissions	Consumption emissions
Japan	4.80	6.60	US	26.42	23.93
UK	3.55	4.22	China	10.19	6.84
France	2.26	3.95	Germany	5.64	4.90
Italy	1.82	3.47	Russian Federation	9.31	3.68
India	2.64	4.33	Canada	2.20	1.96
Brazil	1.00	2.72	Australia	1.20	1.10
Mexico	1.24	1.92	Poland	1.90	0.94
Spain	1.02	1.68	Ukraine	2.35	0.86
Argentina	0.56	1.11	Iran	0.86	0.77
Turkey	0.57	1.08	South Africa	1.33	0.58
High income	57.52	60.45	Romania	0.70	0.31
Lower-middle income	20.71	23.12	Kazakhstan	1.03	0.49
Low income	2.29	4.25	Upper-middle income	19.44	12.19
Non-Annex 1	29.75	34.29	Annex 1	70.21	65.72
EU 27	22.48	25.47	EITs	19.50	9.21

Statistics (IFS), respectively. The remaining data are taken from World Bank's World Development Indicators (WDI). The results for selected countries and groups of countries are shown in Table 1.2. Additionally, the domestic emissions of the countries and groups are also shown.

Table 1.2 suggests that there is a great difference between accumulative consumption emissions and domestic emissions for most countries. For most developing countries with abundant energy endowments, such as China, South Africa, Iran, Russia and most economies in transition (EITs), consumption-based emissions are much smaller than domestic emissions. This also applies to Australia and Canada as they are developed countries with abundant energy endowments. However, for most developed countries such as Japan, the UK, France and Italy, consumption emissions are much larger than domestic emissions. This is because most rich countries consume more emission-intensive goods, whereas most developing countries with abundant energy supplies produce most of these commodities, which is inconsistent with the emissions embodied in trade shown in Figure 1.2 over a much longer period.

The consumption emissions of some rich countries such as Germany and the US are a little larger than domestic emissions. The consumption emissions of developing countries such as India and Brazil are much larger than domestic emissions. This can be explained by two arguments. First, Germany and the US possess the most advanced manufacturing sectors and export a great deal of emission-intensive manufactured goods by importing high levels of energy and resources. Also, there were trade surpluses over a very long period of time between 1950 and 2005, whereas the results in Figure 1.1 are calculated by a static model. By contrast, India and Brazil export a great number of primary products and resources in exchange for foreign manufactured goods. Second, the current input–output analysis may underestimate the difference between consumption emissions and domestic emissions of these countries due to the lack of analysis of the emissions embodied in investment in the production of consumption goods.

From the discussion above, it is fair to determine each country's respective responsibility based on consumption emissions rather than their domestic emissions, since it is not production but excessive and luxury goods consumption that should be given the responsibility for increased emissions. In practice, this would ensure that policies for mitigation do not ignore consumption. For example, if a government wants to adopt a carbon tax, it should use tax instruments to correct people's consumption behaviours and lifestyles, and thus indirectly change producers' behaviour from balancing demand and supply. However, current consumption emissions are analysed using every country's total consumption. It is assumed that consumption emissions are proportional to total consumption, failing to differentiate the structure of consumption. Different consumption goods may have different emission intensities, a topic in need of further investigation.

Developmental rights and mitigation responsibilities

To achieve significant reductions in GHG emissions, a self-enforcing climate agreement needs to broaden participation and increase mitigation goals. Currently, the OECD countries, some major developing countries (China, India and Brazil), Eastern Europe, Russia and Ukraine, account for most of the current and future GHG emissions. An international climate agreement involving only a subset of the world's emitters will lead to carbon leakages in non-participant countries, resulting in emissions in excess of desirable global targets. To guarantee wide-ranging adoption, and break the current North–South climate negotiation impasse, it is important to work out a fair framework for an alliance between the North and South. Since the North itself cannot stabilize the climate without the full commitment of the South, and the South cannot agree with any commitment that will jeopardize its development, an equitable and effective burden-sharing allocation system based on UNFCCC's 'common but differentiated responsibilities' is necessary to ensure the rich countries not only deepen their own mitigation targets for their historically huge emissions, but also do whatever they can to help the poor countries develop, increasing poor nations' capacity to adapt and mitigate emissions for a low-carbon future.

A revised GDR framework

Baer et al (2008) first proposed a new burden-sharing framework – the GDR Framework, which considers both sustainable development and equity issues for less-developed countries, by defining Capacity (C), Responsibility (R) and using a Responsibility and Capacity Indicator (RCI) as a weighting product of the two.

Capacity (C) refers to a measurement of resources to pay without sacrificing necessities, which can be described as:

$$C = P \int_{y_{DT}}^{\infty} (y - y_{DT}) f(y, \bar{y}, G) dy \tag{13}$$

Where, P is the population, y_{DT} is the development threshold, $y - y_{DT}$ is the capacity of an individual with income y, $\bar{y}$ is the per capita income, G is the Gini coefficient, and the income satisfies log-normal income distribution $f(y, \bar{y}, G)$.

While responsibility (R) indicates one nation's contribution to the climate problem considering historical reasons, which can be described as:

$$R = P \int_{y_{DT}}^{\infty} (e(y) - e_{DT}) f(y, \bar{y}, G) dy \tag{14}$$

Where, $e(y)$ is emissions at a given level of income, and e_{DT} is equal to the emissions of a person whose income is precisely equal to the development threshold. The quantity e_{DT} behaves analogously to the development threshold, as the 'emission threshold', such that only emissions above this threshold contribute to R. Therefore RCI can be written as:

$$RCI = R^a \times C^b \tag{15}$$

Where a and b are the weight of responsibility and capacity, respectively, which satisfies the condition $a + b = 1$.

In this chapter, when defining each country's burden share, we adopt Baer et al's (2008) work, but make four main revisions to the calculation of RCI. First, we use more comprehensive historical accumulative GHG emissions data from 1850 to 2005 rather than the emissions data from 1990 to 2005 used by them. It has been convincingly argued that all carbon emissions since the Industrial Revolution accumulatively bear responsibility for the current climate problem. Second, each country's accumulative emissions are represented not by its domestic emissions but rather its consumption-based emissions. Each country's consumption-based emissions are the sum of its emissions from two periods: 1810–1949 and 1950–2005. In the former period, each country's consumption emissions are equal to its domestic emissions because of low-level international trade and missing data for this era. In the latter period, each country's consumption emissions are calculated using the method and preliminary results described above. Third, the responsibility's weight is changed from 0.4 to 0.6 and the capacity's weight is changed from 0.6 to 0.4 to further emphasize the widely agreed principle that whoever emits more GHGs should take more responsibility. Finally and importantly, we do not regard this

Table 1.3 *Global percentage shares of capacity, accumulative consumption emissions, responsibility and RCI for selected countries and groups of countries*

Country and group	Capacity	Consumption emissions	Responsibility	RCI M-GDR	RCI GDR
US	33.7	27.5	38.9	37.4	36.0
Japan	9.4	5.3	6.7	7.8	4.3
EU 27	30.5	29.8	36.3	34.0	27.8
UK	4.7	6.8	8.7	6.9	5.9
Germany	6.1	6.6	8.4	7.5	3.6
France	4.5	4.2	5.4	5.1	1.7
Australia	1.7	1.0	1.3	1.5	8.4
Russia	2.0	3.6	2.2	2.1	1.9
Brazil	2.2	2.2	1.6	1.8	2.9
China	2.3	5.6	1.2	1.6	2.7
India	0.1	3.7	0.0	0.0	0.1
South Africa	0.5	0.6	0.4	0.5	0.8
Annex 1	81.9	71.7	89.6	87.2	81.7
Non-Annex 1	18.1	28.7	10.4	12.8	18.3
High income	84.3	66.2	87.5	86.3	83.3
Upper-middle income	8.7	11.2	7.6	8.0	9.9
Lower-middle income	6.9	19.4	4.8	5.6	6.6
Low income	0.1	3.6	0.1	0.1	0.1

R-GDR framework as a protocol forcing countries to take part in an immediate forced mitigation. Rather, the framework should be seen as a basis on which each country's action blueprint can be described and directed to facilitate a global deal in international finance flow and technology transfer, made and shared among countries. Although each country has the responsibility and capacity to maintain their mitigation burden share in this framework, a supplemental standard is needed to determine whether a country should take part in an immediate forced mitigation.

Based on the revisions above, we can estimate the country's total RCI and compare it with the global total to calculate each country's share. In the R-GDR's calculation we use the 2005 database following Baer et al (2008), excepting the consumption emissions database. The results of this calculation for selected countries and groups of countries are shown in Table 1.3. It can be seen that the US and EU27 contribute the most – 37.4 per cent and 34.0 per cent respectively. Overall, high-income countries or Annex 1 countries currently need to contribute 86.3 per cent and 87.2 per cent respectively. But for most developing countries, such as China, India, Brazil and South Africa, each burden share is very small. Therefore, though lower-middle income and low-income countries possess a quarter of the population and almost a quarter of the income and consumption emissions, their total share of the burden is tiny at not more than 6 per cent.

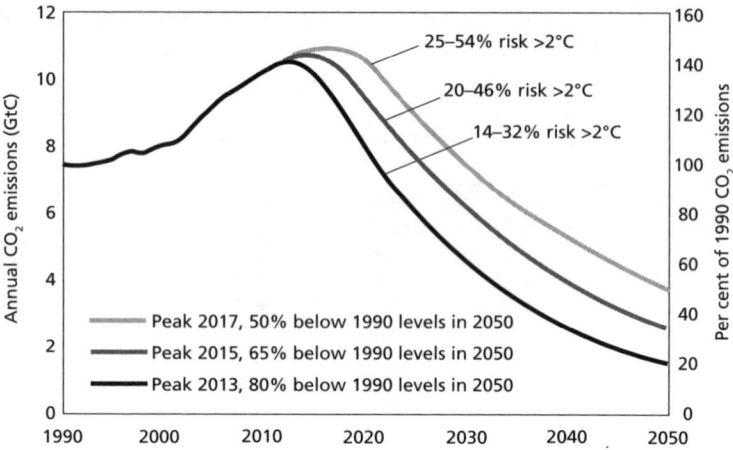

Figure 1.3 *Global emissions pathway for three emergency scenarios, along with each scenario's estimated risk of exceeding the second threshold*

Cap and allocation (and international transfer)

Having calculated the RCI, we can use it to distribute permits to each country under a cap-and-allocate system. This work follows Baer et al (2008), excepting the RCI.

First, the global mitigation requirement is the difference between a global baseline trajectory (constructed as a bottom-up aggregation of national baseline trajectories) and the 2°C emergency trajectory. Figure 1.3 illustrates the global emissions pathway for three emergency scenarios, along with each scenario's estimated risk of exceeding the 2°C threshold. We select the most

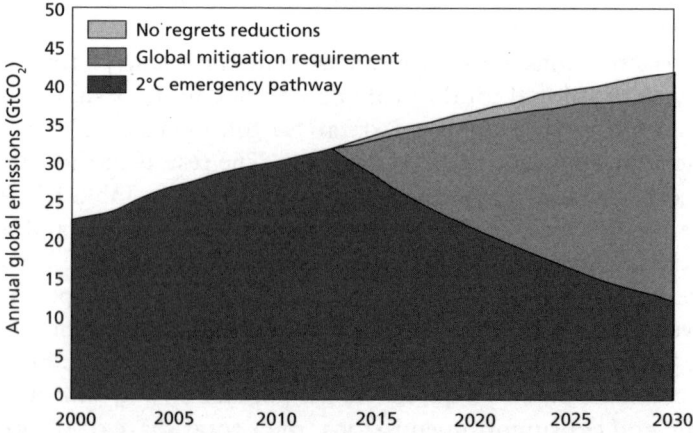

Figure 1.4 *The mitigation gap (mid-grey wedge) between a 'no regrets' baseline (border of mid and light grey) and a second emergency pathway (border of mid and dark grey)*

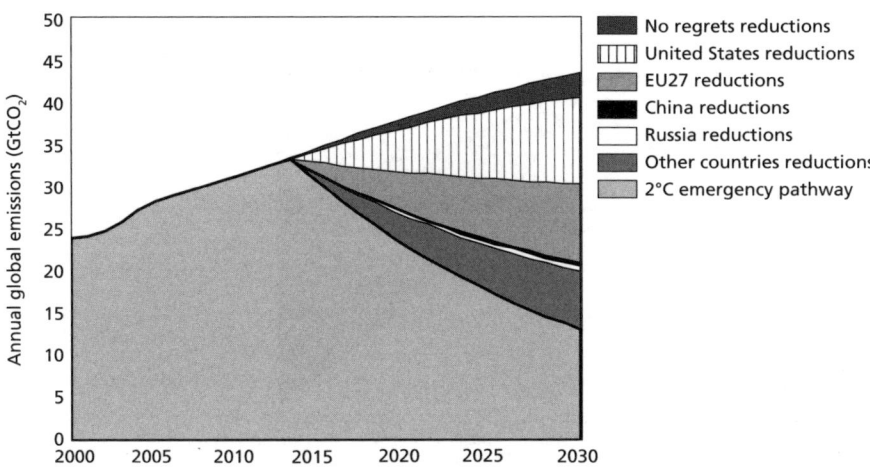

Figure 1.5 *Mitigation requirement, divided into 'obligation wedges' that reflect national/regional shares of RCI*

stringent (the lowest risk in black) 2°C emergency pathway as the goal that the new protocol is to attain. It has emissions peaking in 2013 and dropping off at a constant 5.2 per cent per year, reaching a level of 80 per cent below 1990 levels in 2050. In Figure 1.4, we show two global emissions projections based on a business-as-usual (BAU) trajectory and a 'no regrets' trajectory. The former extrapolates the historical approach to energy conservation, renewables, fossil-fuel subsidies, pollution control and so on, while the latter is a projection if all cost–benefit, negative- and zero-cost emissions reductions were successfully captured. So the 'no regrets' reductions (the light-grey wedge in Figure 1.4) represent free and profitable reductions.

We argue that a country's 'no regrets' trajectory should be adopted as its national baseline. That is to say that all nations should be responsible for capturing their own 'no regrets' reductions first, and that only further reductions, the difference between the 'no regrets' trajectory and the 2°C emergency pathway, should count towards discharging a national mitigation obligation. Given this baseline projection and the 2°C emergency pathway, the global mitigation burden over the period 2013–2030 would amount to 283Gt of CO_2 emissions reductions.

This global mitigation requirement is divided into national mitigation obligations. Each country – however rich or poor it may be – is allocated a portion of the global mitigation requirement in proportion to its aggregate national RCI. Graphically, the global mitigation burden can be divided into wedges, as in Figure 1.5. They show countries and the gigatonnes of reductions they are obligated to pay for. Thus the US's wedge is 37.4 per cent of 283Gt CO_2 or about 106Gt CO_2, while the EU's wedge is 34.0 per cent or about 96Gt CO_2. Russia, a middle-income country, gets 2.1 per cent or about 6Gt CO_2, and China, a developing country, gets 1.6 per cent or about 4.9Gt CO_2.

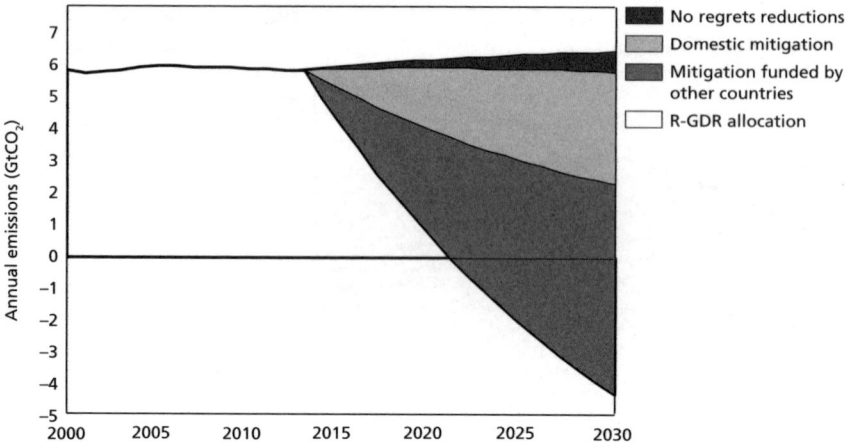

Figure 1.6 *Mitigation requirement of US and the allocation between domestic mitigation and mitigation founded in other countries*

Each country determines how to fulfil its obligation, domestically and internationally. Here, we examine two examples: the US and China. For the case of the US, Figure 1.6 shows US emissions and obligations projected to 2030. If the US reduction obligation were interpreted literally and achieved entirely through domestic mitigation, it would imply reductions of nearly 190 per cent below 1990 levels by 2030, and a US emission level of -4Gt CO_2 by 2030. Obviously, for a mitigation obligation of this magnitude to make sense, the US must not be expected to meet its entire obligation through domestic reductions. Whatever is not accomplished domestically, the US would need to fulfil internationally, by way of reductions in other countries that are 'supported and enabled by technology, financing and capacity-building, in a

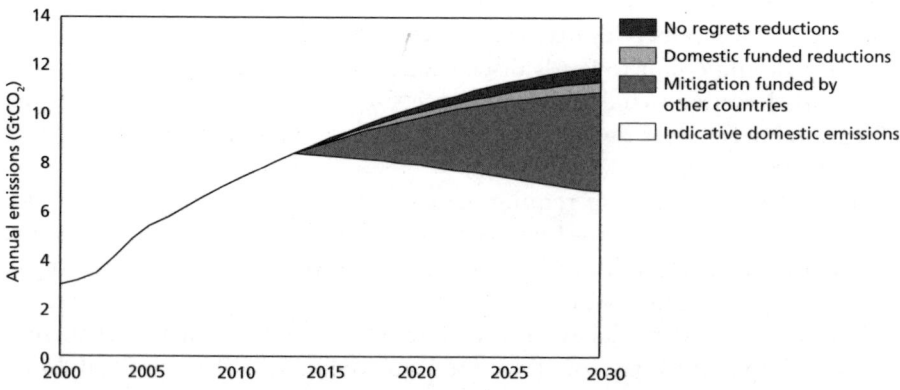

Figure 1.7 *China's domestically funded reductions and mitigation funded by other countries*

measurable, reportable and verifiable manner' (UNFCCC, 2007). Figure 1.6 shows the total US reduction obligation with an indicative division into a domestic mitigation effort and an international mitigation effort. Even if the US reduces its emissions by 5.4 per cent per year starting in 2013, which achieves physical domestic reductions by 2030 of more than 54 per cent below 1990 levels, it satisfies far less than half of the US's total obligation. The remainder must be made in other countries. Therefore, to fulfil the obligation of 108Gt CO_2 by 2030, the US would domestically reduce 38Gt CO_2 and provide funding internationally for the reduction of the remaining 68 Gt CO_2. China, in contrast, is obligated to total reductions of less than 5GtCO_2 by 2030, all of which could be made domestically (see Figure 1.7). At the same time, a much larger quantity of reductions within China, more than 40Gt CO_2 from 2013 to 2030 would be enabled and supported by other countries with higher capacity and responsibility. China's emissions are large and fully exploiting its mitigation potential is essential if we are to keep within the 2°C emergency trajectory.

A gradual step-in framework with graduation threshold

As discussed in the previous section, a gradually implemented project may be preferable in many developing countries. We argue that a reasonable graduation threshold should not be an arbitrary year or income per capita, but the emissions after which the country automatically makes forced mitigation. Adopting time as a threshold makes no sense. Based on the widely agreed principle that whoever emits more GHGs should take more responsibility, we think that income per capita is not as important as emissions and is not a good threshold. Emissions level is a natural threshold. If adopted, it will prompt people to consume less and emit less. Moreover, RCI, combined with income and emissions, may be a more reasonable threshold, though at the current stage this may not be easily accepted by the policy-makers as the post-Kyoto climate architecture is not yet in place. Therefore, we suggest that accumulative consumption emissions be the graduation threshold, so that countries can gradually participate in the fairer regime based on M-GDRs.

We adopt the lowest accumulative consumption emissions 1850–2005 per capita (population in 2005) of Annex 1 countries as the graduation threshold. It can be seen from Table 1.4 that the graduation threshold is 144t CO_2 (Romania). So the accumulative consumption emissions per capita of most developing countries, such as China, India, Brazil and Mexico, are less than this threshold. In contrast, the lowest income per capita in 2005 of Annex 1 countries' is $5583 (Ukraine).

We projected the accumulative consumption emissions per capita and income per capita (population in 2005) in 2015, 2020, 2025 and 2030 under a BAU scenario and 'no-regret' scenario for selected countries to determine when they will be forced to begin mitigation. As can be seen in Figures 1.8 and 1.9, Argentina, Mexico, Turkey and Brazil will need to begin mitigation before

Table 1.4 *Accumulative domestic emissions per capita, cumulative consumption emissions per capita and income per capita for selected Annex 1 and non-Annex 1 countries*

Country	Annex 1=1	Accumulative consumption emissions per capita (t CO$_2$)	Income per capita ($)
Bulgaria	1	228	9328
Croatia	1	263	13,231
Germany	1	848	30,445
Greece	1	314	29,261
Japan	1	435	30,290
Luxembourg	1	550	69,776
Portugal	1	350	19,956
Romania	1	144	9368
Russia	1	262	11,861
Slovenia	1	442	23,010
Switzerland	1	695	35,182
Ukraine	1	190	5583
UK	1	1186	31,371
US	1	966	41,813
Argentina	0	255	10,815
Brazil	0	125	8474
China	0	45	4252
India	0	35	2230
Indonesia	0	61	3209
Korea, Rep.	0	186	21,273
Mexico	0	162	11,387
South Africa	0	136	8478
Turkey	0	128	10,370

2015, China will begin shortly before 2030, but India need not begin even after 2030. Moreover, we argue that if a country takes voluntarily mitigation now, it can delay the date when mitigation becomes compulsory. For example, if China devotes itself to mitigating energy intensity by 20 per cent in the 11th Five-Year Plan and continuing mitigation during 2011–2020, which may mitigate GHG emissions indirectly and can be regarded as voluntary mitigation, it will be possible to delay compulsory mitigation for five to ten more years. Under such conditions, China may not reach the threshold until 2040.

Proposed 'Climate Treaty' regime for post-2012 climate architecture

Based on the analysis above, we suggest a climate treaty for combating future climate change. The main points of this are as follows:

- Each country's responsibility is calculated based on the accumulative consumption emissions.
- Each country's capacity is calculated by considering income level and income distribution, which means some people in each country whose income is above a specific threshold should contribute to mitigation.

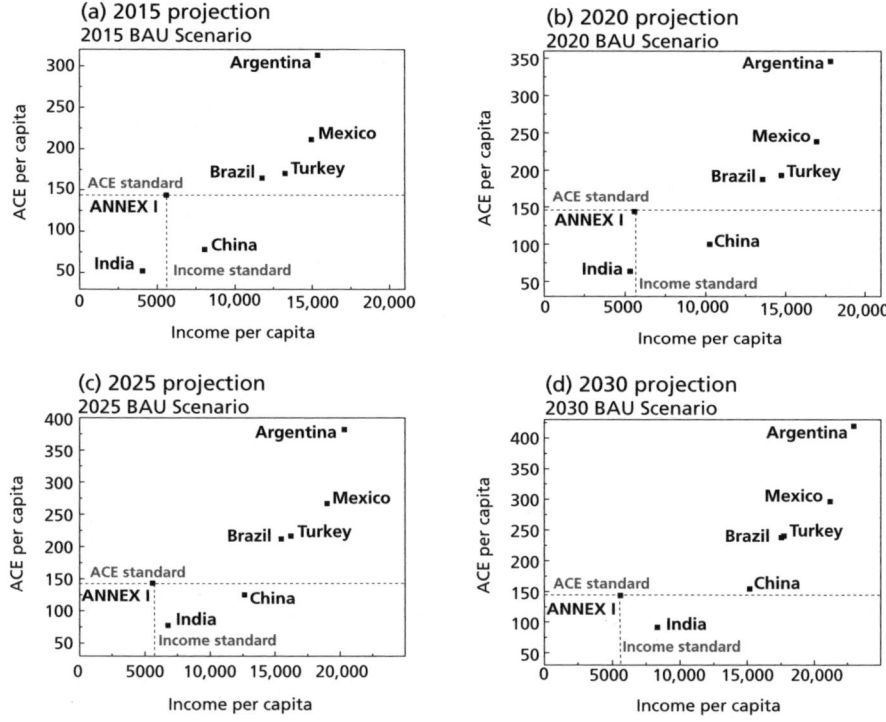

Figure 1.8 *Projection in BAU scenario for selected countries in 2015, 2020, 2025 and 2030*

- Our proposed R-GDR framework, calculated by combining responsibility and capacity, does not propose to force each country to adopt it immediately, but is a basis on which each country's action blueprint can be described and directed, leading towards a global deal in international finance flow and technology transfer that is satisfactorily made and shared among countries, while simultaneously ensuring that each country has a responsibility, capacity and mitigation burden share in this framework.
- The lowest accumulative consumption emissions 1850–2005 per capita (population in 2005) of Annex 1 countries are the graduation threshold, above which the country should automatically face compulsory emissions mitigation.
- Meanwhile, countries under the graduation threshold are not compelled to mitigate but they can engage in voluntary mitigation. This voluntary mitigation will inevitably benefit the nation by delaying the time when the threshold is crossed and mitigation becomes compulsory.
- The countries under the graduation threshold can take part in international carbon trading. They can project their own mitigation trajectory, take part in the 'Inter-country Joint Mitigation Plan' and engage in the mitigation project, technology transfer and financial transfer, together with cooperating developed countries. These reductions in the emissions of countries

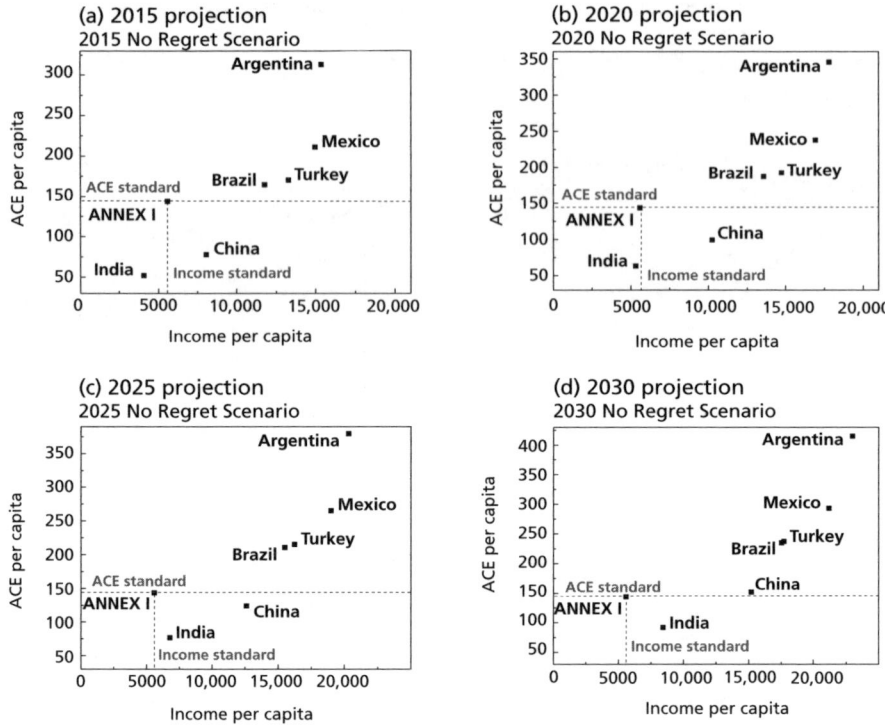

Figure 1.9 *Projection in 'no regret' scenario for selected countries in 2015, 2020, 2025 and 2030*

under the graduation threshold will partially fulfil the cooperating developed countries' obligations under the framework.

- The developed countries joining in the climate treaty should commit to corresponding technology transfers, and financial assistance should be guaranteed.
- If a country under the graduation threshold accepts international technology and financial transfers, it should reduce its graduation threshold correspondingly.

Note

1. The marginal product of a factor of production is the amount that output would be increased by if one or more unit of the factor were employed, all other circumstances being the same.

References

Baer, P., Athanasiou, T. and Kartha, S. (2008) *The Greenhouse Development Rights Framework: The Right to Development in a Climate Constrained World*, Stockholm: Stockholm Environment Institute

Bin Shui and Harris, R. (2006) 'The role of CO_2 embodiment in US–China trade'. *Energy Policy*, vol 34, pp4063–4068

Kainuma, M., Matsuoka, Y. and Morita, T. (2000) 'Estimation of embodied CO_2 emissions by general equilibrium model', *European Journal of Operational Research*, vol 122, pp392–404

Peters, G. P. and Hertwich, E. G. (2008) 'Trading Kyoto nature reports', *Climate Change*, vol 2, pp40–41

UNFCCC (2007) United Nations Climate Change Conference in Bali, Decision 1/CP.13, Bali Action Plan

USEPA (United States Environmental Protection Agency) (2006) 'Climate change – greenhouse gas emissions', www.epa.gov/climatechange/emissions/index.html

Wang, T. and Watson, J. (2007) 'Who Owns China's Carbon Emissions?', Tyndall Briefing No No. 23, Tyndall Centre for Climate Change Research, Norwich, UK

Weber, C. L., Peters, G. P., Guan, D. and Hubacek, K. (2008) 'The contribution of Chinese exports to climate change', *Energy Policy*, vol 36, pp3572–3577

World Resources Institute (2007) *Climate Analysis Indicators Tool (CAIT) Version 5.0*, Washington DC: World Resources Institute

2

Equity Frameworks and a Greenhouse Development Rights Analysis for China

Sivan Kartha

The science and equity challenge

This analysis takes as its key point of departure the overwhelming scientific evidence that action on climate change is urgently needed. Given the widespread and devastating impacts that can be expected from unchecked climate change, combating climate change is a shared global priority. The developed countries certainly must lead, as they have committed to do as Annex 1 signatories to the UNFCCC. But the participation of the developing countries – especially emerging economies such as China – is indispensable. Even if the world's developed countries were to entirely eliminate their emissions, emissions would still need to be curbed significantly in developing countries to keep global concentrations within acceptable levels.

This conclusion is a straightforward consequence of the hard constraints of climate science, and its implications for China are clear. As recently presented in *Nature* by Meinshausen et al (2009),[1] to preserve a reasonable chance of keeping warming below 2°C requires limiting global CO_2 emissions to less than 1000Gt over the first half of the 21st century, for both land-based and fossil fuel-based CO_2 emissions. If heroic efforts are taken to bring deforestation and land degradation to a halt within one decade, then emissions from land could be limited to approximately 60Gt CO_2. This leaves approximately

Table 2.1 CO_2 *from fossil fuels: historic (1850 to 2009)*
and future (2010 to 2050)

	Historic		Future	Total
	1850 to 1999	2000 to 2009	2010 to 2050	1850 to 2050
	Gt CO_2	Gt CO_2	Gt CO_2	Gt CO_2
Global budget	980	280	660	1920
Annex 1 countries	760	140	200	1100
non-Annex 1 countries	220	140	460	820
of which: China	70	60	220	350
rest of non-Annex 1	150	80	240	470

Note: Assumes 1000Gt CO_2 budget for the first half of the 21st century, and Annex 1 emission reduction of 40 per cent by 2020 and 95 per cent by 2050.
Source: Meinshausen et al (2009); CDIAC (2009)

940Gt CO_2 for emissions from the use of fossil fuel, of which we have already emitted approximately 280Gt CO_2. We are therefore left with barely one-third (660Gt CO_2) remaining out of the total budget (1920Gt CO_2) available since the beginning of the industrial age (see Table 2.1).

If the industrialized (Annex 1) countries were to commit to reducing their emissions[2] to 40 per cent below 1990 levels by 2020 and 95 per cent below 1990 levels by 2050, their future emissions would amount to 200Gt CO_2. This would leave 460Gt CO_2 for the non-Annex 1 countries. If we assume that China's portion of this remaining budget is proportional to its share of current non-Annex 1 emissions,[3] its future budget would be 220Gt CO_2.

This combination of scientific facts and simple arithmetic provides a reasonable – and bracing – estimate of the emissions budget available to China. While there is a range of possible emissions paths that would keep China within this budget, all these paths imply bold and ambitious action. For example,[4] China's emissions could peak in 2015 and then decline at a rate of 5 per cent annually (see Figure 2.1). Alternatively, China's peak in emissions could be delayed to 2020, but would then need to be followed by a much more rapid decline of 11 per cent annually. If China's peak in emissions did not occur until 2025, the decline would need to occur at a virtually unattainable rate of 35 per cent every year. And, if the annual 5 per cent rise in emissions continues beyond 2026, the full budget of 220Gt CO_2 would have been expended. While each path reflects unprecedented mobilization, it is clear that the longer that transformation is delayed, the less feasible it becomes. As shown in Chapter 5, the transformation in China is indeed feasible, providing it is launched in the very near future.

There are two further points that must be emphasized about this simple arithmetic exercise.

1 Distinguishing emissions from emission rights

While this simple quantitative exercise tells us that China's emissions must be rapidly curbed to address climate change, it says nothing about how the

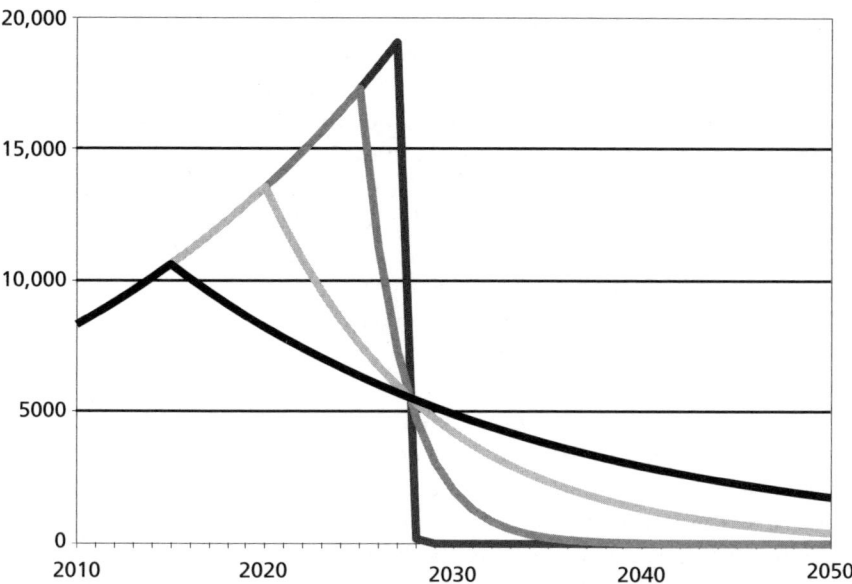

Figure 2.1 *Alternative emission paths for China, consistent with a fossil CO_2 budget of 220Gt CO_2 for the period 2010–2050*

effort required might be shared. Financial and technological resources will be critical for the transformation to occur, and China cannot be expected to bear the burden of generating and providing all those resources on its own. The same, of course, goes for other developing countries.

The allocation of that burden among nations will need to be based on historical and political context, and grounded in the principles of equity that are clearly embodied in the UNFCCC. In other words, the fact that China's emissions must be sharply curbed does not mean that China's *emission rights* should be similarly curbed. The discrepancy between the two must then be filled with a commensurate amount of financial and technological support. One way of providing this support is, of course, a global fund under the UNFCCC. Another (potentially complementary) mechanism is a cap-and-trade system, whereby China is granted emissions rights that it may choose to sell to developed countries that are granted fewer emission rights than they would need even under ambitious pathways of domestic mitigation. In Part I, Fan et al discuss a further arrangement – an 'inter-country joint implementation mechanism' – for mediating cooperation between industrialized and developing countries. Such mechanisms provide China with financial and technological resources to enable a more ambitious emissions trajectory domestically.

. In addition to providing the required financial and technological resources for developing countries, one could also imagine Annex 1 countries undertaking domestic emissions cuts greater than that assumed here (40 per cent by 2020 and 95 per cent by 2050), and thereby making a greater portion of the global emission budget available to the non-Annex 1 countries. If, hypothetically, Annex 1 entirely eliminated its emissions by 2025, it would consume 100Gt CO_2 of the remaining budget (instead of 200Gt CO_2), and leave 560Gt CO_2 to the non-Annex 1 countries, of which China's proportion might be 270Gt CO_2. Given such a budget, China's possible emissions pathways are relaxed somewhat relative to those shown in Figure 2.1. For example, the emission pathway peaking in 2020 could subsequently decline by 7 per cent/year, (rather than 11 per cent/year as shown). This is a modest quantitative difference, which does not qualitatively change the overall picture.

2   **Committing to a risky path**

The above calculation is based on a target that may well be unacceptably risky; the budget of 1000Gt CO_2 entails considerable risks that warming will exceed 2°C, roughly 25 per cent according to the default assumptions of Meinshausen et al (2009). The IPCC would thus describe this scenario as being 'likely', but not 'very likely' to keep warming below 2°C (IPCC WGI, 2007). And, if one makes less optimistic assumptions that are also scientifically defensible, the Meinshausen et al estimate of the chance of exceeding 2°C rises to 42 per cent.

Rockström et al (2009) are among the many research teams that have clearly presented the strong scientific case for adopting a much more restrictive goal, based on the danger of transgressing planetary boundaries 'outside of which the Earth's system cannot continue to function in the stable, Holocene-like state [that] has seen human civilizations arise, develop, and thrive'. Furthermore, delaying climate mitigation increases the need for more ambitious climate targets further down the track and large-scale deployment of yet-unproven mitigation technologies to meet these targets. The key message here from a climate science and risk management perspective is that early mitigation not only reduces the costs but also reduces the risks of delayed climate change action.

Rising to the global climate change challenge will require a concerted, coordinated and committed response from the entire international community. Governments will need to lead the charge in formulating a long-term strategy and creating the institutions, mechanisms, technological alternatives and market environment to correct the greatest global market failure in human history. Promptly establishing these international conditions is necessary to set the expectations of investors who will have to make critical long-term commitments, decision-makers who determine policies at sub-national levels, and citizens whose lifestyles are the ultimate source of emissions.

Burden-sharing frameworks and their implications for China

A global climate regime with the power to induce a sufficiently urgent transformation of the global emission trajectory discussed above has to reflect both equity (taking into account different countries' responsibilities for the climate crises, and capabilities to mitigate emissions and respond to climate impacts) and cost efficiency (reducing emissions and responding to impacts at the least total cost for the global community). Thankfully, the equity and cost-effectiveness objectives do not need to come into conflict, since reductions in one country (the country in which it is cost effective) can be paid for by another country (the country obliged by equitable burden sharing to bear the costs). A number of frameworks for burden sharing have been developed as proposals for allocating obligations within a global climate regime. Each framework applies different distributional principles, which ultimately affect the allocation of emission rights to countries and financial flows between countries.

There are various different principles often put forward in the discussion of allocation rules, such as:

- egalitarianism (same rights to every person);
- ability to pay (capacity to contribute to the mitigation effort);
- polluter pays principle (accounting for responsibility for emissions).

Each of these principles provides a basis for allocating access to a global commons resource. As such, they naturally come into competition with a fourth principle – the principle of sovereignty. Sovereign countries have the inevitable realpolitik tendency to privilege their own national circumstances and to resist joint management of commons resources. This is often expressed (often implicitly) as an appeal to the principle of a 'customary right to emit'. While this is a strong force within the climate negotiations, it provides little basis for a long-term solution to any global commons problem.

Here we provide an overview of some of the burden-sharing frameworks that have been tabled. The comparison takes place on two levels: (1) qualitative comparison – the underlying principles and assumptions are explicated so that the basic structure of a burden-sharing framework can be assessed from the Chinese perspective; (2) quantitative comparison – for several burden-sharing frameworks for which sufficient quantitative detail is available, an assessment is made of the implications from the Chinese perspective.

We reviewed qualitatively key proposals and describe them below and in the columns of Table 2.2, in which the core characteristics of each proposal are indicated.

- **Equal per capita emission rights** is a straightforward approach premised on the equal rights to the atmospheric commons. All countries would be awarded emission allowances in proportion to their population and would

Table 2.2 *Burden-sharing frameworks for allocating emission rights*

	1 Equal per cap	2 Equal per cap (cumulative)	3 Grand-fathering	4 Contraction & convergence	5 Indian Proposal	6.1 GDR	6.2 Modified GDR (three variants)	6.3	6.4
Equal access to *remaining* atmospheric space	X			X	X				
Equal access to entire atmospheric space		X							
Grandfathering: access on the basis of prior emissions			X	X	X				
Capacity: contribution to global effort on basis of available resources						X	X	X	X
Responsibility: contribution to global effort on basis of historical use						X	X	X	X
Consumption basis for assessing responsibility								X	X
Correction for energy resource endowment									X

be free to trade them. The total number of allowances granted globally would steadily decrease along a path consistent with an agreed climate stabilization goal (Agarwal and Narain, 1991).

- **Grandfathering** is a limited interpretation of the 'customary right to emit' whereby countries' allocations are determined by granting all countries emissions in proportion to their prior emissions (constrained by an overall decline in global emissions consistent with the temperature target).
- **Contraction and convergence (C&C)** is a hybrid framework combining grandfathered emission rights with per capita emission rights, with a gradual transition from the former to the latter over a specified number of decades. Countries whose emissions start above the global average would receive allowances that gradually trend down to the global average, while countries whose emissions start below the global average would receive allowances that gradually trend up to the global average (GCI, 2000).
- **Equal cumulative per capita emission rights** extends the concept of equal per capita rights to cover the entire carbon budget, rather than just the portion of the budget remaining for the future. This proposal takes into

Box 2.1 GDRs

The GDR framework aims to ensure global emissions meet a specified climate protection objective, while at the same time explicitly safeguarding a right to development for the world's poor. It straightforwardly and transparently translates the UN notion of 'common but differentiated responsibilities and respective capabilities' into a useful indicator for quantifying national obligations. Specifically, the GDR approach defines an empirically based indicator of national responsibility and capacity, from which a combined RCI enables one to calculate any country's share of the global obligation to solve the climate crisis. *Responsibility* is interpreted as a country's contribution to the climate problem; and capacity as the financial wherewithal to invest in climate solutions. As is typical, responsibility refers to national GHG emissions, and capacity to national income.

However, the GDR approach departs from the conventional definitions in that it defines both responsibility and capacity with respect to a 'development threshold' – a level of well-being that is modestly above a global poverty line. Income below this level does not count towards the measurement of a country's capacity. Nor do emissions that correspond to consumption below this level count towards the measurement of a country's responsibility. And, because the right to development adheres to individuals, not nations, capacity and responsibility are calculated in a manner that accounts for intra-national disparities in income and emissions.

By calculating burden sharing among nations based on capacity and responsibility defined in this manner, the GDR framework shields from all climate costs those individuals who fall below the specified 'development threshold'. People below this threshold are taken as having development as their overriding priority, and thus have no carbon reduction obligations. People above the threshold are taken as having realized their right to development and as being obliged to help preserve that right for others. For the world's comparatively wealthy population living above this development threshold – both in the developing countries and industrialized countries – the GDR framework assigns a share of the global burden to pay for mitigation and adaptation. A reasonable and defensible level at which to set the development threshold is $20/day per capita. This level falls modestly above a global poverty line that corresponds to an empirical assessment of the income level at which the classic plagues of poverty, such as malnutrition, high infant mortality, low life expectancy and lack of access to education and health care, are seen to largely disappear.

Table 2.3 shows the relative amounts of capacity, responsibility and (averaging the two) obligation for a set of key countries and regions.

Table 2.3 *Capacity, responsibility and obligation for key countries and regions, 2010, 2020 and 2030*

	Population (% of global)	GDP per cap ($ purchasing power parity)	2010 Capacity (% of global)	Responsibility (% of global)	RCI (% of global)	2020 RCI (% of global)	2030 RCI (% of global)
EU 27	7.3	30,472	28.8	22.6	25.7	22.9	19.6
EU 15	5.8	33,754	6.1	19.8	22.9	19.9	16.7
EU +12	1.5	17,708	2.7	2.8	2.7	3.0	3.0
US	4.5	45,640	29.7	36.4	33.1	29.1	25.5
Japan	1.9	33,422	8.3	7.3	7.8	6.6	5.5
Russia	2.0	15,031	2.7	4.9	3.8	4.3	4.6
China	19.7	5899	5.8	5.2	5.5	10.4	15.2
India	17.2	2818	0.6	0.3	0.5	1.2	2.3
S. Africa	0.7	10,117	0.6	1.3	1.0	1.1	1.2
Mexico	1.6	12,408	1.8	1.4	1.6	1.5	1.5
Less developed countries	11.7	1274	0.11	0.04	0.07	0.10	0.12
Annex 1	18.7	30,924	75.8	78	77	69	61
Non-Annex 1	81.3	5096	24.2	22	23	31	39

account the fact that some countries (generally, higher-income countries that industrialized earlier) have consumed more than an equal per capita share of the total budget, resulting in a 'climate debt' that is expressed as a negative allocation for the future.

- **The Indian proposal** is a proposal whereby India's per capita emissions would not exceed developed country emissions. Up to that point, its emissions allocation is equal to its unmitigated requirements. As average per capita emissions in developed countries decline, they serve as a cap for India's per capita emissions. This framework is generalized and quantified by interpreting 'developed country' to signify Annex 1 parties, and applying the same allocation rule as India has proposed for itself to all non-Annex 1 countries. (This is an extension of the statement of PM Manmohan Singh (July 2008) across all developing countries.)
- **Greenhouse development rights (GDR)**, as elaborated in Baer et al (2008) is a proposal where the burdens are shared among countries according to capacity and responsibility, with each of these defined with respect to a 'development threshold' defined at an income level modestly above a global poverty line (see Box 2.1 for further description). Three additional variants of the GDR framework named 'Modified GDR' are also considered here and elaborated by Fan et al in Part I.

Figure 2.2 compares these frameworks on equivalent footing. First note the Chinese BAU trajectory (consistent with IEA's *World Energy Outlook 2007*) (solid black line, first from top). Next, note the 2°C pathway for China (solid grey line, fourth from bottom). As discussed above, this is a path in which China's emissions between 2010 and 2050 amount to 220Gt CO_2 (out of a total global budget for that time period of 660Gt CO_2 for fossil CO_2). Compared to China's BAU trajectory, one can see that a considerable mitigation effort will be needed for China to be consistent with a global path that is likely to keep warming below 2°C. Chapter 5 shows one possible way in which emissions in China could be kept within this stringent budget. While it requires extraordinarily ambitious mitigation efforts, it is in fact technically achievable providing efforts commence very soon.

Next, observe in Figure 2.2 the emission rights allocations corresponding to the nine burden-sharing proposals identified in Table 2.2. Clustered together at the bottom one finds equal per capita, C&C and the Indian proposal. Of the burden-sharing frameworks shown, these three place the most demands on China. In fact, in each of these three cases, China's emission rights allocation is less than even its actual emission pathway under the 2°C global trajectory. Such an allocation implies that China would need to reduce emissions domestically to reach the 2°C pathway, and then would need to ensure through some means that further reductions are generated elsewhere in the world. In other words, expressed in terms of a global cap-and-trade system, under these three frameworks China would be a purchaser (rather than a seller) of emission permits. It would need to invest its own resources in reaching the 2°C-consistent pathway

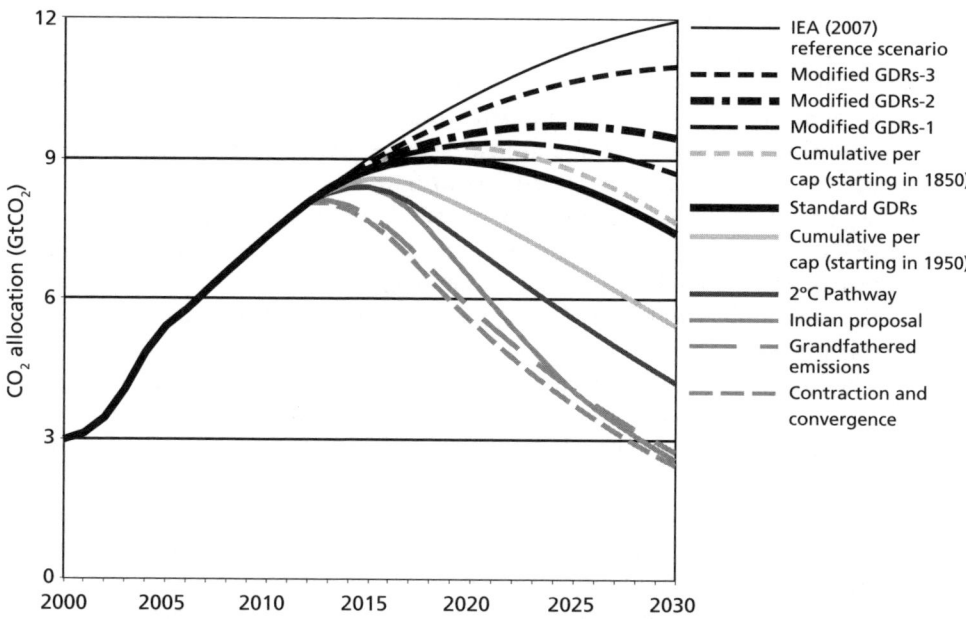

Figure 2.2 *Emission allocations for China under various frameworks*

with no additional international support, and moreover it would need to purchase additional permits from other countries that have excess permits.

Immediately above China's 2°C pathway, one finds an equal cumulative per capita allocation that equalizes per capita emissions since 1850, and another that equalizes per capita emissions since 1950. The choice of year from which cumulative per capita emissions are equalized is a key parameter, and, as would be expected, the later the year the lower China's allocation. As Figure 2.2 shows, both versions allocate fewer emission rights than the IEA BAU scenario for China, but more than would be needed to meet China's 2°C pathway. In other words, China would be expected to invest its own resources in a certain amount of mitigation, but would be able to rely on international financial and technological resources for the remainder. This additional support could be provided, for instance, through a mechanism such as the inter-country joint mitigation plans described by Fan et al in Part I, or by selling excess allocations on an international cap-and-trade market.

Between these two equal cumulative per capita allocations, one finds the GDR allocation. (In this analysis, national responsibility was calculated assuming the rather recent start date of 1990, and the development threshold was set at $20 per capita per day. See Box 2.1.) Like the equal cumulative per capita allocations, the GDR framework also implies an allocation that lies between China's BAU and its actual 2°C pathway, and so would also ensure that China receives partial international financial and technological resources to reduce its emissions to the necessary level.

Above the GDR allocation, one finds the three Modified GDR variants, proposed by Fan et al (in Part I). The first of these (Modified GDR-1) moves the historical responsibility back to 1850 rather than 1990 as in the standard GDR analysis. The second modification (Modified GDR-2) introduces consumption-based accounting, adjusting for emissions embedded in internationally traded goods. The third (Modified GDR-3) corrects the carbon intensity of the domestic energy mix by using an average global carbon intensity. All of these modifications result in allocations for China that are greater than the standard GDR allocation. Taken together, they reduce China's share of the global obligation for reducing global emissions to roughly a third of its standard GDR share.

The Modified GDR approach of Fan et al also proposes that the legal form of climate obligations would evolve over time. Their proposal suggests that countries should automatically face compulsory mitigation once they have reached a 'graduation threshold' for emissions. This threshold is defined as the level of accumulative consumption emissions per capita of the Annex 1 country with the lowest level of such emissions (currently, this happens to be Romania). Countries below the graduation threshold could pursue voluntary mitigation activities but would not be compelled to mitigate. Such voluntary mitigation would delay the point when the threshold is crossed and mitigation becomes compulsory.

Various other proposals that are related to the above could be assessed as well. For example, the *individual targets* proposal, elaborated by Chakravarty et al (2009), would lead to somewhat greater allocation for China than equal per capita emission rights (since it allocates to countries whose average emission rate is greater than the global average a free transfer from countries with emission rates less than the global average). The *One Standard, Two Convergence* proposal by Chen et al (2005), further elaborated by Gao (2007) would allocate somewhat less than the equal cumulative per capita (since its 'double convergence' partially accounts for cumulative emissions).

The equity implications of a burden-sharing regime rely not just on its underlying principles, but also on the mechanism through which it is implemented. Two types of mechanisms in particular can be highlighted. A market-based cap-and-trade system allows permits to be purchased on an open market at a price that reflects a marginal abatement cost. A managed fund, by contrast, would direct resources to mitigation opportunities in different countries based on their actual incremental costs of abatement, rather than global marginal price set by a market. Within a market, some countries will be able to reduce emissions at a cost lower than the global price at which carbon allowances trade. Countries that are net sellers of permits under a global cap-and-trade system would thereby benefit from 'rents' – allowance revenues that exceed their actual costs of emissions abatement. Unless a burden-sharing approach explicitly takes this additional benefit into account, its outcome could differ appreciably from the underlying equity principles on which it is based. Implementing such burden-sharing approaches in part through a fund-type financing mechanism could help avoid the excessive rents that would occur in a carbon market.

Equity within China: Implications for burden sharing

A more disaggregated analysis for China that draws on available Chinese data at the province level has also been carried out to help clarify the burden-sharing implications of a GDR approach implemented at the province level. This is significant given the considerable differences in wealth between provinces. The results can be useful both for an understanding of the distributional impacts of climate policy in China, and to provide information relevant to the design of a national trading system in which provinces would be given allocations. Figure 2.3 shows the wide diversity of income levels and makes

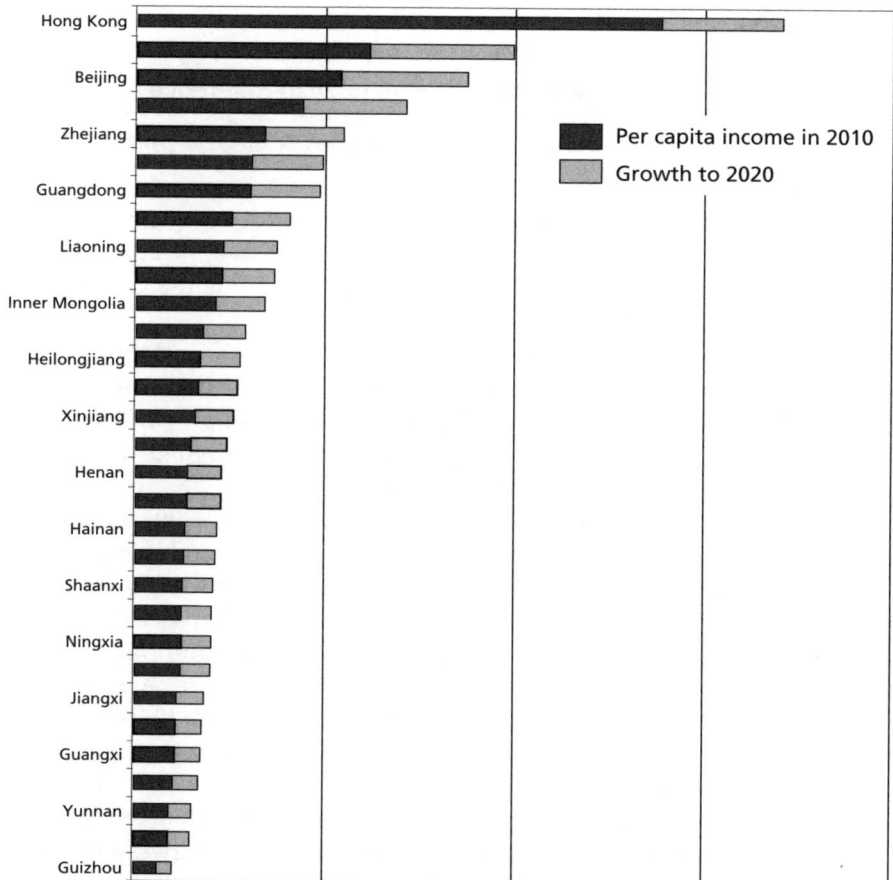

Figure 2.3 *Present and future (purchasing power parity-adjusted) incomes in Chinese provinces*

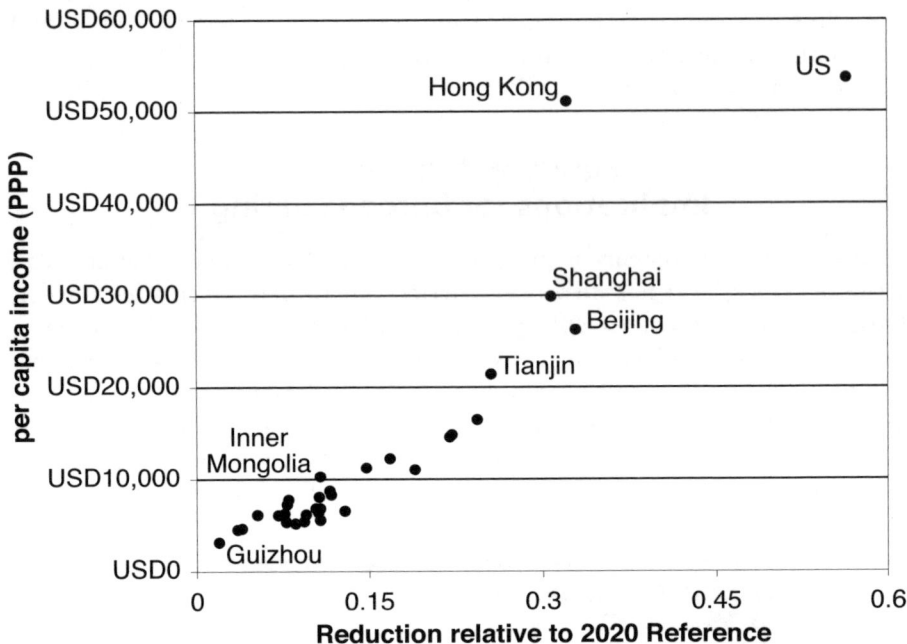

Figure 2.4 *Reduction obligation versus income in Chinese provinces*

evident the importance of intra-national analyses of burden sharing. The dark-grey bar shows incomes (purchasing power parity adjusted) in 2010, and the light-grey segment shows income growth to 2020 (adopting the GDP growth assumptions of the IEA).

To carry out a province-level GDR analysis, we relied on province level incomes from the Chinese Statistical Yearbook, Gini coefficients provided by Professor Wang Xiaolu (disaggregated by rural and urban provincial populations) and emissions figures provided by Professor Cao Jing of Tsinghua University. We are indebted for this assistance.

Figure 2.4 shows the reduction target for each province, plotted against income per capita. The reduction target is for the year 2020 and is measured as a percentage reduction below the 2020 reference (from IEA's *World Energy Outlook 2007*). The deviation from a perfect correlation derives from the fact that difference provinces have different carbon intensities and thus different levels of responsibility. The plot also shows the US for comparison. Despite the fact that its per capita income is not much greater than Hong Kong's, it has a much more demanding reduction target because of its considerably greater historical responsibility.

This suggests that the variation between Chinese provinces' level of obligation in a national climate effort is as great as the variation between disparate countries' level of obligation in a global climate effort. The distributive impacts of mitigation activities will need to be considered, whether climate policy is implemented in terms of a national cap-and-trade system, a carbon tax system (with compensating redistribution of carbon tax revenues), an 'Inter-province Joint Cooperation' mechanism or any other approach.

Conclusions

There are five main conclusions to be reiterated here. The first is that if we are to protect the climate, it is not sufficient for only industrialized countries to reduce their emission. The severity and urgency of the climate crisis puts very serious demands on all countries – rich and poor – to reduce emissions.

Second, equity considerations show that it is not reasonable for countries that are still suffering from underdevelopment to divert their own limited resources to climate protection. Industrialized countries have primary responsibility for the CO_2 pollution contributing to climate change, and they also have the greater capacity for addressing the problem. This recognition is the basis of the agreement in the UNFCCC that 'Parties should protect the climate system … on the basis of equity and in accordance with their common but differentiated responsibilities and respective capabilities'.

Third, there is thankfully no contradiction between the first and second points. The latter simply means that those countries with responsibility and capacity must provide the financial and technological resources to enable and support the necessary mitigation measures in developing countries. Various mechanisms are being considered for channelling this support. A global cap-and-trade system is one option, inter-country joint mitigation plans another and a global climate fund is yet another.

Fourth, various burden-sharing approaches have been put forward for allocating the responsibility for addressing the global climate problem. Several of the well-known approaches (equal per capita emissions, C&C and the Indian proposal) result in extraordinarily stringent allocations in which China must purchase allowances from other countries. The equal cumulative per capita emissions approach and the various GDR-based approaches are much more reasonable from China's perspective. This conclusion is not entirely surprising, as the equal cumulative per capita emissions approach explicitly aims to equitably allocate access to an entire global resource (not merely the small fraction remaining), and the GDR approaches explicitly aim to safeguard a right to development.

Fifth, China itself has considerable disparities in responsibility and capacity, when disaggregated by province. It is important, therefore, for China to consider the distributive impacts of climate policies and ensure that measures are implemented to allocate the costs in an equitable manner. Just as it is true at the global level that a climate solution cannot succeed unless it is equitable, the same is also true at the national level.

Notes

1   The relevant science and its implications have been presented quite clearly by
    Meinshausen et al (2009), who state that 'Limiting cumulative CO_2 emissions over
    2000–2050 to 1,000 $GtCO_2$ yields a 25 per cent probability of warming exceeding
    2°C'. With such odds, the IPCC would thus describe this scenario as being 'likely'
    but not 'very likely' to keep warming below 2°C (IPCC WGI, 2007) An assumption
    of Meinshausen et al is that comparably ambitious efforts to limit emissions from
    non-CO_2 GHGs such as methane, nitrous oxide and halocarbons are concurrently
    undertaken.
2   These reduction levels are at the more stringent end of the ranges presented in the
    IPCC Fourth Assessment Report for Annex 1 (25–40 per cent by 2020 and 80–95
    per cent by 2050, relative to 1990 levels) for emission scenarios consistent with
    stabilization at 450ppm CO_2eq (IPCC WG III, 2007). If Annex 1 countries were
    less ambitious, and their reductions reached only the lower end of these ranges (25
    per cent by 2020 and 80 per cent by 2050), they would occupy a significantly
    greater fraction of the available budget: roughly 305Gt CO_2 (between 2010 and
    2050), leaving 355Gt CO_2 for non-Annex 1 countries to emit. Three independent
    analyses (UNFCCC Secretariat, 2009; IIASA, 2009; AOSIS, 2009) have examined
    the Annex 1 countries' pledges (as of the end of 2009) and found that the aggregate
    implied Annex 1 reductions for 2020 fall shy of even the less ambitious end of the
    25–40 per cent range.
3   If, alternatively, we assume that China's proportion is proportional to its share of
    non-Annex 1 population, its budget would be approximately 110Gt CO_2.
4   The three examples given here are simple indicative paths intended to clarify the
    size of the budget and the significance of timing of reductions. In each path,
    emissions continue along recent trends, growing at 5 per cent/year until the
    specified year (2015, 2020 or 2025), reaching a peak level (approximately 10Gt
    CO_2, 13Gt CO_2 or 17Gt CO_2, respectively), and then steadily declining at the
    stated annual rate (5 per cent, 11 per cent and 35 per cent, respectively). (Note that
    these are rates of decline in emissions, not in emissions intensity.) In each of the
    three paths, China's total emissions over the period 2010 to 2050 would amount to
    220Gt CO_2.

References

Agarwal, A. and Narain, S. (1991) *Global Warming in an Unequal World: A Case of Environmental Colonialism*, New Delhi: Centre for Science and Environment
AOSIS (Alliance of Small Island States) (2009) 'Aggregate Annex-1 reductions for 2020', in *Compilation and Analysis presented by the Alliance of Small Island States to the Ad Hoc Working Group on Further Commitments for Annex 1 Parties under the Kyoto Protocol*, 21 June, Bonn
Baer, P., Athanasiou, T., Kartha, S. and Kemp-Benedict, E. (2008) *The Greenhouse Development Rights Framework: The Right to Development in a Climate Constrained World*, second edition, Berlin: Heinrich Böll Foundation, Christian Aid, EcoEquity, and the Stockholm Environment Institute, www.greenhousedevelopmentrights.org
CDIAC (Carbon Dioxide Information Analysis Center) (2009) *National CO_2 Emissions from Fossil-Fuel Burning, Cement Manufacture, and Gas Flaring:*

1751–2006, T. A. Boden, G. Marland and R. J. Andres, contributors, Washington DC: CDIAC of the United States Department of Energy, http://cdiac.ornl.gov/

Chakravarty, S., Chikkatur, A., de Coninck, H., Pacala, S. W., Socolow, R. H. and Tavoni, M. (2009) 'Sharing global CO_2 emission reductions among one billion high emitters', *Proceedings of the National Academy of Sciences of the United States of America*, vol 106, no 29, pp11884–11888

Chen, W., Wu, Z. and He, J. (2005) '"Two Convergence" approach for future global carbon permit allocation', *Journal of Tsinghua University* (Science and Technology), vol 45, no 6: 130J137 (in Chinese)

Gao, G. (2007) 'Carbon emission right allocation under climate change', *Advance in Climate Change Research*, Article ID: 1673–1719, Supplement 0087–05

GCI (Global Commons Institute) (2000) *GCI Briefing: Contraction and Convergence*, available at www.gci.org.uk/briefings/ICE.pdf

IEA (International Energy Agency) (2007) *World Energy Outlook 2007: China and India Insights*, Paris: International Energy Agency

IIASA (International Institute for Applied Systems Analysis) (2009) *Analysis of the Proposals for GHG Reductions in 2020 Made by UNFCCC Annex I Parties*, analysis and report by F. Wagner and M. Amann, Luxembourg: IIASA

IPCC WGI (Intergovernmental Panel on Climate Change Working Group I) (2007) *Climate Change 2007: The Physical Science Basis. Contribution of Working Group I to the Fourth Assessment Report of the Intergovernmental Panel on Climate Change*, S. Solomon, D. Qin, M. Manning, Z. Chen, M. Marquis, K. B. Averyt, M. Tignor and H. L. Miller (eds), Cambridge, UK and New York: Cambridge University Press

IPCC WGIII (Working Group III) (2007) *Climate Change 2007: Mitigation of Climate Change. Contribution of Working Group III to the Fourth Assessment Report of the Intergovernmental Panel on Climate Change*, B. Metz, O. R. Davidson, P. R. Bosch, R. Dave and L. A. Meyer (eds), Cambridge, UK and New York: Cambridge University Press

Meinshausen, M., Meinshausen, N., Hare, W., Raper, S. C. B., Frieler, K., Knutti, R., Frame, D. J. and Allen, M. R. (2009) 'Greenhouse-gas emission targets for limiting global warming to 2°C', *Nature*, no 458, pp1158–1163

Rockström, J., Steffen, W., Noone, K., Persson, A., Chapin, F. S., Lambin, E. F., Lenton, T. M., Scheffer, M., Folke, C., Schellnhuber, H. J., Nykvist, B., de Wit, C. A., Hughes, T., van der Leeuw, S., Rodhe, H., Sörlin, S., Snyder, P. K., Costanza, R., Svedin, U., Falkenmark, M., Karlberg, L., Corell, R. W., Fabry, V. J., Hansen, J., Walker, B., Liverman, D., Richardson, K., Crutzen, P. and Foley, J. A. (2009) 'A Safe operating space for humanity', *Nature*, no 461, pp472–475 (for more detail see 'Planetary boundaries: Exploring the safe operating space for humanity', *Ecology and Society*, September 2009)

UNFCCC (United Nations Framework Convention on Climate Change) Secretariat (2009) 'Compilation of information relating to possible quantified emission limitation and reduction objectives as submitted by Parties', 12 June, Bonn: UNFCCC

3

Greenhouse Gases and Human Well-Being: China in a Global Perspective

Elizabeth A. Stanton

Climate change and development

Most pollution is an unequivocal social bad – a negative externality – but the relationship between GHG emissions and human well-being is unusually complex. In the long run, there is a strong scientific consensus that GHG emissions will result in higher temperatures and sea levels, and a disruption of historical weather patterns, from heat waves and droughts to more intense storms. The effect of climate change on human well-being will vary greatly from country to country, but if the increase in global average annual temperature exceeds a 2°C threshold, total world food supplies will begin to shrink even as the risk of triggering feedback processes that would accelerate warming grows steeply (Stern 2006; IPCC, 2007).

In the short run, GHG emissions, and the activities that produce these emissions, result in a mixed set of consequences. Industrialized countries have higher emissions, but also more revenue from the sale of industrial products. Countries with more automotive transportation, larger homes, warmer homes in winter, cooler homes in summer, more lights and more consumer electronics have higher residential and transportation-related emissions, but also have lifestyles that are – perhaps – more comfortable, more convenient and more in tune with all of the benefits of the information age.

At the same time, every country suffers the ancillary costs of burning fossil fuels: emissions of SO_x, NO_x and particulate matter, together with a host of attendant health, environmental and aesthetic impacts. And for a small but ever-increasing number of countries, the worst impacts of climate change have already begun. Some countries are especially vulnerable to climatic changes because of their geography, among these: low-lying, small islands; coastal areas in the paths of hurricanes and typhoons; and arid regions where water availability is dropping still lower. Other countries are economically vulnerable; they cannot afford to insulate their populace from the effects of climate change with costly dykes, air conditioning, desalination plants or rigorous building codes.

China and a few other rapidly industrializing countries stand in the middle. On one side are poorer, less industrialized countries with little responsibility for the emissions that cause climate change and few resources with which to combat its effects. On the other side are richer, more industrialized countries with enormous culpability – both past and present – for the problem of climate change and ample funds for adaptation measures to protect human well-being. The case of China, viewed from a global perspective, can illuminate a difficult but not intractable issue of international equity: how can the international community balance each individual's right to an adequate standard of living with the imperative of GHG emissions reductions?

Emissions versus well-being:
The international context

As global emissions of GHGs increase, the impacts of climate change worsen, but these same emissions are essential to maintain high-consumption lifestyles in rich countries and make development possible for poorer and middle-income countries. A statistical examination of this relationship requires good measures of both emissions and human well-being. In this chapter, the primary measure of GHG emissions will be emissions per capita of CO_2 for 2004 (hereafter, simply referred to as emissions per capita).[1] It is worth noting that while each country's total emissions are often emphasized in the Western press and are even at times the subject of international negotiation, any serious analysis of the equity implications of climate change requires a per capita measure of emissions:[2] Luxembourg ranks 90th in terms of total emissions but has the third highest per capita emissions in the world; conversely, China rivals even the US in terms of total emissions but 73 countries have higher per capita emissions.

Establishing a credible measure of human well-being is less straightforward. Two measures are commonly used as proxies for well-being in the development literature: gross domestic product (GDP) per capita adjusted for purchasing power parity (PPP)[3] and the Human Development Index (HDI).[4] PPP-adjusted GDP per capita can be interpreted as average private consumption, but it fails to include access to public goods and other aspects of well-being, both measurable (such as long life or educational attainment) and

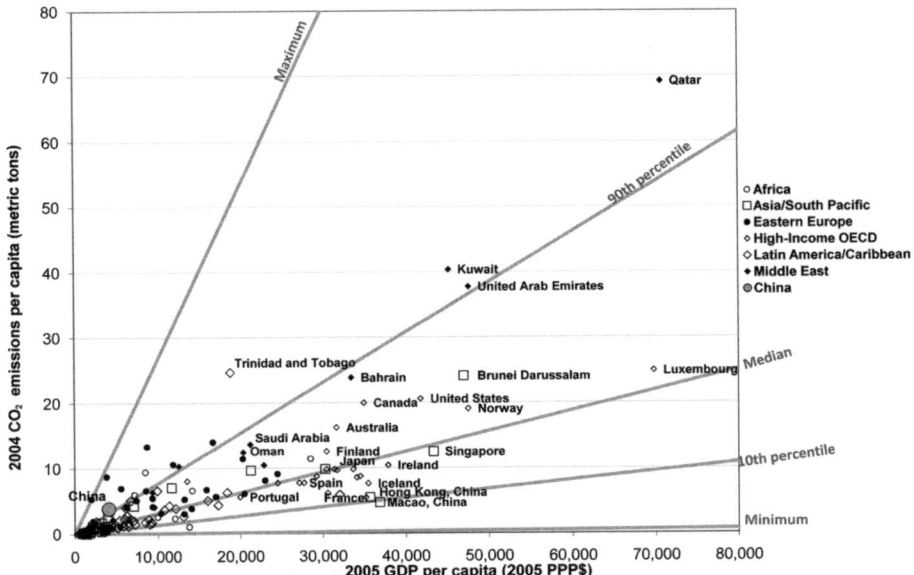

Source: World Bank Development Indicators (2008, 2009, 2010)

Figure 3.1 *Emission per capita versus PPP-adjusted GDP per capita*

immeasurable (such as happiness or the strength of one's community). HDI combines PPP-adjusted GDP per capita with average life expectancy, the literacy rate, and school enrolment;[5] there is a strong correlation between this income measure and the other measures included in HDI. Both PPP-adjusted GDP per capita and HDI are based only on national averages[6] and therefore can impart no information regarding inequalities of ethnicity, gender or region within each nation.

In this chapter, the primary measure of well-being is PPP-adjusted GDP per capita for 2005,[7] but several other development indicators are discussed. Note that GDP per capita alone explains 65 per cent of the variability in emissions per capita, indicating a high degree of correlation between income and emissions. (Hereafter, PPP-adjusted GDP per capita is referred to simply as income per capita.)

Figure 3.1 is a scatterplot of emissions per capita versus income per capita for 174 countries; China, at 3.9t CO_2 emitted annually and $4088 (PPP) per capita, is at bottom left. To give some context to the relative positions of each country: mean annual global emissions per capita is 4.3t (similar to that of Mexico); mean global income per capita is $8700 (similar to that of Kazakhstan).

Emissions intensity is the ratio of per capita emissions to per capita income; graphically, in Figure 3.1 countries aligned along any ray from the origin share the same emissions intensity. The median emission intensity – close to that of Portugal, Japan and Singapore – is 0.3t for every $1000 of income. The 90th

Table 3.1 *Regression analysis for 174 countries*

	Coefficient	Significance[a]		
Per capita GDP (PPP$1000)	0.3819	**	Number of observations	174
Temperature (°C)	0.0713		R-squared	0.803
Transition country (dummy)	3.0240	**		
Per capita oil production ($1000)	0.8249	**		
constant	−2.1259			

Note:
a * indicates significance at the 95% level ** indicates significance at the 99% level
Dependent variable: CO_2 emissions per capita (metric tonnes); independent variables: GDP per capita (PPP 2005 $1000); average annual temperature in country capital (°C); transition country (dummy); and oil production per capita (2005 $1000)
Source: Data are from the World Bank, World Development Indicators online database except for temperatures of capital cities from www.worldclimate.com/

percentile of emissions intensities is 0.8t per $1000, while the 10th percentile is 0.1t. The area of the graph near and above the 90th percentile of emissions intensities is populated almost exclusively by countries of the Middle East, other Organization of Petroleum Exporting Countries (OPEC) members, and the transition economies of Eastern Europe and the former Soviet Union. With the exception of the two Chinese cities for which data are collected in the international dataset – Hong Kong and Macao – countries near or below the 10th percentile are crowded into the bottom left-hand corner of Figure 3.1: only countries with very low incomes have very low emissions intensities. Indeed, only one country in the lowest decile by emissions intensity, Gabon, has an income per capita greater than $3300.

In the regression analysis shown in Table 3.1, four variables explain 80 per cent of the variance in emissions per capita (although average annual temperature was not statistically significant, a point addressed in detail below):

- For every $1000 increase in income per capita there is a 0.38t increase in emissions per capita.
- Eastern European and former Soviet transition countries have 3.0t higher emissions per capita than non-transition countries.
- For every $1000 increase in oil production per capita there is a 0.8t increase in emissions per capita.

Nearly all countries above the 90th percentile in emissions intensity have oil production that is more than 10 per cent of GDP and/or are ex-Soviet transition countries. China has the 8th highest emissions intensity, 0.9t per $1000 income per capita, only surpassed by: Uzbekistan (2.7t per $1000); Turkmenistan (2.3t); Kazakhstan (1.5t); Trinidad and Tobago (1.3t); Ukraine (1.2t); South Africa (1.1t); and Syria (1.0t). Among the top decile by emissions intensity only China and South Africa (with 9.4t and $8400 per capita) are neither highly dependent on oil production nor ex-Soviet transition economies. Uzbekistan, Turkmenistan, Kazakhstan, Ukraine and other transition countries

Table 3.2 *Regression analysis for 127 countries*

	Coefficient	Significance[a]		
Per capita GDP (PPP$1000)	0.3466	**	Number of observations	127
Temperature (°C)	0.0713		R-squared	0.850
Transition country (dummy)	2.3851	*		
Per capita oil production ($1000)	1.0321	**		
Fossil fuel share of energy	0.0348	*		
constant	−4.0132			

Note:
a * indicates significance at the 95% level ** indicates significance at the 99% level
Dependent variable: CO_2 emissions per capita (metric tonnes); independent variables: GDP per capita (PPP 2005 $1000); average annual temperature in country capital (°C); transition country (dummy); oil production per capita (2005 $1000); and fossil fuels as a share of energy production (%).
Source: Data are from the World Bank, World Development Indicators online database except for temperatures of capital cities from www.worldclimate.com/

tend to have outsized industrial infrastructures in relation to current GDP, a legacy of large-scale state-driven manufacturing under the Soviet Union and its satellites.

Table 3.2 adds an additional significant variable to this regression: fossil fuels as a share of energy production (included separately because there are observations for only 127 countries). With the addition of the fossil fuel share of energy, the trend in emissions intensity (that is, the coefficient for income per capita) is slightly reduced: the impact of every $1000 increase in income per capita is a 0.35t increase in emissions per capita. Coincidentally, that is the same as the estimated impact for a 10 percentage point increase in the fossil fuel share of energy production.

In this more limited dataset, only Hong Kong has a low emissions intensity but a high fossil fuel share of energy at 97 per cent. In general, city states (such as Singapore) and the Chinese cities included in the international data have lower than the expected emissions per capita and higher than the expected income per capita. These anomalous results often point to high population density, abundant public transportation and a service-based economy (Leung and Lee, 2000). Like most high-emissions-intensity countries, China produces the vast majority, 84 per cent, of its energy from fossil fuels.[8]

High-income countries with low emissions intensities tend to produce a higher share of energy from renewables and nuclear: France, Iceland, Sweden and Switzerland all fit this pattern. Among high-income OECD countries, Iceland and Sweden have the lowest fossil-fuel reliance; 73 per cent of Iceland's energy comes from geothermal and hydroelectric generation. Sweden generates 13 per cent of its energy from renewables (excluding biomass) and 36 per cent from nuclear.[9]

A third of global emissions are residential in origin[10] – primarily from heating and cooling – but, unexpectedly, the coefficient for average annual temperature is positive and insignificant in these regression analyses. The relationship between emissions per capita and temperature is complex: colder

Table 3.3 *Regression analysis for 174 countries, with interaction term*

	Coefficient	Significance[a]		
Per capita GDP (PPP$1000)	0.2691	**	Number of observations	174
Temperature (°C) × per capita GDP			R-squared	0.813
(PPP$1000)	0.0074	**		
Transition country (dummy)	2.7838	**		
Per capita oil production ($1000)	0.6812	**		
constant	−0.6685			

Note:
a * indicates significance at the 95% level ** indicates significance at the 99% level
Dependent variable: CO_2 emissions per capita (metric tonnes); independent variables: GDP per capita (PPP 2005 $1000); interaction term: average annual temperature in country capital (°C) multiplied by GDP per capita (PPP 2005 $1000); transition country (dummy); and oil production per capita (2005 $1000).
Source: Data are from the World Bank, World Development Indicators online database except for temperatures of capital cities from www.worldclimate.com

countries use more heat and therefore have higher emissions; but countries that are both warm and rich often have high rates of air conditioning use, and therefore very high residential emissions. The regression reported in Table 3.3 replaces the explanatory variable temperature with an interaction term that multiplies temperature by income per capita. A high value for this interaction variable indicates a warm, rich country where air conditioning is likely to be widely used.

The estimated relationship between emissions, income and temperature shown in Table 3.3 (ignoring other variables, and rounding off the estimated coefficients) can be expressed as:

$$\text{Emission/capita} = \text{GDP/capita} \times (0.3 + 0.007 \times \text{Temperature}) \qquad (1)$$

where emissions per capita are measured in metric tonnes and GDP per capita in thousands of PPP-adjusted dollars.

Higher temperature is associated with higher emissions across the full range of countries, but part of the emissions per capita versus income per capita relationship is sensitive to temperature and part is insensitive. When average annual temperature is 0°C, for every $1000 increase in income per capita there is a 0.3t increase in emissions per capita, but when temperature is 30°C, for every $1000 increase in income per capita there is a 0.5t increase in emissions per capita. Differences in emissions based on temperature are real, at high incomes, but are dominated by differences based on income: expected emissions per capita for a 30°C country with $30,000 in income per capita is 15t, compared to 0.5t for a country with the same temperature but $1000 in income per capita. Figure 3.2 represents this relationship as a set of temperature isoquants.[11]

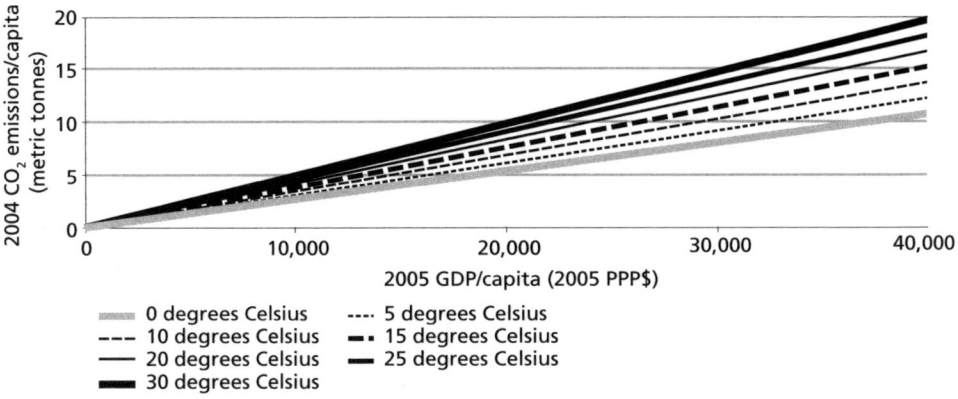

Source: Data are from the World Bank, World Development Indicators online database except for temperatures of capital cities from www.worldclimate.com

Figure 3.2 *Stylistic representation of emissions/capita versus GDP/capita with temperature isoquants*

Chinese provinces in a global context

With 5.3 million people, the least populous Chinese province, Qinghai, has a larger population than 68 of the countries in the international dataset used in this chapter.[12] Figure 3.3 demonstrates the relative emissions intensities of Chinese provinces in a global perspective. Of 30 provinces, 23 are above the 90th percentile by emissions intensity in the international dataset.[13]

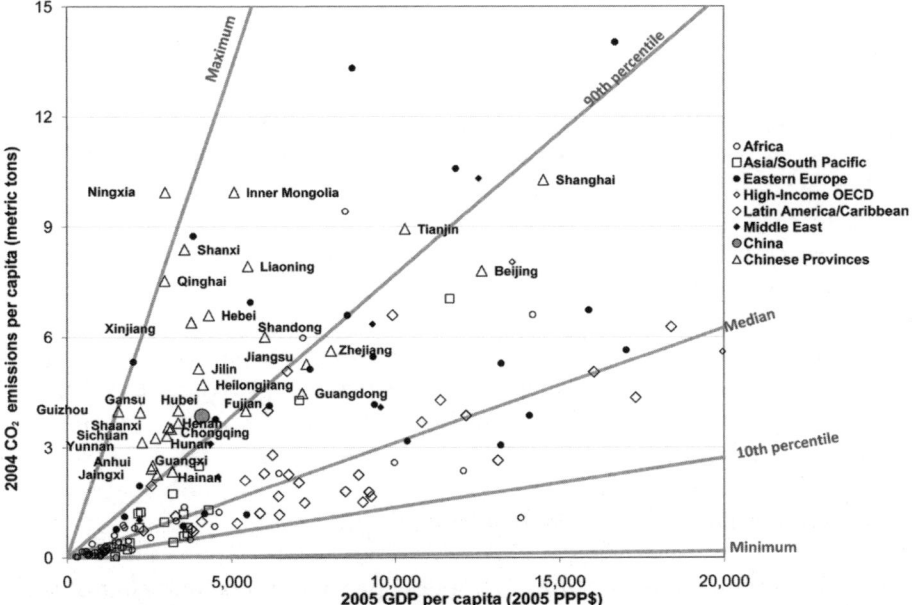

Source: Data are from the World Bank, World Development Indicators online database except for temperatures of capital cities from www.worldclimate.com

Figure 3.3 *Emission/capita versus GDP/capita (PPP): Chinese provinces*

Table 3.4 *Regression analysis for 30 Chinese provinces*

	Coefficient	Significance[a]		
Per capita GDP (PPP$1000)	0.5248	**	Number of observations	30
Temperature (°C)	−0.1891	**	R-squared	0.792
Per capita coal reserves (1000 tons)	1.2862	**		
constant	5.1819			

Note:
a * indicates significance at the 95% level ** indicates significance at the 99% level
Note: Dependent variable: CO_2 emissions per capita (metric tonnes); independent variables: GRP per capita (PPP 2005 $1000); average annual temperature in province capital (°C); and per capita coal reserves (1000t).
Source: Chinese provincial data are taken from the *China Statistical Yearbook 2007*, www.stats.gov.cn/tjsj/ndsj/2007/indexeh.htm.

Few countries share the Chinese provinces' high emissions intensity, and even fewer combine this trait with a similar range of income per capita: South Africa and a handful of ex-Soviet transition countries (see Figure 3.3). Most of the provinces with somewhat lower emissions intensities can be paired with countries that share a very similar emissions per capita and income per capita: Hainan resembles Indonesia; Fujian, Bosnia and Herzegovina; Guangdong, Thailand; Jiangsu, Macedonia; Zhejiang, Bulgaria; and Beijing, Poland. Only Shanghai has no close analogue. It has both the highest emissions per capita (10.3t) and the highest income per capita ($14,500 in PPP-adjusted 2005$) among the Chinese provinces; its closest matches in terms of emission intensity are Iran ($9300), Venezuela ($9900) and Bahrain ($33,500).

As reported in Table 3.4, among Chinese provinces the trend in emission intensity (the coefficient for income per capita) is higher than in the international dataset:

- For every $1000 increase in income per capita there is a 0.5t increase in emissions per capita (compared to 0.38t in the international dataset).
- For every 1°C increase in average annual temperature there is a 0.2t decrease in emissions per capita.
- For every 1000t increase in coal reserves per capita there is a 1.3t increase in emissions per capita.

The relationship between temperature and income per capita is more straightforward in the dataset of Chinese provinces: colder provinces have higher emissions. Taking the constant term into consideration, and ignoring the income and coal reserve effects:

$$\text{Emission/capita} = 5.2 - 0.2 \times \text{Temperature} + ... \qquad (2)$$

The average annual temperatures of Chinese provinces (taken as that of their capital cities) range from 5°C in Heilongjiang to 25°C in Hainan. For a province with an average annual temperature of 5°C, the expected contribution from temperature to emissions per capita is 4.2t; for a province with a temperature of 25°C, the contribution is 0.2t. While no data on air conditioner use and ownership across Chinese provinces are available, it seems plausible to assume that use of this expensive convenience is more limited in China than in countries with higher incomes but similar average annual temperatures. As air conditioning use in China rises along with rapid income growth, residential emissions are likely to increase, especially in warmer provinces.

Per capita coal reserves are an additional significant determinant of the variance in per capita emissions among Chinese provinces. The only provinces below the 90th percentile by emissions intensity in the international dataset – Hainan, Fujian, Guangdong, Jaingsu, Zhejaing, Beijing and Shanghai – all have low per capita coal reserves; the provinces with the highest emissions intensity are among those with the highest per capita coal reserves.

Measuring development in China and abroad

Countries with higher emissions per capita also have higher standards of living, not just in terms of private consumption (or PPP-adjusted GDP per capita) but also as measured by life expectancy and literacy (see Table 3.5). China's emissions per capita place it in the first decile above the median, where the average income per capita among countries in this group is $8500, average life expectancy is 71 years, and the average literacy rate is 89 per cent. (China is a good match to other countries in this group for all data categories considered here; the life expectancy in China is 73 years and the literacy rate is 91 per cent.) For countries below China's emissions per capita, the trend is clear:

Table 3.5 *International dataset: Emissions and development measures by emission per capita decile*

Countries grouped by decile of emissions per capita	Mean CO_2 emissions/ capita (tons)	Mean GDP/capita (PPP)	Mean life expectancy (years)	Mean literacy rate (%)	Mean cumulative emissions/ capita (tons)
Lowest	0.1	841	52	51	2
Second	0.2	1181	54	54	6
Third	0.6	2610	62	75	21
Fourth	1.1	4343	67	85	44
Fifth	1.8	5899	71	87	51
Sixth	3.3	8538	71	89	100
Seventh	5.3	15,425	75	94	240
Eighth	7.6	19,313	75	97	338
Ninth	10.3	24,310	73	95	475
Highest	27.6	38,223	76	95	680

Note: Cumulative emissions are for 1850 to 2005.
Source: Data are from the UNDP Human Development Report, http://hdr.undp.org/, World Resources Institute, Climate Analysis Indicators Tool, http://cait.wri.org/, and the *China Statistical Yearbook 2007* www.stats.gov.cn/tjsj/ndsj/2007/indexeh.htm

Table 3.6 *Chinese provinces: Emissions and development measures by emission per capita decile*

Chinese provinces grouped by global decile of emissions per capita	Mean CO_2 emissions/ capita (tons)	Mean GDP/capita (PPP)	Mean life expectancy (years)	Mean literacy rate (%)	Count of provinces in group
Lowest					0
Second					0
Third					0
Fourth					0
Fifth	2.3	2736	69	92	1
Sixth	3.4	2966	70	87	13
Seventh	5.2	6103	74	90	6
Eighth	7.4	5465	71	91	6
Ninth	9.8	8224	73	89	4
Highest					0

Note: Cumulative emissions are for 1850 to 2005.
Source: Chinese provincial data are taken from the *China Statistical Yearbook 2007* www.stats.gov.cn/ndsj/2007/indexeh.htm

GHG emissions are highly correlated with the level of development. Above China's level of emissions per capita, improvements in life expectancy and literacy slow, but gains in income per capita remain strong.

Table 3.6 sorts Chinese provinces by emissions per capita decile in the international dataset. Only one Chinese province, Jiangxi, has emissions per capita below the world median; Jiangxi has the sixth lowest average life expectancy, but the 11th highest literacy rate among the 30 provinces. Most provinces join China as a whole in the first decile above the median. The four Chinese provinces with the highest emissions per capita are Shanghai (10.3t), Ningxia (9.9t), Inner Mongolia (9.9t) and Tianjin (8.9t). Although the emissions intensity of Chinese large eastern city-provinces – Beijing (7.8t), Shanghai and Tianjin – are much higher than that of cities such as Singapore, Hong Kong and Macao, the basic relationship seen in the global data remains the same within China: cities have higher incomes than larger regions with the same emissions per capita. Chongqing, China's large western city-province, is an exception with much lower income per capita ($3200), emissions per capita (3.5t), life expectancy and literacy than Beijing, Shanghai and Tianjin.

Re-examining the Environmental Kuznets Curve

A key question in understanding the relationship between emissions and development is this: as countries develop and income per capita grows, is a constant emissions intensity maintained? Even with constant emissions intensity, developing countries' contribution to atmospheric concentrations of GHGs will grow in relation to that of industrialized countries as incomes rise. If, however, emissions intensity grows as income grows, future emissions may dwarf BAU projections. Taking an international mandate for poverty reduction as a given,[14] if global emissions are to decrease over time two things will be necessary: first, rich countries must decrease their emissions intensity; and second, low-income and middle-income countries must find a way to increase their income per capita while maintaining or reducing emissions intensity.

By far the strongest determinant of emissions per capita is PPP-adjusted GDP per capita. As noted above, it alone explains nearly two-thirds of the international variation in emissions intensity. One possible form for the relationship of emissions to income is a so-called Environmental Kuznets Curve (EKC), which posits U-shaped pollution levels that first rise and then fall as income per capita increases.[15] The message of the EKC is that negative environmental impacts may increase with development but further development will serve to reduce pollution. In essence, environmentalism is a luxury good that a richer, better-educated populace will purchase either collectively (through a policy response or by importing pollution-intensive goods from elsewhere) or individually.

The linear regression reported in Table 3.1 can only describe a linear relationship between emissions per capita and income per capita; graphically the trend line representing that relationship is nearly indistinguishable from the ray describing the median emissions intensity in Figure 3.1. Table 3.7 reports on a log-log regression, where the natural logarithm of emissions per capita is regressed against the natural logarithm of income per capita. The coefficient of a log-log relationship can be interpreted as a ratio between two percentage changes. Here the estimated coefficient implies that for every 1 per cent increase in income per capita there is a 1.13 per cent increase in emissions per capita.

Far from rising and then falling, the relationship between emissions per capita and income per capita is exponential, with emissions increasing at a slightly increasing rate over the set of countries in the dataset: there is no evidence for an EKC for carbon emissions in this dataset. If the trend of this regression can be said to represent the path of countries as they develop (a

Table 3.7 *Regression analysis*

	Coefficient	Significance[a]		
Log capita GDP (PPP$1000)	1.1303	**	Number of observations	174
Temperature (°C)	−0.0044		R-squared	0.853
Transition country (dummy)	0.6308	**		
Per capita oil production ($1000)	0.0376	**		
constant	−9.2468			

Note:
a * indicates significance at the 95% level ** indicates significance at the 99% level
Dependent variable: natural logarithm of CO_2 emissions per capita; independent variables: natural logarithm of PPP-adjusted GDP per capita; average annual temperature in country capital (°C); transition country (dummy); and oil production per capita (2005 $1000).
Source: Data are from the World Bank, World Development Indicators online database except for temperatures of capital cities from www.worldclimate.com/

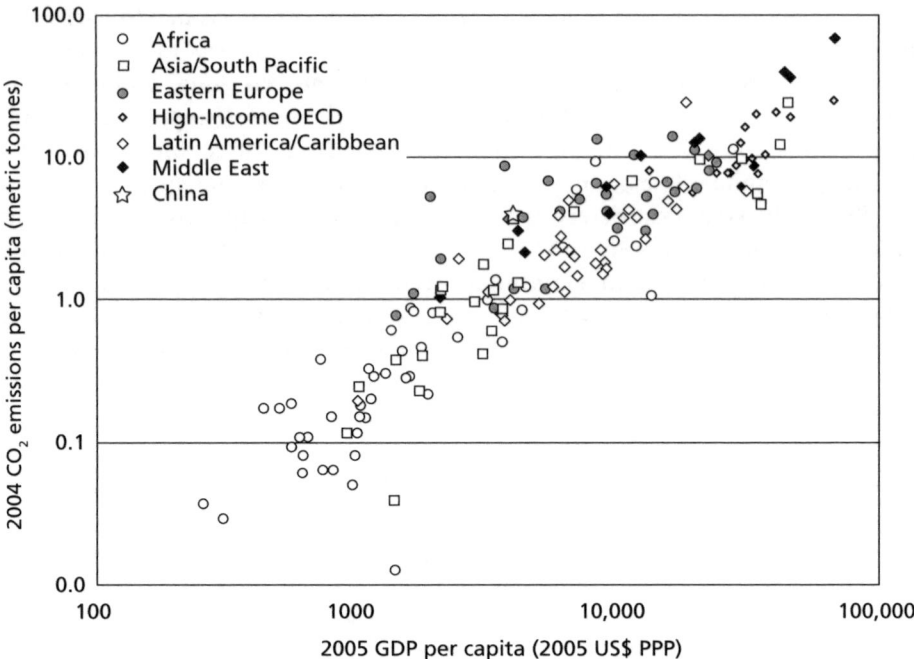

Source: Data are from the World Bank, World Development Indicators online database except for temperatures of capital cities from www.worldclimate.com

Figure 3.4 *Logarithm of emission/capita versus logarithm of GDP/capita (PPP)*

common argument in the EKC literature), than emissions are growing as income per capita grows, and, at least within the range of today's nations, there is no evidence that after some threshold of wealth, emissions will begin to decrease. Figure 3.4 shows the graph of emissions per capita versus income per capita in logarithmic scale.

Avoiding climate catastrophe: Development implications

Stabilization targets of 450ppm CO_2 or lower have the best chance of keeping the change in global annual average temperature below 2°C (Stern, 2006; IPCC, 2007). Figure 3.5 pairs IPCC stabilization trajectories, showing total emissions in the top panel (IPCC, 2007), with their global emissions per capita in the bottom panel, based on median UN population projections (UN, 2006). To hold steady at the 2005 level of total global emissions, per capita emissions will have to shrink from 4.3t to 3.2t over the next century to compensate for expected population growth. Note that the speed with which emissions reductions are begun is just as important as the eventual per capita target: the IPCC trajectories for 550ppm and 500ppm both reach 1.6t per

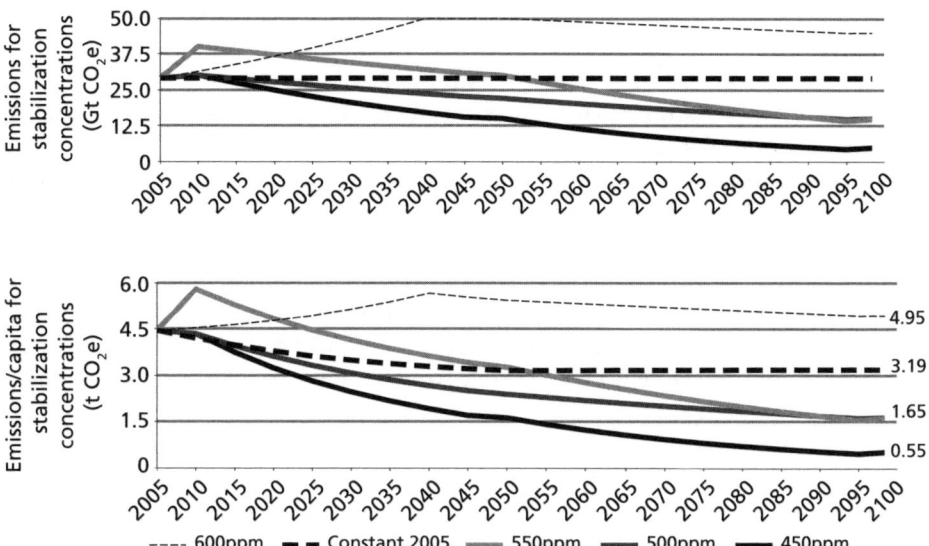

Source: Data are from the World Bank, World Development Indicators online database except for temperatures of capital cities from www.worldclimate.com

Figure 3.5 *Stabilization concentrations:*
Global emissions and emissions/capita

capita by 2100. The 550ppm trajectory projects rising global emission through 2010; in the 500ppm trajectory, emissions are reduced steadily from 2010 onward.

To achieve a 450ppm or lower stabilization concentration, global per capita emissions must decrease steadily throughout this century, dropping well below 1t per capita by 2100. As the world works towards lowering is emissions, there may be both good examples and lessons to be learned among the 57 countries that already have emissions levels lower than 1t per capita.

Today, most of the countries with emissions per capita less than 1t are the poorest countries with the lowest standard of living in the world – with life expectancies as low as 40 years in Zambia and literacy rates as low as 24 per cent in Mali and Burkina Faso – but there are some exceptions to this pattern. The richest countries today with emissions below 1t per capita are El Salvador ($5300 in income per capita) and Swaziland ($4500). The poorest countries in this group are Burundi and the Democratic Republic of Congo (both $300).

The highest life expectancies among this group are found in Nicaragua and Sri Lanka (both 72 years, the same as China, compared to Japan's 82 years). The transition countries have the highest literacy rates among this group, 100 per cent literacy in Georgia and Tajikistan; Samoa is a close third with 99 per cent literacy. Some low-emissions countries with anomalous achievements in development have taken a path that few would choose to follow: reaching near-Western levels of development and than losing it again (as with former

Soviet republics and countries that were highly dependent on economic relations with the Soviet Union). But other examples of low-emissions-intensity development warrant further study with the goal of replicating the success of a Sri Lanka or a Samoa: low emissions per capita but relatively high life expectancy and literacy.

If development is understood not as meeting basic needs but instead as achieving high levels of private consumption, however, examples of high-income countries with very low emissions per capita simply do not exist. To achieve a low stabilization concentration while maintaining high consumption, Western-style lifestyles or increasing levels of private consumption in developing countries over time, the development of new technology and a new global energy infrastructure will be essential. No high-income country has emissions lower than 5t per capita.

Table 3.8 reports the countries in the bottom quintile by emissions intensity. At 1.1t per capita and $14,000 in income per capita, Gabon has a remarkably low emissions intensity; its life expectancy is low at 56 years, but its literacy rate is relatively high at 84 per cent.

In the second decile by emissions intensity (see Table 3.8), Costa Rica and Uruguay stand out as middle-income, middle-development countries with low emissions, less than 2t per capita. Renewables other than biomass contribute a relatively high share of these countries' energy production: 20 per cent for Uruguay and 41 per cent for Costa Rica. Among high-income, high-development countries, only Switzerland has an emissions intensity in the lowest quintile; 18 per cent of Switzerland's energy production is from renewables and 23 per cent from nuclear. Switzerland has a cold climate (limiting air conditioning use), a very small manufacturing sector and a comprehensive system of local and intercity public transportation. Costa Rica, Uruguay and Switzerland all produce no oil and have large service sectors (62, 59 and 70 per cent of value-added to GDP from services, respectively). In comparison, China's energy production is composed of just 2 per cent renewables, excluding biomass and waste, and 1 per cent nuclear, and China's service sector generates only 40 per cent of the value added to GDP.

The development implications of GHG emissions reductions are twofold. First, lower emissions are currently associated with a lower standard of living; extraordinary efforts will be required to change this pattern. A combination of the widest access to today's best technologies and policies, with far-reaching and immediate investment in technology innovation will be necessary to keep global emissions per capita below 5t – much less bring it below 1t by the end of the century – while expanding the number of people at an adequate standard of living. Second, for many developing countries, the cost of emissions reductions will be prohibitive to development; for many transition countries and middle-income countries, such as China, these costs could be devastating to the maintenance of current standards of living and could even cause a reversal in the level of development.

Table 3.8 *Lowest quintile by emissions intensity*

Countries in bottom decile by emissions intensity	CO_2 emissions/capita (tons)	GDP/capita (PPP)	Life expectancy (years)	Literacy rate (%)	Cumulative emissions/capita (tons)
Chad	0.01	1468	50	26	0.7
Cambodia	0.04	1440	58	74	1.5
Mali	0.05	1004	53	24	1.3
Uganda	0.07	846	50	67	1.7
Gabon	1.08	13,821	56	84	108.8
Burkina Faso	0.08	1026	51	24	1.9
Rwanda	0.06	772	45	65	1.7
Burundi	0.03	319	49	59	0.8
Central African Republic	0.06	644	44	49	2.1
Tanzania	0.12	1049	51	69	2.8
Cameroon	0.22	1959	50	68	10.1
Nepal	0.11	960	63	49	1.7
Malawi	0.08	648	46	64	2.6
Lao PDR	0.23	1814	63	69	3.2
Vanuatu	0.42	3226	69	74	13.3
Comoros	0.15	1127	64	57	3.4
Countries in second lowest decile by emissions intensity	CO_2 emissions/capita (tons)	GDP/capita (PPP)	Life expectancy (years)	Literacy rate (%)	Cumulative emissions/capita (tons)
Congo, Dem. Rep.	0.04	267	46	67	3.1
Guinea	0.15	1081	55	30	4.9
Switzerland	5.47	35,182	81	99	324.6
Mozambique	0.11	677	43	39	5.3
Niger	0.09	584	56	29	2.3
Sri Lanka	0.59	3546	72	91	12.9
Costa Rica	1.50	9008	79	95	30
Gambia, The	0.18	1078	59	43	4.5
Sudan	0.29	1679	57	61	5.7
Zambia	0.20	1183	41	68	15.6
Ethiopia	0.11	628	52	36	1.3
Côte d'Ivoire	0.28	1614	47	49	9
Uruguay	1.66	9266	76	97	81.5
Haiti	0.19	1068	60	55	4.5
Madagascar	0.15	834	58	71	3.1
Peru	1.17	6460	71	88	37.9

Source: Data are from the UNDP *Human Development Report*, http://hdr.undp.org/, and World Resources Institute, Climate Analysis Indicators Tool, http://cait.wri.org/

At an average cost of $80 per tonne (see McKinsey & Company, 2009) of carbon eliminated, the cost of China's abatement from current emissions to 1t per capita would be 16 per cent of its GDP. With 20 per cent of the world's population and 17 per cent of world emissions, China's full participation is essential to achieve global emissions reduction goals. For China and many other countries, their ability to reduce emissions will be hobbled by their ability to pay for those reductions. Abating emissions need not be synonymous with paying for abatement; indeed, freeing abatement from the ability to pay for abatement may be the only way to reach emissions levels below 1t per capita without the kind of rapid loss of standard of living experienced in Eastern Europe and the former Soviet Union (Baer et al, 2008).

Conclusion

If China maintains a constant or slightly increasing emissions intensity as its income per capita grows, it will begin to resemble Bahrain, Kuwait or the UAE in terms of emissions per capita and income per capita. There are no examples of highly industrialized, high-income countries with as high an emissions intensity as China. Where China's emission intensity is 0.9t per $1000 of income per capita, India's is 0.6t, US 0.5t, Japan and Germany 0.3t and Brazil 0.2t. If China reaches Japan's income per capita (in PPP terms) while maintaining the same emissions intensity it will emit 32t per capita of CO_2 each year; if China reaches US income per capita at this emissions intensity, its emissions will reach 44t per capita.

The message here is not that a low stabilization trajectory is inconsistent with global development, rather, that today's industrialized countries do not provide an example of how to achieve a high average standard of living while keeping GHG emissions low. Even the best current technology cannot square high levels of consumption with emissions lower than 5t or 6t per capita. Quick action to implement the best existing technology worldwide is essential – and can be especially effective in countries such as China where changes to energy infrastructure are not limited by a lock-in to built structures – but so too are large-scale investments in innovation in the areas of low-carbon electricity production, low-energy transportation, residential and industrial technology, and CCS.

Notes

1   See the World Bank's World Development Indicators online database.
2   For a discussion of per capita emissions rights see Baer et al (2008).
3   PPP adjustments to GDP per capita are an output of the International Comparison Project. According the World Bank, this 'uses a series of statistical surveys to collect price data for a basket of goods and services. For meaningful inter-country comparisons, the International Comparison Project considers the affordability and price level of necessities and luxuries, which exchange rates ignore. Surveys are held every three to five years, depending on the region. The data collected are combined with other economic variables from countries' national accounts to calculate Purchasing Power Parities or PPPs, a form of exchange rate that takes into account the cost and affordability of common items in different countries, usually expressed in the form of US dollars. By using PPPs as conversion factors, the resulting comparisons of GDP volumes enable us to measure the relative social and economic well-being of countries, monitor the incidence of poverty, track progress towards the Millennium Development Goals and target programs effectively' (www.worldbank.org).

4   For more information on the HDI see http://hdr.undp.org/.
5   PPP-adjusted GDP alone explains 61 per cent of the variability in HDI.
6   Or nation-level projections in the case of life expectancy data.
7   World Bank, World Development Indicators online database.
8   Note that China, as defined in this international dataset, does not include Hong Kong and Macao. Data are from the World Bank, World Development Indicators online database.
9   Data are from the World Bank, World Development Indicators online database.
10 World Resources Institute, Climate Analysis Indicators Tool, http://cait.wri.org/, 2000 data.
11 Assuming a non-transition economy with zero oil production.
12 Tibet is excluded from this analysis for lack of data. Hong Kong and Macao are also excluded.
13 To compare Chinese provinces with nations, emissions per capita were imputed by adjusting China's 3.9t per capita in proportion to the ratio of each province's energy consumption per capita to China's in standard coal equivalents; similarly, PPP-adjusted gross regional product (GRP) per capita is China's income per capita weighted by the ratio of each GRP per capita to China's GDP per capita in Yuan. Chinese provincial data are taken from the *China Statistical Yearbook 2007*, www.stats.gov.cn/tjsj/ndsj/2007/indexeh.htm. Of course, this method assumes that the energy production mix and PPP are consistent across all provinces. For a comparison with other sources of emissions per capita data for Chinese provinces see the appendix to this chapter.
14 See for examples the literature on the United Nations' Millennium Development Goals, www.un.org/millenniumgoals/
15 For a good summary of issues surrounding the EKC, see Torras and Boyce (1998).

References

Baer, P., Athanasiou, T., Kartha, S. and Kemp-Benedict, E. (2008) *The Greenhouse Development Rights Framework: The Right to Development in a Climate Constrained World*, Berlin: Heinrich Böll Foundation.
IPCC (Intergovernmental Panel on Climate Change) (2007) *Climate Change 2007: IPCC Fourth Assessment Report*, Cambridge, UK: Cambridge University Press
Leung, D. Y. C. and Lee, Y. T. (2000) 'GHG emissions in Hong Kong', *Atmospheric Environment*, vol 34, pp4487–4498
McKinsey & Company (2009) *Pathways to a Low-Carbon Economy*, available at www.mckinsey.com/clientservice/ccsi/pathways_low_carbon_economy.asp
Stern, N. (2006) *The Stern Review: The Economics of Climate Change*, London: HM Treasury
Torras, M. and Boyce, J. K. (1998) 'Income, inequality, and pollution: A reassessment of the environmental Kuznets Curve', *Ecological Economics*, vol 25, no 2, pp147–160
UN (United Nations) (2006) *World Population Prospects: The 2006 Revision Population Database*, http://esa.un.org/unpp/

Appendix: Comparison of emissions per capita data

Differences among these datasets are generally within 15 per cent and are entirely within 25 per cent with the exceptions of Qinghai and Sichuan provinces.

Table 3.9 *Total CO_2 emissions per capita by Chinese province, 2005–2006*

	Princeton University[a]	Tsinghua University[b]	Author's Calculations[c]
Anhui	2.87	2.95	2.43
Beijing	7.61	9.56	7.80
Chongqing	3.07	3.23	3.51
Fujian	4.00	4.60	4.00
Gansu	3.74	4.70	3.96
Guangdong	4.03	5.87	4.48
Guangxi	2.45	2.61	2.51
Guizhou	4.60	4.56	3.99
Hainan	2.00	2.21	2.35
Hebei	7.42	7.12	6.60
Heilongjiang	4.73	5.64	4.71
Henan	4.16	4.10	3.68
Hubei	3.83	4.15	4.01
Hunan	3.34	3.72	3.33
InnerMong	12.03	11.39	9.94
Jiangsu	6.02	6.37	5.27
Jiangxi	2.41	2.45	2.27
Jilin	6.41	6.66	5.15
Liaoning	7.89	8.40	7.93
Ningxia	10.42	10.57	9.94
Qinghai	4.10	4.08	7.53
Shaanxi	3.75	4.05	3.56
Shandong	6.56	7.21	6.02
Shanghai	10.84	12.87	10.28
Shanxi	9.89	9.01	8.38
Sichuan	2.30	2.25	3.27
Tianjin	9.95	10.32	8.94
Xinjiang	6.06	6.85	6.41
Yunnan	3.43	3.36	3.15
Zhejiang	5.37	6.70	5.63

Source: a Personal communication. Jie Li, Princeton University, 2008. Data are for 2005. b Personal communication. Cao Jing, Tsinghua University, 2008. Data are for 2005. c Author's calculations using data from *China Statistical Yearbook 2007*. Data are for 2006.

4

Carbon Embedded in China's Trade

Frank Ackerman

Introduction

A large fraction of China's carbon emissions are caused by international trade: emissions that occur in China are, in many cases, incurred in order to satisfy final demand in other countries. The responsibility for these emissions is a crucial question of policy: is China responsible, based on the location of production – or is the country that buys carbon-intensive products from China responsible, based on the location of consumption?

This chapter analyses the sectoral composition of the carbon embedded in China's foreign trade. It provides the research basis for Chapter 13, exploring policy options for carbon embedded in trade. The analysis presented here may help to clarify the implications of rapid growth of the Chinese economy. By 2050 or sooner, if current trends continue, China will reach the level of per capita income in the US today. What changes will occur in the magnitude of carbon embedded in trade as the Chinese economy continues to grow and change? To shed light on that broad topic, this chapter examines a more specific question: what would happen to the carbon embedded in current trade patterns if China had the current US level of carbon-intensity in production?

The analysis relies on the Multi-Region Input–Output (MRIO) model developed by Glen Peters and his colleagues, which has been widely used for analyses of carbon embedded in international trade (see, for example, Peters and Hertwich, 2008). Conforming to the data categories of the Global Trade

Analysis Project (GTAP) model of international trade, and building on the GTAP dataset, the MRIO model provides consistent global estimates for domestic emissions and carbon embedded in imports and exports, for 87 countries or regions and 57 economic sectors in 2001. This provides consistency and completeness, at the price of a somewhat dated dataset. Since 2001, China's economy has grown rapidly, with increases in both the trade surplus and carbon intensity of production through 2006 (Raupach et al, 2007).

The carbon intensity of trade: Three scenarios

China is currently less carbon efficient than developed countries. That is, it has higher carbon emissions per dollar of output. As a consequence, exports from China are more carbon intensive than almost all of the imports into the country. Economic growth, however, is likely to bring more advanced, carbon-efficient technology into use, and may therefore narrow the carbon-intensity gap between exports and imports. How would a switch to more carbon-efficient production affect the volume of carbon embedded in trade?

To explore the effect of technological change and convergence on the carbon intensities of exports and imports, consider the following three scenarios for carbon embedded in trade:

1   actual trade flows and emissions in 2001 (MRIO estimates);
2   actual trade volume (that is, dollar value of China imports and exports unchanged from scenario 1), but all imports produced at China's current carbon intensity in each industry;
3   actual trade volume, but all imports and all China exports produced at the current US carbon intensity in each industry.

The second and third scenarios are hypothetical and unrealistic; they serve only to separate the effects of carbon intensity from the effects of trade volumes. Figure 4.1 shows the carbon embedded in exports, imports and the balance of trade in emissions under each of these scenarios.

Under the first scenario, with actual carbon intensities for exports and imports, there is of course a large trade surplus: carbon embedded in exports far exceeds that in imports, so China is a large net exporter of carbon. However, if imports to China were produced at the same intensity as domestic production, the second scenario shows that the emissions trade balance would almost vanish: imports would embody almost as much carbon as exports. This is consistent with findings by other researchers; for example, Peters et al (2007), examining China's carbon emissions, report that 'There is a rough balance between CO_2 emissions from the production of exports and emissions avoided by imports.' Emissions avoided by imports are the emissions that would have resulted from producing the country's imports in China, which is the same as the import emissions in the second scenario in Figure 4.1.

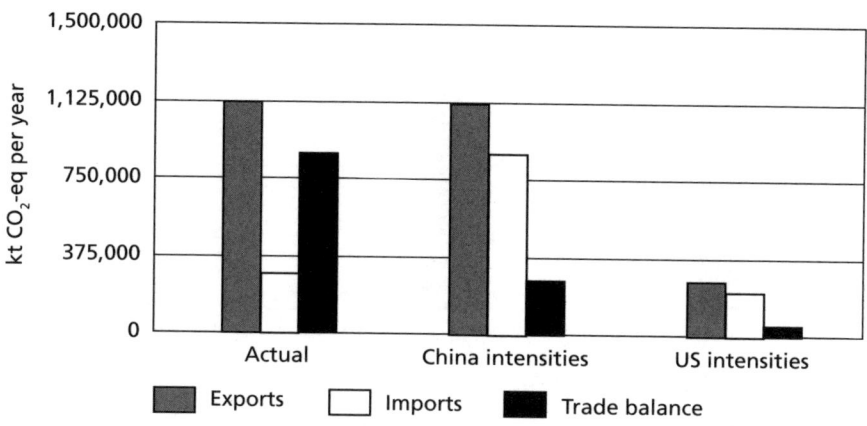

Figure 4.1 *Carbon embedded in China's trade, actual and scenarios*

The third scenario shows that if China reached current US carbon intensities while maintaining its current volume of exports, and its imports were also produced at current US intensities, the carbon trade surplus would be close to zero. This scenario slightly increases the actual carbon embedded in imports, because many Chinese imports come from Japan and Europe, where carbon intensities are below US levels.

The second and third scenarios both imply that, at equal intensities for imports and exports, China's trade surplus in embedded carbon would almost vanish – while by hypothesis, its monetary trade surplus would remain large. Everyone agrees that China has a sizeable trade surplus, but the amount is subject to dispute: the MRIO dataset used in this analysis shows that China had a trade surplus of \$108 billion in 2001; this is larger than the surplus reported by China, but smaller than the sum of the amounts reported by China's trading partners[1] (see Table 4.1 for trade balances for leading export sectors).

The combination, in the second and third scenarios, of roughly balanced trade in carbon with a big trade surplus in monetary terms is possible because at equalized sectoral intensities, imports to China would be more carbon intensive than exports, as shown in Figure 4.2. In the first scenario, with actual carbon intensities, China's exports are far more carbon intensive than its imports. But in either of the other scenarios – with each sector's intensity equalized worldwide either at China's intensities or at US intensities – the current mix of imports to China is on average slightly more carbon intensive than exports from China.

This conclusion is perhaps unfamiliar and deserves some emphasis: China's position as a net exporter of carbon does *not* result from exporting uniquely carbon-intensive products. Rather, China relies heavily on coal and uses energy less efficiently than many of its trading partners. China is a net exporter of many manufactured goods, including both high-technology products such as

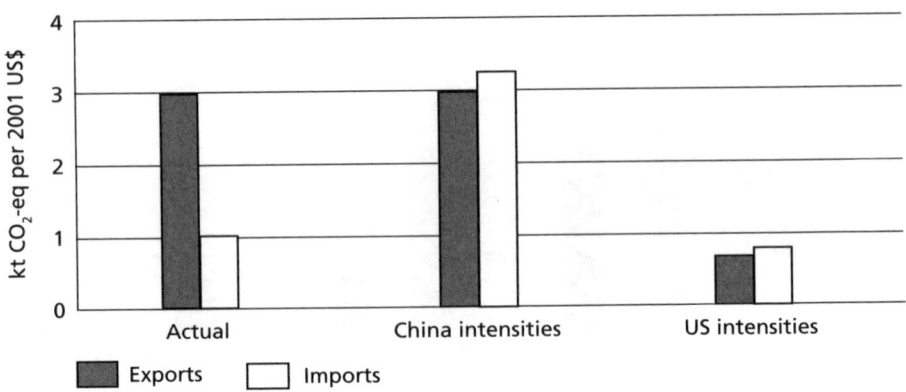

Figure 4.2 *Carbon intensity of China's trade, actual and scenarios*

electronics and machinery and traditional manufactures such as leather goods, apparel and textiles. By contrast, China is a net importer of chemicals, metals, minerals and oil, among other things. At any one level of technology, either Chinese or American, China's imports would be more carbon intensive than its exports. However, China's imports and exports are not produced at the same level of technology and carbon intensity.

Carbon intensity by sector: China versus the US

How do carbon intensities by sector compare in China and the US? China has higher emissions per dollar[2] in almost every industry, but there is no simple relationship between Chinese and American intensities.

Figure 4.3 plots both countries' carbon intensities for each sector, with China on the horizontal axis and the US on the vertical. Figure 4.4 enlarges the lower left corner of Figure 4.3 (note the difference in scale on the axes). The diagonal line represents equal intensities in both countries; only oil seeds fall on the line, while a few other sectors such as paddy rice are close to it. In other sectors, China's carbon intensity is greater, often much greater, than the US intensity. The round dots (except for coal, these are shown only in Figure 4.4) are the sectors that account for more than 3 per cent of China's total volume of carbon embedded in exports; they do not appear to be uniquely high or low in carbon intensity.

The most carbon-intensive sectors in China are energy sectors (coal, gas, electricity) and selected agricultural sectors (rice, meat); note that non-CO_2 GHGs play a large role in agricultural emissions. The sectors named in Figure 4.3 – the only ones in which China's carbon intensity exceeds 10 kilograms (kg) of CO_2-equivalent per dollar of output – together account for less than 1 per cent of the dollar value of exports: 0.81 per cent for coal and a combined total of only 0.11 per cent for the other six industries.

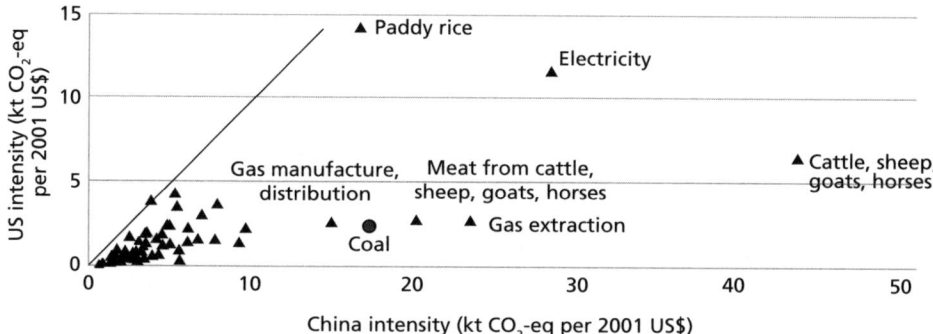

Figure 4.3 *Carbon intensity of industries: US versus China (a)*

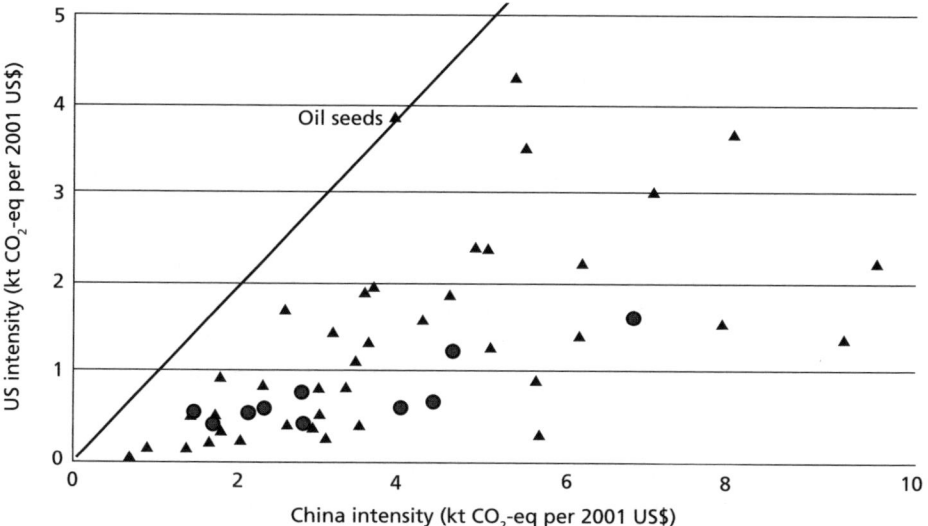

Note: Round dots are sectors accounting for more than 3% of the carbon embedded in China's exports as a whole. From left to right, they are: forestry; apparel; wood products; manufactures not elsewhere classified; textiles; transport equipment (excluding motor vehicles); leather products; metal products; chemicals rubber and plastics; and mineral products; See also the coal industry, shown in Figure 4.3.

Figure 4.4 *Carbon intensity of industries: US versus China (b)*

Revealed comparative advantage and carbon intensity

China has of course been a remarkable success in world trade. And China has very carbon-intensive industries. This section explains why these two facts have little to do with each other: China's most successful export sectors are not its most carbon-intensive ones.

Economists frequently analyse trade in terms of comparative advantage. A country is said to have a comparative advantage in an industry if it has lower costs of production in that industry, relative to other sectors, than its trading partners. A country can usually maximize its income, at least in the short run, by exporting the products in which it has a comparative advantage. (Longer-run questions of development strategy are not addressed by this short-run calculation.)

There is no obvious, direct measure of comparative advantage; instead, it has become common to assume that comparative advantage is revealed by success in trade. 'Revealed comparative advantage' (RCA) is often measured by the Balassa index (first introduced in Balassa, 1965): for China, the RCA for sector j is j's share of China exports, divided by j's share of all world exports. That is:

$$RCA_j = \frac{ChinaExports_j / ChinaExports_{ALL}}{WorldExports_j / WorldExports_{ALL}} \tag{1}$$

If $RCA_j > 1$, then sector j is more important in China's exports than in world exports in general; hence China is said to have an RCA in that sector.

There are 39 sectors that account for at least 0.1 per cent of China's exports. For these 39 sectors, there is no correlation between RCA and China's carbon intensity, as can be seen in Figure 4.5. The same is true if, instead of China's absolute carbon intensity, RCA is compared to China's intensity relative to the US, that is, the ratio of Chinese to American carbon intensity.[3] Using somewhat different data categories, my colleagues and I found a similar pattern in a study of US–Japan trade: comparative advantage between the US and Japan has only a weak, statistically insignificant correlation with greater carbon intensity (Ackerman et al, 2007).[4]

There are only 13 sectors with RCA > 1, i.e. sectors where China has a comparative advantage relative to world trade as a whole. Data on trade, emissions and intensities for these 13 sectors are presented in Table 4.1. As the

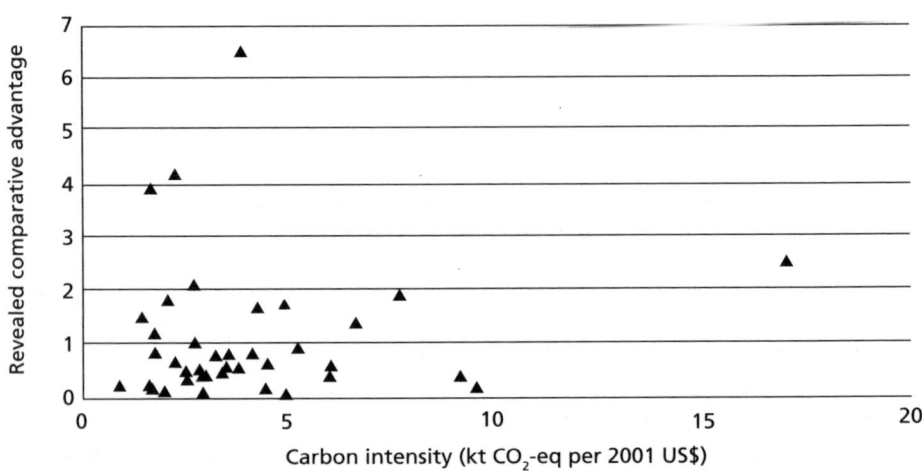

Figure 4.5 *Revealed comparative advantage versus carbon intensity*

Table 4.1 *Sectors where China has a revealed comparative advantage*

	Shares of total exports China	World	RCA	Carbon intensity	Exports	CO_2-eq in exports	Net exports
Electronic equipment	17.9	11.8	1.5	1.4	67,811	98,252	16,207
Machinery and equip- ment nec	15.3	14.8	1.0	2.7	57,874	158,362	10,744
Manufactures nec	11.2	2.7	4.2	2.3	42,573	96,302	39,976
Wearing apparel	8.9	2.3	3.9	1.7	33,650	56,065	31,049
Leather products	8.4	1.3	6.5	3.9	32,021	124,475	29,693
Textiles	6.0	2.8	2.1	2.7	22,660	61,506	5879
Wood products	3.3	1.8	1.8	2.1	12,449	25,816	9977
Metal products	3.2	1.9	1.7	4.3	12,136	51,833	9406
Mineral products nec	2.3	1.6	1.4	6.6	8655	57,503	5534
Coal	0.8	0.3	2.5	17.0	3055	51,976	2983
Animal products nec	0.4	0.2	1.8	4.9	1500	7379	18
Processed rice	0.2	0.1	1.9	7.7	598	4597	492
Fishing	0.1	0.1	1.2	1.8	510	894	441
All sectors with RCA > 1	77.9	41.8	1.9	2.7	295,492	794,960	162,399
All other sectors	22.1	58.2	0.4	4.0	83,975	334,181	(54,163)
Total	100.0	100.0	1.0	3.0	379,468	1,129,141	108,236

Units: Carbon intensity – kg CO_2-eq / US 2001$ of output
Exports, net exports – millions of US 2001$
CO_2-eq in exports – kt CO_2-eq
Note: nec means not elsewhere classified.

table shows, these 13 sectors account for 78 per cent of exports in China, versus 42 per cent worldwide. They also account for 70 per cent of the carbon embodied in China's exports. In terms of net exports, the 13 sectors account for 150 per cent of China's positive balance of trade – since the other sectors of the economy as a whole are substantial net importers.

The 13 sectors, ranked in Table 4.1 according to the dollar value of Chinese exports, suggest a variety of bases for comparative advantage. The top three, representing 44 per cent of exports, are all manufacturing sectors, combining modern (electronics) and unspecified activities. The next three, at 23 per cent of exports, are traditional low-technology manufacturing sectors, namely textiles, apparel and leather products. Note that leather products is the sector with the highest RCA, and apparel is third highest. The next four sectors, accounting for almost 10 per cent of exports, are resource-based sectors: wood, metals, minerals and coal. Three small agricultural sectors, totalling less than 1 per cent of exports, complete the list.

Judging by the list of 13 sectors, China's comparative advantage resides in a combination of advanced and traditional manufacturing, with only a minor role for natural resources. With the exception of the coal industry, none of these sectors are extraordinarily carbon intensive. Indeed, the other economic sectors, where China does not have a comparative advantage, are on average more carbon intensive than these 13.

These findings are consistent with the research literature on China's comparative advantage. Numerous analysts have focused on the role of wages and labour supply, assuming that China has a comparative advantage in manufacturing based on labour costs; this is often discussed in combination with government policies that have protected and promoted key industries. Cai and Wang (2006) focus on the demographic transition and project that China's working-age population could start to decline within the next decade, a trend that could ultimately conflict with specialization in labour-intensive manufacturing. Gerard Adams et al (2006) discuss the role of low-cost labour, exchange rates and a range of other factors in explaining China's competitiveness. Xiaohui Liu and Chang Shu (2003) find that the export performance of Chinese industries is significantly influenced by labour costs, foreign direct investment and firm size. Lee et al (2005) compare China to other rapidly developing East Asian economies, exploring the institutional framework for development and the interplay between 'comparative-advantage-following' and 'comparative-advantage-defying' strategies. Dani Rodrik (2006) argues that China has not simply relied on the comparative advantage produced by cheap labour; rather, government policy has helped to develop competitive industries in areas such as consumer electronics, making and exporting goods that would normally be produced by countries at a higher level of income. In this framework, future growth depends on China's ability to identify and promote higher-income products as it develops.

A rich understanding of Chinese economic success emerges from these analyses, among others; none of them mention energy, carbon intensity or GHG emissions as an important factor in the country's development. China is developing at an unprecedented rate and its economy is currently very carbon intensive – but no one suggests that there is a necessary link between these two important trends.

Towards policy implications

This data analysis suggests two conclusions: first, China's success in world trade is not closely tied to carbon intensity; and second, as China develops, technological change may well eliminate the surplus of embedded carbon in trade, even if China retains a large trade surplus in monetary terms.

This does not mean that embedded carbon can be ignored in policy debate. Assigning responsibility for exported carbon to the consuming country will heighten the interest of the high-income importing countries in aiding the development process. If the US 'owns' the share of China's carbon emissions embedded in its imports from China, then investing in modernizing China will be a much higher priority for the US – and likewise for other developed countries. This will help to promote policies that lead to the joint goals of climate control and development. The fact that success in development could eliminate China's large trade surplus in carbon is only a footnote to the more important objective of adopting low-carbon technologies and creating sustainable paths to development.

The policy implications of carbon embedded in trade, and proposals to establish international prices on embedded carbon, are discussed in Chapter 13.

Notes

1 Chinese data show a trade surplus of $22.6 billion for 2001; reports from China's trading partners imply a surplus of $191.8 billion for the same year. The surplus grew rapidly after 2001, reaching $102.1 billion (China version) or $342.2 billion (partners version) by 2005 (Lum and Nanto, 2007).
2 Since the analysis involves international trade, all monetary amounts are based on market exchange rates, rather than PPP.
3 In both cases, comparing RCA either to China's carbon intensity or to the China/US ratio of carbon intensities, ordinary least square (OLS) regression shows that the adjusted r^2 is negative and $p > 0.5$, indicating *less* correlation between the two data series than would be expected by chance. The same is true in both cases if the coal industry, the extreme outlier in carbon intensity, is removed from the regressions.
4 The relationship between carbon intensity and the balance of trade was closer to significance in the Japan–US case ($p = .064$), but still explained only 2 per cent of the inter-industry variance in the balance of trade.

Acknowledgements

I would like to thank Glen Peters for generously sharing the data and results from his MRIO model. He is not responsible for the content of this chapter, which is my analysis based on his work.

References

Ackerman, F., Ishikawa, M. and Suga, M. (2007) 'The carbon content of Japan–US trade', *Energy Policy*, vol 35, pp4455–4462

Adams, F. G., Gangnes, B. and Shachmurove, Y. (2006) 'Why is China so competitive? Measuring and explaining China's competitiveness', *World Economy*, vol 29, no 2, pp95–122

Balassa, B. (1965) 'Trade liberalization and "revealed" comparative advantage', *Manchester School*, vol 33, pp99–123

Cai, F. and Wang, M. (2006) 'Challenge facing China's economic growth in its aging but not affluent era', *China & World Economy*, vol 14, no 5, pp20–31

Lee, K., Lin, J. Y. and Ha-Joon Chang (2005) 'Late marketisation versus late industrialisation in East Asia', *Asian-Pacific Economic Literature*, vol 19, no 1, pp42–59

Lum, T. and Nanto, D. K. (2007) 'China's trade with the United States and the World', US Congressional Research Service, http://fas.org/sgp/crs/row/RL31403.pdf

Peters, G. P. and Hertwich, E. G. (2008) 'CO_2 embodied in international trade with implications for global climate policy', *Environmental Science and Technology*, vol 42, pp1401–1407

Peters, G. P., Weber, C. L., Guan, D. and Hubacek, K. (2007) 'China's growing CO_2 emissions: A race between increasing consumption and efficiency gains', *Environmental Science & Technology*, vol 41, no 17, pp5939–5944

Raupach, M. R., Marland, G., Ciais, P., Le Quéré, C., Canadell, J. G., Klepper, G. and Field, C. B. (2007) 'Global and regional drivers of accelerating CO_2 emissions', *PNAS*, vol 104, no 24, pp10288–10293

Rodrik, D. (2006) 'What's so special about China's exports?', *China & World Economy*, vol 14, no 5, pp1–19

Xiaohui, L. and Chang, S. (2003) 'Determinants of export performance: Evidence from Chinese industries', *Economics of Planning*, vol 36, pp45–67

5

A Deep Carbon Reduction
Scenario for China

Charlie Heaps

Introduction

This chapter presents an initial technical exploration of how China's energy systems might be altered over the coming four decades to allow China to meet ambitious goals for development and income growth at the same time as keeping GHG emissions within very tight budgets that provide a reasonable chance of keeping global temperature increases below 2°C.

To explore this question two scenarios have been developed. The baseline scenario examines current and historical trends in China's CO_2 emissions and projects CO_2 emissions to 2050, assuming that China continues to develop very rapidly, albeit at a slowing rate compared to the last two decades. The baseline assumes a general continuation of current policies that include some significant efforts to address sustainability and the climate challenge, but it does not foresee any fundamental shifts in energy policy. The net result is that energy sector emissions are expected to continue to grow rapidly in the baseline scenario. Starting from a base of 4.8Gt in 2005, energy sector CO_2 emissions reach 12.4Gt in 2030 and 18.0Gt in 2050 in the baseline scenario, an almost fourfold increase. The baseline scenario relies upon a sector by sector review of historical trends, as well as an examination of how future energy and CO_2 emissions patterns can be expected to evolve as China develops and average income levels increase. Up until 2030 the scenario closely matches trends in energy use and CO_2 emissions foreseen in the IEA's *World*

Energy Outlook 2008 (WEO) for China (IEA, 2008a). For example, the WEO Reference scenario foresees China's energy sector emissions reaching 11.7Gt in 2030. Additional analysis has been done to extrapolate energy and CO_2 emissions patterns out to 2050.

The second DCRS examines the feasibility of massively reducing China's CO_2 emissions in 2050: with energy sector GHG emissions reduced to only 10 per cent of the 2050 levels projected in the baseline scenario or about 85 per cent of the level in 1990. Achieving such a target is made even more difficult because the DCRS attempts to meet these reductions while continuing to assume the same income growth rates as in the baseline scenario.

The DCRS pathway for China is designed to stay within an overall emissions budget for energy sector emissions of about 230Gt CO_2 between 2005 and 2050. This budget was developed by Baer et al (2008) as part of the GDR framework. The GDR framework suggests a global emergency pathway that is consistent with giving the world a realistic chance of keeping global temperature increases below 2°C. This requires that atmospheric CO_2 concentrations peak below 420ppmv and then begin to fall. Clearly, such a pathway is extremely ambitious, but even so it implies considerable risks. Its authors estimate that its probability of exceeding 2°C is roughly 14–32 per cent, which in the language of the IPCC, is 'likely', but not 'very likely' to keep warming below 2°C. Indeed, a growing number of climate scientists now conclude that 350ppmv would be a more prudent goal for concentrations (Hansen et al, 2008).

The GDR global emissions budget is allocated among countries as follows. First, Annex 1 countries (primarily OECD countries) are assigned an ambitious trajectory that ends with emissions roughly 90 per cent below 1990 levels in 2050. For these countries, emissions stay roughly constant until around 2013 and then decline at an essentially constant rate. Non-Annex 1 (NA1) countries, including China, are allocated a budget such that all NA1 countries show the same overall percentage decrease relative to their baselines. That percentage is set at a level that makes the whole world consistent with the overall trajectory.

Given the overall momentum for growth and development in China, and the lack of availability of technologies that can be deployed immediately on a large scale, it is simply inevitable that China's emissions will continue to climb in the next decade even under the most ambitious of mitigation scenarios. This makes the requirements for reducing emissions after 2020 particularly challenging. To stay within the overall 230Gt budget, China's CO_2 emissions need to be reduced to about 10 per cent below their 1990 values. This figure seems challenging enough, but this equates to no more than 10 per cent of what emissions are likely to be in 2050 in a baseline scenario.

It is important to note that the DCRS developed here for China is only a technical feasibility study. It examines whether China's emissions might be cut to a level that is compatible with protecting the planet while also giving enough 'emissions space' for China to continue to develop. It is not a proposal about the level of obligation that China should take on in any international climate negotiations. The issue of what financial burdens different parties should take on for achieving these emissions reductions is an entirely separate question.

The DCRS also is not intended to represent a least-cost development path. The analysis presented here does not examine the economics of different mitigation options and thus it has not attempted to find an optimized pathway that balances the costs of mitigation against the costs of failing to act to avoid climate change damages.

Finally this analysis also should not be read as a proposal for specific energy policies. A range of alternative pathways could potentially yield the same total emissions as the DCRS. The DCRS is intended only as an initial existence proof of whether China's emissions might be reduced sufficiently if the will emerged to do so, both within China and in the rest of the world.

Methodology

The two scenarios described here have been developed using SEI's Long-range Energy Alternatives Planning System (LEAP) energy modelling system: a transparent and user-friendly accounting-based software tool for scenario-based energy analysis and GHG mitigation assessment (Heaps, 2008a, 2008b).[1]

A baseline scenario

The baseline scenario (BLS) examines how China's energy system and its CO_2 emissions might evolve to 2050 in the absence of significant new policies specifically designed to address climate mitigation. The BLS covers energy

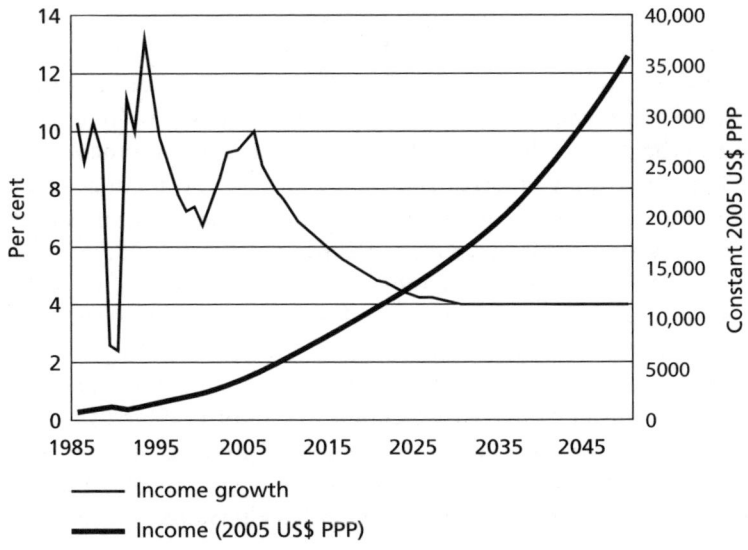

Source: Stockholm Environment Institute/LEAP

Figure 5.1 *Baseline average income and income growth*

consumption and production and related CO_2 emissions going back histori-cally from 1990 to 2006 and projecting forward to 2050 (see Figure 5.1).

The analysis uses a straightforward accounting methodology in which emissions of different pollutants are calculated as the product of fuel combus-tion and an emission factor. Energy consumption is in turn calculated as the product of an activity level measuring the level of energy service provided (for example number of households, passenger-km of transportation, dollars of value added in an industry, etc.) and an energy intensity. Put simply, emissions are calculated as follows:

$$P = A \times E \times F \tag{1}$$

where P is the emission of CO_2; A is a measure of economic activity; E is energy intensity of the activity [TJ/activity]; and F is the CO_2 emission factor [tonnes/TJ]

Levels of activity in each consuming sector (industry, households, services, agriculture, transport) are first projected forward based on overall assumptions about levels of growth of the Chinese economy and how its structure might shift (for example from industry to services, or between heavy and light indus-try) as income levels grow.

The scenario is driven forward by two high-level exogenous assumptions: population and average income levels. For population, the medium variant of the UN population projections is used (UN, 2007), which foresees China's population peaking in 2030 at 1.46 billion people before declining slightly to 1.41 billion in 2050. Average income levels are assumed to continue to grow rapidly in China (although somewhat slower than in the last two decades) as shown in Figure 5.2. Historical income levels are taken from the World Bank World Development Indicators, 2008 (World Bank, 2008) expressed in constant 2005 US$ PPP terms.[2] After 10.1 per cent growth in average income levels in 2007 the scenario assumes that growth declines gradually to 4 per cent in 2030 and thereafter. The assumption is that rapid development will continue but as China's economy matures and the size of its workforce peaks, its rate of growth will also decline. The net result of these assumptions is that average income levels grow enormously from $4062 in 2005 to $16,487 in 2030 and $35,711 in 2050, an increase of almost 800 per cent and a value roughly equal to present-day income levels in richer European countries such as the UK, Germany, France and Scandinavia.

GDP is calculated as the product of population and average income so the above assumptions also imply that China's GDP in PPP terms increases from $5.3 trillion in 2005 to $24 trillion in 2030 and $50 trillion in 2050. By comparison, US GDP in 2005 was about $12.4 trillion. Figure 5.2 shows some sensitivity analyses of GDP projections under alternative assumptions about income growth.

GDP is separated into its sectoral value-added components (industry, services and agriculture) using historical data from the World Bank. Industry is

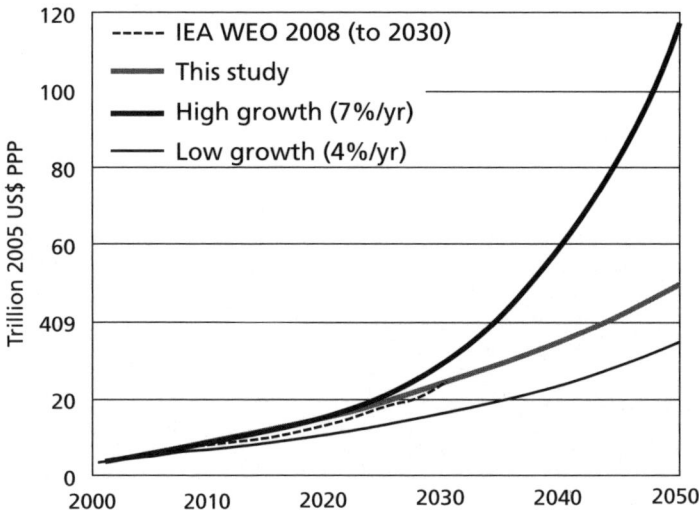

Source: Stockholm Environment Institute/LEAP

Figure 5.2 *China GDP sensitivity to growth assumptions*

further broken down within the manufacturing subsectors. Here data from the United Nations Industrial Development Organization (UNIDO, 2008) is used to calculate historical value added in each major manufacturing sector corresponding to the sector groupings used by the IEA (IEA, 2007b) for its industrial energy use statistics (iron and steel, chemicals and petrochemicals, non-ferrous metals, non-metallic minerals, machinery, food and tobacco, paper pulp and print, textiles and leather, transport equipment, wood and wood products, and 'other').

These historical data are used to calculate the share of value added in each major sector and subsector so that the estimates of GDP can be allocated down to the various energy-consuming sectors and subsectors of China's economy. Future estimates of these activity levels are calculated based on the overall future growth in GDP coupled with cross-country regressions that estimate how the value-added share of GDP from the industry, service and agriculture sectors are likely to change in the future as average incomes increase. In the BLS, the share of GDP coming from services is projected to increase from 40 per cent in 2005 to 52 per cent in 2050. The share from agriculture decreases markedly from 12.6 per cent in 2005 to only 2 per cent in 2050. The balance (the share from industry) stays almost constant going from 47.5 per cent in 2005 to 46.3 per cent in 2050 (see Figure 5.3).

These activity levels are multiplied by energy intensities that are initially calculated from historical data on energy consumption from the IEA (2008b), which contains information on fuel use in all major energy-consuming sectors. The intensity values are calculated by dividing the IEA's total consumption data by each sector's historical activity level. Energy intensities are projected

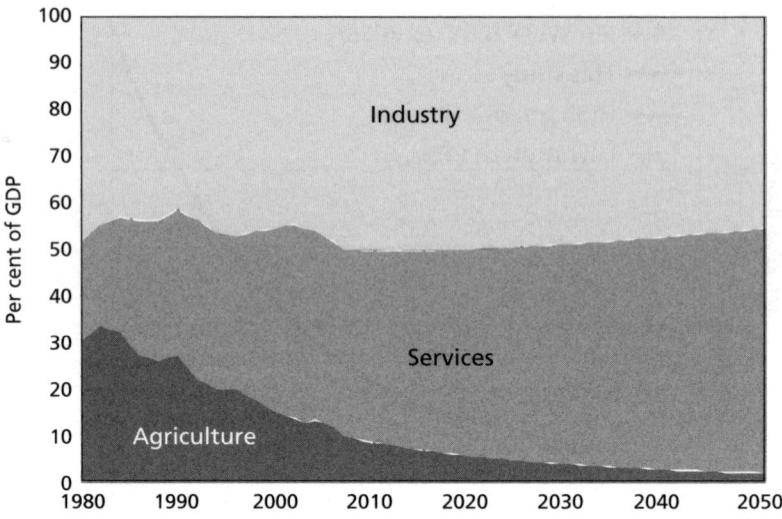

Source: Stockholm Environment Institute/LEAP

Figure 5.3 *Historical and baseline value added trends*

forward based in part on an assessment of historical trends as well as cross-country comparisons of how energy intensities in each sector have evolved as incomes increase, adjusting for temporal improvements in energy intensities that can be expected due to autonomous energy efficiency improvements. For example, Figure 5.4 shows how energy use per capita in the household sector was compared with income levels across countries for the year 2000. These regressions inform a level of energy intensity that China is assumed to gradually converge towards as its average income levels approach those seen in the richer countries. This type of 'convergence algorithm' was used for estimating future intensities in the household, services and agriculture sectors. A key benefit of this approach is that future energy intensities can be driven by overall assumptions on income growth – making it possible to do sensitivity analysis of different assumptions of GDP growth. Future energy intensities in the industry and transport sector are based primarily on historical trends as well as the Chinese government's stated plans for the period to 2010.

In the transport sector (see Figure 5.5), historical data on energy use are again taken from the IEA's energy statistics, which are broken down into passenger and freight energy use and major mode (road, rail, air, water and pipelines). Historical data on passenger transportation demands in passenger-km (p-km) and freight transportation demands in tonne-km (t-km) are taken from the *China Energy Databook* (LBNL, 2008) prepared by the Lawrence Berkeley National Laboratory (LBNL), which in turn is based on the *China Statistical Yearbook* (National Bureau of Statistics of China, 2006). These data are used to calculate historical energy intensities per p-km and per t-km respectively. Activity levels are projected forward using GDP growth rates and the

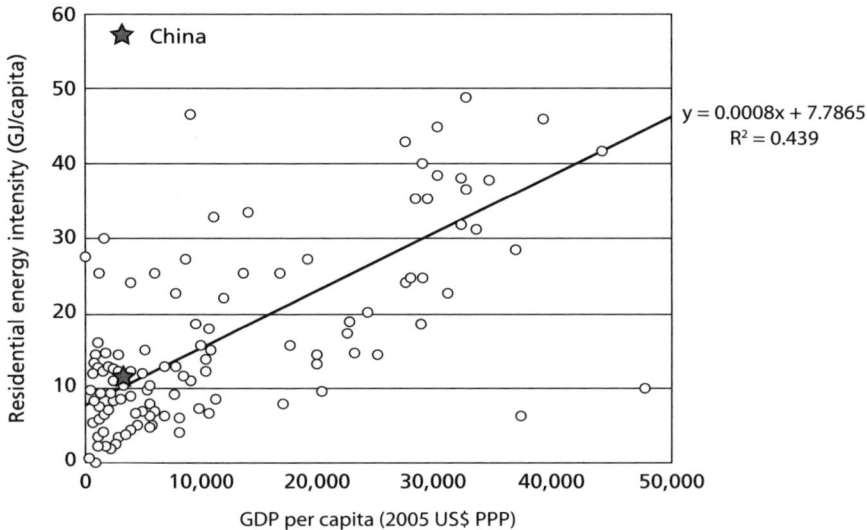

Source: Stockholm Environment Institute/LEAP

Figure 5.4 *Income versus household intensity in 2000 for various countries*

overall elasticities of passenger and freight transport with respect to GDP. Modal shares are projected forward informed by trends in the shares in recent years as well as typical values seen today in OECD countries.

Both passenger and freight transport show huge growth in the BLS when total passenger transport grows by a factor of more than six from 1.7 trillion p-km in 2005 to 11 trillion p-km in 2050. Total freight transport grows by a factor of more than four from 8.0 trillion t-km in 2005 to 34 trillion t-km in 2050. While these levels of growth rates appear extremely high, it is worth noting the trends in transport demand per capita (for passenger travel) and per unit of GDP (for freight) as shown in Figure 5.5.

Passenger transport demand per capita grows from 1325p-km per capita in 2005 to 7710p-km per capita in 2050, which is still lower than current values for all OECD nations (IEA, 2007a). Transport freight requirements per dollar actually decline in this scenario from 1.5t-km/$ in 2005 to 0.68t-km/$, which is similar to the current values in large OECD nations such as the US, Australia and Canada, but still significantly higher than the average for OECD nations of about 0.4km/$. Such a decline in freight transport per dollar of value added is to be expected given the gradual shift from heavy industry to services and lighter industries seen in the scenario. Nevertheless, the huge absolute levels of growth do call into question the plausibility of the baseline assumptions. In other words, it is questionable whether China can really improve its transportation infrastructure sufficiently by 2050 to support the huge increases in transportation projected in these scenarios.

Historical and baseline energy intensity trends in the service and household sectors and in various industrial manufacturing sectors are summarized in

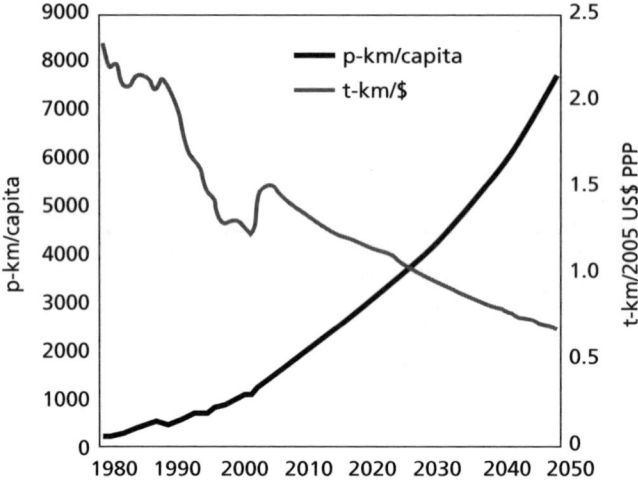

Source: Stockholm Environment Institute/LEAP

Figure 5.5 *Transport demand trends in the baseline scenario*

Figures 5.6 and 5.7. Service and industrial energy intensities are projected to decline to 2050, continuing the historical trend but at a slower rate. However household intensities are forecast to increase in the BLS, a consequence of increasing average income levels.

Within each major sector and most major industrial sectors, fuel shares are calculated for the historical period (1980–2006) and then projected forward to 2050 using an assessment of past trends and a review of the particular circumstances, and suitability and availability of fuels in each sector. For example, in the household sector, increases in average income are likely to be accompanied by a strong shift away from biomass fuels and coal, and increased use of electricity and other modern fuels.

Additional analysis was done to examine trends in two of the most important energy-intensive sectors: iron and steel, and cement. Aggregate historical data on fuel consumption for these sectors from the IEA (2008b) were calibrated against physical production statistics (tonnes of steel, tonnes of cement), data on types of production (basic oxygen furnace (BOF) versus electric arc furnace (EAF) for steel, types of kilns for cement) and data on energy intensities and feedstock fuels for each different type of production from China's national statistics (National Bureau of Statistics of China, 2006) and elsewhere (IEA, 2007b).

Projections of future energy consumption in the baseline sector were based on expected future changes in processes, energy intensity improvements, and projections of future steel and cement production in the country. Here a couple of important trends that have a significant bearing on future energy consumption and GHG emissions are worth noting. First, the high

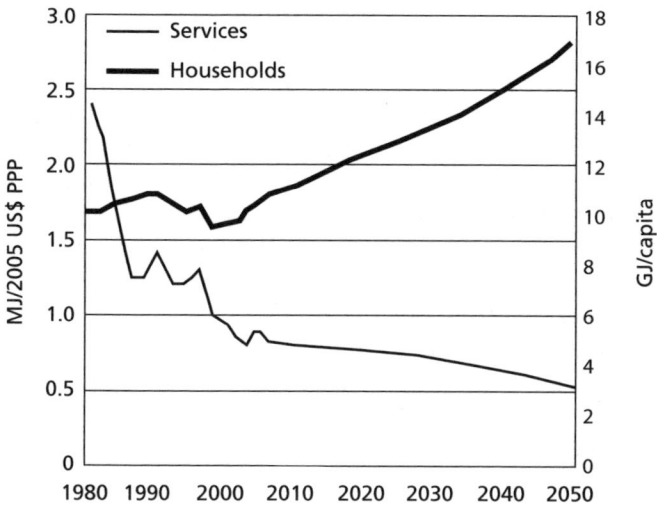

Source: Stockholm Environment Institute/LEAP

Figure 5.6 *Baseline household and service sector energy intensities*

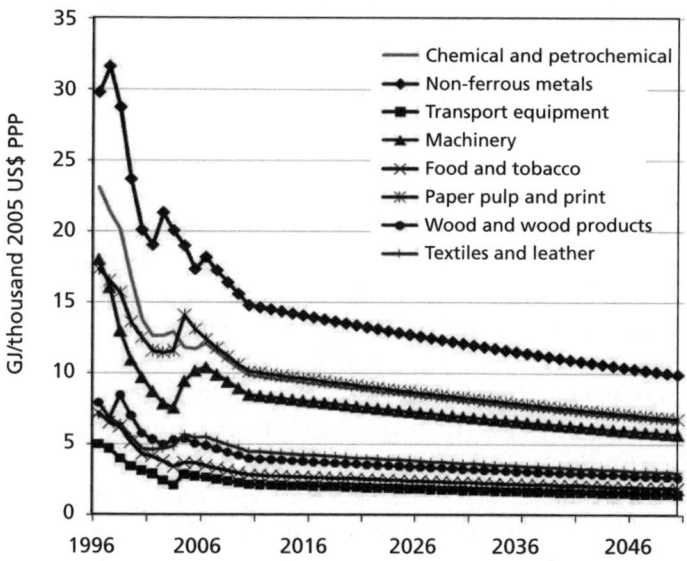

Source: Stockholm Environment Institute/LEAP

Figure 5.7 *Manufacturing energy intensities*

rates of growth of production of steel and cement are assumed to abate after about 2025 with only modest growth seen thereafter (Wang et al, 2007). This agrees with the projections of other researchers in the field (McKinsey & Company, 2009) and reflects an assumption that as China's economy matures its requirements for these basic physical products peaks. Similarly, for iron and steel it is assumed that as China's economy matures so the availability of steel scrap will increase, so that by 2050 China can make much greater use of EAF furnaces, with consequent reductions in the use of the more energy- and carbon-intensive BOF technology. Reflecting this, EAF production as a share of total steel production is assumed to grow from 13 per cent in 2005 to 50 per cent in 2050. EAF production has an energy intensity of around 8 gigajoules (GJ)/t compared to 20GJ/t for BOF/blast furnace production (IEA, 2007b), in part because it can use steel scrap as a feedstock, thus avoiding the need to produce pig iron from ore.

In the cement sector, based on trends to 2030 described in McKinsey & Company (2009), a gradual phase-out of the more energy-intensive types of kilns (wet process and vertical shaft kilns) is assumed, so that advanced dry process pre-calciner kilns achieve a market share of over 90 per cent by 2050. Similarly, a gradual decrease in the clinker content of cement is assumed as clinker quality improves and more high-quality substitutes become available (granulated blast furnace slag and fly ash). Dry process cement has a significantly lower production energy intensity of about 3GJ/t of cement compared to the now largely obsolete wet process kilns (6.4GJ/t) and vertical shaft kilns (5.7GJ/t), the latter of which still accounted for 47 per cent of China's cement production in 2005 (IEA, 2007b). The final energy demands in the BLS are summarized in the following figures showing historical data and projections for energy demand by sector and by fuel for China as a whole and broken down into various key sectors.

Overall final energy demands by fuel and by sector are summarized in Figures 5.8 and 5.9. Notice the increasing importance of electricity and oil products and the continued importance of coal, and the continued dominance of the industrial sector and the rapid growth in the transport sector.

In terms of energy supply, the analysis focuses on likely trends in the electric sector as shown in Figures 5.10 and 5.11. Historical data on conversion technologies are taken from the IEA's *World Energy Balances* (IEA, 2008b) and the *China Energy Databook* (LBNL, 2008). Future trends in terms of generating efficiencies and the expansion of capacity are based the IEA's *World Energy Outlook 2008* (IEA, 2008a) as well as characteristics for various technologies described in the IEA's *Energy Technology Perspectives* report (IEA, 2008c). Potential penetration rates for various renewable technologies are adapted and extrapolated from baseline estimates in McKinsey & Company (2009) and ERI (2008). In the BLS, total capacity expands from 516 gigawatts (GW) in 2005 to 3116GW in 2050 – equivalent to an annual rate of addition of 58GW/year – very high but similar to the rate of additions in the last decade. Over the scenario period coal's share of capacity actually increases slightly from 72.3 per cent in 2005 to 77.3 per cent in 2050. Nuclear power also increases from 1.3 per cent to 2.6 per cent and renewables (primarily wind) increases from 0.2 per cent to 4.8 per cent. Hydropower decreases from a high of 22.7 per cent in 2005 to 10.7 per cent in 2050.

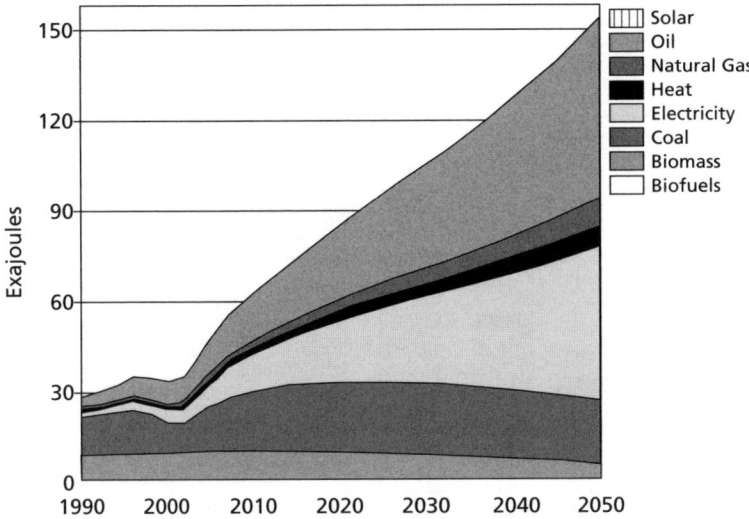

Source: Stockholm Environment Institute/LEAP

Figure 5.8 *Baseline final energy demand by fuel*

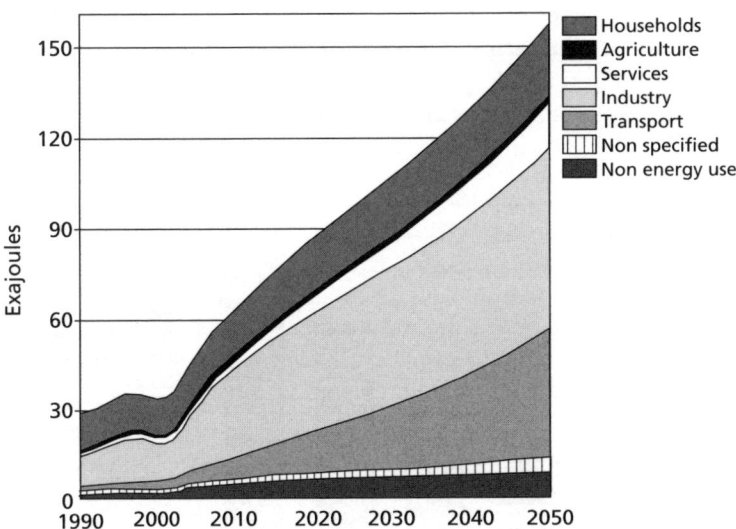

Source: Stockholm Environment Institute/LEAP

Figure 5.9 *Baseline final energy demand by sector*

Significant progress is seen in reducing transmission and distribution losses. New additions to power generation are expected to be significantly more efficient than the current average stock, even in the BLS, and are dominated by more efficient coal power plants as well as new additions of natural gas, hydro and nuclear power plants. Renewable generating

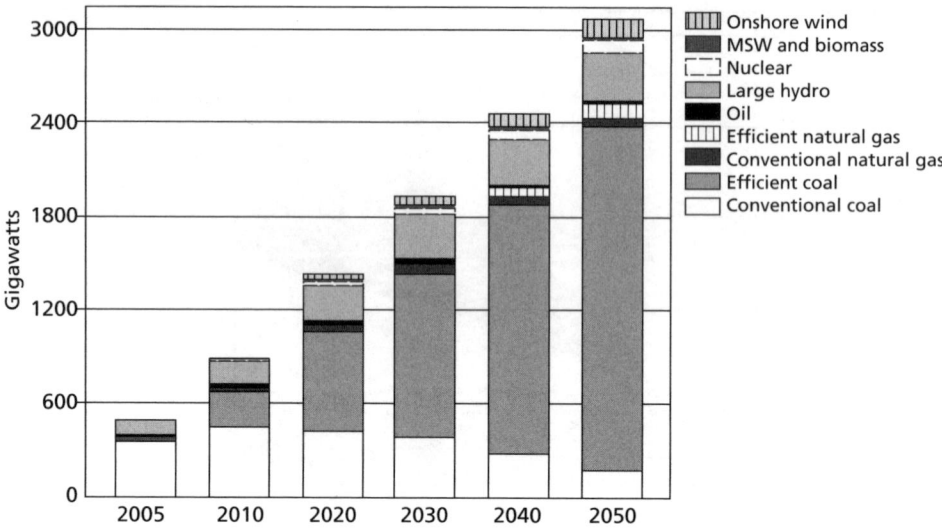

Source: Stockholm Environment Institute/LEAP

Figure 5.10 *Baseline electric generation capacity*

technologies are expected to increase as well, although not enough to gain a substantial share of generation. CCS is not expected to gain any significant share of generation in the BLS.

Primary energy requirements grow from 74.1 exajoules (EJ) in 2005 to 172EJ in 2030 and 254EJ in 2050. Coal continues to be the dominant energy

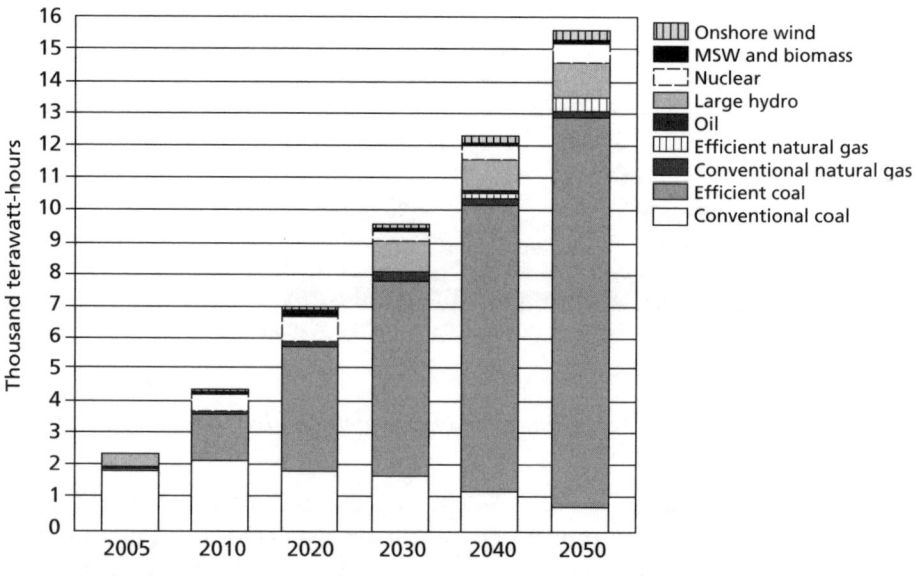

Source: Stockholm Environment Institute/LEAP

Figure 5.11 *Baseline electricity generation*

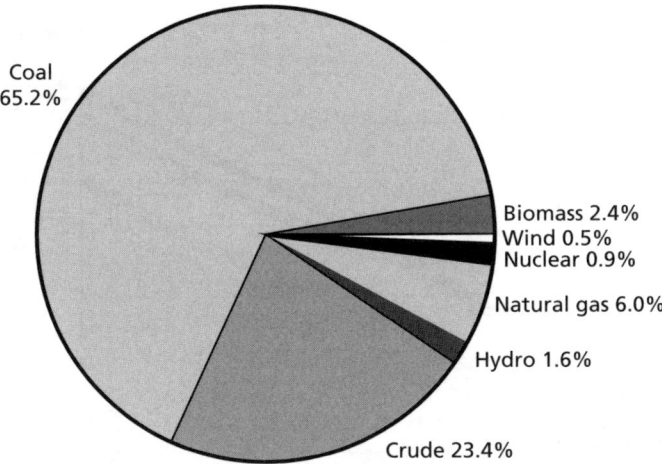

Source: Stockholm Environment Institute/LEAP

Figure 5.12 *Baseline primary energy requirements in 2050*

form with its share of total primary requirements staying almost unchanged at about 65 per cent. The importance of biomass declines as its traditional use in households gradually wanes. Crude oil gains in importance due to the rapid growth of road and air transport.

Emissions of CO_2 are estimated by applying standard IPCC Tier 1 emissions factors wherever a fuel is combusted in the system. The emission factors are specified in terms of metric tonnes of CO_2 per terajoule (TJ) of fuel being combusted. Thus, while the factors are approximate they do capture the variation in the energy content of fuels being combusted in China.

Energy sector CO_2 emissions, shown in Figure 5.13, increase from 4.8Gt CO_2 in 2005 to 12.3Gt in 2030 and 18.9Gt in 2050. By 2050 electric generation is by far the leading source of emissions although industry and transport are also important.

Emissions intensities per capita, shown in Figure 5.14, increase from 3.6t/capita in 2005 to 7.8t/capita in 2030 and 11.1t/capita in 2050, which remains only 58 per cent of US levels in 2005 (19.6t/capita), but higher than those of Sweden in 2005 (5.64t/capita).

Emissions intensities per dollar of GDP continue to decrease although more slowly than in recent decades, declining from 0.9kg/\$ in 2006 to 0.47kg/\$ in 2030 and to 0.31kg/\$ in 2050. These figures can be compared to the equivalent 2005 figures for the US and Sweden of 0.47kg/\$ and 0.18kg/\$ respectively, showing on the one hand considerable improvements over the period but also showing the potential for much greater declines on the other hand.

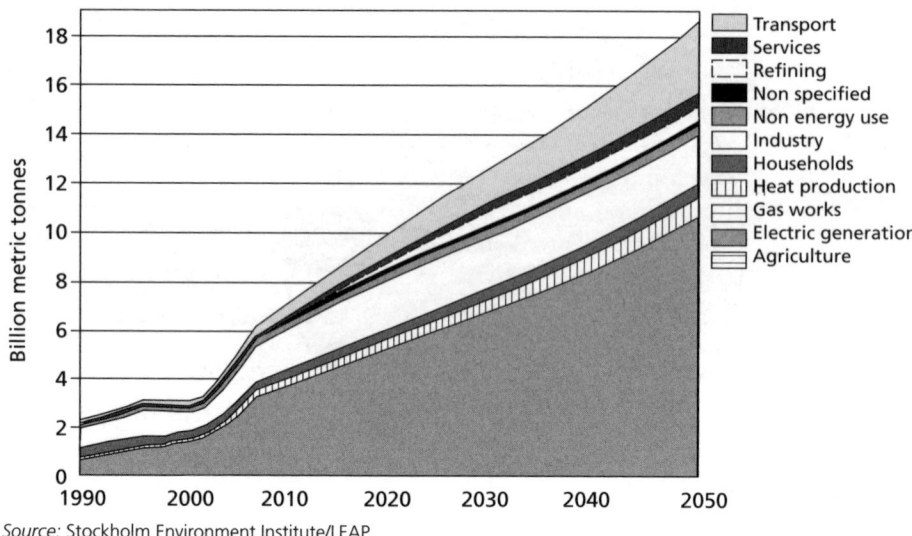

Source: Stockholm Environment Institute/LEAP

Figure 5.13 *Baseline CO$_2$ emissions from energy*

A deep carbon reduction scenario

As mentioned earlier, DCRS is intended to explore the feasibility of massively reducing China's CO$_2$ emissions by 2050: with energy sector GHG emissions reduced to about 32 per cent of their 2005 values or about 90 per cent below the 2050 levels projected in the BLS, with a goal of achieving levels of emissions in China that are compatible with the GDR global emergency

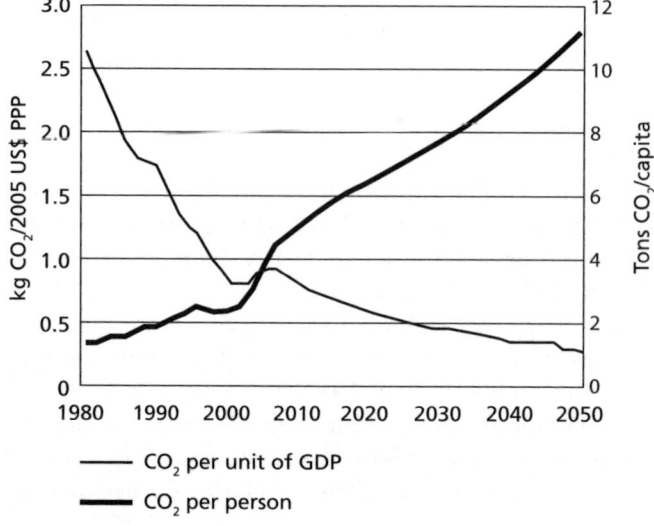

Source: Stockholm Environment Institute/LEAP

Figure 5.14 *Baseline CO$_2$ intensities*

pathway under the same demographic and macroeconomic trends as in the BLS.

In one respect, the DCRS is much more ambitious in terms of its CO_2 reduction goals than other recent global energy scenarios. For example both the Blue Map scenario of the IEA's recent *Energy Technology Perspectives* report (IEA, 2008c) and the recent *Greenpeace Energy [R]evolution* scenario (Greenpeace International and EREC, 2007) both aim for reduction in CO_2 emissions of about 50 per cent versus 2005 levels or about 80 per cent versus the expected 2050 values in their BLSs. The DCRS for China has more ambitious goals because it is designed to be compatible with the overall emissions reductions pathway of the GDR framework that aims for global reductions of about 82 per cent versus 2005 levels.

The measures included in the DCRS that enable it to achieve these ambitious goals are described in more detail below. Estimates of mitigation potential were developed based on a sector by sector review of options available in each sector. Due to the constraints of this study, these drew in large part from existing studies for China, most notably those developed in studies by McKinsey & Company (2009) and the IEA (2008c).

The DCRS focuses primarily on mitigation technologies that are either already commercialized or are expected to become commercialized in the next decade, in time for their deployment to have a significant impact on reducing China's emissions.

Note, however, that particularly for the period after 2030, the DCRS assumes that energy intensities will continue to decline rapidly, which inevitably implies massive and sustained levels of R&D on a global scale.

The main mitigation measures included in the DCRS are described in the following sections.

Buildings

The buildings sector has huge potential for mitigation of CO_2. Not only can energy intensities be reduced dramatically but the sector can also be largely decarbonized by replacing direct use of fuels with greater use of electricity, heat and solar energy. Energy intensity reductions can be achieved through measures including: improved design of new buildings to incorporate passive heating and cooling principals; retrofitting of existing building shells to reduce heating and cooling loads; the installation of more efficient heating, ventilation and air conditioning (HVAC) systems; the use of efficient lighting; the introduction and enforcement of stringent appliance efficiency standards for refrigerators, washing machines, dryers, televisions etc.; and improvements in the construction of buildings (for example to use more sustainable and less energy-intensive building materials). Of these measures, improved new building design is likely to have the largest long-term potential and the best economic benefit–cost ratio. Because China is developing so rapidly and building construction is at such a high level, this area clearly presents a huge opportunity. However, rates of construction are expected to decline in the latter part of the scenario as China's population peaks, its workforce ages and its economy matures. Thus, time is short to implement such measures and avoid large 'lock-in' effects. Implementing efficient new building designs will require wholesale retraining of architects and engineers and policy changes at the highest level to require that these new types of buildings become the norm. Failure to act quickly will mean that greater numbers of buildings will eventually have to be retrofitted for energy efficiency: an approach that is more costly and provides less emissions reduction.

In China, a proper assessment of energy and GHG emissions reduction potential in the buildings sector requires a detailed end-use analysis of energy consumption in both the household and services sectors (see Figures 5.15 and 5.16), considering the most important energy end-uses: space heating and cooling, water heating, cooking, lighting and appliances. This would need to account for the differences between urban and rural households as well as the various climatic regions, which affect heating and cooling demands. In the service sector, the differences between major service sectors (offices, hospitals, shops, restaurants, hotels, government buildings, etc.) also need to be considered. Such a detailed assessment is beyond what was possible in this study. So for the DCRS a simpler approach was used: first adopting an approximate estimate of the potential for energy intensity improvements, and then combining this with a judgement of the potential for fuel use shifts in each sector. A number of studies cite the technical potential for energy efficiency reductions to be as high as 80 per cent in the buildings sector in OECD countries (IEA, 2008c) and given the importance of space heating in China – estimated by the Lawrence Berkeley Laboratory to be over 50 per cent of final residential energy demand (Zhou and Levine, 2008) – and the poor energy efficiency of its housing stock, a similar target is also likely to be possible in China. However, given the lower average income levels of China versus OECD countries, a lower target of 50 per cent reduction in the 2050 energy intensity versus the baseline has been assumed for the DCRS.

As a result, per capita energy intensities for the household sector decline from 10.6GJ/capita to 8.4GJ/capita in the DCRS instead of growing to 16.9GJ/capita in the BLS. This fairly modest rate of reduction of 0.68 per cent/year from 2005 to 2050 reflects the initial low level of energy use per person in China in 2050 compared to the values more typical in OECD countries (between 15GJ/capita for Sweden and 41GJ/capita for the US, normalized to 2700 heating degree days).

In the service sector, energy intensities per dollar of value added decline more rapidly at 2.1 per cent/year from 2005 to 2050 from 0.87 megajoule (MJ)/$ in 2005 to 0.3MJ/$ in 2050. China's intensity in 2005 is very similar to the current OECD average value of about 0.9MJ/$, although these values are very hard to interpret given the extreme difficulties associated with using value added as a measure of activity in this sector, particularly for 2050.

The mitigation benefits of these energy intensity improvements are further magnified through a shift in the fuels used in the buildings sector: with a nearly

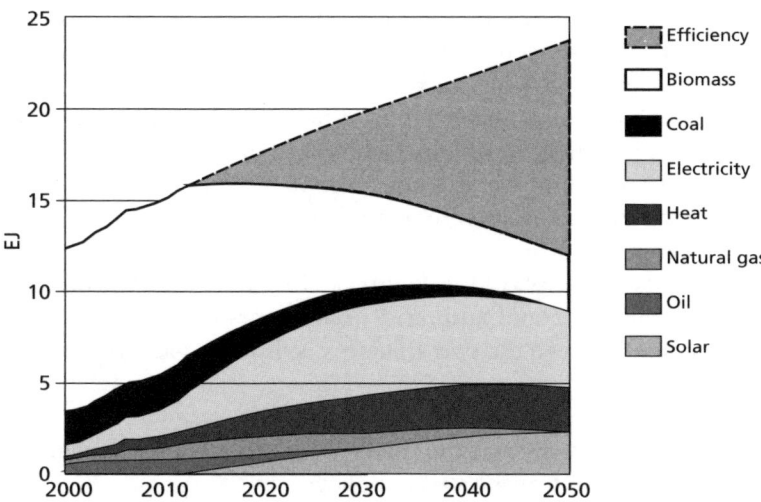

Note: Savings versus baseline shown as 'efficiency'.
Source: Stockholm Environment Institute/LEAP

Figure 5.15 *Household energy demand in the DCRS*

complete shift away from coal, oil and natural gas in favour of electricity, district heating and solar energy (the last of these primarily for hot water production). As will be seen later, this shift away from small-scale combustion of fuels is coupled with a dramatic decarbonization of electricity and heat production (described later). Biomass fuels, which according to IEA statistics in

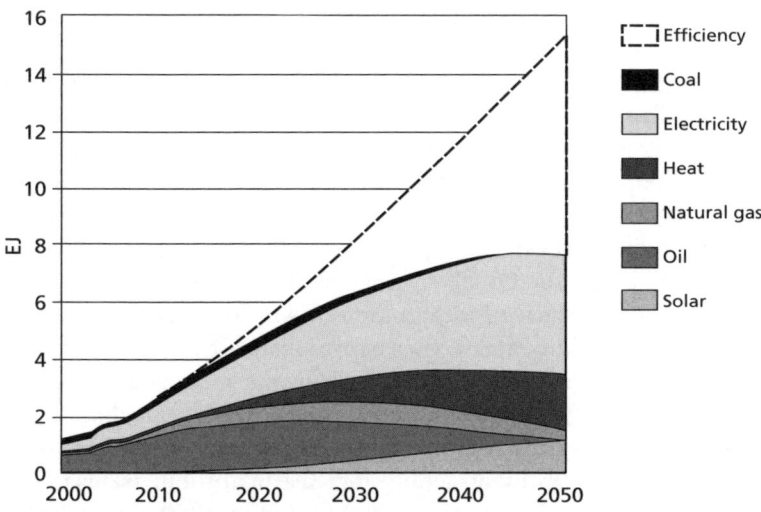

Note: Savings versus baseline shown as 'efficiency'.
Source: Stockholm Environment Institute/LEAP

Figure 5.16 *Services energy demand in the DCRS*

2005 still accounted for more than 60 per cent of final energy consumption, are expected to rapidly decline in the baseline scenario as rural incomes improve. But in the DCRS a significant level of biomass fuel use is retained – reflecting the development of cleaner, more efficient and more convenient ways of utilizing of biomass fuels in rural households. In the service sector the DCRS similarly sees the potential for the near-complete phase-out of fossil fuels, replaced by electricity, heat and (to a smaller extent than in the residential sector) solar energy.

The net results are displayed in Figures 5.15 and 5.16, showing the final fuel demands in the household and residential sectors in the DCRS versus the BLS. The reduction between the two scenarios is displayed as 'efficiency'.

Transport

In terms of passenger transportation, the DCRS reflects the implementation of three substantial shifts in transportation policy:

1   The DCRS reflects policies to slow the rapid overall growth in passenger transportation. In the BLS, the overall demand for p-km grows by a factor of 6.3 from 1.7 to 10.9 trillion p-km. The DCRS assumes a range of policies such as increased fuel pricing, better urban planning to reduce the need for commuting, congestion charging, a revitalization of cycling, parking restrictions, restricted airport developments, etc. Taken together, these policies are assumed to reduce the overall growth in p-km to a factor of 'only' 3.7, resulting in an overall demand for 6.3 trillion p-km in 2050. It is worth noting that apart from their GHG benefits, such policies would also yield innumerable benefits in terms of the liveability of China's cities (reduced pollution, less congestion, fewer traffic fatalities). Indeed it is questionable whether China would be able to develop the infrastructure that would be required to meet the huge levels of transport growth envisaged in the BLS.

2   The DCRS assumes shifts in modal shares relative to the BLS. The baseline assumes a continued shift towards the North American model for passenger transportation – founded upon private cars. In the BLS, road as a share of total p-km increases from 53 per cent in 2005 to 70 per cent in 2050 – a level still well below the OECD average of about 80 per cent. In the DCRS, road transport as a share of total p-km is assumed to stay roughly constant even as the total p-km value grows enormously – by a factor of 3.75. The share of road transport made up by buses is not modelled explicitly but is also assumed to stay roughly constant. Similarly, the rail share of p-km is assumed to decrease only slightly from 34.8 per cent in 2005 to 30 per cent in 2050. Again this represents a huge increase in absolute terms from about 606 billion p-km in 2005 to about 1900 p-km in 2050. This would be an enormously high value for a developed nation but is not unprecedented. In particular in terms of shares it is close to the 29 per cent rail share seen in Japan in 2004. The share for air travel, both domestic and international,

increases from 11.8 per cent in 2005 to 15 per cent in 2050, below the 18 per cent reached in 2050 in the BLS. Figure 5.17 summarizes points 1 and 2, showing the overall decrease in p-km in the DCRS compared to the BLS and transition of modal shares in the DCRS.

3   Perhaps most challenging of all, the DCRS assumes a massive shift away from dependence on oil-based internal combustion engines towards complete electrification of passenger road transport. Such a transition could happen in a number of ways but is likely to follow a gradual path through hybrids and plug-in hybrids to fully electric vehicles. In the DCRS, this is modelled as a gradual transition starting slowly in about 2015, ramping up after 2030 and culminating in 2050 with 90 per cent of all road p-km being delivered by electrical vehicles. While extremely challenging, such a pathway does seem possible if one takes an optimistic view of the development of the necessary technologies (most importantly the ability for advanced battery technologies to come down in cost) and of course of the development of the political will at the highest level (internationally not just within China) to enable such a massive transition. The transition from oil to electricity not only yields a significant decarbonization of transport, it also yields important efficiency benefits. Energy requirements of full electric vehicles per p-km in the DCRS are based on estimates in Mackay (2009) and equal roughly a quarter of the energy requirements of gasoline vehicles.

Similar but much less dramatic transitions are assumed for freight transport. Specifically, because of the assumption of the same levels of GDP growth in both the BLS and the DCRS it is assumed that the overall levels of freight transport in t-km are equal in both scenarios. The decline in the modal share of rail

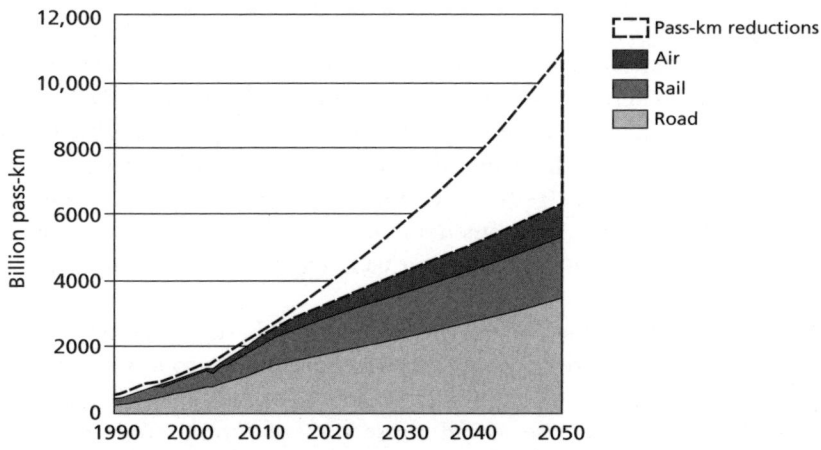

Source: Stockholm Environment Institute/LEAP

Figure 5.17 *Passenger transport modal shares in the DCRS*

freight expected in the baseline is assumed to be arrested in line with policies to promote rail over road transport so that the modal shares for both rail and road stay roughly constant over the period to 2050. Finally, road freight is also assumed to begin a path towards electrification. However, this path is assumed to lag well behind the changes seen for passenger road transportation, so that by 2050 only 30 per cent of road t-km are delivered by electricity.

In addition to these fundamental shifts, the DCRS also assumes modest penetration of biofuels in the air travel and maritime sectors. Due to the current inefficiency of biofuels production, which yields few benefits in terms of life-cycle CO_2 emissions per litre of fuel (with the notable exception of biofuels produced from sugarcane) and the potential for creating competing demands between food and fuel, it is assumed that first generation biofuels do not gain significant market share in the DCRS. However, second generation biofuels produced from woody crops are assumed to become available over the next five to ten years and are assumed to make a modest but measurable contribution in providing substitute fuels for water freight transport and to a lesser degree for air transport. The life-cycle CO_2 emissions profile of these second generation biofuels is still not clear, but for the purpose of this analysis we have tentatively assumed they will emit half as much CO_2 as fossil fuels per GJ.

Finally, in terms of rail transport, both the BLS and the DCRS assume the complete phase-out of coal-fired railways (sorry train spotters!), while the DCRS also assumes a complete transition by 2030 to electric-powered trains. The DCRS also assumes gradual efficiency improvements for rail travel due to the introduction of new technologies such as regenerative braking.

The results of this analysis are shown in Figures 5.18 and 5.19. Figure 5.19 compares net final energy demand for 2030 and 2050 in the BLS and the DCRS, while Figure 5.18 shows the even more noticeable difference in CO_2 emissions between the two scenarios.

Industry

In China, industry is the largest final consumer of energy, accounting for 55 per cent of final energy consumption. As mentioned earlier, industrial energy demands are projected to continue rising in the BLS, although significant growth in energy use in the heaviest industrial sectors is expected to come to an end after the late 2020s as China's economy matures. This trend, combined with steady improvements in energy efficiency and a switching to less carbon-intensive fuels, results in CO_2 emissions declining after 2035 even in the BLS.

The DCRS is assumed to further accelerate these trends through more concerted efforts to improve energy efficiency and switch to lower carbon fuels wherever possible. Two major sectors, iron and steel, and cement, were looked at in some detail, while for the remaining sectors, due to the limitations of this study, the analysis relied on simpler assumptions – primarily that China will continue its current intense efforts to reduce energy intensities. These assumptions need to be confirmed by further research but serve as a placeholder for now.

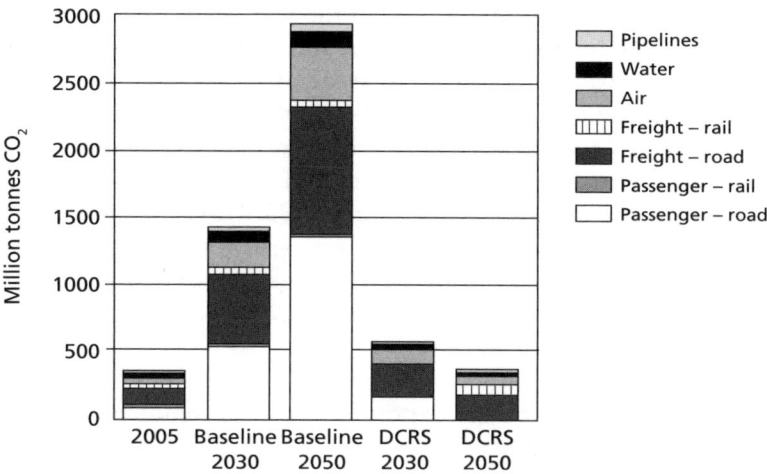

Source: Stockholm Environment Institute/LEAP

Figure 5.18 *DCRS and baseline transport* CO_2

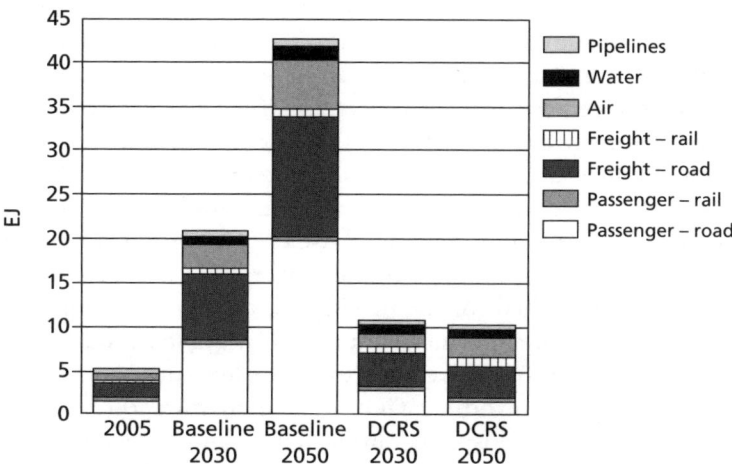

Source: Stockholm Environment Institute/LEAP

Figure 5.19 *DCRS and baseline transport energy use*

Iron and steel

In the iron and steel sector, large emissions reductions can be achieved if the use of EAFs for steel production can be more widely utilized. We assume in the DCRS that by 2050 EAF furnaces achieve a market share of 75 per cent, up from only 13 per cent in 2005. Such a high share is today seen in only a few countries (for example Mexico and Spain), and will only be possible if sufficient scrap steel is available. Scrap supplies are generally seen as a function of the maturity of the economy. Given the current rapid development of China's economy and its expected maturation in the late 2020s, the required supplies of scrap steel may be possible but clearly this is an area that requires further research. Notwithstanding efforts to switch to EAF steel production, coal-based BOF/blast furnace production will inevitably remain important in China given the country's reliance on coal. Here, reductions in emissions will have to rely on energy efficiency improvements and CCS. Based on a review of a variety of industry research (Chatterjee, 1993; ULCOS, 2009), we assume energy intensity improvements of 30 per cent (per tonne of steel produced) by 2050 combined with CCS for the CO_2 generated from 50 per cent of BOF-produced steel by 2050.

Cement
The cement industry is of huge importance for CO_2 emissions and hence for CO_2 mitigation efforts. Energy use in the Chinese cement industry in 2005 produced 203 million metric tonnes of CO_2, roughly 17 per cent of all industrial energy-related emissions. However, the cement making process (calcination) itself emits CO_2 when calcium carbonate ($CaCO_3$) is heated in kilns to form calcium oxide (CaO) and CO_2. These cement process emissions amounted to an estimated additional 518 million metric tonnes of CO_2 in 2005, adding an additional 42 per cent to industrial CO_2 emissions.[3]

In the coming decades a number of options are available to reduce emissions from cement and are reflected in the DCRS. These include a more rapid and complete switch to the less energy-intensive advanced dry process pre-calciner kilns, further reductions in the clinker-to-cement ratio (to reduce the overall need for cement in concrete), the use of CCS for up to 50 per cent of cement kilns by 2050, and the limited use of low-carbon agricultural residues or biofuels for co-firing of kilns. In the longer run a number of new technologies are being researched for the production of new 'eco-cements' that could potentially reduce emissions more dramatically. These include new magnesium (as opposed to calcium) based cements that are less energy intensive to produce and while they also produce CO_2 during manufacture, most of this is reabsorbed from the atmosphere during setting and hardening (Pearce, 2002). Another concept involves the production of cement-like substances from the waste heat and flue gases of fossil-fired power plants in a process that mimics the way marine cement (coral) is produced (Biello, 2008). All these processes are only design concepts at present. They will face significant hurdles before they can be commercialized, not least because cement quality is very carefully regulated for safety due to its use as a construction material. For these reasons, these new types of production process are not included in the DCRS even after 2040. Nevertheless, there is reason to think that cement production could be significantly decarbonized beyond 2050 and perhaps even before.

Other sectors
As mentioned above, in all other manufacturing sectors (chemicals, non-ferrous metals, transport equipment, machinery, food and tobacco, paper,

pulp and print, wood and wood products, textiles and leather, and others) energy consumption in the DCRS is projected using fairly simple placeholder assumptions that will require further in-depth analysis to confirm their feasibility. China is assumed to continue its intense efforts to reduce energy intensities. It has been aiming to reduce its overall energy intensity by 20 per cent between 2006 and 2010, although at present, insufficient data are available to conclude if it will meet that goal. It is also worth noting that the goal is for overall energy intensity across all sectors of the economy and thus not a measure of technical improvements in a particular sector. So for example, a shift in the production of GDP from industry to services would yield a reduction in energy intensity (per dollar of value added) even if Chinese industry continued producing at its current technical intensity. Nonetheless, for the purpose of this analysis, it is assumed that a significant proportion of the intensity reduction goal is met by genuine energy efficiency improvements in the industrial sector and that these reductions (assumed to be 2.5 per cent/year to 2010) will continue, albeit at a lower rate of 1.5 per cent/year thereafter. This can be compared to the assumption of only 0.5 per cent/year decrease in the BLS after 2010. The net result is that industrial energy intensities in 2050 in the DCRS are only 47 per cent of their starting value in 2005. This assumption of a 53 per cent reduction is clearly quite optimistic, but it is not without precedent, especially in China where IEA statistics suggest that in the 25 years since 1980, energy intensities in the major industrial sectors other than cement and iron and steel have been reduced by between 50 per cent and 89 per cent. Of course these reductions may reflect the gathering of 'low-hanging fruit' and the benefits of economies of scale, which may be one-time gains that cannot easily be repeated, but it is at least possible that the reductions assumed in the DCRS can be engineered in the coming 40 years if the international political will emerges to do so.

Coupled with these assumptions on future intensity improvements, the DCRS examines the potential within each sector to switch to lower-carbon fuels based largely on an acceleration of past trends to switch away from coal and oil, and towards greater use of electricity and central production of heat. In two sectors (paper, pulp and print; and wood and wood products) the DCRS also assumes the use of biomass and agricultural residues. CCS is not assumed to be viable in sectors other than the largest and most carbon-intensive sectors: cement, and iron and steel.

Dematerialization
In addition to improvements to energy efficiency, there are a variety of opportunities for China to begin to pursue less material-intensive forms of development. Dematerialization implies that a larger fraction of GDP will come from services and light industry and a smaller fraction from the more energy- and carbon-intensive heavy industrial sectors. However, such shifts might not necessarily imply genuine dematerialization – they might simply imply that production shifts away from China to other perhaps equally

carbon-intensive nations, while material consumption in China continues to increase. Such a transition would in fact be replicating how most OECD nations have grown in recent decades – apparently reducing their energy intensities even while substantial amounts of production have been shifted overseas (much of it to China).

Since the goal in the modelling the DCRS is to reflect a future for China that is genuinely compatible with an assumed global effort at climate protection, the DCRS does not include any structural shifts beyond those seen in the BLS. This is not to imply that dematerialization is not an important option for China. Indeed, many will argue that the task of providing growing welfare to society while consuming less is the key challenge for humanity in the 21st century. Nevertheless, DCRS ignores this question, primarily because the intention with this analysis is to examine whether mitigation targets can be met at the same time as fulfilling China's existing development goals. Of course if China does succeed in meeting its development goals, so that by 2050 its per capita income levels are close to those of OECD nations today, it is also likely that OECD nations will have in the meantime grown considerably themselves. Thus, unless richer countries quickly take up the task of dematerializing their own economies, it is hard to imagine that China will wish to do so.

Results for industry
The net results for industry are displayed in Figure 5.20, showing the final fuel demands in the industrial sectors in the DCRS versus the BLS, and in Figure 5.21, showing industrial sector CO_2 emissions including both energy-related emissions and cement process emissions.

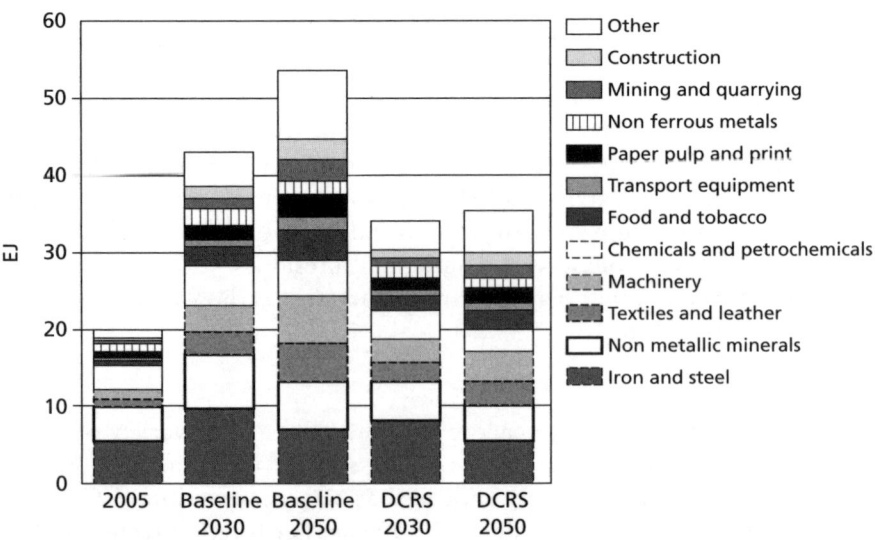

Source: Stockholm Environment Institute/LEAP

Figure 5.20 *DCRS and baseline industrial energy use*

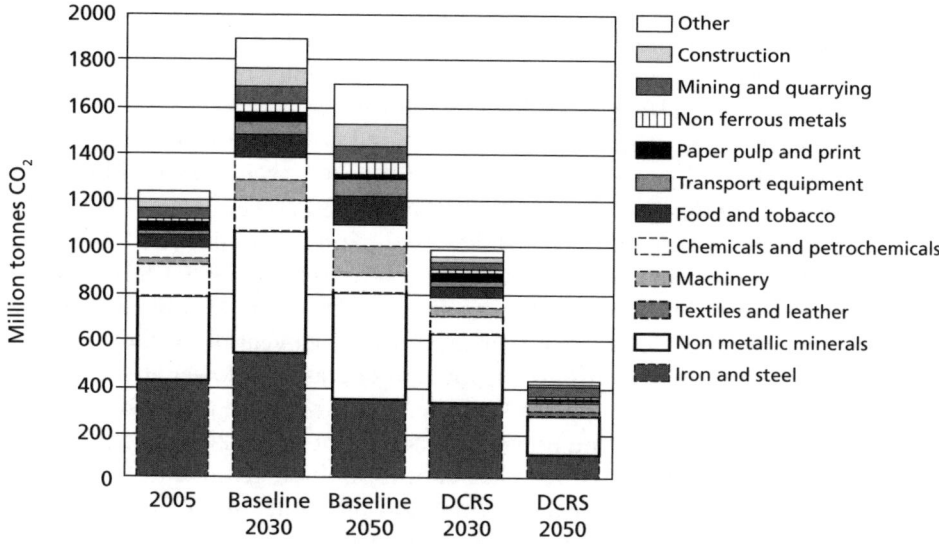

Source: Stockholm Environment Institute/LEAP

Figure 5.21 *DCRS and baseline* CO_2 *emissions
(including cement-process emissions)*

Electricity generation

The fourth and perhaps the most difficult area for policy intervention in the DCRS is China's energy supply and more specifically its electricity generation sector. Here the DCRS models a massive undertaking to decarbonize the supply of electricity and heat to the extent possible given China's resource base. This pathway is designed to work in tandem with efforts on the demand side to eliminate the localized combustion of fossil fuels, replacing them with electricity and centrally supplied heat.

The requirements for electricity in both scenarios are actually quite similar. In the baseline, electricity requirements grow enormously from 2445 terawatt hours (TWh) in 2005 to 15,781 TWh in 2050. In the DCRS, the savings from aggressive energy-efficiency efforts and efforts to reduce transmission and distribution losses are partly counteracted by efforts to switch final consumption away from direct use of fossil fuels and towards electricity and heat, resulting in final requirements of 11,767 TWh in 2050.

The production of China's electricity and heat supplies is currently dominated by coal, which in 2005 accounted for 79 per cent of electric generation. This share even increases over time in the BLS reaching 85 per cent in 2050, albeit with a switch to more efficient types of power plants. The DCRS takes a very different path in four key respects:

1   Early retirement of existing inefficient coal-fired electric generation. The DCRS assumes the accelerated retirement by 2045 of all existing coal-, oil- and gas-fired power plants. Since a large proportion of these plants were

built in the current decade during China's most rapid phase of economic expansion, this represents a significant level of early retirement of capacity – an expensive proposition but essential for keeping China's CO_2 emissions within the overall budget specified by the scenario.

2 Large-scale deployment of efficient coal-fired power with CCS. The DCRS assumes that all new coal plants built from 2010 onwards will be much more efficient than the current stock of coal-fired power plants. The DCRS assumes an average efficiency of new coal plants of 43 per cent versus the roughly 29 per cent average efficiency of the current stock. CCS is assumed not to be available until 2020, but after that date all new power plants are assumed to use a CCS system that captures 90 per cent of the CO_2 emitted by the plants. These plants are assumed to operate at a lower efficiency (39 per cent) due to the energy needs of the CCS process. In addition by 2050, 90 per cent of the existing power plants built between 2010 and 2020 are assumed to be retrofitted for CCS capture (which again entails a loss of efficiency). Smaller amounts of natural gas plants are also constructed and are subject to the same assumptions about CCS.

3 Large-scale development of renewables. The DCRS also assumes that large amounts of new wind (offshore and onshore), solar (concentrating solar panels and solar photovoltaic), municipal solid waste (MSW)/biomass and small hydropower plants are added. The estimates for these were based on a study by McKinsey & Company (2009) for 2030, with the added assumption that the potential capacity of these plants in 2030 could be expanded a further 30 per cent by 2050. Given the growing resistance to large-scale hydropower schemes, the capacity for these plants is assumed to be the same as in the BLS. Finally, the DCRS also foresees the potential for significant increases in nuclear power. While nuclear power is still deeply unpopular in large parts of the world, due primarily to concerns over cost, safety and nuclear proliferation, it has been included in the DCRS due to its potential for CO_2 mitigation and the demonstrated preference for this technology by policy-makers in China.

4 Combined heat and power generation. Apart from operating at greater efficiencies, the DCRS also assumes that all new thermal power plants are designed for the combined production of both heat and power. By 2050, 25 per cent of all energy inputs to thermal and nuclear plants are assumed to be captured as usable process or district heat. This is likely to be hard to achieve, particularly since many power plants will be built away from the potential consumers of the heat. However, this assumption is key in lowering emissions in the DCRS since it allows an essentially zero-carbon resource (waste heat) to be used to meet the increasing need for energy in China's industrial sector as it simultaneously switches away from a reliance on carbon-intensive fossil fuels. Such a scenario is only likely to be plausible if future industries and power plants are developed in a much more holistic fashion.

Figure 5.22 summarizes the capacity expansion path in the DCRS, while Figure 5.23 shows the level of generation from each type of power plant. Total capacity increases from 516GW in 2005 to 3043GW in 2050. Coal capacity triples from 373GW to 1112GW, while various types of renewable electricity (wind, solar, MSW/biomass and small hydro) increase from only 1.06GW to 1247GW so that by 2050 they account for 41 per cent of installed capacity. While this level of capacity is clearly unprecedented, the numbers do at least lie well within the estimates of the available resources in China (ERI, 2008). Nuclear power also increases more rapidly than in the baseline – from 7GW to 237GW, reaching a share of 7.8 per cent in 2050.

Figure 5.24 shows the net results of the electric generation strategy in the DCRS: the sector is largely decarbonized with electric generation CO_2 emissions peaking in 2020 at 4.22Gt/year and declining thereafter to 0.60Gt/year by 2050. The CO_2 emissions remaining in 2050 are largely a result of the CCS process capturing only 90 per cent of emissions, since by that date nearly all fossil-based power generation is assumed to be equipped with CCS. To keep emissions at this level and with this amount of power generation will require the capture of 60.0Gt of CO_2 between 2020 (when CCS is assumed to begin operation) and 2050. By 2050, CO_2 is required to be captured at a rate of 3.78Gt per year. In other words, even if CCS plants were expanded no further, the total amount required to be captured over the next century would be approximately 249Gt CO_2. As yet there are no firm estimates of the potential CO_2 storage capacity in China. Storage options include coal seams, oil and

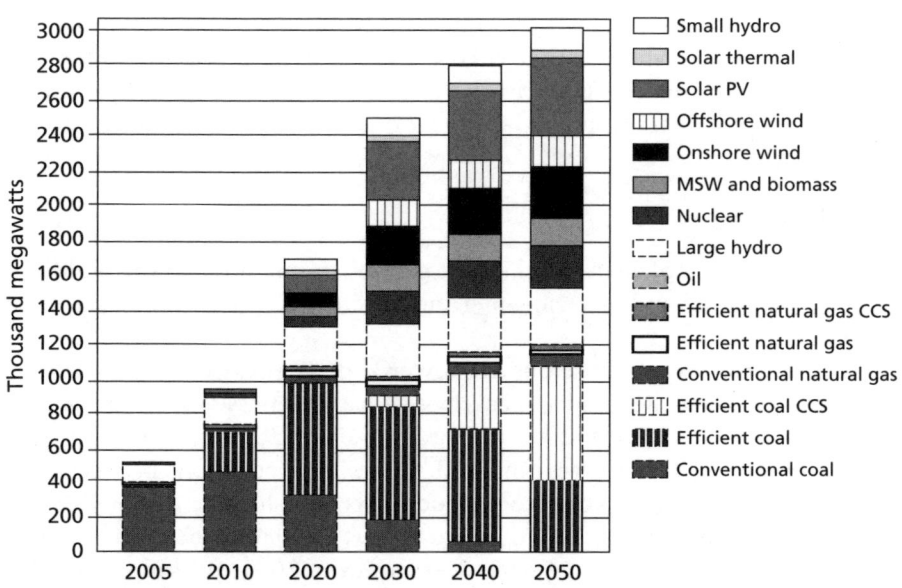

Source: Stockholm Environment Institute/LEAP

Figure 5.22 *DCRS electric generation capacity*

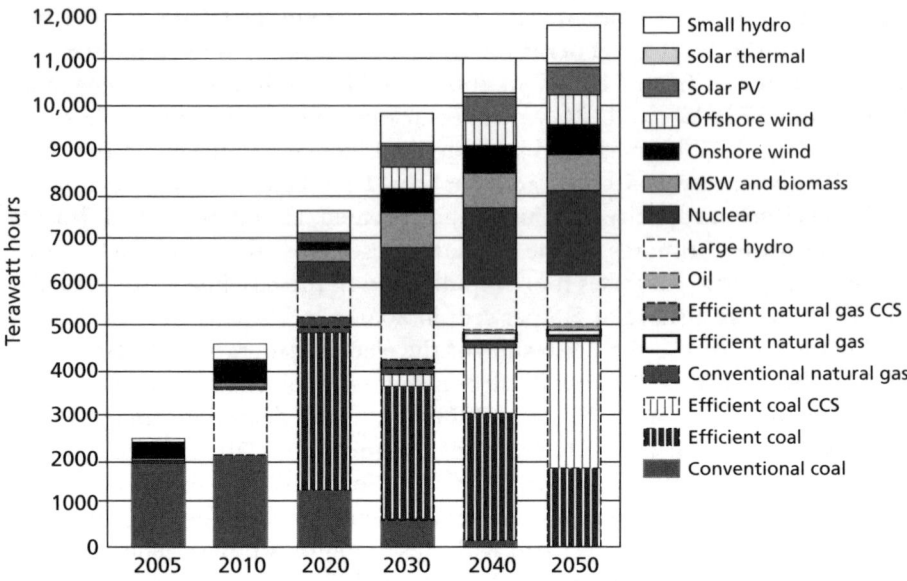

Source: Stockholm Environment Institute/LEAP

Figure 5.23 *DCRS electric generation*

gas fields, deep saline aquifers and ocean storage. A recent estimate by the Chinese Institute of Soil and Rock Mechanics (quoted in Luo et al, 2008) provisionally estimates the total geological storage capacity in China at 196Gt CO_2. If this estimate proves to be reasonably accurate, and assuming that power-generation facilities can be sited conveniently to the storage sites, then the levels of CCS development in the DCRS could potentially be realized. However, CCS appears to be, at best, a stop-gap option unless ocean storage proves viable.

Overall results

The net result of implementing the four groups of measures in the buildings, industry, transport and electric generation sectors in the DCRS is shown in Figure 5.25.

Total energy sector emissions are lowered dramatically compared to the BLS. In 2050, the DCRS results in energy sector CO_2 emissions that grow from a base of 4.8Gt CO_2/year in 2005 and peak in about 2017 at 7.4Gt CO_2/year, before falling to 1.9Gt CO_2/year in 2050. Cumulative emissions between 2005 and 2050 are 229Gt CO_2: equal to the budget set for China by the GDR framework and described in the above. Annual CO_2 emissions intensities show a similar shaped curve, growing from 3.6t/capita in 2005, peaking in 2016 at 5.3t/capita then falling to only 1.3t/capita in 2050.

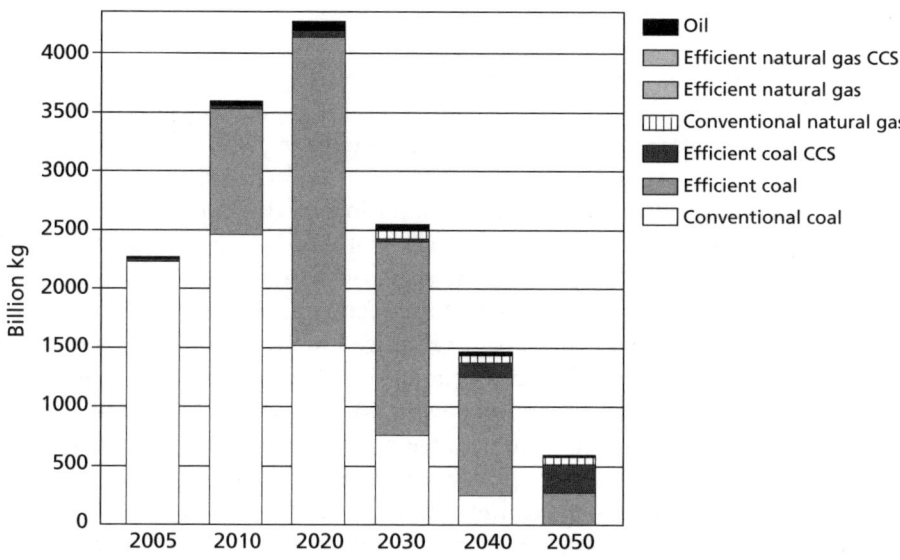

Source: Stockholm Environment Institute/LEAP

Figure 5.24 *DCRS CO$_2$ emissions from electric generation*

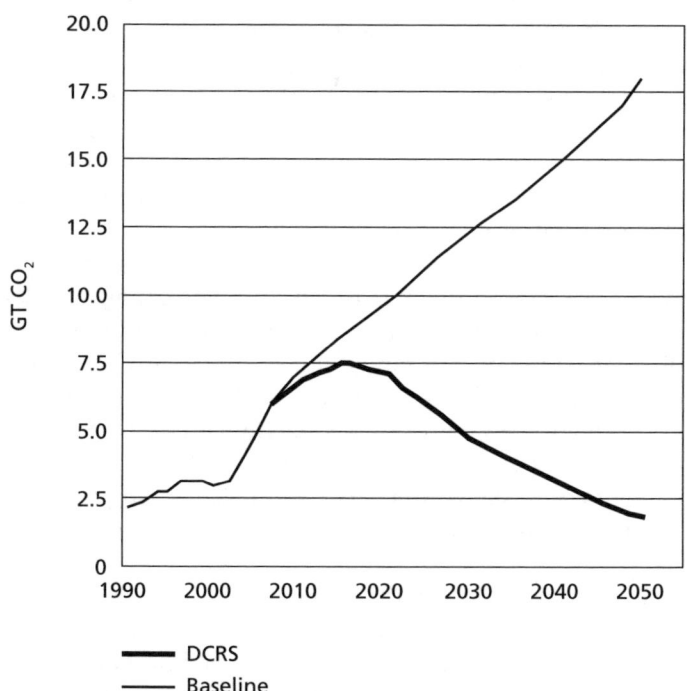

Source: Stockholm Environment Institute/LEAP

Figure 5.25 *Net result of measures implemented in the buildings, industry, transport and electric generation sectors in the DCRS*

Conclusions

The previous sections provided an initial exploration of a scenario that would enable Chinese emissions to be reduced to a level that gives the world a reasonable chance of keeping global temperature increases below 2°C. There are many uncertainties associated with such a scenario that require significant further research. In particular, given the limitations of this study, the author has not yet developed an economic analysis of the costs and benefits of the scenario. In addition, many parts of the scenario require more in-depth study. For example, more detailed analysis of the technical feasibility and the costs and benefits of alternative options are required in all sectors. As mentioned earlier, the dataset associated with this analysis is being made available by SEI and it is hoped that other researchers, especially those in China, will wish to correct and improve upon it in due course and then share their findings with other researchers and the climate policy community.

In spite of the acknowledged limitations of this analysis, two key points seem fairly clear. First, all of the elements in the DCRS need to happen. While the DCRS may be possible, it can only be achieved if all of its elements are not just technically feasible but also economically and politically plausible. The DCRS is barely able to stay within the emissions budgets set through the GDR framework and the loss of any single major element (the electrification of vehicles, the massive deployment of renewables, the complete switch to CCS-based coal-fired generation, huge improvements in energy efficiency, significant changes to passenger transportation modes, etc.) would prevent a plan based on the DCRS from meeting its mitigation goals. It is of course possible that some important options that are available now or will become available in time to be implemented well before 2050 have not been included. The author welcomes feedback on what those options might be.

Second, time is short. Not only do all of the options need to be implemented, but they need to happen quickly. Because climate change is related to cumulative emissions of GHGs, any delay in implementing options will make it almost impossible to meet the overall target budget of 230Gt CO_2. The DCRS is already very optimistic in the dates it assumes for the commercialization and deployment of key technologies. For example, it assumes that CCS starts being used commercially in 2020 and is fully deployed by 2050. Similarly electric vehicles start gaining market share after 2015 and reach a market share of 90 per cent by 2050. Any serious delay in this schedule will make the scenario's goal unattainable.

It is also important to recognize that other development pathways are available that can help meet the same climate protection goals. Such pathways would need to be much less materials intensive and would probably emphasize the provision of welfare more through the delivery of services than through the consumption of goods. For example, they might include less consumption of meat, more consumption of vegetables, better urban planning to reduce the need for transport, and more emphasis on health care and environmental protection. The net result of these measures would be a smaller industrial sector but a larger service sector with a consequent lowering of energy and GHG emissions. Just as with technical options, these new dematerialization options need to be 'demonstrated' before China can be expected to adopt them as its own model for development. This puts the onus on countries in the developed world to investigate and pursue these alternatives.

Notes

1   More information on LEAP is available at www.energycommunity.org. Both LEAP and the LEAP dataset containing these scenarios for China are available for download at the LEAP website or by emailing leap@sei-us.org
2   Unless otherwise stated, all monetary values in this chapter are in units of constant PPP 2005 US$.
3   Estimates of cement process emissions and other non-energy sector sources and sinks have not been studied in detail, but preliminary estimates for the BLS and DCRS are included in the LEAP dataset.

References

Baer, P. Athanasiou,T., Kartha, S., Kemp-Benedict, E. (2008) *The Greenhouse Development Rights Framework: The Right to Development in a Climate Constrained* World, second edition, Berlin: Heinrich Böll Foundation

Biello, D. (2008) 'Cement from CO_2: A concrete cure for global warming?', *Scientific American*, 7 August

Chatterjee, A. (1993) *Beyond the Blast Furnace*, Boca Raton, FL: CRC Press

China National Bureau of Statistics (2008) *China Statistical Yearbook 2008*, Beijing: China Statistics Press, www.stats.gov.cn/tjsj/ndsj/2008/indexeh.htm

ERI (Energy Research Institute of the National Development and Reform Commission) (2008) *China Renewable Energy Development Overview* 2008, Beijing: Energy Research Institute, www.cresp.org.cn/uploadfiles/7/977/2008en.pdf

Greenpeace International and the EREC (European Renewable Energy Council) (2007) *Global Energy [R]evolution: A Sustainable World Energy Outlook*, Amsterdam: Greenpeace and EREC, www.energyblueprint.info/fileadmin/media/documents/energy_revolution2009.pdf?PHPSESSID=0d4943629e3ac0ae42959a7ff9af37fc

Hansen, J., Sato, M., Kharecha, P., Beerling, D., Berner, R., Masson-Delmotte, V., Pagani, M., Raymo, M., Royer, D. L. and Zachos, J. C. (2008) 'Target atmospheric CO_2: Where should humanity aim?', *Open Atmospheric Science Journal*, vol 2, pp217–231

Heaps, C. (2008a) *An Introduction to* LEAP, online document, Stockholm: Stockholm Environment Institute, www.energycommunity.org/documents/LEAPIntro.pdf

Heaps, C. (2008b) *LEAP User Guide*, online document, Stockholm: Stockholm Environment Institute, www.energycommunity.org/documents/Leap2006UserGuideEnglish.pdf

IEA (International Energy Agency) (2007a) *Energy Use in the New Millennium*, Paris: IEA, www.iea.org/textbase/nppdf/free/2007/millennium.pdf

IEA (2007b) *Tracking Industrial Energy Efficiency and CO$_2$ Emissions*, Paris: IEA, www.iea.org/textbase/nppdf/free/2007/tracking_emissions.pdf

IEA (2008a) *World Energy Outlook 2008*, Paris: OECD/IEA

IEA (2008b) *World Energy Balances 2008 Edition*, CD ROM, Paris: OECD/IEA

IEA (2008c) *Energy Technology Perspectives 2008: Scenarios and Strategies to 2050*, Paris: OECD/IEA

LBNL (Lawrence Berkeley National Laboratory) *China Energy Databook Version 7.0*, October 2008, CD ROM, Berkley, CA: Lawrence Berkeley National Laboratory

Luo, Z., Zhang, J. and Burnard, K. (2008) 'Near zero emissions coal: A China-UK initiative to develop CCS in China', *Energy Procedia*, vol 1, (2009), pp3909–3916

MacKay, D. J. C. (2009) *Sustainable Energy – Without the Hot Air*, Cambridge: UIT Cambridge Ltd, www.inference.phy.cam.ac.uk/sustainable/book/tex/cft.pdf

McKinsey & Company (2009) *China's Green Revolution: Prioritizing Technologies to Achieve Energy and Environmental Sustainability*, McKinsey & Company, www.mckinsey.com/locations/greaterchina/mckonchina/reports/china_green_revolution.asp

National Bureau of Statistics of China (2006) *China Statistical Yearbook 2006*, Beijing: National Bureau of Statistics of China, www.stats.gov.cn/tjsj/ndsj/2006/indexeh.htm

Pearce, F. (2002) 'Green foundations: It's time to make the concrete jungle emulate the real thing', *New Scientist*, vol 175.2351, 13 July, pp38–40

ULCOS (Ultra-low CO$_2$ Steelmaking) (2009) *Ultra–Low Carbon Dioxide Steelmaking*, ULCOS, www.ulcos.org

UN (United Nations) (2007) *Word Population Prospects 2006 Revision*, New York: United Nations, http://esa.un.org/unpp/

UNIDO (United Nations Industrial Development Organization) (2008) *INDSTAT 4 2008 ISIC Rev. 3*, CD ROM, Vienna: UNIDO

Wang, K., Wang, C., Lu, X. and Chen, J. (2007) 'Scenario analysis on CO$_2$ emissions reduction potential in China's iron and steel industry', *Energy Policy*, vol 35, no 4, pp2320–2335

World Bank (2008) *World Development Indicators*, CD ROM, Washington, DC: World Bank

Zhou, N. and Levine, M. (2008) *LEAP data set for China*, electronic database, Berkeley: Lawrence Berkeley National Laboratory

Part III

Growth, Opportunity and Sustainability

6

Tax Instruments for Reducing Emissions: An Overview

Ottmar Edenhofer, Robert Pietzcker,
Matthias Kalkuhl and Elmar Kriegler

The challenge of climate change

Climate change is a market externality. Market actors emit GHGs leading to costs in terms of climate change damages that are not borne by the emitters themselves but by others. This results in an overproduction of GHG emissions as compared to the socially efficient choice of an emissions' level that fully accounts for the external costs. Market inefficiencies due to pollution externalities are well characterized in environmental economics, and various instruments to remove such externalities have been proposed. In theory, market instruments internalizing the costs of pollution are suited best to re-establish the socially efficient level of pollution. Two major types of market instruments have been proposed: Pigovian pollution taxes (a price signal) and tradable pollution permits (a quantity signal). The idea of Pigovian taxes is to make the polluter pay the external costs of pollution, thus bringing together the social and private rate of return on the polluting activity and inducing an adjustment of pollution to the socially efficient level. The key difficulty of Pigovian taxes is calculating what level of tax will counterbalance the pollution externality. In contrast, tradable pollution permits will give rise to a price on pollution that reflects the relative scarcity of pollution allowances. The key difficulty here is to set the overall pollution quantity to a socially efficient level. It has been shown that implementing a Pigovian tax or a tradable pollution permit system (cap-and-trade) are equivalent under perfect market conditions and perfect foresight. In practice, however, such conditions are never met. There is therefore a long-standing debate in environmental economics over which of the two instruments is superior in which circumstances. The answer will ultimately depend on the nature of the pollution problem and the specific design of the tax or permit system and the assumptions about the game theoretic information structure that changes the strategic behaviour of agents. Although the focus of this chapter is on tax designs for reducing GHG emissions and limiting climate change, we take up this debate wherever it is appropriate. We show that the debate yields important insights into questions of design for both taxes and quantities.

The climate change problem has several characteristics that require going beyond the standard economic practice for dealing with pollution externalities. Because of those characteristics, climate change has been called 'a market failure on the greatest scale the world has seen' in the Stern Review on *The Economics of Climate Change* (Stern, 2006). Those characteristics include:

- Climate change is a global phenomenon, with the emissions of a single actor affecting all other actors globally (due to the mixing of long-lived GHGs in the atmosphere). This means that no isolated national or regional solution to climate change exists. For the choice of instrument to reduce GHG emissions, this means that instruments at the disposal of a national or regional regulator should not be perceived in isolation but should always be evaluated in terms of how they could be linked to a global effort for GHG abatement.
- Climate change is a stock-pollutant problem due to the fact that the stock of GHGs in the atmosphere, not the GHG emissions themselves, determines the amount of climate change. CO_2, the most important GHG of human origin, is removed from the atmosphere on various timescales. The oceans and the terrestrial biosphere take up substantial amounts of anthropogenic CO_2 entering the atmosphere (between 50 and 60 per cent, but this fraction is declining; Canadell et al, 2007). It is often neglected, however, that even when anthropogenic emissions eventually cease, 15–30 per cent of anthropogenic CO_2 will remain in the atmosphere after its equilibration with the ocean reservoir within approximately 1000 years. The further removal from the ocean–atmosphere system operates on timescales of 10,000 (calcium carbonate compensation) to 100,000 years (silicate weathering; see Archer, 2005), far beyond any meaningful socioeconomic time horizon. This means that climate change is accompanied by a twin problem – ocean acidification – that has received comparatively little attention but may lead to similarly large damages (WBGU, 2006). It can be argued that ocean acidification should be taken into account in any assessment of damages from CO_2 emissions, which implies that the CO_2 stock in the ocean–atmosphere system, rather than in the atmosphere alone, should be counted as pollution stock. Even for climate change alone, the persis-

tence of the climate change damages exceeds ordinary human planning time horizons by far, because the climate and sea level response to an increase in atmospheric GHG concentrations operates on timescales of 100 to 1000 years. Thus, the decay rate of the pollution stock is very close to zero. This means that the evaluation of policies for reducing GHG emissions will inextricably be linked to normative assumptions about how to value, first, the welfare of future generations (time preference) and, second, the accumulation of wealth over time (inequality aversion across generations). It is fair to say that environmental economics has yet to develop a sound understanding of how to choose appropriate normative settings for such long time horizons.

- Climate change was erratic in the past, with strongly non-linear responses to only small changes in orbital forcing. There is no reason to exclude the possibility that climate change may respond in a similar way to the current anthropogenic interference. Recent years have witnessed a growing concern about tipping points in the climate system whose transgression could switch (either abruptly or gradually) important subsystems into a qualitatively new state. Depending on the timescale of the transition, this could entail a strong rise in damage costs for societies across the globe. Candidates for harbouring such tipping points mentioned in the literature (see for example Lenton et al, 2008) are, inter alia, the Arctic, the Indian and West African monsoon systems, the Greenland and West Antarctic ice sheets, El Niño/the Southern Oscillation, the Atlantic Ocean overturning circulation, the Amazon rainforest, the boreal zone and the Tibetan plateau. The existence of such tipping points would have important implications for instruments to reduce GHG emissions. Most importantly, a strict quantity control may be necessary, and the flexibility to adapt to newly incoming information about the non-linear nature of the climate system may be paramount.

- Technological solutions to reduce GHG emissions and limit climate damages involve long-lived and capital-intensive equipment. Thus, GHG emissions cannot be reduced 'overnight', and large amounts of investment will have to be mobilized to achieve technology advances allowing significant emissions abatement. Since many low-carbon technologies are characterized by significant learning potentials, early investments are paramount. This implies that policy instruments need to be able to stabilize long-term investor expectations about the stringency of future climate policy.

- As CO_2 emissions from burning fossil fuels form the main contribution to global warming, successful mitigation primarily has to decrease fossil fuel use and, thus, the extraction of fossil resources such as oil, gas and coal. It is therefore important to consider the supply-side dynamics of fossil resource extraction under strategic behaviour of resource owners in response to climate policy instruments (Sinn, 2008; Kalkuhl and Edenhofer, 2010).

Outline of this chapter

In the following four sections, we discuss the question of tax design and its relation to tradable emissions systems in greater detail. The focus of the discussion is on economic efficiency and distributional aspects. We start from a cost–benefit point of view and explore the important question of the efficiency of taxes versus quantity instruments under uncertainty about costs and benefits of mitigating climate change. Then we take the perspective that our information about future climate damage is too weak for reliably estimating the social costs of carbon and that it is therefore advisable to set a climate protection target, as called for in Article 2 of UNFCCC. This can be translated into a carbon budget constraint. The section investigates several tax designs to meet the carbon budget constraint under explicit strategic resource supply and various other market externalities besides climate change, and how they compare with a cap-and-trade approach. The following section takes a political economy perspective and briefly discusses additional important arguments in the taxes versus quantity debate that were raised in literature. We then investigate existing energy and carbon taxes in Europe and their effects. Finally we summarize the main conclusions that can be drawn from the investigations presented in this chapter.

Taxes versus quantities under uncertain climate change and mitigation costs

Over the course of the 20th century, cost–benefit analysis (CBA) has become one of the most important – albeit controversial – instruments used to assess public and private sector projects. It is used to appraise the total expected costs and benefits of all sorts of decisions, be it investment choices, the development of public policies or infrastructure developments. Due to the widespread use of CBA, we begin this study by investigating the performance of taxes and quantity instruments in the classical cost–benefit approach to pollution control in a world with uncertainties. We briefly describe the static case as analysed by Weitzman (1974); then we present the framework used by Newell and Pizer (2003) to analyse the stock-pollutant problem in a dynamic setting. Following the model description, we present the main results found by the authors. By reproducing their model, we are then able to analyse the dynamics of the model and to perform sensitivity analyses. It is necessary to actually delve into the numerical calculations of the model, as we will show that the results – whether taxes or quantity instruments are deemed more efficient – depend strongly on the parameters used. Finally, we discuss the validity of Newell and Pizer's assumptions and the implications that differing assumptions have on the policy proposal.

The Weitzman argument

In a world in which the impacts of policies and the normal evolution of the economy are known beforehand (perfect foresight), CBA can be used to compare the economic efficiency of different policies by simply comparing the sum of discounted benefits of a policy with the sum of its discounted costs.

Let us consider a situation in which there is a good with externalities, for example pollution. Increasing pollution produces growing damages, while decreasing pollution is costly. Given convex costs and damages, there exists an economically efficient level at which marginal damages equal marginal costs. From basic economic theory, it is widely recognized that in a deterministic world with perfect foresight, a regulator can achieve any desired level of output equally efficiently through price or quantity regulation.

Weitzman (1974) shows in his famous publication that this symmetry breaks down as soon as we introduce uncertainty: once the regulator loses perfect knowledge of the costs of GHG abatement and the damages from climate change, one of the two instruments may become preferable on average.[1]

According to Weitzman, the slope of the marginal cost and damage[2] curves determines which instrument is preferred, as shown in Figure 6.1. If the cost curve has higher curvature, equalling a higher slope of marginal costs, a tax instrument will on average be more efficient (in terms of social costs), while a quantity instrument will be preferred when the damage curve has higher curvature.

Taxes versus quantities in a stock-pollutant problem

Weitzman analysed a static case – costs and damages depend directly on the level of pollution, so there is an equilibrium in which an economically optimal level of pollution can be achieved. As pointed out above, the climate change problem has another structure. While abatement costs are directly dependent on the amount of emissions reductions, the damages from climate change depend on the stock of GHGs in the atmosphere, which depends on the *cumulated emissions*. Emissions increase the stock, but the stock may also show some decay, leading to finite lifetimes. While costs occur only once and can easily be attributed to the abatement of one unit of emissions, damages occur over time,[3] making complicated the valuation of the climate change damage due to a single emission unit.

The question of how Weitzman's argument can be transferred to climate change where benefit and cost curves depend on different variables is therefore important. To answer this question, Newell and Pizer (2003) developed a simple model to compare the economic efficiency of taxes and quantity regulations under uncertainty in the case of a stock externality.

Model description

Their model comprises climate change damages that are quadratic in the CO_2 stock[4] and abatement costs that are quadratic in the amount of abated emissions. Uncertainty is introduced by adding a linear cost term equal to the amount of abated emissions multiplied with a stochastic, autoregressive coefficient.[5] Their uncertainty formulation does not allow for a time-invariant shift in marginal abatement costs or a change of the slope of the marginal abatement cost curve. In their representation of CO_2 abatement, there are no path dependencies, nor are inertia or endogenous experience learning effects considered. Furthermore, they do not account for uncertainties in the climate change damages.

There are several exogenously fixed properties that have important consequences for the model results: the coefficient to the damage function grows by 2.5 per cent per year, while the coefficient to the cost function decreases by 2.5 per cent per year. The former is justified through GDP growth (leading to higher exposure to climate change damages), while the latter is explained through technological development. Another important aspect is the development of CO_2 emissions in the baseline scenario without climate protection: they assume a growth of 1.5 per cent per year indefinitely.[6] Future damages and costs are discounted at a rate of 5 per cent. Concerning the GHGs, they assume that only 64 per cent of the initial emissions reach the atmosphere and are added to the CO_2 stock. The CO_2 stock in the atmosphere decays at a rate of 0.83 per cent per year, which is equivalent to a half-life of 84 years.

The optimal emissions path is calculated from this model by numerically minimizing the expected value of the sum of discounted damages and abatement costs. To analyse the differences between tax and quantity regimes, Newell and Pizer introduce a tax on CO_2 emissions, which results in the abatement of a certain quantity of emissions as firms will abate emissions until their marginal abatement costs equal the carbon tax. They qualitatively argue that while dangerous climate change is still far away, taxes will be the preferred instrument. As time progresses, climate damages rise and abatement options become cheaper; therefore, a quantity instrument will be implemented in the long run. Based on this premise, they analytically derive a comparison of the losses or gains obtained by postponing the switch from a price instrument to a quantity instrument by a certain number of years.

Model results

From the analytic expression for the GDP gains or losses produced by a tax instrument, the following conclusions are drawn. First, as Weitzman noted for the flow pollutant problem, the slopes of the marginal damage and cost curves have a strong influence on whether price or quantity instruments are economically more efficient. A steeper marginal cost curve favours price instruments; a steeper marginal damage curve favours quantity instruments.

Second, in the case of a stock-pollutant, two important observations can be made. Due to the stock nature of the problem, the *one-period marginal benefit* of reducing one unit of emissions is five orders of magnitude smaller than the cost of reduction. However, the slope of the *total marginal benefit curve* used to determine advantages of price or quantity instruments will be increased by:

- a lower discount rate;
- a lower decay rate of the CO_2 stock in the atmosphere;
- a higher correlation between cost shocks;
- a higher damages growth rate.

Consequently, taxes will usually dominate in the first periods when severe climate damages are still far away and are therefore reduced through discounting, while in later periods discounting will have less effect, leading to a dominance of quantities over taxes. This is aggravated by Newell and Pizer's assumption that the benefit curve becomes steeper and the cost curve becomes flatter as time progresses. Therefore, the system switches from a preference for taxes in the first periods to a preference for quantities in the later periods, as illustrated in Figure 6.1.

These analytic insights are then explored with numerical calculations using actual data from different publications.[7] Within their parameter choices, taxes clearly dominate quantity regulations, although the preference for taxes decreases as longer time periods are analysed so that after a century, the preference switches to quantity instruments.

Model dynamics

However, no emissions paths are shown in the publication or the accompanying working paper (Newell and Pizer, 1999), and no comments are made upon the dynamics of the model. We reproduced the model with the numerical optimization software GAMS in order to better understand the assumptions

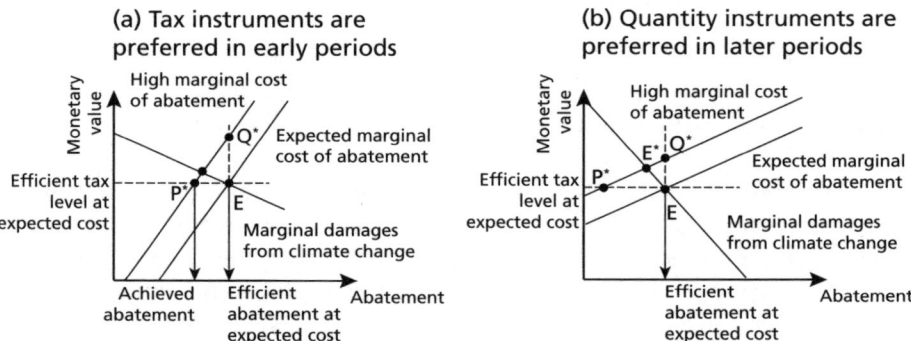

Note: Efficient abatement levels for the expected marginal abatement costs (MAC) are given by point E, which can be fixed by a price instrument (horizontal dotted line) or a quantity instrument (vertical dotted line). If MAC turns out to be higher than expected, the efficient level will be E* instead of E. Figure 6.1a represents early periods in which MAC are steeper than marginal damages. Therefore, the outcome under a price instrument P* is closer to the efficient point E* than the outcome under a quantity instrument Q*. Figure 6.1b represents later periods in which MAC have become less steep than the increased marginal damages. Therefore, the outcome P* is now farther away from E* than Q*, so total welfare is higher with a quantity instrument.
Source: the authors, based on Newell and Pizer (2003)

Figure 6.1 *Comparison of efficiency of tax and quantity instruments for abatement*

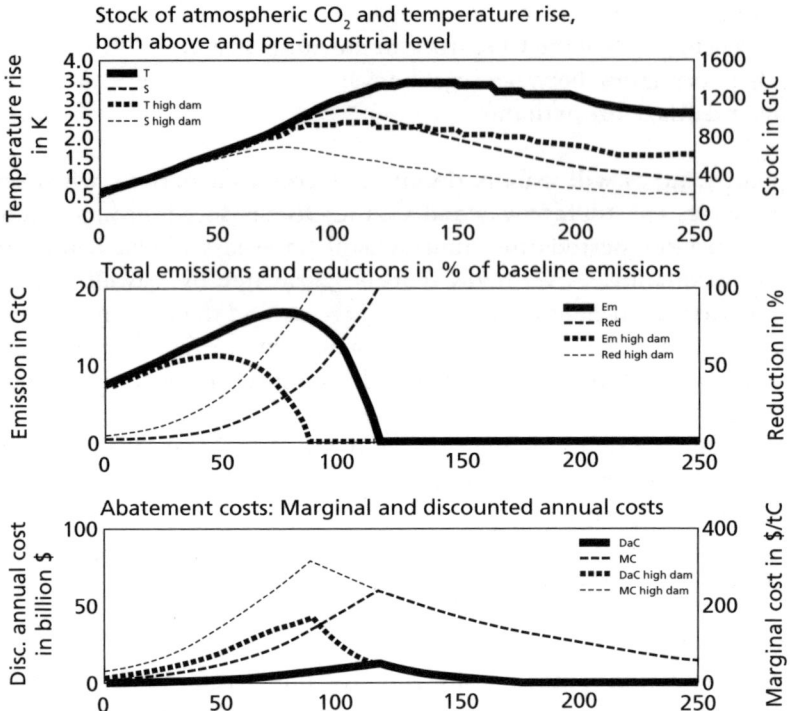

Note: Results of two sample runs of the Newell and Pizer model, one with all values at their default, one with high damage estimates (damages increased by factor 4). The top panel shows the temperature increase ('T') and the CO_2 stock ('S') above pre-industrial levels, the centre panel shows annual emissions ('Em') and emission reductions as a percentage of baseline emissions ('Red'), while the bottom panel displays the marginal abatement costs (MC) and the discounted annual abatement costs ('DAC').
Source: the authors, based on Newell and Pizer (2003)

Figure 6.2 *Results of two sample runs of the Newell-Pizer model*

underlying their parameter choices as well as their policy implications and to analyse the CO_2 dynamics.

The general dynamical behaviour of our implementation of the Newell and Pizer model for most parameter choices made by the authors is depicted in Figure 6.2 and can be described as follows: at time zero, the quantity emitted is practically equal to the baseline case, and there is almost no emission abatement. As time progresses, the relative amount of abatement rises like an exponential or a higher-order polynomial. Together with exponentially rising baseline emissions, this behaviour is equivalent to a slow rise of emissions, which comes to a halt after about 50 to 90 years and then falls in an accelerating manner to zero. Full mitigation is usually reached between year 70 and 150 (115 years using their default parameters).

The CO_2 stock in the atmosphere, the direct driver for damages in the Newell and Pizer model, rises continually until almost all emissions are abated and then begins to fall as the fast decay of atmospheric CO_2 assumed by

Newell and Pizer dominates the remaining annual emissions. Maximum atmospheric stock levels are usually in the range of 300 to 1500GtC[8] (default: 700GtC) above pre-industrial levels before the decay starts to dominate emissions, while the maximum (undiscounted) marginal abatement costs reach $150–800/tC (default: $240/tC) before decreasing again. This general behaviour is robust under quite large parameter variations, while the exact timing of the abatement and therefore maximum stock levels may change. As an example, decreasing the discount rate from 5 per cent to 2.5 per cent results in full abatement already by the year 98, while decreasing the cost reduction rate from 2.5 per cent to 1.5 per cent delays full abatement to the year 150.

Implications of uncertainty for the choice of policy instrument

From the emission paths of the model, several explanations for the calculated dominance of taxes can be deduced, and several problematic assumptions are revealed that question the possibility to transfer the results to the real world. The almost non-existent level of abatement in the first 20 years the model runs can directly be attributed to the extreme difference in marginal cost and marginal benefit curves – the coefficient for the costs is 2×10^5 times larger than that for the damages. This is partly a result of the stock nature of CO_2 emissions and partly a result of the values used to calculate the damage parameters, namely the low climate sensitivity of 2 degrees Kelvin (K) per doubling of atmospheric CO_2 (state-of-the-art climate models use climate sensitivities between 2.1 and 4.4K per CO_2 doubling (IPCC, 2007)) and the damage estimate of 1.85 per cent GDP loss when global mean temperature rises by 3K compared to the pre-industrial level (Nordhaus (2008) increased the damage estimate to 2.5 per cent GDP loss). Although damage estimates vary widely, especially between different scientific disciplines, the values used by Newell and Pizer are at the lower end of the spectrum. A survey among natural scientists gave an average estimate of 13 per cent GDP loss for a 3K temperature increase in 2090, while a similar survey among environmental economists gave an average estimate of 6.6 per cent GDP loss, still 3.5 times higher than Newell and Pizer (Roughgarden and Schneider, 1999). The significance of the damage estimate can be seen in Figure 6.2, where quadrupling the damage coefficient reduces the maximum achieved temperature from 3.4°C to 2.3°C.

In order to test the CO_2 decay rate that Newell and Pizer take as default value – namely 0.83 per cent, equivalent to a half-life of 84 years – we implemented a small climate model into the Newell and Pizer model. To this end, we used the formulae and values given by Kriegler and Bruckner (2004), updated to fit the IPCC Third Assessment Report (TAR) results, and implemented climate change damages depending quadratically on the temperature, not the CO_2 stock.[9] When trying to reproduce the emission paths from the default Newell and Pizer model, we noticed that – once emissions are fully abated – the stock in the Newell and Pizer model decays much faster than the observed temperature decrease in the more realistic climate model.

Since climate change damages are driven by temperature, the stock-pollutant in the Newell and Pizer model should mimic the behaviour of the temperature, not of the CO_2 stock in the atmosphere. Therefore, the fast decay of the pollutant stock in the Newell and Pizer model is equivalent to an underestimation of temporally aggregated climate change damages. To produce similar results in both models, the decay rate in the Newell and Pizer model needs to be set to 0.25 per cent, equivalent to a half-life of 280 years.

According to some recent research in climate physics, the decay rate of the pollution stock might have to be set to even lower values. Matthews and Caldeira (2008) show that the inclusion of reduced ocean carbon uptake due to higher ocean temperatures into the climate models will result in a very long-term stability of increased air temperatures: even 500 years after the last CO_2 emissions, the authors see almost no reduction of temperature. To model such behaviour, the decay rate in the Newell and Pizer model would have to be set to 0.02 per cent or less. The tendency of decreasing the decay rate to better model reality is also reinforced by the fact that we have not accounted for ocean acidification in our damages estimates. As pointed out above, 'decay' of atmospheric CO_2 in the models represents the uptake of CO_2 by oceans, which reduces radiative forcing and temperature rise. If the uptake itself produces damages, it might be possible to aggregate both climate change and acidification into one damage function depending on cumulative CO_2 emissions, having a decay rate near zero. As the decay rate is prominent in deciding whether taxes or quantities are economically more efficient, these insights from climate physics have direct implications for the economic valuation.

Furthermore, the exogenously given exponential decrease in abatement costs leads to the effect that at some point of time, all emissions will be mitigated at low cost. Together with the substantial decay assumed by Newell and Pizer, this leads to the effect that even as high CO_2 emissions pile up in the first years, they can be removed at low cost at a later point of time. This may be exacerbated by the possibility of having negative net emissions in later years to bring the CO_2 stock to zero.[10] Furthermore, this chapter assumes no limit on the rate of increase of emissions abatement, leading to a default scenario in which emissions are reduced from 8GtC (equivalent to 2004 emissions) to zero within ten years – a very daunting task.

These properties all facilitate fast mitigation reactions to the possible event of new research showing that climate change has more severe consequences than expected (for example in the instance of crossing tipping points in the climate system as mentioned above). There is no 'path-dependency' in the model; low emission abatement in the first periods can easily be undone by stronger abatement later. This makes the use of taxes more efficient, as arising cost shocks can be immediately used to abate more in years in which abatement is cheaper and less in years in which it is more expensive.

Real-world experiences seem to contradict these assumptions. Most abatement technologies will require major investments and have a long lifetime, which makes it difficult to economize on short periods of low abatement costs.

Furthermore, it seems improbable that the cost reductions implemented in the model will be realized without prior major investments into abatement technologies, leading to inertia in the abatement sector – if there is no early investment, cost reductions will not happen, making cheap large-scale abatement in the future very unlikely.[11] Also, technology deployment takes a long time in the real world – cycling through the different development stages and achieving cost reductions through learning-by-doing all require significant time to pass. Therefore, the model realizes a world view in which precaution and inertia play a very minor role.

To demonstrate how the above-mentioned observations influence the results, we used a decay rate of 0.02 per cent as implied by Matthews and Caldeira, highly correlated cost shocks[12] as well as a discount rate of 2.6 per cent.[13] Under these assumptions, quantities dominate taxes after model year 45; if we furthermore take the higher estimates of climate sensitivity and climate change damages mentioned above into account, the switch from taxes to quantities will occur already after model year 25.

Many real-world uncertainties are omitted in the implementation by Newell and Pizer. While cost shocks will surely take place, other types of uncertainty and correlations between uncertainties will be equally, if not more, important to the climate change problem.

For example, correlations between uncertainties in abatement costs and uncertainties in climate change damages are not accounted for. Stavins (1996) shows that for the basic Weitzman problem, a positive correlation of uncertainties in cost and benefits increases the efficiency of quantity regulation, while a negative correlation leads to a preference of price instruments. A case can be made for both positive (economic damages will reduce investment opportunities in abatement technologies) and negative correlations (climate damages will mobilize investments in abatement technologies), so it is not clear which correlation will dominate in reality.

More fundamentally, however, it is important to account for uncertainty about the slope and non-linearity of the damage curve. As pointed out above, climate has responded very non-linearly to radiative forcing in the past, and there is a possibility that it will respond in a similar fashion to anthropogenic forcing in the future. According to Newell and Pizer, more non-linear damages, for example from rising sea levels due to the Greenland Ice Shield melting once a threshold temperature is reached, will lead to a preference for quantity instruments.[14] As the concern about such a non-linear climate response grows (Lenton et al, 2008), it becomes increasingly important to intensify research on the economic impacts of climate system tipping points, a topic that has until now received little attention.

To summarize, the CBA of the climate change problem in a world with uncertainties as performed by Newell and Pizer and further examined above offers the following insights:

- In the short run, tax regulation can be welfare increasing, compared to quantity regulation when cost shocks occur.
- In the long run, quantity regulation such as an ETS is more preferable due to rising climate damages that undergo less discounting.
- The estimation of climate change damages and their functional dependence on rising temperatures has a major influence on optimal mitigation pathways. However, it is a topic that has received only limited attention in the past. Much more research is needed to improve the reliability of damage estimates.
- The decision when to switch from tax to quantity regulation is strongly influenced by model parameters such as decay rate, climate sensitivity, discount rate or GDP growth.
- If these parameter values are chosen differently to better incorporate recent findings about climate change, quantity regulation becomes desirable earlier on.
- The informational requirement for deriving the social cost of carbon is extremely demanding. Therefore, estimates of the social cost of carbon cannot provide a robust basis for policy advice.

Taxes versus quantities under supply-side dynamics and multiple market failures

Taxes and permits with a carbon budget constraint

In contrast to the previous section, in which the difference between price and quantity instruments within a CBA were discussed, this section assumes a given and fixed carbon budget and evaluates several instruments to achieve this target with least costs. We focus on this assumption due to methodical and normative difficulties of determining an appropriate damage function – as the social costs of carbon reflect the future damage of an incremental amount of emissions, they are extremely sensitive to future emission paths, carbon cycles, climate sensitivity and available technologies (Stern, 2008). Furthermore, other side-effects of high CO_2 concentrations in the atmosphere, for example on the oceans (acidification), have to be considered. Finally, a valuation of damages is not possible without estimating the potential changes in relative prices for ecosystem services (Sterner and Persson, 2008) and normative assumptions on future generations' needs and preferences (Hepburn and Stern, 2008). Beside the problem of intergenerational justice, the earth's system as a whole may have a value on its own that goes beyond its economically quantifiable value.

Hence, in this section we assume a given mitigation target as a result of a complex political and social decision process. As we do not distinguish between various types of fossil resources, the emissions can be assumed to be proportional to the amount of resources consumed and the problem of climate protection is thus reduced to the problem of fossil resource conservation.[15] Our considerations are based on an inter-temporal general equilibrium model to

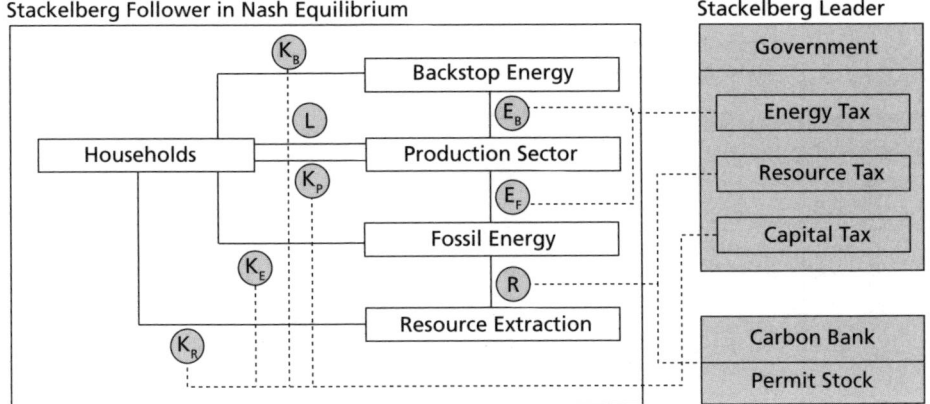

Stackelberg Follower in Nash Equilibrium   Stackelberg Leader

Note: Control variables (grey circles) are influenced by tax paths of the Stackelberg leader. The carbon bank can control the resource extraction directly by issuing permits from a fixed permit stock.

Figure 6.3 *Game theoretic structure of the model*

calculate optimal policy instruments and to illustrate qualitative dynamics numerically. Our model comprises the following features: first, it is an endogenous growth model; second, it comprises multiple sectors to explore the differences between resource and energy regulation; and third, the model explicitly represents the government as the leader in a dynamic Stackelberg game.[16] This approach has the advantage of determining endogenously optimal policy instruments by considering the strategic behaviour of economic sectors. Figure 6.3 gives a schematic overview of the game theoretic structure of our model.

Starting from a given inter-temporal carbon budget, there are two policy design options to achieve emission reduction. *Price instruments* or taxes reduce the demand of several economic factors and thus decrease emissions. In contrast, *quantity instruments* limit emissions directly by restricting the available amount of permits. In our model, we implemented the quantity instrument by a present-value maximizing carbon bank issuing permits that have to be purchased by fossil energy firms for resource consumption.

Price and quantity instruments

In this deterministic framework, price and quantity instruments are equivalent with respect to economic efficiency. If the government has all the necessary information to estimate economic development and no further market failures occur, an optimal resource tax as well as a permit scheme will achieve climate protection at least costs for the society. Both instruments result in the same net resource price[17] to reduce the demand for fossil resources. It should be noted that the net resource price follows a modified Hotelling path (Hotelling, 1931) in an inter-temporal general equilibrium model. The price grows with the interest rate corrected by an extraction cost term (see Figure 6.4).

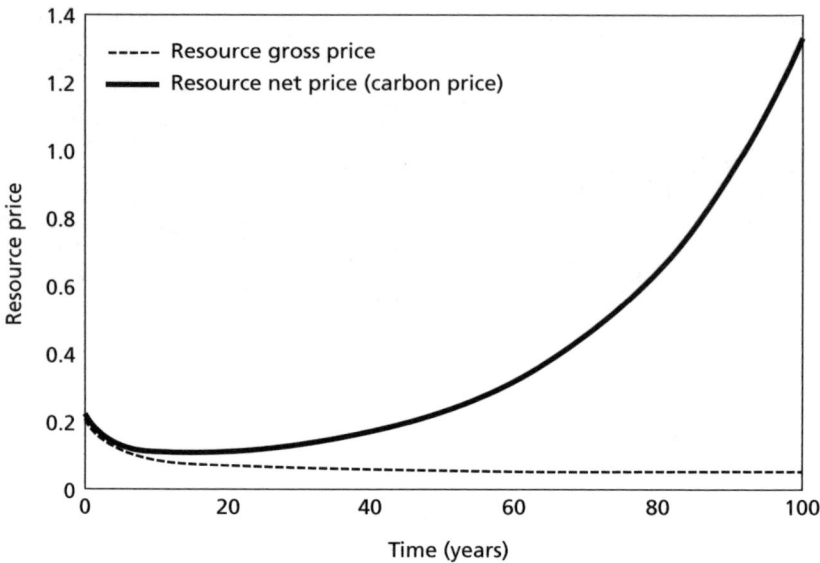

Note: The difference between both prices equals the resource unit tax or the permit price, respectively.

Figure 6.4 *Resource net and gross price*

However, both the optimal tax and the optimal use of the carbon budget through a quantity instrument have different institutional requirements. The optimal tax requires a government that maximizes the welfare of the representative household. Hence, one important implication is that the government has to be able to impose the optimal time path of the tax without any political or institutional limitations. Otherwise the private sector cannot find its intertemporal market equilibrium. In the case of a quantity instrument, the carbon bank has to issue its permits, which can then be traded on future markets.

In any case, the carbon budget creates a scarcity rent for the owners of allowances. In this perspective, creating rents is at the heart of environmental policy. Translating the scarcity of resources into the creation of rents is the reason why economic agents care about the environment. However, it is also a common understanding within welfare economics that rents can be removed from private agents without distorting the allocative efficiency. One feature of a resource tax is that it transfers the rent to the government. In contrast, the quantity instrument without permit auctioning leaves this rent to the permit owners. However, auctioning of these permits, as well as using a resource tax, is symmetric.

Input and output regulation

With regard to Figure 6.5, taxation or quantity regulation can be imposed at different stages of the production processes. Climate policy can be targeted, inter alia, at fossil fuel resources, secondary energy or final output. To achieve an efficient emission reduction, an instrument must be related to the emission or to the economic factor causing the emission. Taxing or restricting secondary energy or output products that are related only indirectly to an emission production process is – in general – not efficient. Although such an instrument could still reach the mitigation target, it fails to set appropriate price signals for a reallocation of economic factors and substitutes. We illustrate this problem by two popular instruments that are imposed on the fossil energy sector to reduce emissions: resource tax and energy tax. The resource tax can be classified as an input instrument (equivalent to an emission tax in our framework) and the energy tax as an output instrument with respect to the transformation process that produces final energy such as heat or electricity. In contrast to the resource tax, the energy tax cannot achieve a significant reallocation in the fossil energy sector compared to the BAU scenario. Although it reduces emissions due to decreases in energy consumption, it has almost no influence on factor allocation and resource substitution within the energy production process (see Figure 6.6). Taxing final consumption products is an even worse way to achieve emission reductions as no reallocation can be achieved either between capital and resources in the fossil energy sector or between energy and capital and labour in the production sector. Output taxation relies on the income effect only while input regulation mobilizes both the income and the substitution effect.

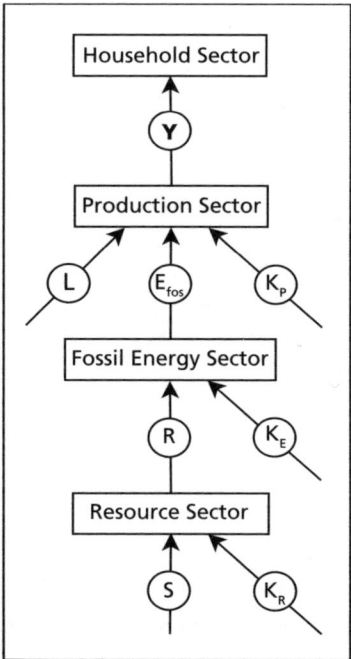

Figure 6.5 *Production chain*

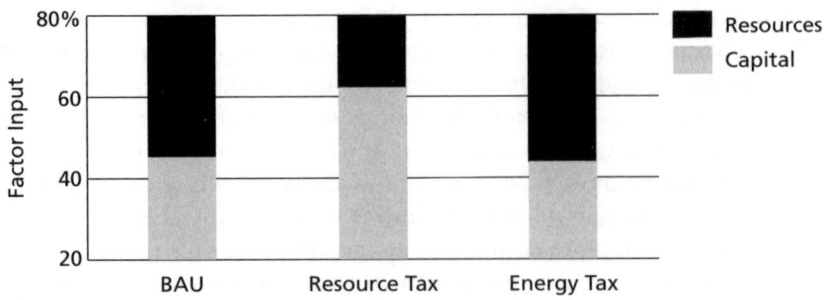

Figure 6.6 *Factor input shares for the input and output pricing instrument*

In general, welfare losses due to output policies depend on the elasticity of the substitution of carbon: the better carbon is substitutable, the more important are appropriate price signals for its substitution and the higher are welfare losses of output instruments that fail to generate these signals (see Figure 6.7 for an illustration of energy tax inefficiency). Only if no substitution is possible, are input and output regulation equivalent.

The lesson from this modelling result is the following: internalizing an externality will be most successful if those factors are addressed that cause an externality because this activates substitution options through the whole production chain. Consumers have the least substitution possibilities; they can only reduce their demand. Production firms can decrease secondary energy use

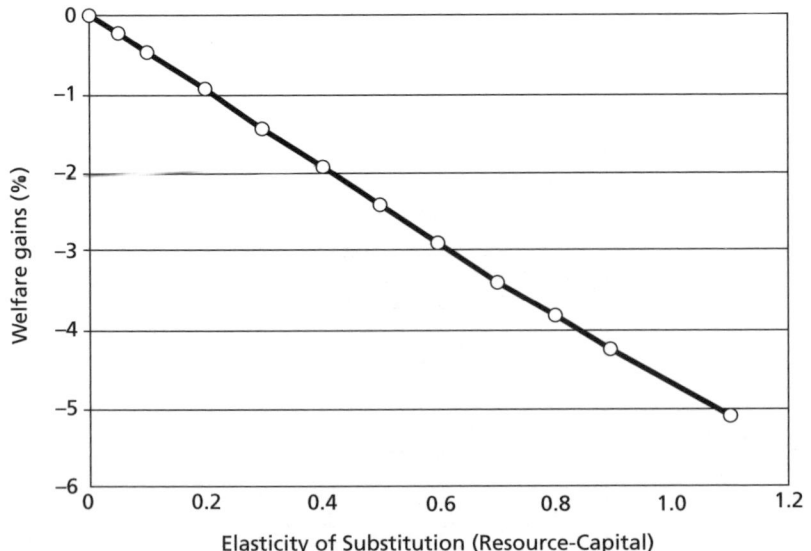

Figure 6.7 *Relative welfare changes of output taxation (energy tax) compared to input taxation (resource tax)*

by switching to less energy-intensive production processes. Power producers have the most substitution possibilities as they can reduce their energy output, increase power plant efficiency or use less-emission intensive options such as natural gas, nuclear power or renewable energy.

Sectoral coverage

It is worth mentioning that such a regulatory instrument has to cover all relevant sectors, i.e. all relevant fossil resource flows through the economy (Hargrave, 2000). This can be done by an upstream system where the fossil resource extracting sector is regulated, or by a downstream system where higher sectoral levels have to pay taxes or buy permits for the fossil resource use of their factor inputs. In an idealized world of complete sectoral coverage and zero monitoring and transaction costs, these approaches are equivalent.

In real-world policy implementations, it is commonly observed that individual sectors are exempt from tax or quantity regulations (Rupp and Bailey, 2003; Bach, 2005). This decreases sectoral coverage, thereby reducing substitution possibilities and strongly increasing total cost. Hence, exempting sectors from the regulation will lead to suboptimal social welfare.

To sum up, it can be said that in a deterministic inter-temporal world, taxes and quantities are symmetric when all institutional requirements are fulfilled. The institutional requirements are a government that can adjust taxes freely for the tax instrument and a complete set of future permit markets for the quantity instrument. In addition, lump-sum transfers are also required to ensure that both regulatory devices are equivalent. Moreover, input regulation is preferable, even if all emissions can be covered in an output regulation system without transaction costs. The general intuition is that policy instruments should be imposed on the production process where more options are available to avoid harmful activities.

Supply-side dynamics and green paradox

With regard to the strategic behaviour of resource owners, Sinn (2008) concludes in a cost–benefit framework that increasing resource taxes accelerates extraction and, hence, worsens global warming instead of slowing extraction. His analysis relies fundamentally on the assumption that resource owners take into account only the resource budget given by nature in their dynamic optimization problem: resource owners will extract the entire resource stock over the infinite time horizon and resource taxes are only capable to transform the time-path of extraction without changing the total (cumulative) extracted amount of resources. Increasing resource taxes, such as a common Pigovian tax, devalues resources in the ground for future extraction and, thus, gives an incentive to accelerate extraction. Within this framework, an asymmetry of price and quantity instruments arises, since a permit system enforces the extraction path on the resource owners and will preclude the possibility to extract the entire resource stock. The examination of the green paradox shows that there is almost no practically feasible option to use price instruments for internalizing the climate externality into the private resource extraction problem, for example by using stock-dependent Pigovian taxes (Kalkuhl and Edenhofer, 2010). Policy instruments suffer from serious credibility problems, or high transaction costs or imply huge transfer payments to resource owners that are not politically feasible.

In our model, however, both the resource tax and the permit scheme will impose the permit cap determined by the climate target onto the resource owners' extraction problem.[18] One important difference for Sinn (2008) is that the mitigation target is not formulated endogenously as CBA but as a resource budget where the concept *internalization* gets a completely new meaning. Price as well as quantity instruments lead to a transformation of the resource scarcity rent into a climate rent that protects the atmosphere as a global common. As it is extremely difficult to achieve climate protection without devaluating the resource owners' scarcity rent, successful policy instruments have to comply with two objectives: first, they have to devaluate the resource owners' scarcity rent and, second, they have to establish an optimal resource price by a public authority that governs the global common on behalf of humankind.

The quantity instrument steers the transformation of the resource rent into a climate rent directly by announcing a credible fixed permit budget. Thus, resource owners anticipate that the scarce permit stock has already devalued their (now) abundant resource stock and that there is almost no room left for rent-making extraction. A price instrument, however, obscures the politically set carbon budget because the resource owners only take into account the tax path and might ignore the fact that the government will impose the tax according to the fixed carbon budget. Moreover, it obscures the devaluation of the resource rent. If resource owners do not anticipate the tax only as a function of time, they might *believe* that they *could* make rents by changing the extraction path. The quantity instrument and the tax are only equivalent if the resource owners anticipate the correct time-path of the tax and believe that the public authority is committed to fixing the carbon budget. If the resource owners perceive the tax only as a function of time, they will change their extraction path in a non-optimal way.

Institutional aspects for optimal carbon pricing

Optimal resource taxing

As already mentioned, a socially optimal carbon price follows the Hotelling rule equating present-value shadow prices of fossil resources at each time step. As cumulative demand for resources has to equal the carbon budget, the calculation of the efficient resource price requires an exact estimation of demand and extraction costs for the entire time horizon. The informational requirements for calculating optimal resource prices are highly demanding and probably beyond the computational capacity of a real-world government or research institution,

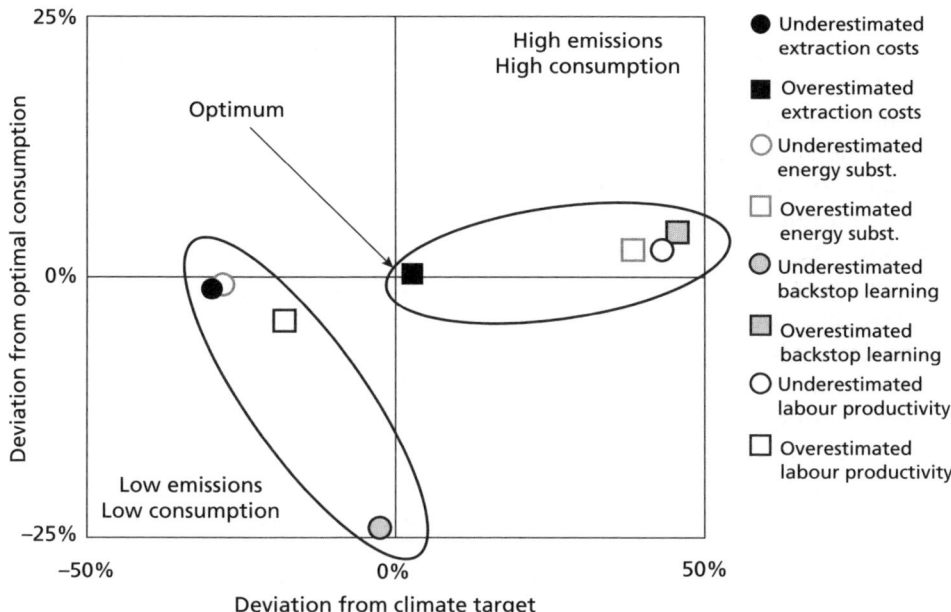

Figure 6.8 *Deviation from climate target and optimal consumption resulting from inaccurate estimates by the government on various parameters in order to calculate the optimal carbon tax*

especially as future low-carbon technologies have to be assessed. If the regulator errs in predicting resource demand, he fails either to achieve the mitigation target (accompanied by over-consumption) or the optimal consumption path due to too restrictive climate protection (see Figure 6.8).

Optimal permit issuing

In contrast to direct resource pricing, an ETS with a fixed carbon budget restricts cumulative fossil resource use and, hence, cannot lead to a violation of the climate target. However, the carbon bank has to decide upon the inter-temporal allocation of permit use (i.e. the timing of permit issuing). Now the carbon bank faces the same uncertainties about the future demand that influence future resource prices. A wrong estimation of economic parameters leads to suboptimal timing and causes welfare losses.

If the carbon bank allows for banking and borrowing of permits, it will shift these uncertainties about future demand to the private sector. Given the full inter-temporal flexibility of permit use, private firms have to decide *when* to use their permits according to their estimation of future permit prices.

In order to deal with the same uncertainties that a government is faced with, private agents might invent institutions such as insurance schemes and hedging strategies. These institutions might respond in a much more effective way to parameter changes than a government with only a limited capability to fine tune due to the nature of the political decision-making process. Experience shows, however, that markets are not always superior but can also suffer from failure.

In principle, an inter-temporal permit market with a fixed carbon budget follows the same rules as the market of exhaustible resources with negligible extraction costs (Rubin, 1996; Kling and Rubin, 1997). Using one permit now reduces the totally available amount of permits at all times and this is why expectations regarding future demand are essential. Hence, for a successful functioning of such a market, it is necessary that the future prices are already known with certainty or that traders believe that they can predict future prices (Dasgupta and Heal, 1979). Future markets can be distorted by insecure property rights, imperfect information, limited access to markets in the future or uncertainty about the regulator's future policies. The collapse of permit prices within the EU ETS's first trading period was, for example, caused by an over-allocation of permits and the absence of banking (see Chapter 8). As the government tries to coordinate the inter-temporal plans for private agents on the market, this type of policy could be called the management of expectations.

As the goal of a successful ETS is to cover all relevant economic sectors and activities, the permit market will be highly fragmented into different economic sectors with different abatement options. However, private agents of one sector might hardly be able to estimate long-term abatement plans of other sectors that influence future permit prices. Furthermore, the assessment of future markets needs information and research that are always affected by costs. Hence, in a completely deregulated permit market, only economically powerful enterprises would provide exhaustive private market research and information collection. According to the Greenwald-Stiglitz theorem (Greenwald and Stiglitz, 1986), markets will not be efficient if all relevant information is not provided as a public good for all market participants. Therefore, institutions are required that provide such information about long-term abatement options and their costs and risks to all market participants.

As the permit price follows the Hotelling dynamics, optimal timing crucially depends on well-functioning capital markets that determine discount rates. There is, however, qualified concern about this assumption. The debate on discounting in the context of global warming emphasizes reasons why market discount rates probably exceed social discount rates (Weitzman, 1998; Portney and Weyant, 1999; Heal, 2008). Furthermore, capital markets could become instable or even collapse, as the recent financial crisis has dramatically shown. Hence, allowing for full inter-temporal flexibility with potential efficiency gains should be balanced against the consequences and risks of capital market imperfections. Existing ETSs consider these aspects by allowing for banking but prohibiting borrowing in order to prevent excessively high pollution levels for present time periods. Another possibility to influence the time-path of permit use is trading ratios that determine the value (in terms of the amount of allowed emissions) of one allowance banked to the next period (Leiby and Rubin, 2001).

In order to stabilize expectations of firms and to regulate the permit market, some institutional requirements are important. The carbon bank endowed with a carbon budget manages permits by maximizing the net present value of its permit stock according to the Hotelling rule for exhaustible resources. Furthermore, it can define trading ratios to influence the time-path of mitigation if market discount rates differ from social discount rates. The bank guarantees property rights of permits, monitors transactions, provides banking and borrowing, offers transparent information and creates credibility regarding the fixed amount of permits over the time horizon and hence provides planning security for economic actors. Being an independent institution like a central bank, the carbon bank reduces regulatory uncertainty about future policies that might be exposed to political pressure (elections or public finance). Nevertheless, it should react flexibly to new insights in the climate system. This will contribute to reducing regulatory uncertainty by only letting pass natural (climate-system-related) uncertainty to private actors that have to be considered anyway by the society.

These considerations show that a careful examination is required of what role private sectors could perform in managing the carbon budget and what independent institutions or governments should administrate. The tax instrument demands a computational capacity that is beyond the limits of what we can expect from a real-world government. The government would have to be able to implement a resource tax that makes cumulative demand compatible with the carbon budget. However, as an additional new market is created by an ETS, additional institutional substitutes are required that reduce the room for possible market failures. The institutional requirements for price and quantity regulation are daunting. A precise assessment of both options would require a detailed understanding of the transaction costs, which is beyond the scope of this chapter.

Multiple market failures and complementary instruments

In this section we discuss further market failures and the policies required to correct them. We focus on three market failures that are related to fossil resource extraction and global warming: oligopolistic market power, high discount rates due to insecure future markets, and technological spill-overs.

Oligopolistic market power

Oligopolistic market power in the resource sector increases the resource price above the optimal level and thus leads to a more conservative resource extraction path. Although this might contribute to climate protection, it is not an economically efficient way to reduce emissions as the oligopolist provides resources on a suboptimal level in order to generate substantial rents. Furthermore, it does not guarantee the attainment of the carbon budget in the long run as the resource owner may extract his/her whole resource stock if it is profitable (see Figure 6.9a).[19] Hence, market power in the resource sector cannot replace a climate policy instrument. Quite the contrary, in the case of climate protection, it can be welfare-improving if governments not only reduce emissions but also regulate oligopolistic resource owners.

If the mitigation goal is implemented by a resource tax, such a tax can address the two market failures at the same time: the climate protection target (which is not anticipated by resource extractors) as well as monopolistic or oligopolistic market power. This was shown by Misiolek (1980) in a static framework: while a decreased resource price (i.e. a subsidy) corrects the monopolistic undersupply, an increased resource price internalizes the social costs of resource extraction. Hence, our model transfers these insights to a dynamic context by calculating an optimal resource tax that lies quantitatively between the pure emission-reducing tax in a competitive economy and the pure oligopoly output subsidy in a BAU economy.

Under a permit scheme, defeating oligopolistic market power depends crucially on the structure of the permit market. If allowances are issued to an oligopolistic resource or energy sector, extraction again will be delayed in order to augment resource prices (see Figure 6.9b). Although the mitigation goal is achieved, suboptimal timing of resource extraction causes welfare losses. A decreasing permit (or resource) subsidy then can enforce the socially optimal net resource price. If, by contrast, permits are issued to a competitive market or if they are auctioned by a carbon bank – i.e. the permit price equals the competitive Hotelling price – oligopolistic market power in the resource sector will clearly be reduced by devaluating the resource scarcity rent.

To sum up, with respect to the oligopolistic market distortion in the resource sector, a resource tax can always achieve the socially optimal extraction while the performance of the permit instrument depends on the good functioning (i.e. competitiveness) of the permit market.

Expropriation risk and suboptimal timing

As already outlined when we discussed the optimal permit price path, the assumption of a complete set of future markets is crucial for a first-best optimum. If resource owners expect that their property rights are insecure,[20] they will change their resource extraction (Sinn, 2008). Uncertain property rights imply a higher discount rate of resource owners leading to accelerated extraction and, thus, contributing to global warming (see Figure 6.9a). In the presence of price or quantity instruments that remove the resource rent, however, uncertainty about property rights of resources is removed jointly. The former resource tax could even be used without modifications as the discount rate of resource owners becomes irrelevant if the private Hotelling dynamic is defeated.

Nevertheless, a similar problem evolves for permit owners that face regulatory uncertainties about future trading ratios or permit caps. Although the carbon budget is always held, timing is suboptimal because effectively higher discount rates are used to compensate for uncertainty (see Figure 6.9b).

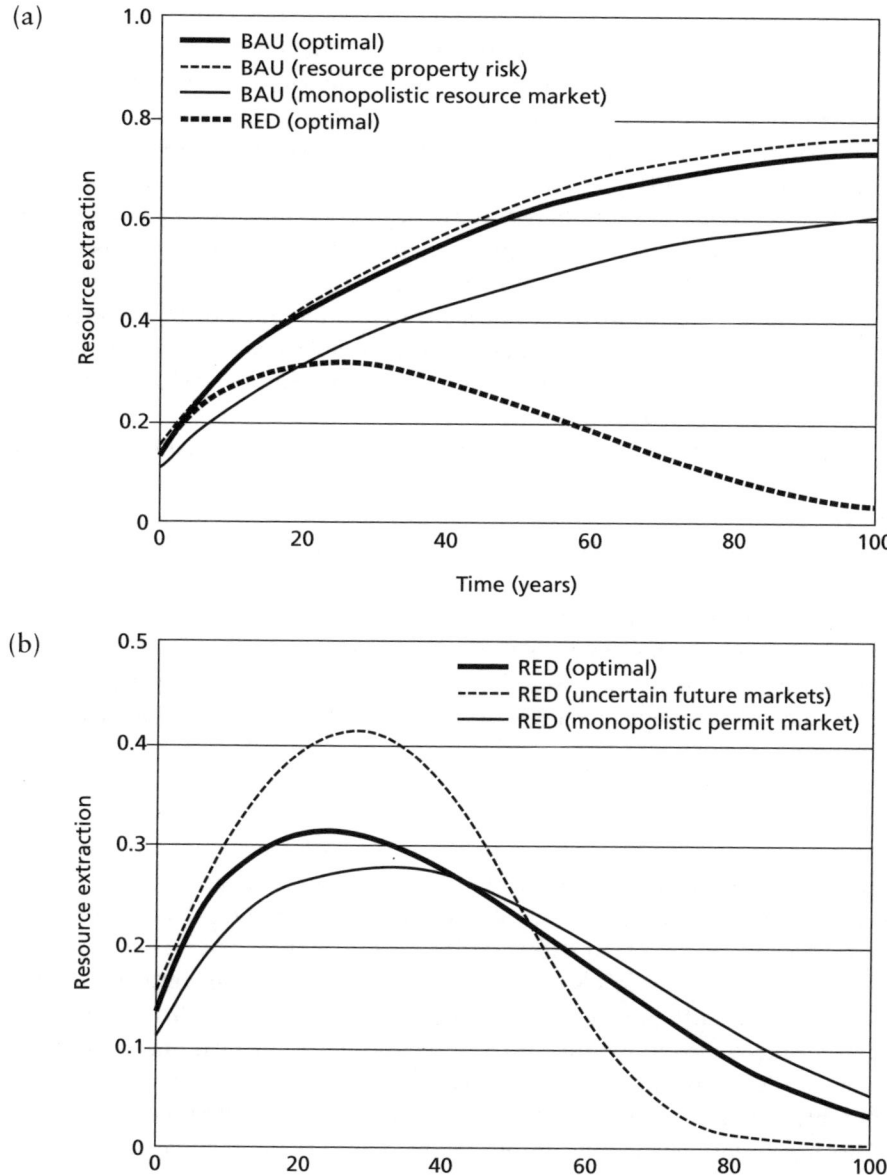

Note: RED = reduction scenario. Resource property risk as well as uncertain future markets augment discount rates in the resource and permit sectors, respectively, and accelerate resource extraction. Monopolistic resource or permit markets slow down resource extraction.

Figure 6.9 *Impact of several market imperfections on the resource extraction under (a) BAU and (b) the quantity instrument within the reduction scenario*

In the context of his resource supply analysis, Sinn (2008) proposes a capital source tax that is imposed on the resource-extracting countries, thereby decreasing their discount rate and shifting extraction into the future. Here arises the problem of the limited feasibility of economy-wide capital taxes due to their economic distortions. Another possibility to remove the effect of uncertain future markets and, thus, to flatten the permit price path is to implement an additional tax on (grandfathered or auctioned) permits that decreases with time and turns into an increasing subsidy. This makes present permit use more unattractive than future use and, hence, leads to an optimal timing. Nevertheless, distribution of rents and transfer incomes differs under the capital source tax and the permit tax/subsidy because a subsidy shifts income from the government to the permit owners. Furthermore, decreasing taxes cannot be implemented credibly.

The problems of these additional tax instruments addressing the uncertainty about future markets reveal the importance of the institutional framework that creates or removes such uncertainties. The problem of uncertain future markets for fossil resource owners could be eliminated by a resource tax or a permit scheme that both remove resource rents. Elimination by imposing a permit market shifts the problem of future markets to the permit owners or the carbon bank. In contrast, the resource tax removes uncertainty from the private sector completely by socializing the extraction problem. Hence, an appropriate institutional framework is needed to stabilize long-term expectations and decrease risk premiums of effective discount rates for permit markets.

Technological change

Is carbon pricing the only important thing that a government should do in order to avoid dangerous climate change? Conventional economic wisdom would say 'yes', as *The Economist* (2008) recently did when subsidies for clean technologies were criticized. Admittedly, a high carbon price gives stronger incentives for investing in clean technologies. However, carbon prices alone will fail to optimally push clean technologies if there are additional market failures connected to innovation-driven technologies. Typical market failures can emerge due to the nature of knowledge as a (partly) public good with non-excludable and non-rival characteristics. Endogenous growth theory describes the chronic underinvestment in innovations due to these positive externalities (Jones and Williams, 2000). It recommends subsidies for investments that are related to spill-over effects (for example the learning-by-doing model of Romer, 1986) or R&D expenditures in order to augment the productivity of several production factors (Jones, 1995; Popp, 2004; Edenhofer et al, 2005).

Biased technological change is an important aspect in the context of global warming (Edenhofer et al, 2006). The investors not only have the option to invest in clean technologies but can also invest in enhancing labour productivity, which might be a much more productive option. Hence, technological change not only describes the speed of growth but also the direction of growth, i.e. which sectors benefit from the productivity increase.

Our model shows that it is important to apply further instruments to the price or quantity instrument, in particular: R&D expenditures both for energy efficiency and renewable energy technology and investment subsidies to internalize spill-overs of learning-by-doing effects within the renewable energy sector (Kalkuhl and Edenhofer, 2010). The R&D investments in overall energy efficiency compensate basically for the decline of the marginal capital productivity due to decreasing energy supply. The learning-by-doing subsidies boost private investment in renewable energy capacity building by equating private and social rates to return. Finally, R&D investments in the backstop technology create the public knowledge stock independently from built capacities. The last two instruments decrease production costs of backstop energy in the course of time and hence form a backstop price that competes with the fossil energy price. The rising carbon price increases the fossil energy price until the backstop price is reached. After that break-even point, renewable energy is fully competitive.

Although underinvestment in clean technology markets can be addressed by specific technology subsidies, one may ask if an increase of the carbon tax could induce sufficiently higher investments. Assuming that clean technology markets are immature markets that have much higher social returns to R&D investments – and are therefore more undersupplied than productivity-enhancing technology markets – an augmentation of the carbon tax above the Pigovian level should improve welfare (Hart, 2008). Calculations within our model show that an augmented resource or emission tax, however, can only marginally internalize technological spill-overs and remains a second-best option.

Thus, both a quantity and a price instrument require additional technology policies. The funds necessary to implement such technology subsidies can come either from the carbon tax revenues or from the revenues from permit auctioning. A grandfathered permit system without auctioning, however, would not raise the needed funds to push technology development, but rather need additional state funds for this purpose. This strengthens our main argument for an auction-based permit system and against a grandfathered permit system.

Assessment of tax and quantity regulation

The considerations above lead to the following assessment of price versus quantity instruments. The price instrument can correct the climate externality under multiple externalities such as oligopolistic extraction and high discount rates due to uncertain future markets. It furthermore can partly correct underinvestment in clean technology markets (though it cannot replace specific technology instruments). The clear advantage of the tax lies in its robustness against market failures in capital, permit or resource markets that might become even more important for long-term emission reduction.

The quantity instrument, by contrast, guarantees the achievement of the carbon budget per definition. In contrast to the tax, an estimation of the resource demand is forwarded from the government to the private sector, which has to assess future demand and hedge against uncertainties. While the size of the carbon budget is fixed, the depletion of the permit stock (i.e. timing) can be privatized by allowing for banking and borrowing. Thus, the quantity instruments allow for more market mechanisms and market dynamics in order to achieve the carbon budget. This may provide higher cost effectiveness but also higher welfare losses if markets (especially capital and permit markets) are not well functioning. The major advantage of the quantity instrument lies in its environmental effectiveness, which even holds when taxes are implemented incorrectly.

The lesson learnt from the green paradox is that climate policy implicitly or explicitly has to devaluate the resource owners' scarcity rent if stock-dependent Pigovian taxes or subsidies for the resource stock in situ are ruled out due to their high transaction costs and undesirable distributional effects, respectively. If climate policy instruments cannot propagate this devaluation credibly, resource owners will provide resources at suboptimal high prices despite increasing resource taxes. Devaluation is only anticipated correctly if resource owners expect zero profits due to taxation (i.e. zero shadow prices for resources) at the end of the time horizon.

Hence, the challenge for climate policy is to overcome the green paradox by a stable and credible expectation building of a now-and-forever devaluation of the resource scarcity rent. We think that a quantity instrument implemented by an ETS with a cumulative carbon budget will make this devaluation more explicit for resource owners than a price instrument in terms of a resource tax that only implicitly contains a carbon budget.

Political economy of taxation instruments

We follow our theoretical analysis with a qualitative assessment of several arguments in the prices versus quantity debate that have been raised in literature (see for example the discussion of Nordhaus, 2008, Chapter 8). While the focus on economic efficiency in first-best worlds was already broadened above to include effects of market failures and externalities, we go a step further here and take a political economy perspective, discussing additional issues relevant for real-world policy-making. These concern in particular the setting of long-term expectations, the problem of political support and the institutional design of taxes and emission allowances trading systems.

Raising revenues and preventing perverse incentives

An important feature of internalizing the emissions externality is the raising of revenues that the regulator could redistribute in a socially efficient way, for example to foster research, development and deployment of GHG mitigation technologies or that can be used to compensate households and industries that are disproportionally affected by the policy. Raising the full costs of the externality from the polluters will also prevent perverse incentives, for example increasing pollution to obtain a greater share of allowances. As already mentioned above, a carbon tax exhibits this feature, while it is not guaranteed for an ETS. If the emissions allowances are handed out to firms, no revenues will be raised by the government and perverse incentives for polluters may be induced. A tax regime would be favourable compared to such a type of ETS in terms of distributional equity. However, if the full amount of emissions allowances is auctioned on a regular basis, this disadvantage of an ETS will disappear. Like a tax, the auctioning will raise revenues for the government and avoid perverse incentives for polluters.

Institutional requirements and transaction costs

Both a carbon tax and an ETS for carbon permits require minimum standards for monitoring, reporting and verification (MRV) of carbon entering the economy or leaving it in the form of CO_2 emissions. The point of reporting requirements may be chosen upstream, for example at the level of producers and importers of fossil fuels, or downstream, for example at the level of combustion or industrial facilities. The choice of the reporting requirement will depend upon the point of regulation that is chosen for pricing a specific GHG. The more upstream the point of reporting and regulation is chosen, the easier it is to implement MRV with broad coverage and minimized transaction costs. Independently of the choice of instrument, it is important to advance and enforce MRV standards economy-wide at an early stage. Given that climate change requires a coordinated global response, such MRV standards should be harmonized internationally.

If MRV is in place, a carbon tax can be implemented using the existing institutions for tax collection. The institutional requirements of an ETS are more complicated. Emission allowances need to be allocated or auctioned to market actors and registered, and institutions such as exchanges or clearing houses need to be established to allow a proper functioning of the carbon market. Firms may not behave efficiently during the initial phase of an ETS during which new information needs to be gathered and carbon trading internalized in the firms' operations. In summary, taxes are favoured from the perspective of institutional and accounting requirements and related transaction costs.

Political requirements

In some countries, imposing taxes is politically difficult. An ETS might be easier to implement if participants can be compensated with free emissions allowances, but this might compromise the environmental effectiveness due to over-allocation of emission rights resulting from lobbying activity. If emission allowances are auctioned (and there are good reasons for this, as mentioned above) political opposition against a carbon tax and an ETS may be similar on average. Past experiences with energy and carbon taxes (see below) and the

ETS implemented by the EU show that such opposition routinely leads to a dilution of the policy in terms of exemptions for energy-intensive and politically well-organized industries – resulting in high economic and environmental costs due to arising inefficiencies. However, as an ETS creates a new market, there will also be incentives for some market actors such as the finance and insurance industries to favour its introduction. This will not be the case with taxes.

International harmonization of carbon prices

Since climate change can only be tackled globally, a meaningful effort will have to rely, at least in the medium term, on the implementation of carbon pricing mechanisms in most regions of the world. In order to avoid distortionary effects between regions and to ensure overall cost efficiency of the global mitigation effort, their carbon prices should be harmonized. It is a clear advantage of emission trading systems that mechanisms (1) creating an integrated international cap-and-trade system, and (2) establishing an opportunity cost for emissions in regions without an emissions cap by means of an appropriately designed carbon credit system (as attempted by the CDM) are conceivable that would lead to the emergence of a globally harmonized carbon price (see Chapter 11). Although the implementation of such mechanisms may still be a daunting political task, it is far less conceivable that carbon taxes raised in the jurisdiction of individual nation states would be harmonized internationally (although some propose such an approach; see Aldy et al, 2008; and Cooper, 2008). The environmental tax reforms in the European Union (see below) have shown that it is notoriously difficult to harmonize taxes across nation states even when they are aligned in a close union. It may have been this advantage of ETS over taxes that gave rise to a focus on cap-and-trade systems and carbon credit markets in international climate policy negotiations.

International burden sharing

Another advantage of implementing carbon markets rather than a multitude of national carbon taxes is that international burden sharing of the costs of climate change and emissions abatement can be easier achieved by adjusting regional caps and allowing for inter-regional carbon credit and permit trade (see Chapter 11). In the case of national carbon taxes, it is difficult to conceive how national tax revenues would be redirected to international burden-sharing funds to a sufficient degree. However, some authors argue that international burden sharing is not necessary to reach an international agreement, and therefore the tax revenues can remain with the nation that raised them (Cooper, 2008).

Closely linked to the question of burden sharing is the question of baseline setting: while all evidence speaks in favour of auctioning permits on a national level, how should the permits be distributed between the states participating in an ETS? The possibility of changing this distribution by setting different baselines allows for international burden sharing, but at the same time it creates a very difficult negotiation topic: as it is necessary to set an individual baseline for each country, each country will try to influence the negotiations to increase its own baseline.[21] A tax, in contrast, does not necessarily create this debate; setting an equal tax without tax exemptions can be appealing due to its simplicity and perceived equal treatment of all parties. Whether this difference is seen as an advantage or disadvantage of a tax compared with an ETS depends on the assumptions about the political process leading to an international agreement and the negotiation position of the different nations involved.

Stabilizing market expectations

For inducing large-scale and long-term investments in carbon abatement technologies, stable expectations about a substantial future carbon price need to be created. A credible tax policy that may include the announcement of a yearly increase in the tax rate (as now implemented in the Canadian province of British Columbia) can create such conditions. Nevertheless, unlike for an ETS, carbon taxes may need to be adjusted if they deliver technology advances and associated emission reductions below expectations for working towards the fulfilment of a domestic or international reduction target. As mentioned above, an ETS will face different difficulties in stabilizing investor expectations. Since carbon prices are formed on the market as a function of supply and demand, investors are subject to volatility of spot prices as well as uncertainty about future prices and future caps. Both effects will suppress investments substantially if volatility and uncertainty cannot be effectively mitigated. Several measures have been proposed. Price floors and caps could limit price volatility to a corridor. Above, we discussed banking and borrowing of emission allowances as a means to smooth price volatility across trading periods (see also Fell et al, 2008), but this depends on functioning capital markets. Under these circumstances, governments could endow a carbon bank with a fixed amount of allowances that could then be used to control the liquidity of the market. In addition, credible and mature ETS should deliver forward contracts that enable hedging against uncertainty of future carbon prices. Despite all these proposals, investor uncertainty of future carbon prices will not be completely eliminated from an ETS. A key question is whether inefficiencies due to incomplete future markets under a permit system can push carbon prices to a level at which the cap may be loosened due to rising political pressure. The circumstances under which such a possibility could be exploited strategically by energy producers also need to be investigated.

In summary, taxes may send a more stable price signal that could convert into an advantage for mobilizing the necessary investments. This will only materialize, however, if governments are able to implement a tax policy that adequately reflects their long-term climate protection commitment. As shown above, this will be a daunting task.

Given the various advantages and disadvantages of taxes versus emissions trading systems, hybrid systems combining features of both instruments have been advocated. These systems come in two flavours: first, ETS with price floor and price caps that effectively limit the carbon price range that firms have to cope with – if the allowance price exceeds the price cap, firms will buy allowances at the maximum price from the regulator, which is equivalent to raising a carbon tax; second, carbon taxes with the possibility to opt out by joining an ETS established in parallel.

Countries that want to move quickly on carbon pricing should ask the question which instrument, or combination of instruments, can induce innovation in energy and emission reduction technologies within years, and at the same time will retain the possibility of linking the domestic effort to global carbon markets as they develop.

Analysis of existing energy and CO_2 taxes in the EU

In the previous sections, we have dealt with qualitative assessment and economic modelling of tax and quantity instruments. In this section, we review empirical studies of the implementation of environmental taxes: what are the effects of regulation and what general conclusions can be drawn for future regulations? Real-life governments, firms and consumers do not necessarily behave rationally and in a welfare-maximizing manner. Rather, the taxes result from a complicated political process in which environmental targets are weighed against national interests and budget restrictions, government decisions are influenced by lobby groups and reactions to regulation can depend on whether prices are explicit or implicit. Therefore, the reviewed taxes cannot be assumed to reflect Pigovian taxes that fully internalize the pollution externality.

Price elasticities

The steering effect of a tax is directly linked to the price elasticity of demand for the taxed good. The closer a price elasticity is to zero, the lower the reduction in demand in response to the price signal sent by a tax. Therefore, a very high tax has to be implemented to yield the expected reductions in demand. Although in a quantity regime the price of the permits would reach the same level as the tax required to achieve the emission reduction targets, the explicit nature of the tax may have implications for political feasibility: raising a very high tax may lead to strong opposition, while reduction limits in a quantity regime or stricter technology standards for energy efficiency do not expose the price signal as directly during negotiations, making it more difficult to raise public protest. This difference would not exist for fully rational interest groups with complete information, however, real-life experience has demonstrated repeatedly that firms and interest groups react differently to explicit and implicit pricing, thereby showing irrational behaviour. It should be noted, though, that protest may be raised later, potentially leading to a weakening of the cap ex post.

It is important to bear in mind that the short-term and long-term price elasticities of energy demand are quite different (Sterner, 2003). Due to the fact that much of the energy demand is determined by the technology used and the infrastructure (which can increase or decrease individual possibilities to switch to less energy-intensive alternatives),[22] price changes might lead to only minor immediate changes in energy use that can be achieved through changed behaviour such as reduced heating or reduced car usage for leisure. However, in the medium and long terms, high energy prices will focus new investments into more energy-efficient technology and infrastructure, for example public transport, more efficient buildings and appliances, yielding a higher decrease in energy demand.

Empirical studies confirm the timeframe-dependent behaviour of energy price elasticities: Cooper (2003) analyses the short- and long-run price elasticities of crude oil from 1971 to 2000, finding short-run price elasticities of -0.02 to -0.08 and long-run elasticities almost ten times as large, with values between -0.1 to -0.6 for most countries. Bernstein and Griffin (2006) find only a minor difference between short- and long-term price elasticities for residential electricity demand in the US (-0.2 to -0.32), but a factor of five between short-term and long-term elasticity of commercial electricity demand (-0.21 to -0.97). This finding highlights the importance of fully exposing commercial electricity users to energy taxes, as they show a stronger reaction to price signals.

It has to be noted, though, that the calculation of energy elasticities is a controversial topic as the actual results are strongly influenced by the model used to calculate the elasticity and the assumptions, for example concerning major technological or cultural changes that occur with little causal relation to energy prices. To exemplify this, Hang and Tu (2007) find a positive price elasticity of electricity in China. This counter-intuitive result (higher prices should lead to a decrease, not increase in demand) might be explained by a general shift towards higher electrification over recent decades.

The matter is further complicated by the effect of income on energy use: several studies estimate income elasticities in the range of 0.5 to 1.2, with higher values usually found in developing countries (Sterner, 2003). This has pronounced implications for the steering effect of price mechanisms: total reductions achieved through price signals can easily be overcompensated by the effect of growing GDP (resulting in growing incomes). Therefore, tax levels either have to rise continually to offset growing GDP, or taxes have to be combined with additional policy instruments such as efficiency standards or R&D subsidies to increase substitution possibilities and thereby the cost elasticities.

If one accepts the plausible observation that long-term energy price elasticities are much higher than short-term elasticities, it is crucial to create credible long-term pricing signals if any significant reduction in energy use is to be achieved. The expectation of firms that the tax will be stable, if not continually increasing, therefore has merit in itself. It is an important factor necessary for any successful abatement policy and should not easily be sacrificed for higher flexibility. Therefore, a tax policy that from the point of its inception envisages later adjustments after more information on mitigation costs becomes available may not foster the needed planning security and confidence leading to large-scale investments in carbon-saving technologies. Therefore, if we assume risk-averse firms and a government responsive to policy instrument failure,[23] the real-world outcome of an inflexible implementation with strong commitment to a time evolution path with rising taxes may be more efficient than the flexible implementation.[24]

Environmental tax reforms in the EU

The European Commission promoted the idea of implementing taxes on energy and pollution in 1993 when the president of the European Commission, Jacques Delors, presented the 'White Paper on Growth, Competitiveness and Employment' (European Commission, 1993). The paper mentions the main principle behind environmental tax reforms: to tax environmentally damaging activities such as pollution or resource use and to decrease the tax burden on labour and other wanted activities. This notion of revenue-neutral price signals to direct consumer behaviour received wide support from both academia and politics.

However, taxation decisions in the EU require unanimity instead of the qualified majority voting needed for most other decisions. As all member states regard taxation to be a field of high national interest, it proved very difficult to reconcile the different positions and develop a consensus over tax rates. It therefore took until 2003 to formulate the 'Taxation of Energy Products Directive' (European Commission, 2003) that laid the groundwork for a common energy resource tax policy. The directive requires all member states to levy a minimum tax on energy resources, including coal, oil, gasoline, natural gas and electricity.[25] As most member states already levied higher taxes on most energy resources, the directive had only minor impact on the total energy taxation and led to little harmonization.

In view of the difficulties of harmonizing energy taxes in the EU – especially given the complications of varying existing taxes and subsidies – achieving global harmonization of regional CO_2 taxes will be a major hurdle should tax instruments be chosen as an alternative to a global carbon market (Aldy et al, 2008). On a national level, however, many different energy tax schemes have been implemented over the last two decades or are in the process of implementation, especially in the new member states. Two different approaches are described in more detail: the Norwegian CO_2 tax that is close to being a resource tax and the German eco-tax that is mostly an end-user energy tax.

Case study: The Norwegian CO_2 tax

Although most of Norway's electricity production is based on hydropower, Norwegian industries and households rely heavily on fossil fuels, mostly mineral oils. As a reaction to scientific research on climate change, Norway implemented a CO_2 tax in 1991 and has adjusted the tax several times over recent years. Current rates are between €20 and €43 per ton of CO_2, depending on the fuel taxed. In addition, Norway levies taxes on electricity use as well as basic energy taxes on oil and natural gas used for heating. Coal was taxed in earlier years but has been exempt from taxation since 2003. Tax exemptions exist for pulp and paper industry, fisheries, cement and metal processing industries and air transport. Norway implemented an important tool to assure that the steering effect of the CO_2 taxes is not undermined by inflation on longer timescales – the tax is indexed to inflation, that is, it automatically increases each year according to last year's inflation (Speck et al, 2006).

A large share of emission-intensive sectors is exempt, so the carbon taxes cover only about 64 per cent of total CO_2 emissions. Although the tax exemptions greatly increase the national cost of emission abatement – the exempted sectors would have contributed to the emission reductions, for example by investing in energy-efficient technologies – they can be explained by considering the political decision-making process: energy-intensive sectors are often well organized and have therefore considerable political influence that they use to shift the burden to other, less organized sectors.

Evaluation studies confirm that Norway's energy and CO_2 taxes lead to a decrease in CO_2 emissions compared to the baseline; however, exact values of the tax effect vary (Speck et al, 2006). Ex-post analyses of the CO_2 reductions due to energy resource taxes are difficult as the data need to be corrected for resource price changes, general economic trends, changes in other taxes and market-induced industrial restructuring over an extended period of time (see comments on short- and long-term elasticities above). Bruvoll and Larsen (2004) performed an ex-post modelling of the Norwegian economy with a disaggregated general equilibrium model. They calculate that out of the total 19 per cent decrease of CO_2 emissions per GDP from 1990 to 1999, only 2 percentage points are attributable to the carbon tax. They attribute this low abatement effect to the many differentiations and exemptions from taxation.

The most visible effect of the Norwegian CO_2 tax was the early deployment of CCS technology at the Sleipner gas field in the North Sea by StatoilHydro. Having started in 1996, Sleipner is the oldest CCS project in operation. CO_2 is a waste product of the field's natural gas production and the gas contains more CO_2 than is allowed into the natural gas distribution network. Therefore, the CO_2 needs to be stripped from natural gas and is usually vented into the air. To avoid paying carbon taxes, Statoil decided to dispose of this CO_2 in a deep saline aquifer. Since 1996, Sleipner has been storing about 1 million tonnes of CO_2 a year.

Concluding, it can be stated that Norway introduced a hybrid between an energy and a resource tax. At first glance, it has the appearance of a resource tax, covering all basic energy resources, adding only a minor tax on electricity. At second glance, however, the exemptions and differentiations in the tax lead to the result that many resource-intensive industries have to pay only a small share of the normal tax rate, so in effect the tax burden is mostly borne by the consumers. Therefore, the disadvantages of a pure energy tax (see above) are at least partly reproduced: cheap opportunities for substitution of CO_2-intensive fuels and increase of efficiency in resource-intensive industries are not tapped, increasing the total abatement cost.

Case study: The German eco-tax

After several economic studies supported the economic advantage of introducing an environmental tax reform, the German government introduced an end-user eco-tax on electricity, transport fuels, natural gas and light and heavy oils in 1999, and repeatedly raised the rates until 2003.[26] A major goal of eco-taxation was to enhance energy efficiency by increasing the costs of energy, while reducing the costs of labour by offsetting part of the labour costs with the tax revenues. These effects are often called the 'double dividend of environmental tax reforms' – environmental targets are achieved and, at the same time, existing negative distortions in other markets are reduced.

From the start of the eco-tax, manufacturing industry, agriculture and forestry were given large tax provisions and the possibility of claiming tax refunds when total taxation was higher than a base amount. This 'flat base amount rule' leads to a decreasing marginal tax rate as energy use increases, clearly favouring larger companies in comparison to small- and medium-sized firms. Although the reduction of CO_2 emissions was one of the objectives of the eco-tax, electricity generated from renewable energy sources was taxed the same as from fossil fuels – therefore the tax posed no incentive for electricity-generating companies to switch to renewable energies.

The tax generated a total revenue of €18 billion in 2004 (about 0.9 per cent of GDP), with €10 billion coming from transport fuel taxes, €6.5 billion from the electricity tax and the rest from natural gas and heating oils (Speck, 2007a). The exemptions and tax refunds for the manufacturing industry were estimated at €4 billion in 2003 (Bach, 2005). The revenues generated by the eco-tax were in the first years exclusively used to reduce employers' and employees' social security contributions. In later years, roughly 10 per cent of the revenues were transferred to the general national budget, while 1 per cent was used to fund the promotion of renewable energies.

The effects of the tax have mostly been described as positive, both from the economic and environmental perspective: most companies are experiencing net benefits from the tax, social security contributions are estimated to be about 10 per cent lower than without energy tax and fuel consumption has decreased (Bach, 2005; Speck, 2007a). Ex-post analysis revealed that total transport fuel consumption decreased by about 14 per cent from 2000 to 2005, although this effect cannot solely be attributed to the eco-tax – while the tax rose from €0.50 to €0.65 per litre on gasoline, rising oil prices further increased the end-user price of gasoline by €0.6 per litre until 2005. Fears of reducing companies' competitiveness have not materialized; rather, most companies actually gained from the introduction of the eco-tax.

Most criticism focuses on three issues: first, the tax is criticized as being too limited in order to have a large impact on the ratio of labour versus energy productivity growth rates (Beuermann and Santarius, 2006). Second, the tax is criticized as inefficient since it sets a price on energy, instead of directly on fossil fuel carbon (see above). Third, the major tax exemptions mentioned above strongly reduce the taxes' cost effectiveness. As a further result of the tax exemptions and the existence of a flat base amount payable by all companies, the majority of the tax burden is borne by small enterprises and people without work,[27] while almost all larger companies have actually profited from the eco-tax as the reductions of social security contributions were higher than the levied energy taxes (Rupp and Bailey, 2003; Bach, 2005).

Important results for tax design

The pattern of exemptions and differentiation leading to reduced steering effects of energy taxes can be found in most energy tax implementations during recent decades: energy-intensive industries (or even most industries) have to pay much lower energy tax rates than do households. While this may be explained by the fear of losing international competitiveness, studies show that often other factors such as infrastructure, wage levels, education and proximity to growing markets have a higher influence on competitiveness and choice of location than environmental regulation (Sijm et al, 2004; Speck, 2007b). Also, most energy taxes recycle the revenues for reductions of social security contributions or income taxes – the so-called 'second dividend' – thereby leading to a net reduction of firm-level taxes (Bach, 2005). Furthermore, the ongoing shift towards energy taxes in many countries will lead to a level playing field again, reducing the competitive pressure on energy-intensive firms as competitors face similar increases in energy prices.

It is important to note that the efficiency of environmental taxes is weakened by the presence of additional externalities besides the pollution externality. In particular, infrastructure lock-in and technological spill-overs reduce the flexibility to substitute the polluting activity with cleaner alternatives. This is indicated by the large difference in short- and long-run price elasticities that reflect timescales of socioeconomic transformations. It is therefore necessary to supplement any price mechanism with other policy instruments such as R&D subsidies or efficiency standards to increase substitution possibilities.

As tax exemptions increase the total cost of achieving abatement targets, the main conclusion that is drawn in most tax evaluations is that there should be as few tax exemptions and differentiations as possible. Also, it is important to index resource taxes to inflation to ensure their long-term steering capacity. This is emphasized when incorporating political feasibility concerns: repeated explicit tax increases may face major political opposition and therefore create political risk that has to be carried by firms planning long-term investments, whereas automated adjustments to inflation create more stable planning conditions.

Summary and conclusions

It is widely accepted that a price on CO_2 and other GHG emissions is required for successful climate protection. This can be achieved either through price instruments such as taxes on emissions or quantity instruments such as emission allowances. In this chapter we discussed the effects of and the issues surrounding the implementation of carbon and energy taxes and contrasted them with quantity instruments based on tradable emissions allowances where appropriate.

While climate protection is a must, debate arises over how much effort should go into climate protection measures. Theoretically, this is a question of weighing the costs of emissions reduction with the benefits from avoiding additional climate change to find a socially efficient level of emissions. In practice, this would require the valuation of all mitigation costs and climate change damages today and in the future to identify the social costs of carbon. This is clearly an unrealistic requirement. Uncertainties about mitigation costs and climate change damages abound, and expectations of future social and technological changes have a dominant influence on the evaluation of the social costs of carbon.

We took up the debate on appropriate choice of instruments for carbon pricing in such an idealized cost–benefit setting in the second section of this chapter. If the uncertainties were quantified reasonably well, we could calculate the emission path which – on average – results in the highest expected social welfare. The stock nature of climate change seems to favour taxes in the short run, while emission permits become the dominant policy instrument half a century from now or later. When we make more realistic assumptions about GHG decay, inertia of mitigation measures and the extent of climate change damages, quantity instruments become the favourable choice before 2050.

However, if we assume – as many scientists do – that the uncertainties in mitigation costs, but even more so in climate damage costs, are significant and extremely difficult to quantify, the concept of CBA evaluating the social cost of carbon is weakened. Based on the concern that there are tipping points in the earth's climate system whose transgression could increase climate change damages (and the uncertainty about them) dramatically, policy-makers may decide to avoid dangerous interference with the climate system (as implemented in Article 2 of the UNFCCC). This decision would most likely involve setting a climate protection target, for example in terms of a maximum temperature or GHG concentration limit. Such a target can be converted to a total maximum carbon budget that may be used without incurring an unacceptably high probability of violating the climate protection goal. Once the carbon budget is set, the question remains how to cost-effectively allocate its usage over time. This case was discussed in the third section.

The preference for tax or quantity instruments in such a setting hinges on the assessment of whether governments or markets are better suited to bear risks and make predictions about the future. A price instrument puts the risk of misjudging the right tax rate on the government. Possible consequences of predicting the wrong mitigation costs are economic losses (if taxes are too high) or non-compliance with the carbon budget (if taxes are too low). A quantity instrument moves the risk to the economic agents, with profit losses as a consequence of wrong predictions of future allowance prices. Although markets are often seen as more capable in information-collecting than a centralized authority, this will entirely depend on the implementation of an efficient carbon market including mature future markets or other institutions for stabilizing future price expectations. Furthermore, quantity instruments are robust against the green paradox that predicts backlashes of strategic resource owners against resource taxes resulting in suboptimal resource extraction.

Finally, it is not sufficient to assess policy instruments exclusively from an economic efficiency point of view, but also with regard to institutional requirements and political support. We therefore took a political economy perspective in the fourth section of the chapter and briefly discussed arguments raised in literature. Taxes are easier to implement with existing institutions and may involve smaller transaction costs. In contrast, national or regional emission allowances trading systems are easier to link into a global effort with appropriate burden sharing and harmonized carbon prices. Both instruments need to send a credible long-term carbon price signal to investors, reallocate scarcity rents of carbon sensibly and assure a broad coverage of GHG-emitting sectors.

The necessity to include institutional requirements and political support into the analysis was reinforced by our analysis of existing implementations of carbon and energy taxes in the last section. It showed that in almost all cases political constraints led to the dilution of the tax measures. Economic inefficiencies were introduced due to exemptions for powerful industries and high transactions costs from choosing downstream points of regulation.

The following conclusions can be drawn from the combination of theoretical analysis, numerical models and analysis of existing carbon and energy taxes:

- Economic theory is indecisive on whether a tax or a quantity instrument is better suited for putting a social price on carbon. Although taxes seem to be the better short-term option, preference for one instrument hinges strongly on the exact parameterization of the global climate system, economic growth, technological progress, discount rates and damages from climate change and the uncertainty about these factors.

- With regard to the supply-side dynamics of fossil resources, successful policy instruments have to devalue resource owners' scarcity rent in order to propagate the political constraint on the resource stock depletion. A quantity instrument in terms of a fixed permit stock makes the carbon budget explicit and, hence, more transparent and credible for resource owners. A resource tax obscures the politically set carbon budget and may lead to suboptimal resource prices if resource owners do not anticipate future tax paths and their impacts on resource demand correctly.

- The more the daunting task of estimating cumulative demand is shifted from the government (tax) to the private sector (ETS), the more crucial becomes the good functioning of capital markets in order to estimate, valuate and hedge against future mitigation costs. An efficient ETS with full inter-temporal flexibility (i.e. with banking and borrowing) crucially relies on functioning permits and capital markets, i.e. there should be no uncertainty on future caps, no market power in the permit market and no imperfections in the capital market such as liquidity constraints or risk premiums due to regulatory uncertainties. A tax is more robust against these permit market-related problems but it leaves the government with the enormous challenge of exactly estimating resource demand for the entire planning horizon with regard to the optimal resource price.

- It is important to stabilize investor expectations of the stringency of future carbon constraints by setting a credible, long-term and clear signal of future carbon prices. For a carbon tax, this requires a credible long-term commitment to a mitigation goal that is to be reached by increasing resource taxes in real terms. It also requires that the tax path is chosen appropriately (not too high, not too low) so that market actors do not expect surprising future adjustments to the tax provision. For an ETS, this requires reducing volatility of spot prices for emissions allowances and creating stable expectations of future caps. In the third section, we presented the idea of a carbon bank that regulates the allowance market, thus increasing price stability and preventing additional market failures.

- It is also important that the government raises the revenue from introducing a scarcity value of carbon to the economy. This avoids perverse incentives for permit owners and allows the revenues to be used in order to (1) offset distortionary taxes, (2) subsidize abatement technologies to offset other market externalities from technological spill-over, or (3) counteract the regressive effect of the carbon constraint (distributional equity). While carbon taxes will directly deliver annual revenues to the government, the cap-and-trade system will require the auctioning of permits to raise a comparable revenue stream. In the context of a cap-and-trade system, auctioning allowances is therefore strongly preferred to handing out allowances for free.

- Emissions taxes or input taxes on fossil fuels are better than output taxes on secondary energy (such as electricity) because they directly address the pollution externality. We showed in the third section that output taxes do not lead to an efficient readjustment of factor shares in the fossil energy sector, i.e. they do not lead to a reduction of its resource intensity but only to a reduction of overall energy intensity of the economy.

- Taxes and tradable allowance systems could be implemented both upstream (at the point where carbon enters the economy, i.e. at the level of fossil fuel producers and importers) or downstream (for example at the level of utilities and firms). It is widely recognized that upstream systems are much better than downstream systems because they allow for a wider coverage of sectors with less transaction costs. An upstream tax system would amount to a tax on coal at the mine mouth, crude oil entering the supply chain and natural gas entering the pipeline system. An upstream cap-and-trade system would only involve companies in the fossil-fuel producing sector. It needs to be thoroughly checked whether this limitation still allows for the formation of an efficient market or whether such a carbon market would be prone to oligopolistic distortions.
- Since climate change is a global problem, the effort to reduce GHG emissions must be global. The long-term goal therefore should be the international harmonization of carbon prices. As experience with failed tax harmonization projects suggests, this will be more difficult to achieve with a system of national or regional carbon taxes than with a global system or regionally interlinked systems of emissions trading. However, international integration of regional carbon markets is only possible based on harmonized standards for MRV. To implement an efficient tradable allowance system, it is necessary to have good information on emissions. Collecting data and developing institutions is a process that may well take five to ten years, therefore the process should be started now. As an additional advantage, information on energy and CO_2 wastage is collected, showing no-regret options to improve efficiencies.

Besides the above mentioned points, it is also important to note that the emissions externality is not the only externality leading to socially inefficient outcomes. There are various other externalities that are relevant for the evaluation of carbon pricing instruments, and we discussed some of them above:

- *Technology spill-overs* – technology markets generally suffer from under-investment due to spill-over effects and positive externalities, but this problem is even greater for highly innovative and strongly growing new technology markets, such as the clean technology market. Although high carbon prices can at least stimulate investment to some extent, they cannot internalize the externality completely and thus have to be complemented by subsidies and R&D expenditures.
- *Oligopolistic resource markets* – imperfect resource markets suffer from an undersupply of resources that aim at raising resource prices and therefore resource rents. In standard economic theory, an output subsidy for resource owners leads to an efficient extraction path. In the presence of a

climate target, imperfect resource markets can be corrected by a carbon tax slightly below the Pigovian level. A common ETS that devalues the owners' scarcity rent also reduces their market power.

- *Uncertain future markets* – if property rights on resources or, more generally, on permits due to uncertain future cap setting are insecure, owners incorporate these risks by increasing their discount rate. High discount rates lead to an inefficient timing of abatement and investment and hence penalizes future generations. This problem could be removed by implementing a capital tax that only addresses the affected sector. Expropriation risk in the resource sector is removed automatically by price or quantity instruments mitigating climate change.
- *Asymmetric information and irrational market actors* – existing energy markets are not efficient. According to the McKinsey & Company abatement cost curve (Dinkel et al, 2009), huge efficiency potentials are not tapped although they are economically attractive. This can be explained through a lack of information and irrational behaviour, both of which can be targeted through efficiency standards, for example for buildings and appliances.

Based on the above-mentioned findings, the following conclusion can be drawn: in the long run, the efficient policy instrument for climate protection will be a globally harmonized emissions trading system due to expected sharply increasing climate damages. Furthermore, a quantity instrument is more robust against the supply dynamics of strategically acting resource extractors. Full banking and borrowing, however, should only be implemented if the good functioning of capital markets is guaranteed. Otherwise, the problem of appropriate timing of mitigation should be solved by the regulator, just as under the tax instrument.

In any case, it is important to give a credible signal of long-term commitment to climate protection – both for domestic investment behaviour and international cooperation for climate protection – and to encourage major investments in low-carbon technologies.

Notes

1 Weitzman introduces uncertainty as an additive linear term in the cost and benefit functions, resulting in a vertical shift of the marginal cost and marginal benefit functions.
2 Following the naming conventions in economics, both the terms 'damages' and 'benefit' (of avoided damages) are used to describe the climate change damages.
3 This recurring nature of the damages may be explained either through climate change damaging the capital stock, leading to a continued reduction of GDP, or through the effect that a changed climate leads to less productivity in a certain sector, for example agriculture.
4 In reality, the damages occur due to the increased temperature and the resulting climate variations, but the model would not be analytically solvable if the independent variable were not the same.

5 Newell and Pizer use an autoregressive coefficient to model their view that cost shocks are correlated in time, as business cycles and macroeconomic effects span more than one year.

6 This leads to the peculiar fact that in their baseline the emissions rise above the total amount of carbon stored in fossil fuels quite quickly: an amount equivalent to the conservative estimate of total fossil fuels (4000GtC) is emitted by the model year 140, while the amount equivalent to a high estimate of carbon stored in fossil fuels (9000GtC) is emitted by year 195. In year 300, their model shows total baseline emissions of 44,000GtC.

7 Newell and Pizer calibrate their climate damages and abatement cost curves with data from Nordhaus (1994), as reported in Roughgarden and Schneider (1999) and the Energy Modeling Forum 16, as reported by Weyant and Hill (1999). Furthermore, they conduct a one-dimensional sensitivity analysis by varying one parameter while keeping all others at their default value.

8 Throughout this section, CO_2 is measured in 'tons of carbon' (tC). To convert to the metric 'tonne of carbon dioxide' (tCO$_2$), all weights have to be multiplied by 3.67.

9 For comparability, we chose a climate sensitivity of 4.5 and calibrated the damages so that 3°C temperature increase yielded 1.85 per cent GDP damage, the value used by Newell and Pizer.

10 Newell and Pizer do not explicitly state whether they allow negative emissions or not, but their framework makes it plausible and they do not explicitly exclude it.

11 When trying to reproduce the externally given annual cost reduction of 2.5 per cent through a learning curve, a learning rate of 30 per cent was needed to yield similar cost reductions. When compared to commonly observed learning rates of 10–20 per cent, this value seems very high, making the initial assumption questionable.

12 We expect that a misjudgement of abatement costs is closer to reality than highly fluctuating cost shocks. Therefore, we implemented a temporally constant stochastic linear factor to the cost curve.

13 This is equivalent to a pure rate of time preference of 0.1 per cent rather than the 2.5 per cent chosen by Newell and Pizer.

14 This claim is supported by Pizer (2002) who numerically models different exponents in the damage curve. The model he uses differs greatly from the Newell and Pizer model, so no numerical results can be transferred.

15 Equating the atmospheric concentration target with the amount of fossil resources extracted is based on the physical law of conservation of mass and the omission of further processes in the global carbon cycle.

16 A Stackelberg game assumes a hierarchical asymmetry between players in the sense that one player (Stackelberg leader) makes his decision before the other players (Stackelberg followers) by considering secured information about the reaction of the followers after his move.

17 The net resource price is the price the fossil-energy firms have to pay for fossil resources. It is the sum of the gross resource price and the unit tax or permit price, respectively.

18 Technically speaking, the difference lies in the way policy instruments influence the transversality condition of resource owners. A more formal discussion of the explicit assumptions and technical implementations of specific policy instruments can be found in Kalkuhl and Edenhofer (2010).

19 The complete extraction of the resource stock in the absence of climate policy depends on some basic assumptions about the substitutability of fossil resources and

the dynamic of extraction costs as well as on the time horizon considered. One example for a set of assumptions leading to a complete extraction is provided by Sinn (2008).

20 One example of insecure property rights might be authoritarian regimes of oil-exporting countries that are under a certain threat of losing control of their oil resources.

21 This 'haggling' procedure would be prevented if the baseline could be set according to some general metric, for example equal emission rights per capita.

22 This lock-in effect can be illustrated using the example of the US over recent decades. The reduced availability of public transport and the increase of suburban living with the resulting long commuting distances lead to a substantial reduction in the short-term price elasticity of gasoline. People living in suburban areas simply find it impossible not to use their cars any more as they totally depend on them to go to work and for other daily activities (Hughes et al, 2008).

23 If firms expect the government to possibly abolish a tax if it shows no effect in the first years, they may decide to heavily invest in non-carbon-free technologies to reach a lock-in state in which they can force the government to abolish the tax or at least make exemptions.

24 The long-term stability of the German feed-in tariffs that guarantee a fixed price for 20 years and thereby reduce the investment risk almost to zero is often stated as one of the major factors accounting for the success of the feed-in tariffs in achieving capacity expansion.

25 For current tax rates, see 'Excise Duty Tables, Part II: Energy Products and Electricity' (European Commission, 2008).

26 The eco-tax was formulated as a unit tax without inflation indexing, making future adjustments for inflation necessary. Also, the tax was added on top of previously existing mineral oil taxes.

27 Jobless people, students and pensioners have to pay the higher energy prices but get no benefit from the reduction of social security contributions.

References

Aldy, J. E., Ley, E. and Parry, I. W. H. (2008) *A Tax-Based Approach to Slowing Global Climate Change*, Discussion Paper RFF DP 08-26, Washington, DC: Resources for the Future

Archer, D. (2005) 'The fate of fossil fuel CO_2 in geologic time', *Journal of Geophysical Research*, vol 110, C09S05

Bach, S. (2005) 'Be- und Entlastungswirkungen der Ökologischen Steuerreform nach Produktionsbereichen' Band I des Endberichts für das Projekt: 'Quantifizierung der Effekte der Ökologischen Steuerreform auf Umwelt, Beschäftigung und Innovation' Forschungsprojekt im Auftrag des Umweltbundesamts, FuE-Vorhaben Förderkennzeichen 204 41 194

Bernstein, M. A. and Griffin, J. (2006) *Regional Differences in the Price-Elasticity of Demand for Energy*, NREL/SR-620-39512, Santa Monica: Rand Corporation

Beuermann, C. and Santarius, T. (2006) 'Ecological tax reform in Germany: Handling two hot potatoes at the same time', *Energy Policy*, vol 34, no 8, pp917–929

Bruvoll, A. and Larsen, B. M. (2004) 'GHG Emissions in Norway: Do carbon taxes work?', *Energy Policy*, vol 32, no 4, pp493–505

Canadell, J. G., Le Quéré, C., Raupach, M. R., Field, C. B., Buitenhuis, E. T, Ciais, P., Conway, T. J., Gillett, N. P., Houghton, R. A. and Marland, G. (2007) 'Contributions to accelerating atmospheric CO_2 growth from economic activity, carbon intensity, and efficiency of natural sinks', *PNAS*, vol 104, no 47, pp18866–18870

Cooper, J. C. B. (2003) 'Price elasticity of demand for crude oil: Estimates for 23 countries', *OPEC Review*, vol 27, no 1, pp1–8

Cooper, R. N. (2008) 'The case for charges on greenhouse gas emissions', discussion paper, Harvard Project on International Climate Agreements, Belfer Center for Science and International Affairs, Cambridge, MA: Harvard Kennedy School

Dasgupta, P. S. and Heal, G. M. (1979) *Economic Theory and Exhaustible Resources*, Cambridge, London and New York: Cambridge University Press

Dinkel, J., Enkvist, P.-A., Nauclér, T. and Pestiaux, J. (2009) *Pathways to a Low-carbon Economy: Version 2 of the Global Greenhouse Gas Abatement Cost Curve*, McKinsey & Company

The Economist (2008) 'Green, easy and wrong. Why a verdant New Deal would be a bad deal', *The Economist*, 6 November

Edenhofer, O., Bauer, N. and Kriegler, E. (2005) 'The impact of technological change on climate protection and welfare: Insights from the model MIND', *Ecological Economics*, vol 54, pp 277–292

Edenhofer, O., Carraro, C., Koehler, J. and Grubb, M. (eds) (2006) 'Endogenous technological change and the economics of atmospheric stabilisation', Special Issue, *The Energy Journal*, vol 27

European Commission (1993) 'Growth, competitiveness, employment: The challenges and ways forward into the 21st century – White Paper', *Bulletin of the European Communities*, Supplement 6/93, Brussels: European Commission

European Commission (2003) *Taxation of Energy Products Directive*, Directive 2003/96, Brussels: European Commission

European Commission (2008) 'Excise duty tables, part II: Energy products and electricity', Brussels: European Commission, http://ec.europa.eu/taxation_customs/taxation/excise_duties/energy_products/rates/index_en.htm

Fell, H., MacKenzie, I. A. and Pizer, W. A. (2008) 'Prices versus quantities versus bankable quantities', discussion paper RFF DP 08-32-REV, Washington, DC: Resources for the Future

Greenwald, B. C. and Stiglitz, J. E. (1986) 'Externalities in economies with imperfect information and incomplete markets', *The Quarterly Journal of Economics*, vol 101, pp229–264

Hang, L. and Tu, M. (2007) 'The impacts of energy prices on energy intensity: Evidence from China', *Energy Policy*, vol 35, no 5, pp2978–2988

Hargrave, T. (2000) *An Upstream/Downstream Approach to Greenhouse Gas Emissions Trading*, Washington, DC: Center for Clean Air Policy

Hart, R. (2008) 'The timing of taxes on CO_2 emissions when technological change is endogenous', *Journal of Environmental Economics and Management*, vol 55, pp194–212

Heal, G. (2008) 'Climate economics: A meta-review and some suggestions for future research', *Rev Environ Econ Policy*, vol 3, pp4–21

Hepburn, C. and Stern, N. (2008) 'A new global deal on climate change', *Oxf Rev Econ Policy*, vol 24, pp259–279

Hotelling, H. (1931) 'The economics of exhaustible resources', *The Journal of Political Economy*, vol 39, pp137–175

Hughes, J. E., Christopher, R. K. and Sperling, D. (2008) 'Evidence of a shift in the short-run price elasticity of gasoline demand', *Energy Journal*, vol 29, no 1, pp93–114

IPCC (Intergovernmental Panel on Climate Change) (2007) 'Climate models and their evaluation', Randall, D.A., Wood, R. A., Bony, S., Colman, R., Fichefet, T., Fyfe, J., Kattsov, V., Pitman, A., Shukla, J., Srinivasan, J., Stouffer, R. J., Sumi, A. and Taylor, K. E., in *Climate Change 2007: The Physical Science Basis*, contribution of Working Group I to the Fourth Assessment Report of the Intergovernmental Panel on Climate Change, S. Solomon, D. Qin, M. Manning, Z. Chen, M. Marquis, K. B. Averyt, M. Tignor and H. L. Miller (eds), Cambridge, London and New York: Cambridge University Press

Jones, C. I. (1995) 'R&D-based models of economic growth', *Journal of Political Economy*, vol 103, pp759–784

Jones, C. I. and Williams, J. C. (2000) 'Too much of a good thing? The economics of investment in R&D', *Journal of Economic Growth*, vol 5, pp65–85

Kalkuhl, M. and Edenhofer, O. (2010) 'Prices vs. quantities and the intertemporal dynamics of the climate rent', CESifo Working Paper Series no 3044, Munich: CESifo

Kling, C. and Rubin, J. (1997) 'Bankable permits for the control of environmental pollution', *Journal of Public Economics*, vol 64, pp101–115

Kriegler, E. and Bruckner, T. (2004) 'Sensitivity analysis of emissions corridors for the 21st century', *Climatic Change*, vol 66, no 3, pp345–387

Leiby, P. and Rubin, J. (2001) 'Intertemporal permit trading for the control of greenhouse gas emissions', *Environmental & Resource Economics*, vol 19, pp229–256

Lenton, T. M., Held, H., Kriegler, E., Hall, J. W., Lucht, W., Rahmstorf, S. and Schellnhuber, H. J. (2008) 'Tipping elements in the Earth's climate system', *PNAS*, vol 105, pp1786–1793

Matthews, H. D. and Caldeira, K. (2008) 'Stabilizing climate requires near-zero emissions', *Geophysical Research Letters*, vol 35, L04705

Misiolek, W. S. (1980) 'Effluent taxation in monopoly markets', *Journal of Environmental Economics and Management*, vol 7, pp103–107

Newell, R. G. and Pizer, W. A. (1999) 'Regulating stock externalities under uncertainty', RFF Discussion Paper 99-10 (Revised), Washington, DC: Resources for the Future

Newell, R. G. and Pizer, W. A. (2003) 'Regulating stock externalities under uncertainty', *Journal of Environmental Economics and Management*, vol 45, no 2, pp416–432

Nordhaus, W. D. (1994) 'Expert opinion on climate change', *American Scientist*, vol 82, pp45–51

Nordhaus, W. D. (2008) *A Question of Balance: Economic Modeling of Global Warming*, New Haven, CT: Yale University Press

Pizer, W. A. (2002) 'Combining price and quantity controls to mitigate global climate change', *J. Pub. Econom. Management*, vol 85, no 3, pp409–434

Popp, D. (2004) ENTICE: Endogenous Technological Change in the DICE model of global warming', *Journal of Environmental Economics and Management*, vol 48, pp742–768

Portney, P. R. and Weyant, J. P. (eds) (1999) 'Discounting and intergenerational equity', Washington, DC: Resources for the Future

Romer, P. M. (1986). 'Increasing returns and long-run growth', *The Journal of Political Economy*, vol 94, pp1002–1037

Roughgarden, T. and Schneider, S. H. (1999) 'Climate change policy: Quantifying uncertainties for damages and optimal carbon taxes', *Energy Policy*, vol 27, pp415–429

Rubin, J. D. (1996) 'A model of intertemporal emission trading, banking, and borrowing', *Journal of Environmental Economics and Management*, vol 31, pp269–286

Rupp, S. and Bailey, I. (2003) 'New environmental policy instruments and German industry', Working paper, Plymouth: University of Plymouth

Sijm J. P. M., Kuik, O. J., Patel, M., Oikonomou, V., Worrell, E., Lako, P., Annevelink, E., Nabuurs, G. J. and Elbersen, H. W. (2004) 'Spillovers of climate change', Report no. 500036002, Utrecht: Netherlands Research Programme on Climate Change

Sinn, H. W. (2008) 'Public policies against global warming: A supply side approach', *International Tax and Public Finance*, vol 15, pp360–394

Speck, S. (2007a) 'Differences in ETR between CEEC and Germany / UK', PETre working paper, London: Anglo-German Foundation

Speck, S. (2007b) 'Overview of environmental tax reforms in EU member states', in Andersen, M. S., Barker, T., Christie, E., Ekins, P., Fitz Gerald, J., Jilkova, J., Junankar, J., Landesmann, M., Pollitt, H., Salmons, R., Scott, S. and Speck, S. (eds) *Competitiveness Effects of Environmental Tax Reforms (COMETR)*, final report to the European Commission, DG Research and DG TAXUD, National Environmental Research Institute, University of Aarhus, pp19–83

Speck, S., Andersen, M. S., Nielsen, H. O., Ryelund, A. and Smith, C. (2006) *The Use of Economic Instruments in Nordic and Baltic Environmental Policy 2001–2005*, TemaNord, 525, National Environmental Research Institute, Denmark

Stavins, R. N. (1996) 'Correlated uncertainty and policy instrument choice', *Journal of Environmental Economics and Management*, vol 30, no 2, pp218–232

Stern, N. (2006) *The Economics of Climate Change: The Stern Review*, Cambridge, UK: Cambridge University Press

Stern, N. (2008) 'The economics of climate change', *American Economic Review, Papers & Proceedings*, vol 98, no 2, pp1–37

Sterner, T. (2003) *Policy Instruments for Road Transportation. Policy Instruments for Environmental and Natural Resource Management*, Washington, DC: Resources for the Future Press

Sterner, T. and Persson, U. M. (2008) 'An even Sterner review: Introducing relative prices into the discounting debate', *Rev Environ Econ Policy*, vol 2, pp61–76

WBGU (German Advisory Council on Global Change) (2006) *The Future Oceans – Warming Up, Rising High, Turning Sour*, Special Report, Berlin: WBGU

Weitzman, M. L. (1974) 'Prices vs. quantities', *Review of Economic Studies*, vol 41, no 4, pp477–491

Weitzman, M. L. (1998). 'Why the far-distant future should be discounted at its lowest possible rate', *Journal of Environmental Economics and Management*, vol 36, pp201–208

Weyant, J. P. and Hill, J. (1999) 'Introduction and overview', in 'The costs of the Kyoto Protocol: A multi-model evaluation', Special Issue, *The Energy Journal*, vol 20, pp8–44

7

Exploring Carbon Tax in China

Cao Jing

Introduction

Climate change as attributed to increased concentrations of GHGs may have significant environmental and economic impacts on the welfare of the entirety of the human race and world society. The impacts of such may be unequally distributed while primarily and most severely affecting developing countries (Cline, 2007). While there is much existing literature suggesting China may have agriculture gains from climate change, more recent studies such as Cline and those made by Chinese academics and institutions point to the opposite. For instance, the latter show how the estimated effects on agriculture output may be significantly adverse. This is particularly true if carbon fertilization does not materialize in practice.

The explosive economic growth of the past three decades and massive urbanization have seriously exacerbated China's huge energy demands while increasing the exigency for environmental protection. As in many other countries, finding solutions that adequately address these energy and climate issues, without compromising economic development goals and the welfare of the population, is a compound problem of the utmost importance. In this light, the Chinese government has already put forth concrete efforts to address climate change issues domestically, and installed an extensive body of regulations and policies aimed at improving energy efficiency and sustainable development. For example, the Chinese government promulgated an aggressive 20 per cent energy intensity reduction target under its 11th Five-Year Plan, installed a portfolio of environmental laws for renewable energy promotion, and committed to generating at least 10 per cent of its electricity from renewables by 2010. However, with the country's very high dependence on coal and a currently imbalanced economic structure,[1] it will be a long tough road before China may be able to make the full transition to a low-carbon economy.

At present, China is facing particular political pressure to make binding commitments to more comprehensive and tangible environmental goals. With the recent US cap-and-trade plan and the looming climate change summit in Cancún, it is of the essence that China formulates a definite set of climate policies and actions regarding the nation's transition to a low-carbon economy while at the same time reconciling the sustainability of economic growth and social welfare.

Before discussing the specific policies the Chinese government ought to consider, it is necessary to address four prominent features of the Chinese energy and economic structures. First, China is highly dependent on coal. Though renewables have been taking root at a faster pace in recent years, their relative share in the energy structure remains small. Second, there has been an overall declining trend in the nation's energy intensity, however, this trend has slightly reversed or flattened out in recent years. Third, China has an imbalanced economy that is heavily reliant on industrial sectors while having an immature service sector. Lastly, an increase in dependence on automobiles and a heavier reliance on refined oil may significantly reshape the future energy structure if Chinese people follow the same lifestyle pattern as developed countries in their industrialization processes.

Total energy consumption by source from 1980 to 2007 is presented in Figure 7.1. China is currently the world's largest producer of coal – the most carbon-intensive fossil fuel. Coal supplied over 71–76 per cent of the country's commercial energy consumption from 1980 until 1997; this declined to 66–69 per cent over the last decade, while the shares of oil and natural gas rose with the rapidly increasing demand for private automobiles and for domestic heating and cooking in urban centres. Even so, due to a legacy of tremendous coal dependence and investment in recently built coal-fired power plants, coal use is expected to double in less than 20 years.[2] IEA projections estimate that China needs to add more than 1300GW to its electricity-generating capacity. This figure is greater than the present total capacity in the US. Meanwhile, coal remains the dominant energy for future power generation (IEA, 2007).

Though China's enormous economic growth has been accompanied by a fast decline in energy efficiency improvement (see Figure 7.2), the level of energy intensity has been somewhat flat since 2000 and even showed a slight increase between 2003 and 2005. The absolute values of energy and GHG intensities are still higher than in most developed countries. For example, China's GHG intensity for 2005 was 3.6t of CO_2 equivalent per $1000 GDP, which is roughly seven times that of Germany and five times that of the US for the same year (McKinsey & Company, 2009). Current energy intensity decomposition literature suggests that the most important factors in the decline of energy intensity before 2000 were rooted in technological change.

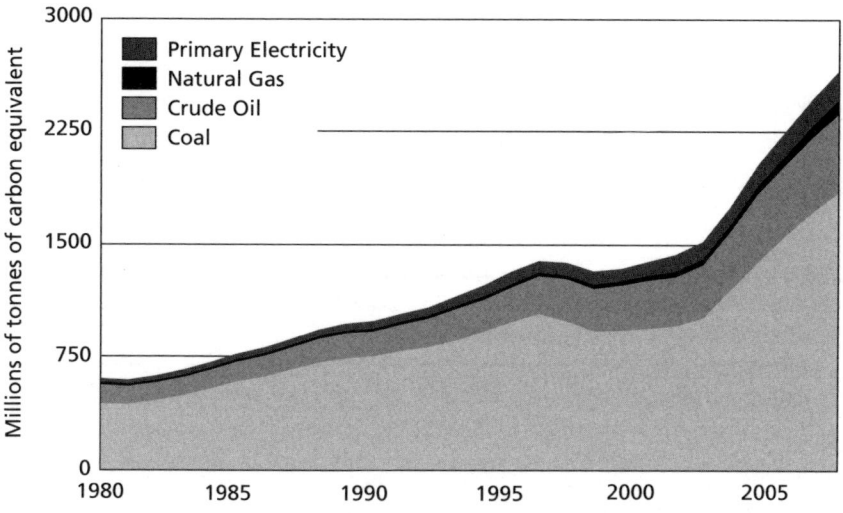

Source: LBNL (2008)

Figure 7.1 *Primary energy consumption by fuel source, 1980–2007*

There is some disagreement over the role of economic structural change. Many find that structural changes only played minor roles in reducing energy intensity. Garbaccio et al (1999) and Ma and Stern (2007) even found that structural change actually increased energy intensity. The flattening-out of energy intensity in recent years may reflect a boom in the real estate and steel industries; however, it is still a mystery why this fast decline has not been a

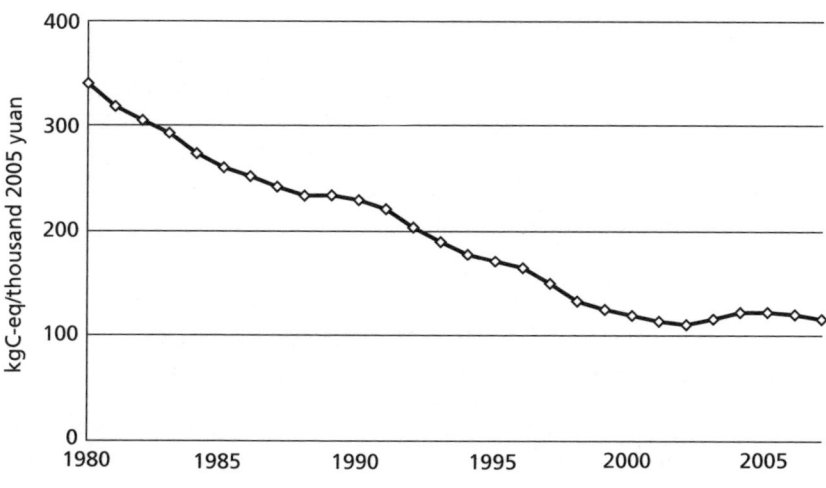

Source: LBNL (2008)

Figure 7.2 *Energy consumption per unit GDP, 1980–2007*

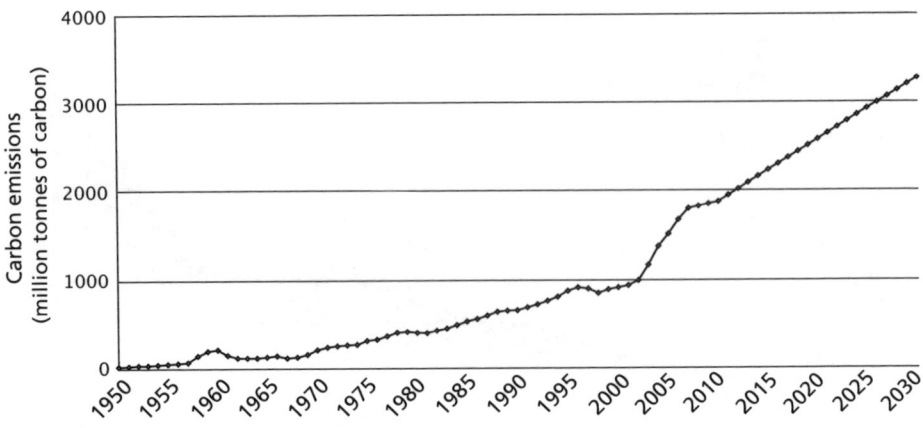

Source: EIA (2008)

Figure 7.3 *CO_2 emissions in China: historical emissions and projections, 1950–2030*

continuing trend. Questions also remain as to how this change may affect future forecasts on energy use and consumption even with the same GDP growth forecasts.

Many researchers have projected China's future CO_2 emissions. Figure 7.3 shows a projected annual growth rate for 2005–2030 of about 3.3 per cent based on our Chinese Computational General Equilibrium (CGE) model, with respect to an annual GDP growth rate at 6.4 per cent. We can see that even with substantial energy intensity improvement from 2005 to 2030, absolute emissions will still more than double by 2030. This will be the equivalent to roughly 28.4 per cent of total global emissions.

In Table 7.1, we provide a summary of different projections from previous studies. The estimate of Cao et al (2009) is roughly the median value between the lower limit estimated by Yang and Schneider (1997) and the upper limit by Auffhammer and Carson (2008), and similar to ERI (2004), Jiang and Hu (2006) and Fridley (2006). Of all the projections, Auffhammer and Carson (2008) obviously propose the most rapid growth, which even exceeds the GDP growth rate.

An answer to global climate change: Carbon tax versus cap-and-trade in China?

In 2007 China issued a white paper on climate change, highlighting the use of economic incentive-based instruments to foster activities for carbon abatement. However, how to formulate a detailed design of appropriate climate policy has yet to be cogently addressed. In particular there still needs to be a discussion of what kind of climate policies, such as price- or quantity-based

Table 7.1 *Comparison of carbon emissions growth rates*

	Date begun	Horizon end	Annual growth rate (%)
EIA (2008)	2005	2030	3.3
Cao et al (2009)	2002	2030	5.03
Auffhammer and Carson (2008)	2000	2010	11.05–11.88
Yang and Schneider (1997)	2000	2025	1.93–3.10
IPCC (2000)	2000	2010	2.58–4.82
ERI (2004)	2000	2010	4.18
Fridley (2006)	2000	2010	5.00–5.02
Jiang and Hu (2006)	2000	2010	4.2

instruments, should be implemented considering both efficiency and political economic criteria. In this section, we discuss the design of China's carbon pricing policies and focus on the debate between a future carbon tax or a cap-and-trade policy in China.

Criteria for policy design

Policy design is a complex process. It must not only be considered within economic, technical and social contexts but also determined by which criteria the policy-maker values the highest. For policies set to tackle climate change, the instrument(s) chosen by an individual country will vary from other countries that have their own characteristics. For instance, a certain country may already have an environmentally related tax on consumption or energy, or it may already host a mature emissions trading market, making it very different from other states. Therefore, a central task for any country addressing climate change is to select and define the appropriate policy instrument(s) for the abatement of GHG emissions. Additionally, the criteria on which such an instrument(s) will be judged must also be considered:

- *Efficiency or cost-effective criteria* – for economists, the efficiency of any instrument plays the most important role in theory. In reality, benefits are difficult to measure, particularly environmental and climate benefits. Thus, cost-effective criteria become more practical a choice in crafting realistic policies. It is very important to choose instruments that are capable of achieving targets without unnecessarily wasting resources.
- *Enforcement and penalty design* – free-riding is a big problem for countries trying to deal with climate change issues in a coordinated manner. Therefore, policy designs on enforcement, in particular penalty mechanisms, should be designed to provide sufficient incentives for the prevention of carbon leakage across countries and policy failure within individual countries. For example, many command-and-control technology mandates have been implemented in China and other countries for decades. However, it is often observed that though firms abide by these mandates, policy failures often exist, for example, many Chinese plants in

reality often leave their end-of-pipe equipment idle and only operate it when they are being inspected.
*   *The political economy* – finally, political economic perspectives must be taken into account when setting climate change policy. Indeed, the problem of climate change is a global public goods issue. Therefore, compliance is the key to the success of any climate policy. In this light, many practical issues need to be considered such as political feasibility, interest group effects, distribution effects and tax incidences, coordination of local- and national-level governance and so forth.

To mitigate the future climate crisis, each country needs to consider, from its own national perspective, what are effective and efficient policy instruments appropriate to its own situation. Specifically, each nation needs to consider whether it should opt for a price-based instrument such as a national carbon tax or a quantity-based instrument such as a cap-and-trade emission permit policy.

The current debate: Carbon tax or cap-and-trade?

For issues that fall under the umbrella of global public goods such as climate change, there are three potential policy/instrument approaches: command-and-control policies; price-based instruments such as carbon taxes; and quantity-based instruments such as cap-and-trade. In general, when considering economic efficiency and incentives on technological change, the last two options are more cost-effective than command-and-control policies (though in some special cases, command-and-control policies can be more efficient; see for example Cole and Grossman, 1999).

In the classical debate between price-based and quantity-based regimes, Weitzman (1974) highlights uncertainty properties as central determinants of climate change policy. As is well known, if the marginal benefit curve is flatter relative to the marginal cost curve, then price-type regulations such as carbon taxes are more efficient and vice versa. When putting the static Weitzman framework into a dynamic extension, carbon abatement benefits become related to the stock of GHGs, which are only slightly influenced by the emissions in one period of time, while the costs are directly related to the carbon flow in this period. Thus, the marginal costs of carbon abatement are very heterogeneous: a total of 3Gt CO_2-eq can be abated for negative or zero cost; but for much larger abatement, the price rises up to 50, 100 or even more $/t CO_2-eq, as the recent McKinsey & Company China report suggests (2009). This implies that price-based instruments will be favoured, as much literature suggests that the climate damage function tends to be linear (Nordhaus, 2006). Edenhofer et al (2008) point out that tipping points may exist with a strong increase in damage costs when climate change becomes worse. Such a non-linear damage function might shift policy from those favouring taxes to those employing quantity instruments half a century from now or perhaps later.

Newell and Pizer (2003) compare the efficiency of taxes and quantity regulations using numerical models. In general, they suggest that tax-based instruments dominate quantity-based regulations. Still, within their parameter choices, preferences shift from tax to tradable permits in later periods. With no experience with tradable permits in China, we believe that a certain amount of time will be required to learn how to build a working permit market. In the short run, then, a carbon tax will be the domestically favoured option in China.

As Cooper (2008) points out regarding developing countries, a cap-and-trade scheme might be more politically difficult than a carbon tax since there may be resistance to any absolute emissions ceiling over the short term in developing states where immediate economic growth is most probably an overriding concern. Cooper (2008) also pays attention to the possibility that a cap-and-trade scheme is likely to foster rampant corruption in developing countries when allocating emission permits to domestic firms or residents. This notion is partly supported by Anger et al (2008) in their empirical study of the first trading phase of the EU's ETS and of cross-sectional data on German firms. Their study found that industries represented by relatively powerful lobbying groups face lower regulatory burdens, thus suggesting that governments consider the preferences of sectoral interest groups when allocating emission permits.

When considering China's domestic situation and characteristics, we find a carbon tax to be more politically feasible and superior to a carbon cap-and-trade system. Currently, the Ministry of Finance is interested in reforming the resource tax; perhaps looking to do so in the short term. In the near future, this tax might gradually be converted to an energy tax or carbon tax. Given this, a carbon tax may match well with China's fiscal reform at a time when there is an increasing exigency for environmentally conscious sustainable growth. In particular, carbon tax revenue can be recycled to mitigate the current distortions within China's fiscal structure, for instance, improving the efficiency of capital and value-added taxes, removing highly distorted subsidies on energy prices, further reforming electricity deregulation and so forth. Although an auction-based cap-and-trade system for carbon is likely to act as a hidden tax, it seems politically difficult to implement for the time being.

Beijing is currently facing political pressures from developed and some developing countries, particularly after the Obama administration introduced the new US climate cap-and-trade policy regime. Domestically, China has mobilized CN¥4 trillion for infrastructure stimulus spending to alleviate some of the worst of the global financial crisis. However, such measures may very well come at the expense of rapidly increased energy use and carbon emissions if investments in manufacturing sectors are still the major stimulus. In this light, a carbon-charging policy may be a solution that tempers the climate externalities of the stimulus spending.

While end-of-pipe technologies may be the most effective in controlling single pollutants such as sulphur emissions, which can be removed by flue gas desulphurization, such equipment is rarely utilized though it has been installed in most power plants throughout the country, as mentioned above. Therefore, an integrated strategy aiming at raising energy prices may very well be the most effective way to control GHG emissions and reduce residual conventional pollutants. Additionally, even if this policy cannot be implemented immediately, the formulation and promulgation of the plan and a timetable will provide incentives and fair warning for households and firms to make adjustments to their behaviour. By making policy commitments to ameliorate climate change, China will be better able to engage in global climate debate.

A Chinese carbon tax: Economic effects and ancillary environmental benefits

Numerous studies have suggested that even a very modest carbon tax can bring about substantial carbon emission reductions in China. For example, Cao et al (2008) used an integrated economic CGE model and an electricity sector engineering model to show that a carbon tax that increased the price of coal by 14 per cent could collect 2.0–2.4 per cent of total tax revenue, and reduce carbon emissions by almost 10 per cent. A later presentation at the CICERO-Tsinghua Workshop of the 2002 base year CGE model suggested that a $10–20 carbon tax could reduce carbon emissions by 9–16 per cent (Cao, 2008).

A carbon charge will affect emissions from two sources. First, in industries that use coal as heavy inputs such as in the power sector, a tax stimulus might shift energy input from coal to oil, natural gas or renewable energy sources. In this way, the composition of energy use will move towards a less carbon-dependent energy structure. This new structure will become more prominent in the long run as consumption behaviours are deeply embedded in the current economic structure and the related elasticities have little response to price increases. Second, when facing a higher price for electricity or other energy-intensive goods, households will change their behaviour to prefer more low-carbon commodities, or will be more likely to take measures such as better insulating their homes. Energy-intensive sectors will, therefore, shrink in response to changes in demand.

A carbon tax regime will not only work towards the abatement of carbon emissions, but also have substantial effects on the macroeconomy, as well changing energy prices and rebalancing economic structures and energy mix structures. In addition, there are varieties of revenue recycling regime that will affect revenues and thus may ultimately affect growth, investment and consumption, as well as potentially impact on both the domestic and international distributions of income and commodities.

Under reasonably set carbon tax rates, we applied a perfect foresight dynamic Chinese CGE model to examine the effects of carbon tax on the Chinese economy, and our simulation analysis shows that the macroeconomic costs are actually very small. For example, when increasing the carbon tax incrementally from $0 to $10/tC, our results suggest that a double dividend result may occur when the tax rates are low and distortions are less than other pre-existing taxes. Therefore, swapping a carbon tax with other taxes will improve economic efficiency. Tables 7.2 and 7.3 show the impact of the unit carbon tax on carbon emissions, public health, energy use, consumption, investment, real GDP and the ratio of carbon tax revenue to the total tax revenue in the 1st and 40th years of the simulations. Compared to the base case, the changes in consumption, utility and real GDP in the simulations are positive. Although there might be a negative shock on investment in the first year, the impact on investments will be positive in the long run if the carbon tax rates are lower than $8/tC.

It should be noted that in this model Chinese labour supply is held to be fairly rigid. This is a plausible assumption, if considering the aggregate labour market without differentiation in the urban and rural markets. Under the assumption of an inelastic enter–exit type of labour supply, the tax interaction effects are negligible in our counterfactual simulations. One possible reason behind China's double dividend is that China's capital tax is highly distorting (the sales tax and value-added tax are small but still distorting) so that imposing a less distorting environmental tax policy can partially offset these tax distortions and improve economic performance as a whole.

While the benefits of an improved climate would ideally be enough of an incentive for the implementation of a carbon tax, it is also widely acknowledged that the CBA regarding climate change is perhaps the most complex issue facing our global society. Even if the costs of mitigating or solving climate change may be estimated, the long-term outlook is muddled at best. Coupled with tremendous heterogeneity in greenhouse emissions sources and the variety of potential climate impacts on different areas and peoples, the picture becomes even more clouded.

Even if the abatement costs of climate mitigation can be estimated, it is still difficult to address uncertain long-term climate benefits. The matter is further complicated by the heterogeneity of GHG sources and the diversity of governments, peoples and geographies within even a single country such as China. In addressing potential climate mitigation policy, it has been acknowledged that the ancillary costs and benefits of reducing GHGs have important policy implications, since these ancillary benefits are relatively *short term* and most of the benefits are *local*, i.e. 'they affect the communities relatively close to the source of the policy or program' (Krupnick et al, 2000). Therefore, the fact that the ancillary benefits of climate change mitigation may offset the costs in the short term and locally is an especially important aspect for increasing political support for climate change policies within developing nations such as China.

Regarding ancillary benefit estimation in China, Cao (2007), Cao et el (2008) and Ho and Jorgenson (2007) suggest that both a carbon tax and a fuel tax (imposed on fuel use and charged not on carbon content but as 30 or 40 per cent of marginal health damage from fossil-fuel combusted pollutants) would substantially reduce conventional pollutants, such as particulate matter and sulphur dioxide, and would thus bring about significant health benefits by abating global climate change. Table 7.4 shows that the ancillary benefits are quite substantial even at a low 14 per cent ad-valorem carbon tax.[3] The tax would bring in ¥14–37 billion ancillary benefits, roughly 0.10–0.18 per cent of China's GDP.

Unlike a Pigovian tax in an ideal world, where an environmental tax is the only tax, the carbon tax policy we propose is actually a second-best tax coexisting with other pre-existing taxes. When economic efficiency is considered, we recommend using a revenue recycling regime to swap a carbon tax with other pre-existing more distorted taxes. Through revenue recycling, energy-intensive firms may continue to operate instead of shutting down, as may be the case if there were no decrease in capital or value-added taxes. Furthermore, incentives to reduce carbon emissions will not be undermined at the margin due to increased energy prices. Rather these would serve as signals to firms and households to shift their preferences towards less carbon-intensive production and commodities.

When considering the climate and environmental benefits, a carbon tax on fossil fuels is indeed a very effective tool. What's more, it comes at only a modest cost to consumption and GDP growth. Though there will be short-term costs and consequences due to necessary adjustments in labour and some industries, these are easily outweighed by the substantial improvement in public health over the short and long terms.

Regarding distribution effects over industry sectors, a carbon tax will inevitably increase the price of fossil fuels and indirectly raise the price of energy-intensive goods and services. Thus, the price changes will disproportionately affect industries based on their respective demands for specific energy inputs. Based on a Ramsey-type perfect-foresight CGE model, we examined the sectoral distribution effects of a carbon tax at roughly $9/tC. This level may achieve a 20 per cent reduction of health damages compared to the BAU scenario at the benchmark year of 2000. The simulation results are shown in Figure 7.4. Coal mining and processing, gas production and supply, and other energy-intensive sectors will be negatively impacted while light industries such as food, textiles, and apparel and leather will experience the opposite. The results also suggest that environmental tax reform would stimulate the industrial structure to shift from heavy to light industry. This would help to correct China's current distorted manufacturing structure.

When considering household tax implication, the main burden of a carbon tax will clearly fall on those in the coal industry. This problem may be easily addressed through income compensation and fiscal rebate policies. Therefore, negative short-term transitional effects *can* be mitigated. While carbon and/or other energy-related taxes such as a gasoline tax may have regressive distribution effects on certain income classes, both empirical and numerical studies have shown that this is not an absolute or universal rule for all cases. For example, Sterner (2008) shows that gasoline taxes in South Africa and India are likely to be progressive as only rich people can afford automobiles. This also holds for other developing nations.

Table 7.2 Effects of carbon tax in China (in the first year) (%)

Tax rate	Carbon emission	SO_2 emission	PM_{10} emission	NO_x emission	Health damage	Energy use	Consumption	Invest-ment	Utility	Real GDP	Carbon tax revenue/total revenue
t = $1/tC	−0.62	−2.13	−1.45	−0.17	−2.19	−0.34	0.01	−0.01	0.01	0.01	0.35
t = $3/tC	−1.82	−6.07	−4.14	−0.52	−6.24	−1.00	0.03	−0.04	0.03	0.01	1.00
t = $5/tC	−2.96	−9.64	−6.59	−0.86	−9.92	−1.63	0.05	−0.08	0.05	0.01	1.59
t = $8/tC	−4.60	−14.42	−9.86	−1.37	−14.83	−2.52	0.07	−0.16	0.07	0.01	2.40
t = $10/tC	−5.64	−17.28	−11.82	−1.70	−17.77	−3.08	0.07	−0.23	0.07	0.00	2.89

Source: Cao (2007)

Table 7.3 Effects of carbon tax in China (in the 40th year) (%)

Tax rate	Carbon emission	SO_2 emission	PM_{10} emission	NO_x emission	Health damage	Energy use	Consumption	Invest-ment	Utility	Real GDP	Carbon tax revenue/total revenue
t = $1/tC	−2.01	−6.30	−3.95	−0.57	−6.60	−0.86	0.04	0.03	0.04	0.03	0.66
t = $3/tC	−5.76	−16.36	−10.27	−1.69	−17.17	−2.41	0.07	0.05	0.07	0.06	1.82
t = $5/tC	−9.19	−24.07	−15.13	−2.78	−25.27	−3.77	0.09	0.04	0.09	0.06	2.77
t = $8/tC	−13.84	−32.76	−20.64	−4.35	−34.43	−5.56	0.07	−0.02	0.07	0.04	3.96
t = $10/tC	−16.67	−37.25	−23.51	−5.36	−39.18	−6.62	0.05	−0.07	0.05	0.01	4.64

Source: Cao (2007)

Table 7.4 *Ancillary benefits of national-level carbon tax policy*

	Carbon emission reduction (million tC)	Ancillary benefits (¥ billion 2000)	Total ancillary benefits (¥/tC)	Ancillary benefits/ real GDP (%)
2000	73.1	14.5	198.2	0.158
2001	79.5	16.3	205.1	0.164
2002	82.9	18.4	222.1	0.17
2003	85.4	20.5	240.1	0.175
2004	87.4	22.5	256.9	0.177
2005	89.1	24.4	274.5	0.179
2006	90.2	26.4	293.1	0.18
2007	90.7	28.3	312.2	0.18
2008	90.6	30.0	330.7	0.178
2009	89.8	31.4	349.4	0.175
2010	88.9	33.0	371.6	0.172
2011	87.2	34.4	395.1	0.169
2012	84.9	35.5	418.6	0.164
2013	82.0	36.4	443.6	0.159
2014	78.5	36.9	470.7	0.153
2015	74.4	37.3	500.6	0.146
2016	69.7	37.2	534.1	0.139
2017	64.5	36.8	570.9	0.131
2018	58.8	36.1	614.7	0.123
2019	52.5	35.0	667.8	0.114
2020	45.8	34.0	740.8	0.105

Note: The carbon tax is imposed at roughly $9/tC, which can achieve a 20 per cent reduction in health damages compared to the BAU scenario.
Source: Cao et al (2008)

In fact, even for developed countries, a portion of a gasoline tax may well be progressive. For instance, West (2004) concludes that a gasoline tax is actually progressive across lower-income groups and only regressive for upper-income groups. The reason for this is that lower-income households do not own automobiles in the first place while having a greater responsiveness to gasoline price changes. Additionally, these progressive and regressive results significantly depend on how the tax revenue is utilized. For example, West and Williams (2004) suggest that a gasoline tax may easily be made progressive by using the resulting revenue to finance cuts in regressive taxes.

Metcalf (2007) calculates how a $15 carbon tax would affect retail prices and how such a burden can be transferred to households of different income brackets. The results suggest that such a tax is mildly regressive, yet this may be reduced by flat income tax rebates for all workers and social security recipients. In this way, the rebate can adjust the tax incidence and change a mildly regressive tax to a mildly progressive one. Therefore, tax incidences are in all actuality an empirical matter with the extent to which a gasoline tax is either regressive or progressive fully depending on its design and its relationship with pre-existing taxes. In principle, then, tax incidences may be mitigated by different types of compensation regimes.

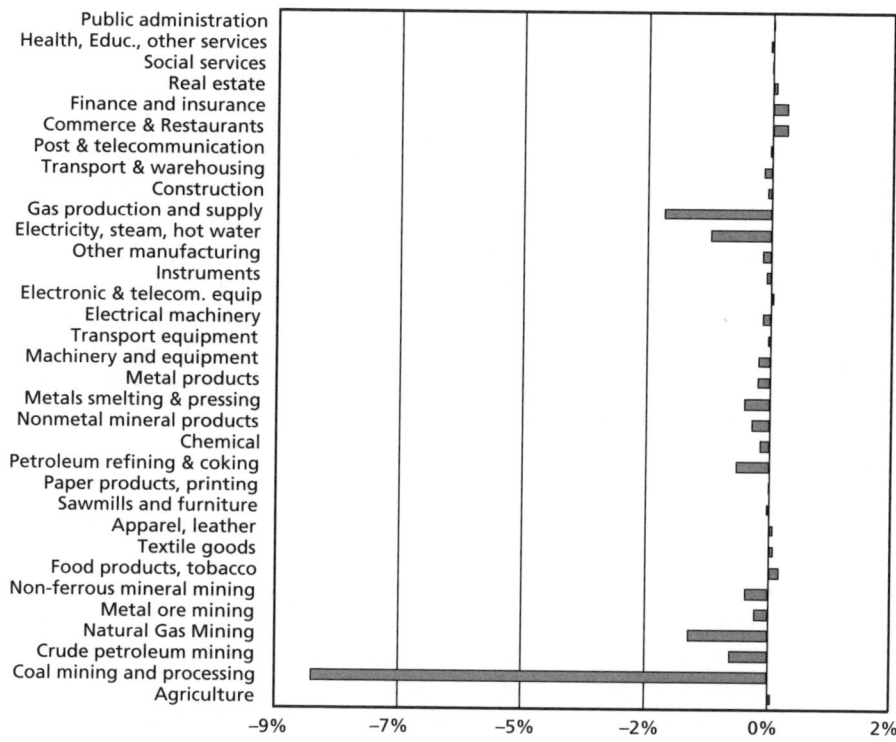

Source: Cao (2007)

Figure 7.4 *Sectoral distribution effects of carbon tax (first year) (20 per cent reduction in environmental health damage)*

Practical issues that need to be addressed

Before the implementation of a national-level carbon tax in China, practical issues must be taken into consideration that address tax bases and rates; potential schemes for tax recycling, enforcement, and review; and reconciliation with international regulations and the transnational cap-and-trade market.

Ideally, a carbon tax should cover all significant GHGs including methane and nitrous oxide. However, practical difficulties often arise in levying a charge on methane produced in the agricultural sector as well as in the measurement of non-CO_2 GHGs (Cooper, 2008). Therefore, it is suggested that initially only fossil fuels are taxed by their carbon content. This would compel the reform of China's current resource tax and significantly raise tax rates to constrain increasing carbon emissions in the country. Though Cooper suggests imposing an across-the-board tax on all GHG emissions worldwide, we must settle for a second-rate solution that takes political considerations into account. Therefore, a lower and less extensive carbon tax is more likely to be accepted by, and implemented in, China in the immediate term with gradual increases over time for the rectification of the country's imbalanced economic structure.

Once the nation's growth rates and welfare levels converge to those of developed nations, the relevant tax rates may be raised to the levels of the latter.

Within the Chinese fiscal system, a carbon tax may interact with other pre-existing taxes imposed on fossil-fuel use such as resource taxes and consumption taxes on gasoline and kerosene. Under current policy, the purposes of these pre-existing taxes are clearly not to constrain carbon emissions. The problem is that current tax rates are too low to achieve meaningful carbon abatement in China. For example, when the domestic oil price was around ¥3000/t in 2008, the resource tax was around ¥14–30/t or only 0.5–1.0 per cent. The consumption tax on gasoline and kerosene was also very low in 2008: only ¥0.2 per litre for gasoline, and ¥0.1 per litre for kerosene. Since 1 January 2009, Beijing has increased the consumption tax to ¥1 per litre for gasoline and ¥0.8 per litre for kerosene. However, most of this increase has only been used to replace existing road tolls. Thus, the overall burden has not been substantially increased. In addition, coke and other coal products are excluded under current resource and consumption tax categories. Therefore, a new carbon tax needs to integrate the current resource tax and fuel consumption tax, substantially increase the tax rate based on carbon content, and cover all industrial sectors and commodities. Furthermore, deepening the existing resource tax policy is likely to remove the political obstacles to raising energy prices and gradually converge to a carbon tax policy that is not based on resource use itself but on carbon content for GWP in the future.

An initial carbon tax would be set by the national-level Ministry of Finance. It should be high enough to affect behaviour significantly, and should include measures such as increasing the price of coal by at least 10 to 20 per cent. However, the tax should not be so high as to lead to substantially adverse transitional impacts on coal mines and refineries. The tax rate should be subject to periodic review, since it is difficult to know how much the carbon tax will reduce carbon emissions prior to actually implementing it. With trial and error, the carbon tax regime can be adjusted along its experience curve. Over time, it will be possible to learn more about the economics of climate change, its full impact, newly developed and perfected technologies, as well as how firms and households respond to carbon taxes. Thus, a carbon tax regime can be gradually increased and adjusted at fixed time intervals thereafter.

Another issue arises in respect to how the new carbon tax may be integrated with pre-existing taxes such as value-added tax, capital tax, labour tax and, in some cases, subsidies on energy and household income. Cao et al (2008) choose a simple revenue recycling mechanism that proportionally reduces all other pre-existing taxes and obtains positive impacts on the Chinese economy. Their study suggests that when a carbon tax is imposed at an appropriate level, tax swapping can be efficiency improving and reduce distortions within the Chinese fiscal system. This is particularly true as China's capital tax remains distorted. Obviously, even with revenue recycling possibly mitigating macroeconomic contractions, an increase in energy prices may stimulate structural inflation and thereby hurt certain income groups. This issue has become the number one political concern regarding any potential regressive energy tax reform. China's debate on gasoline tax reform has lasted over 15 years and has not led to any concrete policies coming into force. This foreshadows the likelihood of political difficulties of imposing a carbon tax in the future.

To complicate matters further, a carbon tax would increase the relative price of energy-intensive products, which may reduce the international competitiveness of certain commodities in the global market. As previously discussed, even without a domestic carbon tax, energy-intensive products may still face border taxes, in which case it would be impossible for the Chinese government to collect the revenue. In contrast, a domestic carbon tax retains revenue from carbon taxing within China. Additionally, the main idea behind international climate policy is to promote common action for the abatement of carbon emissions. One primary instrument that all nations working for climate protection have implemented, or are planning to implement, is the substantial raising of the price of energy derived from fossil fuels through either a tax or a cap-and-trade regime. Therefore, if China raises its own energy taxes, it will make it far easier for other countries to do the same without fear of lost competitiveness to Chinese industry. Also empirical studies have thus far found no evidence to support the notion that China has a comparative advantage in energy-intensive products, other than labour-intensive products. The so-called 'pollution haven' hypothesis is not supported by empirical evidence (Wenhua, 2004).

In this chapter we have focused on carbon tax policy. It should be noted that in practice carbon tax reform should be applied in combination with and consideration of other regulatory policies and instruments. An example of this would be a hybrid policy based on an ETS with a safety valve, as suggested by Portney and Stavins (2000). In this case, if permit prices reach an established ceiling, the government relieves the upward pressure on costs by issuing more permits, which causes prices to fall. Another example used to avoid hot spots is a the tax or permit regime combined with command-and-control ambient mandates for local pollutants.

Finally, it is necessary for policy to take into account the differences between sulphur dioxide and particulate matter. There are currently no economically available end-of-pipe technologies capable of substantially reducing carbon emissions. CCS technologies remain uncertain in their capabilities and financial feasibility for diffusion and expansion, and little is known about their long-term effects and risks. In the future we do not know whether or not CCS or renewable technologies can be adapted to face the challenges of climate change so as to solve the entire climate change problem. But what we do know is that the revenue from a carbon tax can stimulate increased R&D in these important areas, increasing our chances to combat climate change without significant welfare losses in the future. In addition, in order to avoid constraints imposed by certain infrastructure choices, it is important to make immediate changes in our lifestyles and energy-use behaviour. These changes will be better adopted and more quickly adopted by signalling the implementation of a carbon tax. Only through immediate efforts can we win the battle against climate change and overcome the challenge of energy security.

Notes

1   China's imbalanced economic structure refers to overcapacity in producing manufacturing commodities, over-saving with limited domestic consumption, and imbalanced trade with other countries.
2   Forecast coal use is based on our China CGE model.
3   The ad-valorem carbon tax roughly corresponds to a unit carbon tax at \$6/tC, using the 2000 ¥–US exchange rate.

References

Anger, N., Böhringer, C. and Oberndorfer, U. (2008) 'The political economy of allowance allocation in the EU emissions trading scheme: Theory and evidence', presented at the 16th European Association of Environmental and Resource Economists, June, Gothenburg, Sweden

Auffhammer, M. and Carson, R. T. (2008) 'Forecasting the path of China's CO_2 emissions using province level information', *Journal of Environmental Economics and Management*, vol 55, pp229–247

Cao, J. (2007) 'A perfect foresight dynamic CGE analysis of environmental tax reform in China', chapter in 'Essays on environmental tax policy analysis: Dynamic computable general equilibrium approaches applied to China', PhD thesis, Cambridge, MA: Harvard University

Cao, J. (2008) 'Costs and benefits of a carbon policy for China', CICERO-Tsinghua Workshop, 15 and 16 October, Beijing: Tsinghua University, www.cicero.uio.no/workshops/china2008/oct15/CAO_Jing.pdf

Cao, J., Ho, M. and Jorgenson, D. (2008) *The Co-Benefits of Mitigating the Global Greenhouse Gases in China – An Integrated Top-Down and Bottom-Up Modeling Analysis*, EfD working paper series, Beijing: EfD

Cao, J., Ho, M. and Jorgenson, D. (2009) 'The local and global benefits of green tax policies in China', in *Review of Environmental Economic and Policy*, vol 3, no 2, pp189–208

Cline, W. (2007) *Global Warming and Agriculture: Impact Estimates by Country*, Washington, DC: Center for Global Development and Peterson Institute for International Economics

Cole, D. and Grossman, P. (1999) 'When is command-and-control efficient: Institutions, technology, and the comparative efficiency of alternative regulatory regimes for environmental protection', *Wisconsin Law Review*, vol 5, pp887–938

Cooper, R. N. (2008) *The Case for Charges on Greenhouse Gas Emissions*, Harvard Project on International Climate Agreements, Discussion Paper 08-10, Cambridge, MA: John F. Kennedy School of Government

Edenhofer, O., Pietzcker, R., Kalkuhl, M. and Kriegler, E. (2008) 'Taxation instruments for reducing greenhouse gas emission, and comparison with quantity instruments', draft paper prepared for the midterm review of the project 'China Economics of

Climate Change', 14–15 December, Beijing

EIA (Energy Information Administration) (2008) *International Energy Outlook 2008*, Washington, DC: US Department of Energy

ERI (Energy Research Institute) (2004) *China's Sustainable Energy Future: Scenarios of Energy and Carbon Emissions*, Technical Report, Berkeley: Lawrence Berkeley National Laboratory

Fridley, D. (2006) *China's Energy Future to 2020*, Technical Report, Berkeley: Lawrence Berkeley National Laboratory

Garbaccio, R. F., Ho, M. S. and Jorgenson, D. W. (1999) 'Why has the energy-output ratio fallen in China?', *Energy Journal*, vol 20, pp63–91

Ho, M. and Jorgenson, D. (2007) 'Policies to control air pollution damages', in Mun, S., Ho, M. and Nielsen, C. (eds) *Clearing the Air: The Health and Economic Damages of Air Pollution in China*, Cambridge, MA: MIT Press

IEA (International Energy Agency) (2007) *World Energy Outlook 2007 – China and India Insights*, Paris: IEA

IPCC (Intergovernmental Panel on Climate Change) (2000) *Emissions Scenarios*, Cambridge, UK, Cambridge University Press

Jiang, K. and Hu, X. (2006) 'Energy demand and emissions in 2030 in China: Scenarios and policy options', *Environmental and Policy Studies*, vol 7, pp233–250

Krupnick, A., Burtraw, D. and Markandya, A. (2000) 'The ancillary benefits and costs of climate change mitigation: A conceptual framework', paper presented at Expert Workshop on the Ancillary Benefits and Costs of Greenhouse Gas Mitigation Strategies, Washington, DC, 27–29 March

LBNL (Lawrence Berkeley National Laboratory) (2008) *China Energy Databook Version 7.0*, CD ROM, October, Berkeley, CA: Lawrence Berkeley National Laboratory

Ma, C. and Stern, D. I. (2007) 'China's changing energy intensity trend: A decomposition analysis', *Energy Economics*, doi:10.1016/j.eneco.2007.05.005

McKinsey & Company (2009) *China's Green Revolution – Prioritizing Technologies to Achieve Energy and Environmental Sustainability*, McKinsey & Company

Metcalf, G. (2007) *A Proposal for a U.S. Carbon Tax Swap: An Equitable Tax Reform to Address Global Climate Change*, Washington, DC: Brookings Institution

Newell, R. and Pizer, W. (2003) 'Regulating stock externalities under uncertainty', *Journal of Environmental Economics and Management*, vol 45, no 2, pp416–432

Nordhaus, W. (2006) 'After Kyoto: Alternative mechanisms to control global warming', *AEA Papers and Proceedings*, vol 96, no 2, pp31–34

Portney, P. and Stavins, N. R. (eds) (2000) *Public Policies for Environmental Protection*, Washington, DC: Resources for the Future

Sterner, T. (2008) 'Gas taxes: Progressive or regressive', paper presented at EFD annual meeting, 3–8 November, Beijing

Weitzman, M. (1974) 'Prices versus quantities', *Review of Economic Studies*, vol 41, no 4, pp477–491

Wenhua, D (2004), 'Pollution abatement cost savings and FDI inflows to polluting sectors in China' chapter in *'Essays in Environmental Economics and Policies'* PhD thesis, Cambridge, MA: Harvard University

West, S. (2004) 'Distribution effects of alternative vehicle pollution control policies', *Journal of Public Economics*, vol 88, pp735–757

West, S. and Williams III, R. (2004) 'Estimates from a consumer demand system: Implications for the incidence of environmental taxes', *Journal of Environmental Economics and Management*, vol 47, pp535–558

Yang, C. and Schneider, S. (1997) 'Global carbon dioxide emissions scenarios: Sensitivity to social and technological factors in three regions', *Mitigation and Adaptation Strategies for Global Change*, vol 2, pp373–404

8

Domestic Emissions Trading Systems

Steffen Brunner, Christian Flachsland,
Gunnar Luderer and Ottmar Edenhofer

Introduction

The EU ETS is the centrepiece in Europe's climate policy architecture and the largest cap-and-trade system in the world. After having failed to introduce a union-wide carbon tax during the 1990s, the European Commission presented a green paper in 2000 that proposed the use of emissions trading. Launched in January 2005, the EU ETS established a uniform carbon price for specific heavy-industry activities in all EU member states. It covers CO_2 emissions from over 10,000 installations, including power and heat generators, oil refineries and factories for ferrous metals, cement, lime, glass and ceramic materials, and pulp and paper. Together, the covered sources account for roughly 40 per cent of total EU GHG emissions.

This chapter reviews the characteristics and experience of domestic emissions trading as an instrument of climate policy. The rationale of emissions trading and its general principles are briefly summarized first. Then we highlight the basic design elements of cap-and trade, and then review the main lessons from the EU ETS. Finally, we summarize the conclusions.

Rationale of emissions trading

At the very heart of emissions trading lies the notion that if 'factors of production are thought of as rights, it becomes easier to understand that the right to do

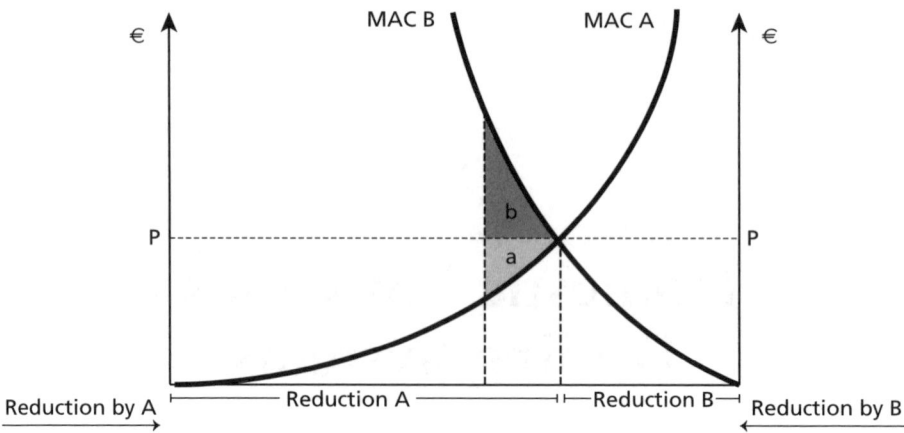

Figure 8.1 *Efficiency-enhancing effect of emissions trading*

something which has a harmful effect ... is also a factor of production' (Coase, 1960). The cost of exercising such a pollution right should be equivalent to the loss that is suffered elsewhere in consequence of this action. Making these rights explicit and transferable enables the market to value and trade them. As a result, emission rights can flow to their highest value use and reductions can occur wherever they are cheapest. Figure 8.1 illustrates this effect. Imagine polluter A with a MAC curve A, and polluter B with a relatively steeper MAC curve B. Given a certain emission limit (left dotted line) and without emissions trading, both polluters would have to reduce emissions until they comply with their individual abatement obligation. Polluter A would face relatively low cost, polluter B relatively high cost for a given quantity. With emissions trading, A will want to increase and B decrease abatement until marginal abatement costs converge at price P. The efficiency gain of A is illustrated by area a, which comprises the benefit from increasing abatement and selling freed up allowances to B. The efficiency gain of B is illustrated by area b, which represents the amount of saved abatement cost net of the money paid to A.

After the groundbreaking work of Coase in the 1960s, scholars quickly began to point out the instrument's usefulness for air pollution control. Tradable emission permits have been successfully used in the US since 1995 to tackle the problem of acid rain by reducing sulphur dioxide and nitrous oxide emissions.

Two broader categories of domestic ETS can be differentiated: systems that limit emissions at relative or at absolute levels (Reinaud and Philibert, 2007).[1] Relative targets can either apply to country or sector/firm level. It is also possible to combine relative targets for some sectors with an absolute target at country level. Relative targets can be based on emission intensity per unit of output or other indexes and have been proposed as a means to address cost concerns as they better encapsulate the notion of decoupling economic growth from emissions (Marschinski and Lecocq, 2006). Ellerman and Sue Wing

(2003) contend that intensity targets can reduce the uncertainties associated with abatement costs under uncertain economic conditions. Pizer (2005) finds that intensity targets may be more appropriate if the short-term objective is to slow, rather than halt, emissions growth.

Systems based on absolute targets cap emissions from covered entities and allocate emission allowances to participants that can then trade them among each other. Under cap-and-trade, compliance is established by comparing actual emissions with the amount of allowances a particular participant holds. Dudek and Golub (2003) argue that absolute caps have more certain environmental results and lower transaction costs for emissions trading, thereby creating stronger incentives for technological change. IPCC (2007a) concludes that absolute targets are more effective than intensity targets if the aim is to meet certain emission reductions, but they may be less effective at addressing costs when the economic outlook is uncertain. The majority of currently existing, announced and proposed schemes apply cap-and-trade and the remainder of this chapter therefore focuses on this type of emissions trading.

Elementary design features of cap-and-trade

In theory, there exists a wide range of different cap-and-trade system designs. Table 8.1 in the appendix to this chapter summarizes the major characteristics of the EU ETS, the Regional Greenhouse Gas Initiative (RGGI), and proposals for systems in California, Australia and New Zealand. While they differ widely in their specific rules, they all build on a limited number of design elements that enable policy-makers to strike a balance between a programme's environmental effectiveness, cost effectiveness, distributional impact and institutional feasibility.

Cap and coverage

The coverage identifies the scope of an ETS in terms of emission sources and GHGs. Emissions sources are usually covered at sector level, where minimum thresholds are defined to exclude very small sources. This can avoid that transaction costs of including additional sources exceed the environmental benefits.[2] The cap defines the maximum level of GHGs these sources may emit during a certain period of time. Several studies have found that full coverage of emissions sources within an ETS is superior to partial coverage because it ensures that marginal costs are equalized across the entire economy. Böhringer and Löschel (2005) find that in the EU significant cost savings would result from economy-wide coverage compared to exclusion of certain sectors. Given the substantial abatement options in different economic sectors such as buildings and transportation (Enkvist et al, 2007), it is important to make these options accessible for the market by covering all relevant sectors by the cap. This can be done upstream in order to minimize transaction cost. The EU ETS, however, excludes buildings and transport from coverage, leaving their abatement potential inaccessible for the trading sector.

Complementary measures such as insulation and fuel efficiency standards should be employed in order to exploit the abatement potential of non-covered sectors. In fact, the presence of negative abatement costs indicates that market failures exist that impede the exploitation of options that would be beneficial even in absence of climate policy. However, it needs to be taken into account that as long as such activities are not subject to a binding constraint on overall emissions, for example through a cap, rebound effects will occur. As higher efficiency reduces the costs of an activity (for example transport or heating), this will lead to an expansion of the activity level. Thus, emission reductions may not be as large as suggested by initial assessments of reduction potentials from efficiency measures that do not take into account rebound effects. Moreover, even if a certain sector previously excluded becomes covered, complementary policies continue to be necessary in order to address the pre-existing market failures indicated by negative abatement costs. Both an absolute cap on emissions and complementary instruments are required to reduce emissions in a cost-efficient manner.

Broad coverage can also widen the potential for carbon market liquidity and efficiency (Baron and Bygrave, 2002). Introducing more sources and sectors to a market can also reduce the impact of economic shocks in any one sector on the trading system as a whole. However, this comes at the expense of partly transferring shocks into other sectors via the permit market. In addition, the larger and more liquid a market, the harder it is for large players to affect the overall price level. However, sectors for which emissions cannot be quantified and monitored with certainty (for example N_2O and CH_4 emissions from agriculture) should be managed with alternative measures and remain excluded from the cap until sound methodologies of measurement and monitoring have been developed. Otherwise, the inclusion of unverifiable emissions could undermine the cap's environmental integrity (Herold, 2008).

Another key aspect in designing ETS is identifying the appropriate incidence of regulation. The point of obligation defines the liable party for surrendering emissions permits. In an upstream design, the point of obligation can for example lie at the level of importers and producers of fossil energy.[3] Small energy-related sources, for example in the transport sector, can be effectively covered by moving the point of obligation upstream. Importers and producers of fossil resources would then pass on the price signal to resource users, who will adapt their economic activities according to this change of input prices. In a downstream design the point of obligation can be the emitting installation or the final consumer of fossil energy. A hybrid system combines both approaches, where some sectors are included upstream and others downstream (Hargrave, 1998).

While CO_2 related to the combustion of fossil fuels accounts for the bulk of man-made emissions, several other sources and compounds also contribute significantly to global warming. Currently, CO_2 release due to deforestation, mostly in tropical countries, accounts for almost 20 per cent of the emissions (IPCC, 2007b). In terms of CO_2 equivalents, emissions of methane and nitrous oxide from agriculture make up 14 per cent of the global total. Agricultural emissions have risen continuously and are particularly difficult to reduce. Other significant sources are nitrous oxide and fluorinated gases from industrial processes. Many of these emissions can be reduced at low costs. A major concern often voiced is that including low-cost abatement opportunities and offset credits (see below) will, in the short term, crowd out other mitigation activities and thus delay the structural energy sector transformation that is necessary for the long-term achievability of ambitious climate stabilization targets. This problem is mitigated by setting appropriately tight short- and medium-term reduction targets in order to ensure sufficient incentives for low-carbon energy investments and inter-temporal cost optimality, while taking advantage of low-cost emission reduction opportunities.

Generally, governments need to be as clear as possible about the future development of cap and coverage, as they are the major determinants for emission prices. For making long-term investments with confidence, market participants require predictability. Clear signals of support for stringent future caps from governments will enhance predictability. This requires in particular the definition of a long-term reduction trajectory. After all, GHGs are stock-pollutants. Their climate effect depends primarily on the long-term emission budget rather than reduction targets for certain points in time.

Importing offset credits

ETS coverage can also be extended by allowing for the import of offset credits. Emission reductions in sectors or countries not covered by the ETS can be certified and redeemed to offset emissions under the cap. The CDM, for example, is the largest offset credit programme in the world with a market value of $13 billion in 2007 (World Bank, 2008). Allowing the import of offsets will lower compliance costs if the marginal costs of creating credits are lower than the allowance price in the cap-and-trade system. This might be sensible for sectors in which emission levels and reductions are difficult to measure for all sources, such as forestry, agriculture and waste. Partially covering these sectors in the ETS might lead to intra-sectoral distortions, whereas allowing them to generate project-based credits can provide incentives for mitigation without generating distortions (Garnaut, 2008). The planned Australian ETS, for instance, allows for domestic forestry offsets in order to provide incentives for this sector before it becomes eventually covered under the scheme. The EU is also planning a new domestic offset mechanism from 2013 onwards (EC, 2008), and the RGGI system also includes a domestic offset credit system (RGGI, 2007).

There are, however, risks connected to the import of credits. First, the excessive reliance on offsets to meet abatement obligations may jeopardize a programme's environmental effectiveness. This concern arises when the additionality of offset credits cannot be fully guaranteed.[4] Offsets are rewards for reductions in emissions measured against an assumed baseline: 'what emissions would have occurred in the absence of the credit mechanism?'.

Determining a counterfactual baseline can be arbitrary to some extent, resulting in the risk that schemes do not promote genuine abatement. Second, the import of relatively cheap abatement options via offset credits drives down allowances prices under the cap, which in turn reduces the financial incentive for investing in domestic emission reductions. This can be a problem if a certain level of domestic abatement and certain incentives for investments are an ETS policy objective. From an environmental point of view, however, it does not matter where GHG emissions are reduced internationally as they equally contribute to global warming. To address these previous concerns, an ETS can limit the amount of credits that can be used. The EU, for example, limited the import of Certified Emission Reductions generated under the CDM into the EU ETS to 13.4 per cent. Limiting the import of credits does, however, not address the problem of additionality (except when limiting the import quota to zero), and can only contain the scale of the problem. Additionality concerns are best addressed by applying rigorous additionality procedures, and/or reforming the credit scheme accordingly. Discounting offset credits is another option to address uncertainty over project additionality.

Allowance allocation

After the cap has been set, the ETS operator issues allowances for units of emissions (for example metric tonnes CO_2[5]) where the total number of permits equals the cap. This step is a contended issue in virtually every cap-and-trade programme as it critically determines the policy's distributional impact. Allowances can either be sold, usually through auction, or allocated for free. The key question is, who receives the scarcity rent created by capping emissions?

In principle, there are two options for free allocation: grandfathering and benchmarking. Grandfathering allowances means that permits are distributed in proportion to past emissions, measured for one or several years. Grandfathering can either be a one-off allocation to existing sources or be regularly updated with new emissions data. Benchmarking constitutes a method where allowances are distributed on the basis of an average or expected performance benchmark. For instance, it can provide fixed allocation quantities based on expected output. Under free allocation, emitters receive the full value of allowances for free and pass on the opportunity costs of allowances to consumers wherever possible. The scarcity rent of allowances is thereby appropriated by the private sector, generating windfall profits for emitters and higher output prices for consumers.

There are a number of reasons why free allocation may distort incentives for reducing emissions (Hepburn et al, 2006). First, the expectation that the baseline year upon which allocations are based will be updated may encourage sources to invest in dirty technology or refrain from investing in clean technology in order to increase or maintain emissions levels and thus receive more free allowances in future. One needs to be cautious to avoid this kind of perverse incentive, for example by choosing a baseline year in the past or by combining baselines with benchmarks. Second, free allocation may present a barrier to market entry. If existing polluters receive allowances for free but new entrants are obliged to buy them, free allocation directly disincentivizes market entry, reducing competition. At the same time, it may present a barrier to market exit. The requirement that an installation must be kept open in order to receive free allowances may prevent the closure of inefficient plants, freezing emissions at higher levels than otherwise necessary. Third, grandfathering leads to increased lobbying activities because emission allowances, which have a monetary value, are given out for free. Lobbying of powerful producer groups may put governments under considerable pressure – a time-consuming and costly procedure.

Auctioning offers several advantages over free allocation. First, it puts upfront costs on polluters. This will tend to enhance management's awareness of carbon cost, leading to more efficient decisions. Second, auctioning enables governments to use revenues to address equity issues through reductions in taxes or other distributions to low-income households. Governments can also use auctioning revenues to invest in the development and deployment of cleaner technologies, or provide finance for other countries' efforts for climate change mitigation and adaptation. This is analogous to the 'double dividend' feature of carbon taxes. Goulder et al (1999) demonstrate that recycling revenues from auctioned allowances can have economy-wide efficiency benefits if they are used to reduce distortionary taxes. Dinan and Rogers (2002) show that free allocation of tradable permits may be regressive because it leads to income transfers towards higher-income groups (i.e. shareholders) at the expense of poorer households (i.e. consumers with high income shares of energy expenditure). Third, Tietenberg (2006) contends that auctioning provides stronger incentives for technological innovation. Under grandfathering some sources are buyers and others sellers. Sellers have an incentive to behave strategically and keep prices high by avoiding technological innovation. In auction markets, all emitters are buyers. Hence, all sources benefit from low-carbon technologies via decreased marginal abatement cost and permit prices.

Allocation methods for sources entering or leaving the scheme represent a contentious issue. New entrants and closure provisions define whether new entrants have to either buy/receive additional allowances from a new entrant's reserve or be obliged to buy permits on the market (Ahman et al, 2006; Ellerman, 2006). Free allocation would be more favourable to encourage new investments, as the obligation to buy allowances represents an additional barrier to market entry and might impede the renewal of facilities. However, if free allocation to new entrants is based on actual emission intensity, it provides a disincentive to install more efficient technology, contrary to the overall aim of carbon pricing. Grubb and Neuhoff (2006) therefore suggest basing free allocation for new entrants on the basis of technology-specific benchmarks.

Overall, auctioning can avoid many of the problems associated with free allocation and offers distinct advantages. But good design is necessary to avoid new inefficiencies. Small frequent auctions may be more effective in limiting the market power of large bidders (Hepburn et al, 2006). They may also encourage learning processes, help players to adjust bids and promote price stability. In contrast, one large auction at the beginning of each trading period may minimize administrative costs. However, it may also enable large polluters to buy the bulk of permits and then use them to extract oligopoly rents on the secondary market. The optimal design for auctions in cap-and-trade systems is subject of ongoing research (Neuhoff and Matthes, 2008).

Banking and borrowing

Economic theory states that the efficiency gains from emissions trading stem from equalizing marginal abatement costs among spatially distributed sources with different abatement cost curves. In addition, however, there is also a temporal dimension to equalizing marginal abatement costs. Due to economic cycles/shocks, resource price hikes, mild winters and so on, allowance prices can be low in one particular period of time and high in another. However, it would be efficient to reduce emissions when this is cheapest. Theoretically, if the cumulative emission budget remains constant and it does not matter in which period emissions are reduced, this will enhance the efficiency of meeting a predefined emissions target. In cap-and-trade systems, temporal flexibility is introduced by banking and borrowing provisions.

Banking means that instead of using an emissions permit at an early date, it is saved for later use because it is anticipated that its value in later periods will be higher than now. As a result, abatement options that are cheaper now than the net present value of a future allowance will be utilized, thus increasing the amount of early abatement relative to the case without banking. Allowance banking can help to equilibrate present-value prices across trading periods. Ellerman and Pontero (2005) analyse the use of banking in the US Acid Rain Program. They find that sources banked 30 per cent of permits in phase one (1995–1999), probably due to the expectation of tighter caps in later phases. As a result, emission reductions in phase one were twice as high as needed to meet the cap. Fell et al (2008) demonstrate that banking reduces price volatility and eliminates about 20 per cent of the cost difference between price and non-bankable quantity instruments. The degree of these reductions, however, depends on the persistence of price shocks. Due to these advantages, banking is a feature found in most ETS designs.

The advantages of borrowing, in contrast, are less unanimous. Borrowing means the ability to buy allowances from a future trading period during an earlier trading period. While in theory, borrowing can significantly enhance economic efficiency of achieving a cumulative cap, concern over the instruments' impact on policy consistency exists. As excessive borrowing undermines early abatement, it is likely to put upwards pressure on future compliance costs, unless substantial technological change occurs in the meantime. Regulators, faced with considerable political pressure from polluters who refrained from investing in low-carbon technologies early on, might be pressed to cancel out borrowed quotas by increasing future allocations.[6] Sticking to announced allocation budgets might thus become politically untenable. Assuming that borrowed quotas will not be repaid, the dominant strategy from the point of view of an emitter is to refrain from early abatement, create pollution lock-in and bet on loosened caps in the future. Excessive borrowing from future compliance periods could therefore undermine long-term policy integrity while also depressing short-term allowance prices and hence financial incentives to invest in low-carbon technologies for all other players. Nonetheless, some proposed emission trading programmes include borrowing as a means of cost containment.[7] The conditions under which borrowing can be in line with a stringent trading system and its environmental effectiveness remain subject to further research.[8]

Price ceiling and floor

As noted earlier, under uncertainty and in the near term, carbon taxes tend to be more economically efficient than emissions trading (Newell and Pizer, 2003). Yet, the introduction of carbon taxes has proven to be politically difficult, resulting in a greater emphasis on emissions trading. In order to combine the economic efficiency benefits of carbon taxes with the political feasibility advantages of emissions trading, some scholars have investigated how quantity controls can be reconciled to efficient price policies via the application of price bounds (Roberts and Spence, 1976; Newell et al, 2005).

A price ceiling, also called 'price cap' or 'safety valve', can help to limit cost uncertainty by defining a fixed price at which additional allowances are made available in excess of the cap (Pizer, 2002; McKibbon and Wilcoxen, 2002; Jacoby and Ellerman, 2004). A price floor, in contrast, defines a minimum charge on emissions. Price floors stabilize expectations for developers of low-carbon technologies as they are assured that abatement technologies below the price floor will be marketable. Most ETS designs do not include a price floor. One exception is the RGGI, which plans to define a minimum 'reserve price' for allowance auctions. Although this establishes a minimum price for allowances during auctions, it does not guarantee a minimum price during trading periods.

The underlying reasoning of price ceilings is that the atmospheric stock of emissions over the long term needs not necessarily be affected by a short-term deviation from the reduction path. There may therefore be less concern about short-term increases in CO_2 as long as the overall trajectory of CO_2 emissions is downward over an extended period (Newell and Pizer, 2004). Price ceilings can cap expected cost and hereby reduce economic uncertainty. However, the uncertainty would only be shifted from the cost side to the benefit side of reducing emissions. Philibert (2008) estimates that implementing price ceilings and floors in the global climate architecture can reduce mitigation cost by two-thirds (from \$10,671 billion to \$3456 billion for the 10.5Gt CO_2 scenario) without greatly increasing global warming (from 2.44°C median increase to 2.49°C by 2050).

Price ceilings, however, can put a cap's long-term environmental effective-ness at risk, especially where they are combined with banking. Suppose market participants expect the emissions cap to be tightened in the near future (for example due to new information on the severity of climate change). As a conse-quence, they also expect a higher price ceiling under the new cap. Market participants would want to buy as many allowances as possible at the current lower trigger price in order to bank them for later use. An unrestricted safety valve mechanism could thus put much stress on the stringency of long-term emission reduction targets (Stern, 2007). In addition, price ceilings generate additional administrative complexity, especially in regard to linking different trading systems. The lowest price ceiling among linked trading systems would set the international price ceiling (Flachsland et al, 2008). In order to safeguard the stringency of long-term caps, Murray et al (2008) propose to limit the quantity of additional allowances available in the safety valve. This 'allowance reserve' would stipulate both a ceiling price at which cost relief is provided and a maximum number of allowances available to provide this relief.

Monitoring and compliance

The provision of robust information is a necessary condition for ETS price stability (Stranland et al, 2002). Stringent monitoring, reporting and verification (MRV) procedures are required to reveal the basis of market supply and demand in a reliable manner, excluding later data adjustment. The price crash in the first trading period of the EU ETS (see below) resulted from the fact that prior to the implementation of the system no reliable emissions data were avail-able. MRV rules ensure that a tonne of carbon emitted by one source is equal to a tonne of carbon reduced by another source. Hence, they are essential for securing the environmental credibility and the financial fungibility of emission permits. The installation of stringent MRV standards may pose a challenge, in particular for countries with weak legal traditions and institutions.[9] Tietenberg (2006) states that emissions trading systems introduce an incentive for cheating regarding the provision of emissions data. Under grandfathering, sources have an incentive to register more historic emissions than actually occurred. Depending on method of allocation, non-registered emissions do not need to be paid for (auctioning) or can be sold (free allocation). For ensuring compliance, excess emission penalties should be set at levels substantially higher than the prevailing allowance price (Swift, 2001). An additional incentive for compliance is provided by public access to data because third parties such as traders or non-governmental organizations (NGOs) can screen the available information for inconsistencies (Kruger et al, 2000).

Avoiding carbon leakage

In an ideal world, global emissions trading would establish a uniform carbon price for all countries and all sectors. But in reality, carbon pricing is only applied partially and unevenly, so that firms facing a cost on carbon might compete with firms who do not. Asymmetric carbon pricing gives rise to concerns over carbon leakage that occurs when domestic production is reduced as a result of carbon costs but (at least some of) the reduced domestic production gives rise to additional production and emissions elsewhere. Trade-exposed energy intensive industries might emigrate to non-carbon jurisdictions and/or carbon-constrained sectors might lose shares in the international market. Both outcomes reduce the environmental effectiveness of cap-and-trade. Concerns exist that both effects will also negatively impact competitiveness and employment of carbon-constrained sectors without necessarily benefiting the global environment.[10]

For assessing the risk of carbon leakage, one needs to distinguish between direct and indirect price effects generated by emissions trading (Grubb and Neuhoff, 2006). Direct costs are caused by emissions originating from the production process itself (which include energy and process emissions). Indirect costs are caused by, for example, higher electricity prices: electricity generators pass on opportunity costs of allowances to firms, which in turn will result in higher prices for consumption goods. The same indirect cost effect may apply to other intermediate inputs to production that may become more expensive due to pass-through of CO_2-related costs. The German government (UBA, 2008) assessed both effects for sectors that may potentially be exposed to distortions in competitiveness. The analysis is based on the concept of value at stake, which is defined as the sum of potential direct and indirect costs in relation to the gross value added (GVA) of a given industrial sector. The sectors with relevant value at stake contribute little more than 2 per cent to overall GDP. In other words, most economic sectors are not at risk of carbon leakage.

Grubb and Neuhoff (2006) argue that for most industries, the net value at stake due to adverse competitiveness impacts are lower than 1.5 per cent of sector value added, and that for EU industries that are affected the most (cement, and iron and steel) import penetration from outside the EU is rather low. This is in line with the findings of the World Bank (2007), which concludes that industrial competitiveness is not seriously impaired by domestic emissions reductions and that the remaining adverse effects can be offset by well-designed policy packages. Further, Reinaud (2008) finds limited risk of carbon leakage. Production factors other than carbon costs also influence investment and production decisions: cost and qualification of employees, exchange rates, infrastructure, existence of technology clusters and so on. With regard to most industrial sectors, carbon cost influences investment decisions only marginally.

For the limited number of sectors that are prone to carbon leakage, one countermeasure is free allocation of emission allowances. However, for those sectors for which the potential value at stake depends mostly on the increase of electricity prices (indirect CO_2 costs), a change in the allocation method would not prevent a potential production cost increase (UBA, 2008). For those sectors with high direct CO_2 costs, it should be noted that the maximum value at stake is based on the assumption of full auctioning of emissions allowances. Direct costs could be lower if part or all of the allowances were allocated for free, depending on whether, and to what extent, the concept of opportunity costs is applied. Without a functioning international carbon market, partial or totally free allocation for a limited number of industries prone to carbon leakage may be necessary to maintain a level playing field among international competitors.

Another measure that features in the discussion surrounding carbon leakage is trade sanctions, i.e. imposing a tariff or border tax on each imported good that is directly proportional to the carbon emitted during its production process. However, border tax adjustments hardly seem feasible in practice, as they would require a complete life-cycle assessment for each single product that is imported. Otherwise, it seems highly likely that imposing uniform carbon tariffs on all goods from all countries that oppose climate protection could be seen as 'arbitrary' and 'discriminatory' under WTO rules. To circumvent this last problem, Ismer and Neuhoff (2007) propose that border tax adjustment should focus on primary materials incorporated in imported products and, employing detailed data on production processes, assume that these products have been produced using the best available technology (BAT). However, it is not entirely clear how this proposal could be implemented in practice, what the implications for WTO law are, and how effective in reducing carbon leakage it could turn out to be.

Key lessons from the EU ETS[11]

The EU ETS runs in phases. Phase I (2005–2007) was planned as a test phase during which experience with the programme could be developed and banking of allowances was effectively not allowed. Phase II (2008–2012) coincides with the five-year Kyoto commitment period. In line with the EU-wide Kyoto target, each member state has its own national emissions target as defined by the EU burden-sharing agreement. Each member state is required to develop a National Allocation Plan (NAP) that addresses the national target in two steps. First, it splits the national reduction burden between ETS and non-ETS sectors, hereby setting the trading sector's cap. Second, it specifies how the allowances will be allocated to the sources in the trading sector.

Setting the cap

The EU-wide caps of Phases I and II where determined in a decentralized negotiation process between the EU Commission and member state governments. The aggregation of national caps as specified in NAPs resulted in the EU-wide cap. The Commission had the powers to review NAPs and reduced 15 NAPs in Phase I by together 290 million tonnes (Mt) annually. Although this cut aimed at enforcing scarcity in the system, Phase I caps were still too lax. NAPs were based on projected BAU emissions, many of which were overestimated due to the influence of firm lobbying and information asymmetries between companies and regulators. Covered sources in Phase I received emission permits for 2080Mt CO_2, while actually emitting 2020Mt CO_2 annually on average (EEA, 2007). Hence, the number of allocated emission allowances exceeded verified emissions by roughly 60Mt CO_2. This figure only reflects allowances allocated for *free* to *existing* installations. Not included are allowances that were auctioned or part of member states' new entrants' reserves. The quantity of excess allowances during Phase I was therefore even higher. When data on verified emissions for 2005 was released in April 2006, it became clear that allocated allowances exceeded emissions by at least 3 per cent. As a result, the price of Phase I EU Allowances (EUAs) crashed immediately and eventually fell below €1 per tonne of CO_2 in 2007.[12]

The extent of cap mismatch in regard to emissions varied from country to country. From 2005 to 2006, only five countries faced a net shortage of allowances: UK (−83Mt CO_2), Italy (−33Mt), Spain (−25Mt), Ireland (−6Mt), and Austria (−1Mt) (Convery et al, 2008). The net position is here defined as the difference between allocation and emissions balance. In contrast, in 11 countries, 8 of which were new member states, allocations exceeded emissions by more than 10 per cent. The differences in allowance allocation had distributional impacts, transferring money from countries with tight caps to countries with lax caps. Given the overall surplus in Eastern Europe, Convery et al (2008) estimate that facilities in Western countries acquired allowances for 41Mt CO_2, worth €700 million.

As a result of the over-allocation in Phase I, the European Commission tightened the carbon constraint for Phase II. NAPs were downsized by the Commission, in some cases to a considerable extent. In relative terms, the Baltic States had to reduce their proposed caps by the highest percentages: Estonia −48 per cent, Latvia −56 per cent and Lithuania −47 per cent. In absolute terms, the caps for Poland (−76Mt CO_2), Germany (−29Mt) and Czech Republic (−15Mt) were cut most. Overall, the Commission fixed the EU-wide cap at 2.08 billion tonnes of CO_2 annually for the period of 2008–2012. That is 9.5 per cent lower than the cap in Phase I.

Setting the post-2012 regime of the EU ETS, the EU adopted a revised EU ETS Directive in December 2008 that attempts to incorporate many of the lessons learnt during Phases I and II (EC, 2008). The directive for Phase III (2013–2020) includes several changes. Total EU trading sector emissions in 2020 will be reduced by 21 per cent below 2005 levels, resulting in a cap of 1720 million tonnes CO_2.[13] To arrive at this quantity, the centrally set EU-wide cap, which replaces the 27 NAPs, decreases annually by 1.74 per cent from 2013 onwards (taking 2010 as basis year). This linear reduction factor is set to apply beyond the end of the third trading period and may be reviewed by 2025 at the latest.

The scheme's coverage will be extended to new sectors and new gases. It will cover CO_2 from emissions from aviation, petrochemicals, ammonia and aluminium, as well as N_2O emissions from the production of nitric, adipic and glyoxylic acids, and perfluorocarbons from the aluminium sector. Road transport and shipping remain excluded, although the latter is likely to be included at a later stage. Agriculture and forestry are also left out. To lower administrative costs, smaller installations, emitting under 10,000t CO_2 per year, will be allowed to opt out from the ETS, provided that alternative reduction measures are put in place. Industrial GHGs prevented from entering the atmosphere through the use of so-called CCS technology are to be treated as not emitted under the new scheme.

Allowance allocation

The current primary allocation method in the EU ETS is to grandfather allowances to sources for free. Most member states made no or only little use of the possibility of auctioning up to 5 per cent of allowances in Phase I and 10 per cent in Phase II. Only four countries used auctioning in Phase I, accounting for 0.13 per cent of total allowances. More allowances are being auctioned in Phase II, though the expected quantity is still well below the 10 per cent limit. Grandfathering was mainly based on historical emissions although benchmarking had been strongly advocated. Convery et al (2008) assume that governments decided against benchmarks because they would have resulted in allocations too far below recent emissions to gain stakeholder acceptance.[14] The use of benchmarks has spread in Phase II, mainly in the power sector, but they are typically fuel specific, that is, different for power plants generating electricity from natural gas and coal.

The power sector has been allocated almost all of the aggregate ETS sectors' emission reduction burden in Phase I, resulting in a net short position in emission allowances (Convery et al, 2008). A sector's net position is here defined as the difference between gross long (allowances received) and gross short (actual emissions) position. The industrial sector, in particular iron and steel, was long in allowances. The privileged treatment of industry can be explained by the general assumption of the sector's relatively lower abatement potential and exposure to non-EU competition.

All member states created reserves that grandfathered allowances to new entrants. Most required closed installations to forfeit post-closure allowances. This measure was adopted for political reasons. The EU as an investment location should not be disadvantaged and the incentive to close facilities and move production elsewhere should be eliminated (Convery et al, 2008). The main effect, however, was to preserve pre-policy incentives to invest in existing (i.e. relatively polluting) facilities. Even though the perverse incentives of this provision were widely recognized, it was apparently not possible to resist political demands.

The decision to grandfather permits generated substantial windfall profits (i.e. additional revenues) for incumbent firms, especially in the power sector. Auctioning allowances would recover the scarcity rent from private companies for public use without impacting electricity prices relative to the free allocation case. On these grounds, auctioning will be the principal allocation method in Phase III. The power sector will face full auctioning of permits from 2013 onwards (this may be delayed until 2020 in some Eastern European countries), while auctioning in all other sectors is to be phased in incrementally, starting with 20 per cent in 2013 and reaching 100 per cent auctioning by 2025 at the latest. Industries that are at risk of carbon leakage may be exempted from this rule. For these industries, free allocation will be based on sector-specific benchmarks. Benchmarks will be based on the top 10 per cent most efficient facilities in the relevant sector. The European Commission determined in December 2009 which sectors are exposed to the risk of carbon leakage.

The total quantity of allowances that member states will auction will be determined at the EU level, but auctions will be organized at country level. The Directive foresees that 88 per cent of the total quantity of allowances to be auctioned will be distributed according to the relative share of 2005 emissions in the EU ETS. For reasons of 'fairness and solidarity', 12 per cent of the total quantity of allowances to be auctioned will be redistributed to poorer member states.

Did abatement occur?

In light of the excess supply of allowances in Phase I, it appears questionable whether the EU ETS was able to encourage CO_2 abatement. For the year 2007, when the EUA price dropped to a negligible level, the answer to this question is probably negative. Prior to this period, however, the EUA price consistently stayed above €6/t CO_2 and abatement should have occurred. Ellerman and Buchner (2008) estimate that for 2005 and 2006, the EU ETS reduced EU-23 emissions by between 50 and 100Mt CO_2/year. This is 2.5 per cent to 5 per cent less than the emissions expected in a scenario without EU ETS. The authors argue that the finding is valid in spite of the over-allocation that effectively existed in some countries and sectors. It is not the initial allocation that causes a source to reduce emissions but the price it must pay for its emissions, even if in opportunity-cost terms. The authors, however, emphasize that this estimate should be treated with care because there exist many potential pitfalls in constructing counterfactual scenarios. A strong argument can be made that the baseline data used for establishing their scenario without EU ETS is biased. Other studies show lower estimates for induced abatement. Delarue et al (2008) constructed a computer model in order to explore the potential of short-term CO_2 abatement in the European power sector. They think that fuel switching reduced EU-23 emissions by 35–64Mt CO_2 in 2005 and 19–25Mt CO_2 in 2006, with more weight being given to the lower bounds. The European Environment Agency (EEA, 2007) expects that reductions induced by the EU ETS by 2010 will be at least 150Mt CO_2 annually in the EU-15 and contribute 3.4 percentage points to their Kyoto-target of −8 per cent relative to 1990 levels. For the new member states, estimates differ greatly depending on the method used to calculate the baseline. The mid-case scenario predicts EU-12 emission reductions of approximately 25Mt CO_2/year by 2010. Overall, the EU ETS is expected to reduce 2010 EU-27 CO_2 emissions by 200Mt CO_2/year relative to BAU.

Besides its extent, researchers also investigate the nature of abatement. Convery et al (2008) found that abatement often occurred where it was not expected. In Germany, the power sector shifted generation from higher-emitting lignite to lower-emitting hard coal. In the UK, there was a noticeable improvement in the carbon efficiency of coal-fired power plants, probably via increased use of biomass or improved energy efficiency. Analysts, however, predicted more reductions from substituting coal with natural gas. The authors argue that this did not happen in the magnitudes expected, largely because of the relatively high natural gas prices during 2005 and early 2006. In contrast, little attention was given to either the German intra-fuel switch or the improved CO_2 efficiency observed in the UK. This underlines that market instruments may trigger abatement in areas where it was not anticipated by regulators.

Price effects

Looking at the interaction of carbon and electricity prices, there is no universal answer on how the EU ETS has affected electricity prices (Reinaud, 2007). Many other factors affect generation prices such as high natural gas prices in 2005. Gas-fired power generation is increasingly the choice of investors in volatile markets due to short lead times and flexibility advantages. With gas thus being the marginal fuel for peak production, the interaction between gas and electricity prices is particularly close. The potential use of market power by electric utilities further complicates the picture. Moreover, as no data can be gathered on the bidding strategies, and the marginal supplier to the market, determining the precise level of pass-through of CO_2 prices in electricity prices is not possible. Nevertheless, several studies attempted to provide estimates of pass-through rates: Sijm et al (2006) find rates ranging from 39 to 73 per cent for Germany and the Netherlands for the period January to July 2005 and from 60 to 80 per cent for the same countries between January to December 2005. For the reasons cited above, however, these estimates need to be treated with care.

Carbon market governance

The development of the European carbon market poses new challenges to institutions and governance. Two general areas of activity can be differentiated here: the monitoring and management of carbon markets.

Monitoring is necessary to ensure that allowance markets work efficiently. In the EU ETS, more than 10 million allowances are traded weekly on average, resulting in a market worth several billion euros. The legal nature of these allowances is, however, unclear. Some countries consider them to be financial instruments whose trading is supervised by the financial service authority, while other countries consider them to be normal commodities and only their derivatives are viewed as financial instruments (EEA, 2007). In order to avoid market manipulation and inside trading, it is important to consider how to apply the rules for financial markets to emission allowances. Furthermore, publication of market sensitive information by the Commission and member states should be strictly and clearly regulated, since the release of market sensitive data impacts allowance prices. The same rules that regulate the dissemination of stock market sensitive information should apply here. In the future, carbon-market governance bodies might be responsible for collecting information on market transactions, analysing their impact on the economy and industrial sectors, and disclosing this information in a predictable and transparent manner.[15] They would also act to uncover and abolish market manipulations (for example by coordinating national and regional market control authorities), organize permit auctions, and administer the overall allocation process.

The second, more controversial area of activity regards the intervention in carbon markets. The Australian ETS, for instance, will feature an 'Independent Carbon Bank' that would be in charge of regulating the carbon market based on the rules provided by the legislator (Australia, 2008).[16] Depending on the specific proposal, the carbon bank may adjust the limits on the use of offsets, extend the possibility of banking/borrowing, set the interest rates for borrowed allowances, or even sell/buy additional allowances. Though these measures may provide near-term flexibility and cost relief, the long-term reduction trajectory and emission cap may be hampered. For this reason, relief measures should only be employed incrementally after in-depth consideration of market conditions and be accompanied by a clear repayment schedule for borrowed allowances. The advantages of an intervening carbon bank can be questioned. It can be argued that the existence of a carbon bank creates uncertainty over allowance quantities if it has the power to temporarily adjust caps and borrowing modalities. Carbon prices and investments in low-emission technologies become less certain than, for example, in a fixed safety-valve approach. Further, market participants might decide to delay abatement actions because they speculate on a loosening of the carbon constraint by the intervening bank. The danger also exists that borrowed quantities are cancelled by an eventual adjustment to the cap. Nevertheless, carbon banks might satisfy politicians' and investors' demands for flexible carbon constraints and therefore represent a possible measure to provide short-term cost relief. The optimal layout and operating principles of carbon banks are the subject of ongoing research.

Complementary measures

Environmental policies are often directed at internalizing externalities. The standard theory of externalities indicates that only one instrument is needed to internalize one externality. In some instances, however, multiple externalities and market failures exist, and it is very unlikely that one instrument can be used to optimally address several market failures simultaneously. Therefore, several complementary instruments are required.

Market failures that lead to negative abatement costs were discussed above, mostly related to energy efficiency measures. It was concluded that full sectoral coverage along with complementary regulatory measures are needed to tap into this low-cost mitigation potential. Moreover, other market failures related to the specific circumstances of technological innovation exist that prevent the large-scale uptake of low-carbon technologies such as renewable energy. For this reason, it is of utmost importance to understand that even a well-designed ETS is not sufficient on its own to encourage the fundamental energy system transformation we aim for when giving carbon a price. Although many renewable energy technologies, given stringent carbon constraints, are likely to be profitable in the long term, most of them fail to attract investments because of high initial costs. This is in spite of their significant cost-reduction potential due to learning processes and economies of scale (IEA, 2000).

The first source of failure in technology markets can be found in the public good nature of knowledge. The positive side-effects of introducing new technologies are not limited to the innovating firm because technology development typically also creates benefits for others, for example in the form of knowledge spill-overs. In the energy sector in particular, technology spill-over to competitors is large and, as a result, firms underinvest in R&D (Azar and Dowlatabadi, 1999). Without additional policies addressing this externality, investments remain lower than socially desirable even in the presence of an optimal carbon price.

Second, learning and network effects introduce multiple equilibria to the energy-economy system. In the status quo, marginal additional investments in many low-carbon technologies do not achieve profitability even though a price on carbon exists because many alternative technologies are still at the beginning of their learning curves. This state is a stable equilibrium. However, once a technology reaches a higher point on the learning curve, another stable state may be reached, one with high deployment of low-carbon technologies that is cost-optimal even if R&D and learning investments are taken into account. This bistability creates a strong path-dependency and potential technology lock-in that need to be addressed, for example via feed-in tariffs or renewable energy quotas (see Box 8.1).

Third, energy infrastructure investments compete for capital with other investment opportunities, and thus will be considered attractive only if they provide satisfying returns on investment in the near term. The benefits of research in new energy technologies, however, may not be realized for two to three decades, which is beyond the planning horizons of even the most forward-looking companies (Anderson and Bird, 1992).

Fourth, large uncertainties exist that concern the future development of energy and climate policy, availability and prices of fossil fuels, as well as the speed of innovation in low-carbon technologies. Investors, being risk averse, fear the threat of stranded investments and consequently tend to delay future engagements. This is particularly damaging for many low-carbon technologies, since they are rather capital intensive and require substantial upfront expenditures.

BOX 8.1 RENEWABLE ENERGY SUPPORT POLICIES IN EUROPE

Among renewable energy support policies one can distinguish between investment support (for example capital grants, tax exemptions or reductions on the purchase of goods) and operating support (for example price subsidies, green certificates, tender schemes, tax exemptions or reductions on the production of electricity). In the EU, operating support plays a far more important role and can either work through prices or quantities. Although, in theory, both price- and quantity-based instruments should yield the same economic efficiency, practical experiences with renewable support policies in Europe draw a different picture.

A study prepared for the European Commission assessed renewable electricity support policies in various EU member states and concludes that for onshore wind, for example, in 2006 well-adapted feed-in tariffs were typically more effective at a relatively lower producer profit (Ragwitz et al, 2005). In other words, it can be observed that quota systems achieved a rather low effectiveness at comparably higher profit margins. However, it should be noted that quota systems are relatively new instruments and there is at present little knowledge of how the certificate prices will develop over time. Nevertheless, quota systems generally involve a higher price uncertainty, which leads to high risk premiums and limited economic efficiency from a social-planner perspective. They also are less capable of differentiating between technologies. For these and other reasons, feed-in tariffs, though they too have disadvantages, achieved better results in promoting renewable energies in Europe than quota-based policies.

Overall, none of these market failures is adequately addressed by an emissions trading system alone. Complementary measures and policies such as feed-in tariffs or quotas for renewable energy sources are required to support the transformation towards a low-carbon economy. At the same time, existing subsidies for unsustainable and carbon-intensive energy sources such as coal need to be abolished as they represent negative carbon prices. This may free up resources that can then be spent on low-carbon technologies. In some circumstances, complementary measures may be transitional because although they may be necessary to address a specific failure in the short to medium terms, they are not expected to be helpful in the longer term. They can support and drive research, development and demonstration of new technologies where the investors are unable to capture the full benefits of their investment and address other market failures, such as non-price barriers.

Conclusion

Emissions trading is expanding worldwide. Cap-and-trade systems effectively establish a price on carbon by setting an absolute quantitative limit for GHG emissions. The allowance cap determines the carbon price and the instrument's environmental effectiveness. But uncertainty over the carbon price makes it difficult to estimate the policy's economic cost. Trading of permits enables participants to reduce emissions where this is cheapest and leads to the formation of a price signal reflecting marginal abatement costs across covered installations. Looking at the EU ETS experience, one can draw several conclusions in regard to general ETS design:

ETS coverage, both in terms of sectors and gases, should be as broad as possible in order to maximize market liquidity and the range of potential abatement options. Sectors with highly dispersed emission sources (for example transport) should be covered upstream in order to keep transaction costs low. Sources that are difficult to measure and monitor (for example agriculture) should be excluded from coverage to safeguard the cap's environmental integrity. In such sectors, domestic offset credits may be used to smooth the transition towards broader trading regimes by allowing experience learning with monitoring emissions. Setting and communicating long-term trajectories for cap and coverage is of paramount importance for enhancing predictability and investor confidence.

The allocation of allowances affects the policy's distributional impact and cost effectiveness. The EU ETS allocation process demonstrated that free allocation can significantly distort incentives but may ease the transition to emissions trading. Increasing the use of auctioning is likely to generate benefits in terms of cost effectiveness, distribution and public finances. Allocation procedures can be used for redistributing wealth to underprivileged regions. New entrants' provisions should be carefully balanced to set reasonable incentives while avoiding shielding incumbents against new competitors. Different allocation methods can apply to different sectors, mirroring industries' diverse vulnerabilities to carbon pricing and international competition.

Banking provisions can provide significant temporal flexibility, reducing cost and volatility. The benefit of borrowing provisions is more ambiguous as concerns over time-consistency exist. No existing trading system allows for borrowing and further research needs to clarify the conditions under which it might represent a sensible design choice.

Price ceilings limit the potential range of allowance prices. On the one hand, this reduces cost uncertainty while, on the other hand, it can diminish the incentive for investors to develop low-carbon technologies. In addition, price bounds complicate linking up with other trading systems.

MRV of emissions is critical for ensuring a trading system's transparency, integrity and credibility. Reliable historic data are essential for determining caps at appropriate levels. Stringent verification rules can avoid market-distorting ex post corrections and enhance investor trust in carbon markets. Institutions need to be put in place that oversee the carbon market and ensure that it is functioning efficiently.

Overall, carbon pricing alone is not enough to mitigate climate change. The standard theory of externalities indicates that only one instrument is needed to internalize one externality. In some instances, however, multiple externalities and market failures exist, and it is very unlikely that one instrument can be used to optimally address several market failures simultaneously. Policy-makers need to implement complementary measures that address other market failures such as underinvestment in technological innovation.

Appendix

Table 8.1 Comparison of different cap-and-trade programmes

	EU ETS	RGGI	California	Australia	New Zealand
Implementation Stage	Up and running since January 2005	Legislation process in progress	2007 Proposal by MAC expert commission	2007 Discussion paper by PM Task Group expert commission	2007 Final Decisions on the core design and on detailed design features on the government level
(Envisaged) Start Date	1 January 2005	1 January 2009	1 January 2012	Scheduled in 2011 or 2012	Stage 1: 1 Jan 2008 Stage 2: 1 Jan 2009 Stage 3: 1 Jan 2010 Stage 4: 1 Jan 2013
Ratification of Kyoto Protocol	Yes	No	No	Yes	Yes
Participating Sub-regions	27 EU Member States	10 US states: Connecticut, Delaware, New Jersey, New York, Maine, Maryland, Massachusetts, New Hampshire, Rhode Island, Vermont	California	Australian Commonwealth	New Zealand
Regulated Sectors	Electricity, refining, iron and steel, cement, glass, ceramics, pulp and paper	Electricity generating facilities $\geq$ 25MW primarily fired by fossil fuels (coal, natural gas, oil), feeding more than 10% of their generated electricity into the grid	Pr 1:[17] electricity, refining, cement, other processes, non-CO_2 Pr 2: like 1, + transport Pr 3 & 4: like 2, + small industry, commercial, residential sources	Electricity and industrial processes emitting more than 25kt CO_2-eq per annum, transport, waste to be discussed	S1: Forestry S2: Liquid fossil fuels S3: Stationary energy; industrial processes S4: Agriculture; waste
Regulated Emissions	CO_2 only	CO_2 only	CO_2, N_2O, HFCs, PFCs and SF_6	CO_2, CH_4, N_2O, HFCs, PFCs and SF_6	CO_2, CH_4, N_2O, HFCs, PFCs and SF_6
Point of Regulation	Downstream	Downstream	Upstream and downstream	Upstream and downstream	Upstream and downstream

Table 8.1 *continued*

	EU ETS	RGGI	California	Australia	New Zealand
Covered Emissions (Mt CO$_2$eq)	~2.000Mt	149Mt (2003)	Pr1: 193Mt Pr2: 356Mt Pr3&4: 409Mt	~ 300Mt (according to calculations by PM Task Group)	S1: 21.8Mt (expected for the period 2008–2012) S2: 15Mt (2005) S3: 22.8Mt (2005) S4: 39.2Mt (2005) Total: 77.6Mt (excluding S1)
Share of Economy-wide Emissions (CO$_2$eq)		~ 24%	Pr1: 39% Pr2: 72% Pr3&4: 83%	~55%	S1: N/A S2: 19% S3: 30% S4: 51%
Number of Covered Entities	~10.000	~630	Pr1: ~450 Pr2: ~480 Pr3: ~490 Pr4: ~150	~ 900	S1: ≥ 1000 S2: ~5 S3: ~ 80 (~45 stationary energy; ~ 35 industrial processes) S4: ≥ 35 agriculture (Point of Regulation for agriculture not decided); ~ 60 waste
Total Emissions (incl. non-energy emissions) in Mt CO$_2$-eq	4979.4 (in 2004, EU25)	624.9 (in 2003)	494.3 (in 2004)	525.4 (in 2005)	77.2 (in 2005) + deforestation
Energy Mix by Sectors	Industry: 28% Households: 41% Transport: 31%	Industry: 17.5% Commercial: 25.9% Households: 27.5% Transport: 29%	Industry: 23% Commercial: 18% Households: 18% Transport: 40%	Industry + Commercial: 50% Households: 30% Transport: 20%	Industry: 30% Commercial: 9% Households: 13% Transport: 44% Agriculture: 4%
Energy Mix by fuel	Oil: 37% Solid Fuels: 18% Natural Gas: 24% Nuclear: 15% Renewables: 6%	Petroleum products: 48.9% Coal: 8.6% Natural Gas: 23.2% Nuclear: 11.7% Renewables: 7.1%	Petroleum products: 46% Coal: 8% Natural Gas: 29.5% Nuclear: 5% Renewables: 11.5%	Oil: 35% Coal: 41% Natural Gas: 19% Nuclear: - Renewables: 5%	Oil: 38% Coal: 13% Natural Gas: 20% Nuclear: - Renewables: 28%

Historical Emission Trends	EU25: 8% below 1990 levels in 2003; EU15: 1.7% below 1990 levels in 2003	+7.4% during 1990–2003 period	+14.3% during 1990–2004 period	+4.5% during 1990–2005 period	+23.4% during 1990–2005 period
Future Projections	4.7% above 1990 levels in 2030	12% above 1990 levels in 2019 (electricity only)	40% increase in 1990–2020 period	27% above 1990 levels in 2020	30% above 2005 levels in 2030
Reduction Goals	- 8% below 1990 levels in 2008–2012 period - 20% (or 30%) below 1990 levels in 2020 - 60%-80% below 1990 levels by 2050	- 2009 cap: 5% above 2005 levels, will remain until 2015 - 10% reduction below this cap by 2019	- 2000 levels in 2010 - 1990 levels in 2020 - 80% reduction below 1990 levels by 2050	8% above 1990 levels in 2008–2012 period	Carbon neutrality: - Electricity by 2025 - Stationary energy by 2030 - Transport by 2040
Unit of Measurement	1 metric tonne CO_2-eq	1 short tonne CO_2-eq (1 short tonne equals 0.90718474 metric tonnes)	1 metric tonne CO_2-eq	1 metric tonne CO_2-eq	1 metric tonne CO_2-eq
Cap	Future levels not specified. Bottom-up emergence of cap through NAP negotiations. Plans for centrally set cap	Annual cap of 170.6Mt CO_2 between 2009 and 2014; annual reduction of 2.5% between 2015 and 2018	Not specified	Not specified	To be linked to NZ commitments under the Kyoto protocol (309.5Mt in 2012) and an international post-2012 regime, respectively
Allocation Method	Grandfathering, benchmarking, maximum 10% auctioning	Auctioning minimum 25%; decision over remaining 75% left to individual states	Auctioning and benchmarking	Free allocation and auctioning	S1: Free allocation S2: Auctioning S3: Some free allocation for industrial processes S4: 90% free allocation for agriculture
Banking	From 2nd period on: Unlimited	Unlimited	Unlimited	Limit proposed as long as price cap applies	Unlimited
Borrowing	Rejected	Rejected	Rejected	Rejected	Rejected
Trading Period	3 years in first period, 5 years in second	3 years	No specification	10 years	2008/9/10–2012 2013–2020
Compliance Period	1 year	3 years	3 years recommended	1 year recommended	1 year (S1: initially 2 years)

Table 8.1 *continued*

	EU ETS	RGGI	California	Australia	New Zealand
Price Cap	Rejected	Two-stage safety valve arrangement: 'Credits Trigger Event' if spot price for emissions exceeds $7 for a period over 12 months; 'Safety Valve Trigger Event' if spot price for emissions exceeds $10 for a period over 12 months	Rejected	Proposed; further specifications to be discussed	Rejected in principal, but considered if no international climate policy agreement post-2012
Price Floor	No	Yes	Encouraged for consideration	No	No
Penalty System	Delivery of the non-delivered allowances + €100 penalty per tonne (2008–2012)	Three times the non-delivered certificates to be delivered at next compliance date	Not specified, but non-delivery shall be made up + penalty	Emissions fee proposed setting a price cap	Delivery of the non-delivered allowances (can be extended to two times of the non-delivered allowances) + NZ$30–60 penalty per tonne
MRV	Updated in 2007	Guicelines for continuing measurement based on CRF 40 Part 75 (Acid Rain Program regulation) that demands a maximum uncertainty of 10%. Verification by regulating authority	To be developed, building on existing MRV infrastructure of California Climate Action Registry	To be developed	To be developed

Registry	Community Independent Transaction Log overseeing communications between national registries	No common registry for RGGI	To be developed, building on existing CCAR infrastructure	Under development; rules clarified under the Australia National Greenhouse and Energy Reporting Act in legislation since September 2007	To be developed building on existing infrastructure established under the Climate Change Response Act 2002
Domestic Credit Programme	No	Credits are accepted from programmes in RGGI states or any other US state or jurisdiction	To be developed using RGGI experience; programmatic approach proposed	To be developed especially focusing on forestry and agriculture	Discussed
Eligibility of CDM/JI	Yes	Generally no; in case of a safety trigger event credit allowances may be awarded for the retirement of allowances or credits from international trading programmes	Most MAC members recommend eligibility of CDM	Yes	Yes + assigned amount units (AAUs)
Import Quota for Credits	Varying from country to country according to set of criteria. Average (weighted) EU quota for CDM/JI import in trading period II: 13.4%	3.3% of a facility's emissions can be covered by credits; the number rises up to 10% in case of a safety valve trigger event	Most MAC members recommend unrestricted credit import	No	No

Source: Adapted from Flachsland et al (2008)

Notes

1 The relative merits and drawbacks of both types are subject of an ongoing academic debate (see for example Quirion, 2005; Sue Wing et al, 2006; Newell and Pizer, 2006).

2 The EU ETS covers more than 10,000 different installations, 7 per cent of which account for 60 per cent of total emissions. In contrast, the 1400 smallest sources account for less than 0.14 per cent of emissions (EC, 2008). Reducing the number of covered sources can cut a programme's cost without significantly affecting its environmental effectiveness, but only if there is no leakage from larger to smaller sources. Alternatively, upstream coverage automatically includes all sources within a sector.

3 Because the amount of CO_2 emitted from burning a unit of coal, oil or gas is largely invariant to the process in which it is oxidized, emission factors can be calculated easily on the level of fossil primary energy carriers. Indeed, observing coal use and calculating emissions from it rather than observing emissions in the flue stack is the standard procedure for calculating emissions, for example for large-scale users of coal in the EU ETS, and also in national emissions inventories under the UNFCCC.

4 See for example Wara and Victor (2008) or Schneider (2007), who provide a critical analysis of the CDM's environmental performance.

5 The RGGI system uses short tons.

6 Garnaut (2008) argues that good governance and stringent law enforcement can avoid this problem. However, the assumption that governance is good and law enforcement is stringent at all times and everywhere can be questioned. After all, if the political agenda of a legislating body is changed following general elections, former decisions on caps etc. may be challenged and eventually altered.

7 For example, the US 'Bill to Reduce the Economic Impact of Climate Legislation' proposes the establishment of a 'Carbon Market Efficiency Board', which would intervene in a future US carbon market when allowance prices exceed a certain threshold. The first measure is to expand companies' ability to borrow permits. The second measure, to be used if high prices are not relieved by the first measure, is to add permits to the market. This temporary increase would be compensated for by reducing available allowances in a later year.

8 A special case of borrowing represents intra-period borrowing. This is the case in the EU ETS, where facilities receive allocations for two compliance years prior to the first actual compliance date. That is, until the very last compliance year in a trading period, there are always two allocation tranches in the market at the date of compliance. Intra-period borrowing does not dilute environmental effectiveness.

9 Several authors have concluded that tradable permit programmes may be less appropriate for developing countries due to their lack of appropriate market or enforcement institutions (Blackman and Harrington, 2000; Bell and Russell, 2002). For China, Wang et al (2004) suggest that a strengthened monitoring and enforcement capacity would be required to implement emissions trading.

10 Reinaud (2008) correctly points out that most discussions of competitiveness effects neglect industries that benefit from climate policy. This is fallacious as climate policy potentially enhances research, development and deployment of lower-emitting technologies and also offers positive competitiveness and employment effects.

11 The following discussion is largely based on the findings provided by Convery et al (2008).
12 Note that in the presence of banking between trading periods, the price would not have crashed to zero since over-allocated allowances could have been used in Phase II or Phase III. However, banking was not allowed. This illustrates the role of banking in containing price volatility.
13 The numbers are based on the scope of the ETS as applicable in Phase II.
14 In the EU ETS, choice of allocation method rests with national governments.
15 Several companies have emerged providing market data, projections and background information on the EU ETS (see for example. www.pointcarbon.com).
16 The Carbon Market Efficiency Board mentioned in Footnote 7 is another example of a carbon market oversight institution.
17 The Market Advisory Committee has proposed four programme options (abbreviated Pr here) with differing coverage and points of regulation.

References

Ahman, M., Burtraw, D., Kruger, J. and Zetterberg, L. (2006).'A ten year rule to guide the allocation of EU emission allowances', *Energy Policy*, vol 35, no 3, pp1718–1730

Anderson, D. and Bird, C. D. (1992) 'Carbon accumulations and technical progress: A simulation study of costs', *Bulletin of Economics and Statistics*, vol 54, no 1, pp1–27

Australia (2008) *Carbon Pollution Reduction Scheme*, Green Paper by the Department of Climate Change, Canberra: Commonwealth of Australia

Azar, C. and Dowlatabadi, H. (1999) 'A review of technical change in assessment of climate policy', *Annual Review Energy Economics*, vol 24, pp513–544

Baron, R. and Bygrave, S. (2002) 'Towards international emissions trading: Design implication for linkages', paper presented at the 3rd CATEP Workshop on Global Trading, Kiel Institute for World Economics, 30 September–1 October

Bell, R. and Russell, C. (2002) 'Environmental policy for developing countries', *Issues in Science and Technology*, Spring, pp63–70

Blackman, A. and Harrington, W. (2000) 'The use of economic incentives in developing countries: Lessons from international experience with industrial air pollution', *Journal of Environment and Development*, vol 9, no 1, pp5–44

Böhringer, C. and Löschel, A. (2005) 'Climate policy beyond Kyoto: Quo vadis? A computable general equlibrium analysis based on expert judgements', *KYKLOS*, vol 58, no 4, pp467–493

Coase, R. H. (1960) 'The problem of social cost', *Journal of Law and Economics III*, October, pp1–44

Convery, F., Ellerman, D. and de Perthuis, C. (2008) 'The European carbon market in action: Lessons from the first trading period, Interim Report, March, Cambridge, MA: Center for Energy and Environmental Policy Research

Delarue E., Ellerman, A. D. and D'haeseleer, W. D. (2008) *Short-term CO$_2$ Abatement in the European Power Sector*, Cambridge, MA: Centre for Energy and Environmental Policy Research

Dinan, T. and Rogers, D. L. (2002) 'Distributional effects of carbon allowances trading: How government decisions determine winners and losers', *National Tax Journal*, vol 55, no 2, pp199–221

Dudek, D. and Golub, A. (2003) '"Intensity" targets: Pathway or roadblock to preventing climate change while enhancing economic growth?', *Climate Policy*, vol 3, no S2, ppS21–28

EC (European Commission) (2008) *Directive of the European Parliament and of the Council amending Directive 2003/87/EC so as to improve and extend the GHG emission allowance trading system of the Community*, COM(2008) 16 final, Brussels: European Commission

EEA (European Environmental Agency) (2007) 'GHG emission trends and projections in Europe 2007', Copenhagen: EEA

Ellerman, A. D. (2006) 'Are cap-and-trade programs more environmentally effective than conventional regulation?', in J. Freeman and C. Kolstad (eds) *Moving to Markets: Lessons from Thirty Years of Experience*, New York: Oxford University Press, pp48–62

Ellerman, A. and Buchner, B. (2008) 'Over-allocation or abatement? A preliminary analysis of the EU ETS based on the 2005–06 emissions data', *Environmental Resource Economics*, vol 41, pp267–287

Ellerman, A. and Sue Wing, I. (2003) 'Absolute v. intensity-based emission caps', *Climate Policy*, vol 3, supplement 2, ppS7–S20

Ellerman, A. and Pontero, J. (2005) 'The efficiency and robustness of allowance banking in the US Acid Rain Programme', Working paper 0505, Centre for Energy and Environmental Policy Research, Cambridge, MA: MIT

Enkvist, P-A., Nauclér, T. and Rosander, J. (2007) 'A cost curve for GHG reduction', *McKinsey Quarterly*, vol 1

Fell, H., MacKenzie, I. A. and Pizer, W. A. (2008) *Prices versus Quantities versus Bankable Quantities*, Washington, DC: Resources for the Future

Flachsland, C., Edenhofer, O., Jakob, M. and Steckel, J. (2008) 'Developing the international carbon market: Linking options for the EU ETS', Report to the Policy Planning Staff in the Federal Foreign Office, Potsdam: Potsdam Institute of Climate Research

Garnaut, R. (2008) *Garnaut Climate Change Review*, Final Report, Canberra: Commonwealth of Australia

Goulder, L., Parry, I., Williams, R. and Burtraw, D. (1999) 'The cost effectiveness of alternative instruments for environmental effectiveness in a second best setting', *Journal of Public Economics*, vol 72, no 3, pp329–360

Grubb, M. and Neuhoff, K. (2006) 'Allocation and competitiveness in the EU Emissions Trading Scheme: Policy overview', *Climate Policy*, vol 6, pp7–30

Hargrave, T. (1998) *US Carbon Emissions Trading: Description of an Upstream Approach*, Washington, DC: Center for Clean Air Policy

Hepburn, C., Grubb, M., Neuhoff, K., Matthes, F. and Tse, M. (2006) 'Auctioning of EU ETS Phase II allowances: How and why?', *Climate Policy*, vol 6, pp137–160

Herold, A. (2008) 'The significance of monitoring capabilities to decide about the scope of a carbon market', presentation at 1st Global Carbon Market Forum on Monitoring, Reporting, Verification, Compliance and Enforcement Session II, 19 May

IEA (International Energy Agency) (2000) *Experience Curves for Energy Technology Policy*, Paris: IEA

IPCC (Intergovernmental Panel on Climate Change) (2007a) *Climate Change 2007: Mitigation of Climate Change*, Contribution of Working Group III to the Fourth Assessment Report of the Intergovernmental Panel on Climate Change, B. Metz, O. R. Davidson, P. R. Bosch, R. Dave and L. A. Meyer (eds) Cambridge and New York: Cambridge University Press

IPCC (2007b) *Climate Change 2007: The Physical Science Basis*, Contribution of Working Group I to the Fourth Assessment Report of the Intergovernmental Panel on Climate Change, S. Solomon, D., Qin, M. Manning, Z. Chen, M. Marquis, K. B. M. Tignor and H. L. Miller (eds), Cambridge and New York: Cambridge University Press

Ismer, R. and Neuhoff, K. (2007) 'Border tax adjustment: A feasible way to support stringent emission trading', *European Journal of Law and Economics*, vol 24, pp137–164

Jacoby, H. D. and Ellerman, A. D. (2004) 'The safety valve and climate policy', *Energy Policy*, vol 32, no 4, pp481–491

Keohane, N., Revesz, R.L. and Stavins, R.N. (1998) 'The choice of regulatory instruments in environmental policy', *Harvard Environmental Law Review*, vol 22, pp313–367

Kruger, J., McLean, B. and Chen, R. (2000) 'A tale of two revolutions: Administration of the SO_2 trading program', in R. Kosobud (ed) *Emissions Trading: Environmental Policy's New Approach*, New York: John Wiley & Sons

Marschinski, R. and Lecocq, F. (2006) 'Do intensity targets control uncertainty better than quotas? Conditions, calibrations and caveats', Kyoto: World Congress in Environmental and Resource Economics

McKibbin, W. J. and Wilcoxen. P. J. (2002) 'Climate change policy after Kyoto: A blueprint for a realistic approach', Washington, DC: Brookings Institution

Murray, B. C., Newell, R. G. and Pizer, W. A. (2008) *Balancing Cost and Emissions Certainty*, Washington, DC: Resources for the Future

Neuhoff, K. and Matthes, F. (2008) *The Role of Auctions for Emissions Trading*, Cambridge, UK: Climate Strategies

Newell, R. and Pizer, W. A. (2003) 'Regulating stock externalities under uncertainty', *Journal of Environmental Economics and Management*, vol 45, pp416–432

Newell, R. and Pizer, W. A. (2004) 'Uncertain discount rates in climate policy analysis', *Energy Policy*, vol 32, pp519–529

Newell, R. G. and Pizer, W. A. (2006) *Indexed Regulation*, Discussion Paper 0632, Washington, DC: Resources for the Future

Newell, R. G., Pizer, W. A. and Zhang, J. (2005) 'Managing permit markets to stabilize prices', *Environmental and Resource Economics*, vol 31, no 2, pp133–157

Philibert, C. (2008) *Price Caps and Price Floors in Climate Policy – A Quantitative Assessment*, Paris: IEA

Pizer, W. A. (2002). 'Combining price and quantity controls to mitigate global climate change', *Journal of Public Economics*, vol 85, no 3, pp409–434

Pizer, W. A. (2005) 'The case for intensity targets', *Climate Policy*, vol 5, no 4, pp455–462

Quirion, P. (2005) 'Does uncertainty justify intensity emission caps?', *Resource & Energy Economics*, vol 27, no 4, pp343–353

Ragwitz, M., Schleich, J., Huber, G., Resch, G., Faber, T., Voogt, M., Coenraads, R., Clejine, H. and Bodo, P. (2005) *Analysis of the EU Renewable Energy Sources' Evolution up to 2020 (FORRES 2020)*, Karlsruhe: Fraunhofer IRB Verlag

Reinaud, J. (2007) CO_2 *Allowance and Electricity Price Interaction: Impact on Industry's Electricity Purchasing Strategies in Europe*, Paris: IEA

Reinaud, J. (2008) *Issues Behind Competitiveness and Carbon Leakage: Focus on Heavy Industry*, Paris: IEA

Reinaud, J. and Philibert, C. (2007) *Emissions Trading: Trend and Prospects*, Paris: IEA

RGGI (Regional Greenhouse Gas Initiative) (2007) *Overview of RGGI CO_2 Budget Trading Program*, New York: RGGI

Roberts, M. and Spence, M. (1976) 'Effluent charges and licenses under uncertainty', *Journal of Public Economics*, vol 5, no 3–4, pp193–208

Schneider, L. (2007) 'Is the CDM fulfilling its environmental and sustainable development objectives? An evaluation of the CDM and options for improvement', Freiburg: Öko-Institut

Sijm, J., Neuhoff, K. and Chen, Y. (2006) 'CO_2 cost pass through and windfall profits in the power sector', *Climate Policy*, vol 6, no 1, pp49–72

Stern, N. (2007) *The Economics of Climate Change: The Stern Review*, Cambridge and New York: Cambridge University Press

Stranland, J., Chavez, C. and Field, B. (2002) 'Enforcing emissions trading programs: Theory, practice, and performance', *Policy Studies Journal*, vol 303, no 3, pp343–361

Sue Wing, I., Ellerman, A. D. and Jaemin, S. (2006) *Absolute vs. Intensity Limits for CO_2 Emission Control: Performance Under Uncertainty*, Report no 130, MIT Joint Program on the Science and Policy of Global Change, Cambridge, MA: MIT

Swift, B. (2001) 'How environmental laws work: An analysis of the utility sector's response to regulation of nitrogen oxides and sulfur dioxide under the clean air act', *Tulane Environmental Law Journal*, vol 14, pp309–424

Tietenberg, T. (2006) *Emissions Trading: Principles and Practice*, second edition, Washington, DC: RFF Press

UBA (Umwelt Bundesamt) (2008) 'Impacts of the EU Emissions Trading Scheme on the industrial competitiveness in Germany', Dessau-Rosslau: Umwelt Bundesamt

Wang, J., Yang, J., Ge, C., Cao, D. and Schreifels, J. (2004). 'Controlling sulfur dioxide in China: Will emission trading work?', *Environment*, vol 46, pp28–39

Wara, M. and Victor, D. G. (2008) 'A realistic policy on international carbon offsets', Program on Energy and Sustainable Development Working Paper #74, Stanford, CA: Program on Energy and Sustainable Development

World Bank (2007) *International Trade and Climate Change: Economic, Legal, and Institutional Perspectives*, Washington, DC: World Bank

World Bank (2008) *State and Trends of the Carbon Market 2007*, Washington, DC: World Bank

9

Emission Reduction and Employment

Cai Fang, Du Yang and Wang Meiyan

Economic growth, employment and emission reduction

There has been growing unassailable evidence on the existing and possible severe consequences of global climate change and the relationship between human economic activities and the global warming (see for example Stern, 2007). The Kyoto Protocol and Bali Road Map put forward either compulsory or moral requirements for all economies, including China with the largest population size, fastest growth rate and second highest total GDP in PPP term. As Thomas (2007) estimates, assuming the ratio of CO_2 emission over GDP remains at the level of 2001, total global emissions will be as high as 25 billion metric tonnes by 2018. While world CO_2 emission will have increased by 69 per cent during the period, emission in China will increase to 9 billion metric tonnes with growth of 218 per cent, exceeding all other countries in terms of total emission.[1]

The reason that emission reduction has been given so much attention economically and politically is because there exists a trade-off between it and economic (employment) growth. Most research focuses on the impacts of emission reduction on economic growth rather than employment. The reasons are that people always believe economic growth will no doubt bring employment growth, thus GDP growth has tended to be the priority while employment growth has been overlooked. However, when we talk about emission reduction, employment should always be put first rather than GDP growth.

The impacts of emission reduction on employment determine its feasibility and sustainability in terms of putting people's livelihoods first. China has abundant labour resources. The purpose of keeping a specific economic growth rate is to keep stable employment growth. The main consequence of a decline in GDP growth is the loss of employment. Though the balance between labour supply and labour demand has improved in China, any shocks to employment will decrease urban and rural incomes, enlarge income inequality and bring social contradictions.

Emission reductions may cause unacceptable costs in terms of GDP loss. However, employment losses do not have to be so large. This is because there is no fixed relationship between GDP and employment. Economic growth has not always brought employment growth, since the employment effects of economic growth differ, depending on which sectors grow. If emission reduction is primarily caused by lower output in capital-intensive sectors, a fall in GDP will bring a minimum of employment loss and in this sense the cost of emission reduction is brought down.

Figures 9.1 and 9.2 show the relationship between employment per unit of value added and energy consumption per unit of value added by sector. We can see that there has been a strong negative relationship between the variables since 2001. Evidence from China shows a clear relationship between emissions and employment.

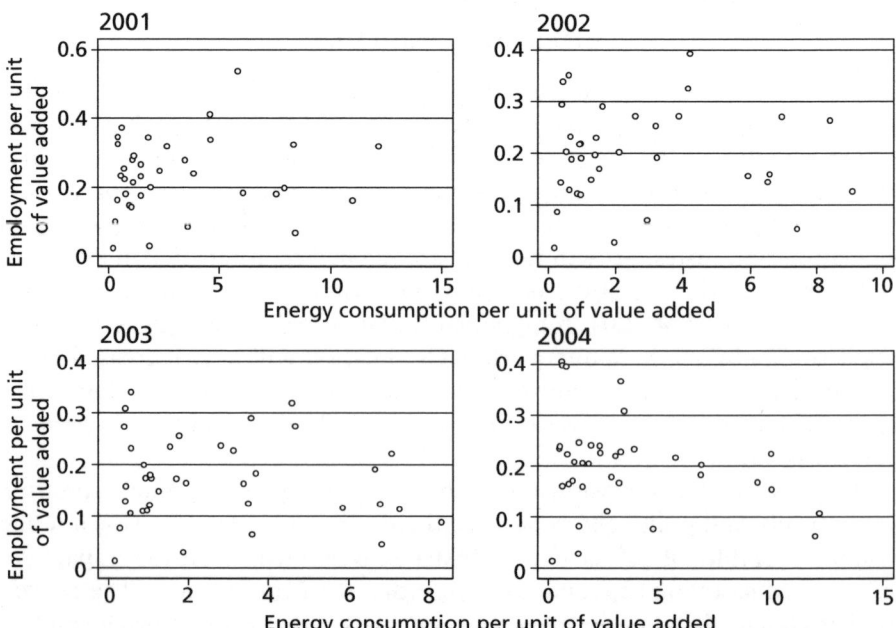

Source: National Bureau of Statistics (NBS) (various years) *China Statistical Yearbook*, Beijing: China Statistics Press

Figure 9.1 *Relationship between employment per unit of value added and energy consumption per unit of value added by sector, 2001–2004*

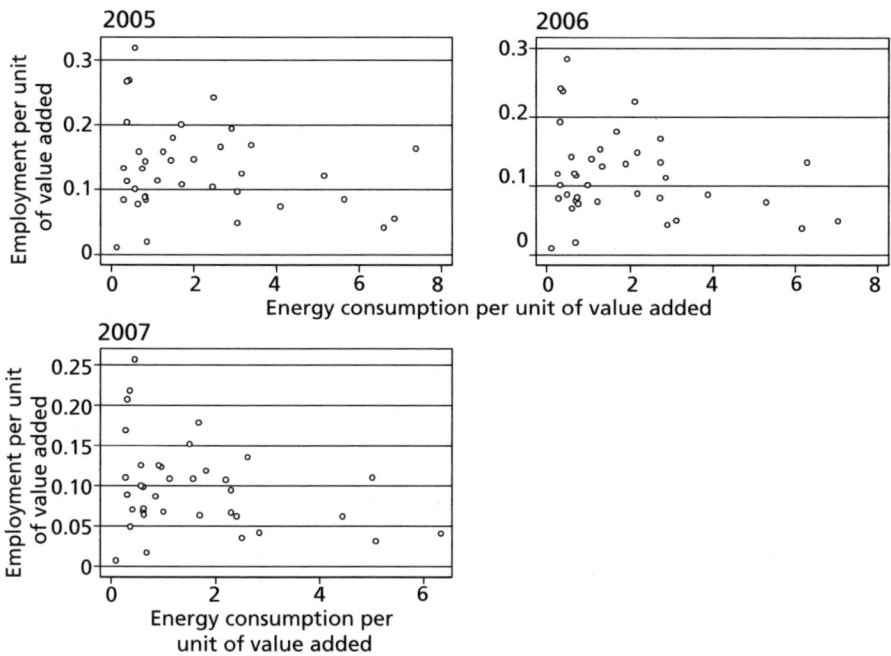

Source: NBS (various years) *China Statistical Yearbook*, Beijing: China Statistics Press

Figure 9.2 *Relationship between employment per unit of value added and energy consumption per unit of value added by sector, 2005–2007*

The environmental problems facing China are the result of its economic growth pattern, characterized by over-reliance on labour and capital inputs rather than on productivity enhancement, though this growth pattern suits China's particular stage of development. The Chinese stimulus of economic growth is unique because it combines both central and local governments to form a developmental state. This is the primary characteristic of government in its economic development function. Consequently, while individual enterprises inherently tend to respond to growth pattern changes resulting in emission reductions, the role of the government is vital to guarantee the fulfilment of the emission reduction strategy. The coming new development stage can help the government shift its policy orientation and put in place the relevant regulations to act as incentives among stakeholders.

The political economy of emission reduction relates to the different reactions of all relevant parties. First, for the central government, decision-making must be based on the recognition of the importance of emission reduction as required by the shift in China's stage of development. Second, for local governments, incentives to shift the growth pattern from input driven to productivity driven and the willingness to sacrifice short-term growth in GDP and fiscal revenue for sustainable growth are becoming more relevant in response to the central mandate. Third, for enterprises, assumed to care more about profits than externalities, internal problems requiring a solution in the form of a transition from input-based expansion to productivity-based expansion can be tackled by changing the attributes of production. Finally, for people, stronger demand for environment quality ultimately comes from the transition from subsistence-dominated livelihoods to more holistic human development, a result of China's development. In the final analysis, a strategy to deal with environment can only be implemented effectively by combining the quartet of capacity, responsibility, obligation and incentive.

The change in development stage requires the transformation of the growth pattern accordingly. The economic growth pattern can be described by the ways in which factors of production are allocated at both micro and macro levels and it can usually be classified by the sources that economic growth is based on. With China's rapid demographic transition, the period of reform and opening-up has been characterized by the adequate supply of labour and a high savings rate. This demographic dividend has resulted in a productive population structure, a market-led resources allocation mechanism and participation in the global economy. The favourable population structure has provided China's economy with a window of opportunity and thus, the phenomenon of diminishing return to capital has been avoided (Cai and Wang, 2005). Economic growth in transition China has heavily relied on inputs of production factors other than on productivity improvements. After a short rise in total factor productivity (TFP) and in its contribution to growth in the early reform period, China's TFP performance has been unsatisfactory since the 1990s (see for example Zheng and Hu, 2004; Kuijs and Wang, 2005). This has contributed to the growth pattern characterized by heavy pollution, high resource depletion and low efficiency (Kaneko and Managi, 2004).

The same analysis was made with regard to the Four Tigers when they created the so-called (disputable) East Asian miracle. At that time, Krugman (1994) noted that East Asian economic growth was merely fuelled by inputs of labour and capital, and the growth of TFP and its contribution to economic growth were insignificant. Krugman believed the miracle was doubtful because of neoclassical growth theory, which assumes diminishing return to capital due to a limited supply of labour, which was not true in those economies. Subsequently, once unlimited labour supply came to an end, their economies had to and did transform their economic growth patterns from input driven to TFP driven, and they have sustained their growths (Bhagwati, 1996). In China, after three decades of extraordinary economic growth largely fuelled by the demographic dividend, the population is aging and the reservoir of surplus labour in the rural areas is running dry. Thus, the scenario of labour supply and demand has fundamentally changed in China since 2004. With the approach of a Lewis turning point, the conditions are maturing under which economic growth is becoming increasingly reliant of productivity improvements rather than of expansion of inputs (Cai, 2008).

Increases in per capita income induce greater demand for security and quality of life, leading to calls for a better environment. A decade ago, the

World Bank (1997) estimated that in 1995, the losses resulting from air and water pollution amounted to $54 billion or 8 per cent of total GDP in China. Since then, especially during 1995 to 2006, the real income per capita of urban households increased by 131 per cent, and real income per capita of rural households increased by 74.8 per cent. As a result of the much faster rate of the income growth for upper-income groups, the richest 20 per cent earned 4.6 times more than the poorest 20 per cent. Per capita income level is the decisive factor in both of the widely used approaches to estimating the loss caused by environment damage – namely, the human cost and willingness-to-pay approaches. In particular, the upper-income group has the strongest bargaining power and ability to impact environmental policy decision-making. Improvements in the income levels of China's residents must play a role in increasing demands for environmental improvements. The frequent environmental incidents of recent years have demonstrated that citizens and press respond to environmental disasters in both a timely and enthusiastic manner (Hayward, 2005).

The concerns of Chinese residents, scholars, policy-makers, and to a lesser extent enterprises, over environmental quality have been translated conceptually into the central policy documents and into protocols of the 11th Five-Year Plan. All those documents repeatedly call for transformations in growth patterns, while the 11th Five-Year Plan stipulates restrictive criteria for emissions reduction. As a response partially to these regulations and partially to the increase in raw materials prices and wages, Chinese manufacturing enterprises have successfully improved their efficiency when using intermediate inputs and improved labour productivity (Kim and Kuijs, 2007).

An evolutionary or shock path

Labour markets are continually adjusting and unemployment is normal. In most cases, a small amount of unemployment can be resolved by the labour market alone. Most economies, however, try to avoid negative shocks to their labour markets since they cannot afford the social instability and welfare loss to individuals when large-scale unemployment happens.

An example is the labour market dislocation that took place in the 1990s when the Chinese labour market was affected by the Asian Financial Crisis and restructuring of state-owned enterprises (SOEs), which brought about massive lay-offs and unemployment in the urban labour market. It took the Chinese government several years to deal with the outcome by initiating active labour market programmes and constructing social safety nets in urban areas. The recent financial crisis also caused a negative labour market shock, as evidenced by a large flow of migrants back to rural areas. Employment recovery is one of the essential components of the recovery plan.

Considering the social and economic costs of labour market shocks, it is necessary to take into account employment consequences when planning for emissions reduction. Actions to reduce emissions could cause structural unemployment. Nevertheless, structural unemployment is not necessarily the result of a shock. On the contrary, if emission reduction is pursued in a gradual way, it is easy to harmonize the labour market and economic restructuring. It is evident that, under the pressure for emission reduction, an evolutionary path is helpful in order to stabilize the economy and the labour market.

The first thing to note is that most emission reduction plans decrease employment and slow down economic growth. Within the confines of the reduction obligations made by government, policy-makers still have scope to select different means to achieve the reduction. In all cases, taking into account the effects on employment is helpful in order to harmonize adjustment of the labour market.

Among the emissions reduction options, one minimizes employment loss. Figure 9.3 depicts the relationship between employment and emissions. Energy consumption is an intermediate input for production. In most cases, the greater the rate of energy consumption and emissions, the more employment results. In other words, if emissions are reduced, the price of unemployment will have to be paid. In addition, Figure 9.3 shows how the relationship between employment and emissions is not constant – the marginal costs of emissions reduction go up with increasing emissions reduction. The emission reduction is the same from E_0 to E_1 as from E_0' to E_1', however, the employment loss is S_0–S_1 and S_0'–S_1' respectively. It is evident that the latter incurs high costs in terms of employment.

Emissions reductions are realized through interventions in economic activities. Although both market signals and administrative means can be used to achieve reductions, these play different roles in the labour market. In general, emission reductions take an evolutionary path or a path that leads to economic shocks.

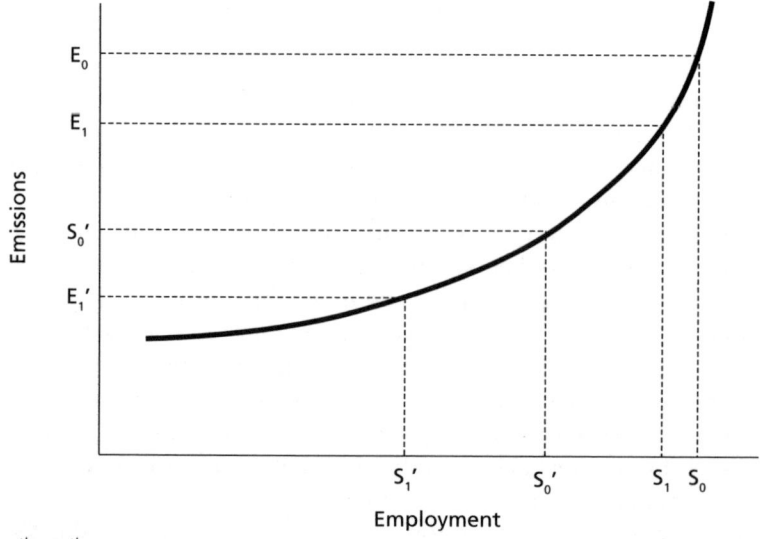

Source: the authors

Figure 9.3 *The relationship between emissions and employment*

In order to reduce energy consumption and emissions, the government has chosen administrative measures, shutting down enterprises characterized by high energy consumption and high pollution. Such enforcement echoes, to some extent, the measures that were typically employed in the planned economy and that are familiar in China. For instance, in the 1980s the central government demanded the closure of many township and village enterprises (TVEs) that were blamed for high energy and raw-materials consumption and were responsible for inflation and macroeconomic imbalance. As a result, the development of SOEs was contained. A similar situation occurred in the 1990s – when the macroeconomy was overheated, policy-makers insisted on the closure of some small businesses with high energy consumption and low efficiency, such as small-scale coal pits and small cement factories. Although these measures achieved their goal, they caused negative shocks to both economic development and employment. By contrast, large enterprises, in particular, the SOEs, have bargaining power, making small- and medium-sized enterprises the target of emission reduction policies.

The evolutionary path largely uses market mechanisms and price signals. This way differs from the above in terms of implementation, process and outcomes. By adjusting the carbon price, evolutionary policy changes the relative price between inputs with high emissions and others, and contains the growth of demand for high-carbon inputs. In the long run, enterprises have stable expectations and knowledge with regard to the high price of carbon, leading them to apply low-carbon technology and achieve the goal of emissions reductions. A carbon tax is a typical means to change the price of high-carbon inputs. In addition to inducing technology changes, the evolutionary path is easy to enforce because all the enterprises face the same carbon price, which makes implementation transparent and fair. Since governments do not have to inspect each enterprise, this path saves on implementation costs and helps eradicate the corruption associated with enforcement.

Finally, enterprises will reduce production when faced with high carbon prices, but as the emissions reduction process is incremental, this helps avoid massive unemployment caused by the shutting-down of enterprises. It is evident that the evolutionary path helps eliminate labour market shocks and facilitate social stability.

Figure 9.4 depicts the two alternative paths. The top curve describes the initial relationships between emissions and employment. Through shutting down the firms with high emissions, path I is a route that leads to economic shocks. In practice, the government may set up a permissible level of emissions and shut down those firms with emissions above that level, shown as the dashed line on the curve. However, this policy is only effective regarding the targeted enterprises if those that are untargeted (the solid part of the curve) do not change their behaviour. In this scenario, employment is affected and its loss is S_0–S_1.

Path II differs from path I. The price signal is intended to be effective for all enterprises and induce them to reduce emissions through adopting new low-

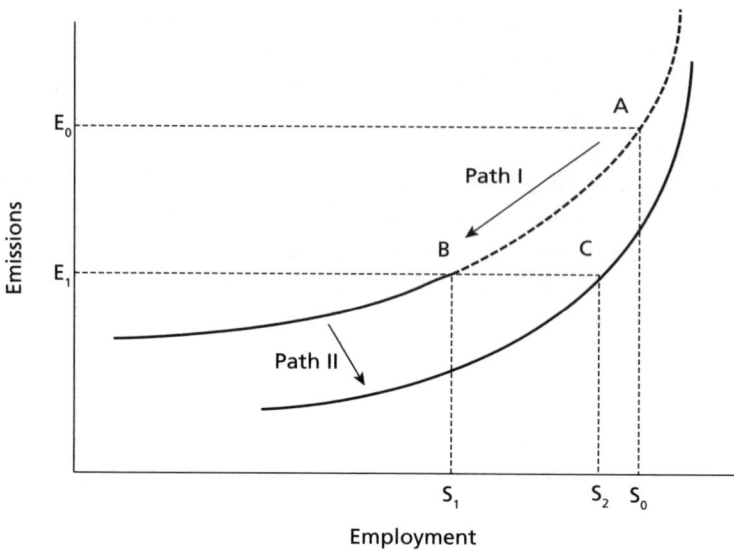

Source: the authors

Figure 9.4 *Two paths for emission reductions*

carbon technology. Therefore, the curve shifts to the right. When the emission level is set at E_1, corresponding employment will be S_2. The change from A to C reflects the effects of path II and the employment loss is S_0–S_2. Path II therefore results in limited employment loss.

In summary, these two paths to emissions reduction have different implications for employment. Path I leads to greater employment loss than path II. In addition, regardless of the path chosen, the effect of the reduction depends on the industries or regions that are targeted. In other words, when the government intends to intervene in economic activities in order to reduce emissions, how policies target industries, regions and enterprises makes a difference.

Sectoral priority: Energy consumption or energy efficiency?

The 11th Five-Year Plan of National Economy and Social Development clearly announces the goal of saving energy and reducing emissions. By the end of 2009, energy consumption per unit had been reduced by 15.61 per cent (see Table 9.1). To reach its goal, the Five-Year Plan needs to be implemented very effectively in 2010. Sparing no pains in the pursuit of energy saving and emissions reduction is likely to sacrifice not only economic growth but also employment expansion.

For a long time, the economic growth pattern characterized by extensive physical inputs has been a major factor driving increases in energy consumption. Transformation from an input-driven economic growth pattern to a

Table 9.1 *Energy consumption per unit of GDP (metric tonnes of standard coal equivalent (SCE)/¥10,000)*

| | Current price | | 2005 price | |
	Production energy	Total energy	Production energy	Total energy
2005	10.99	12.26	10.99	12.26
2006	10.42	11.62	10.80	12.04
2007	9.20	10.23	10.43	11.60

Source: NBS (2008)

productivity-driven pattern requires a relatively long period of time; however, the accomplishment of the objective of energy saving and emissions reduction is set out in the 11th Five-Year Plan and thus aims to fulfil the objective within a relatively short period of time. This contradiction results in calls for administrative tools, with strict regulations imposed on key sectors that dominate energy use and emissions.

To suit different purposes, industries or sectors can be categorized by various criteria. In addition to the most common categorization that divides sectors into primary, secondary and tertiary, the divide between light and heavy industries helps in the examination of energy consumption and emission of GHGs. In general, heavy industry is the number one energy consumer and polluter, and it also does not do well in creating employment. In the US, for example, the six biggest energy-consuming industries use 81 per cent of total energy consumption in manufacturing but only contribute 4 per cent of the nation's GDP and 2.5 per cent of total employment (see Chapter 5). Correspondingly, because these industries have small employment elasticity (see Figure 9.5), they are the sectors in which the emission costs of employment can be comparatively low.

The classification of industries into heavy and light does not, however, directly and sufficiently reflect energy consumption and emission level by sector, and also obscures differences within sectors. Moreover, such categorization usually generates no operational policy measures. For instance, at an aggregated level, one cannot simply recommend policy measures based on shock therapy in order to reduce the proportion of heavy industry. However, the following is applicable. First, the share of heavy industry can be depressed in a gradual way so that emission levels can be reduced. Second, considering endowments, costs of production, security of resources and potential market risks caused by overcapacity, China should not seek to become the global centre of heavy industry. Therefore over time heavy industry development has to be tempered. Since China has no comparative advantage in energy-intensive industries (see Chapter 4), it will not be harmed by the adjustment of energy prices.

In what follows, we empirically divide industries into two subsectors in accordance with their performance in energy consumption and energy efficiency, and examine the related characteristics accordingly. Energy is used

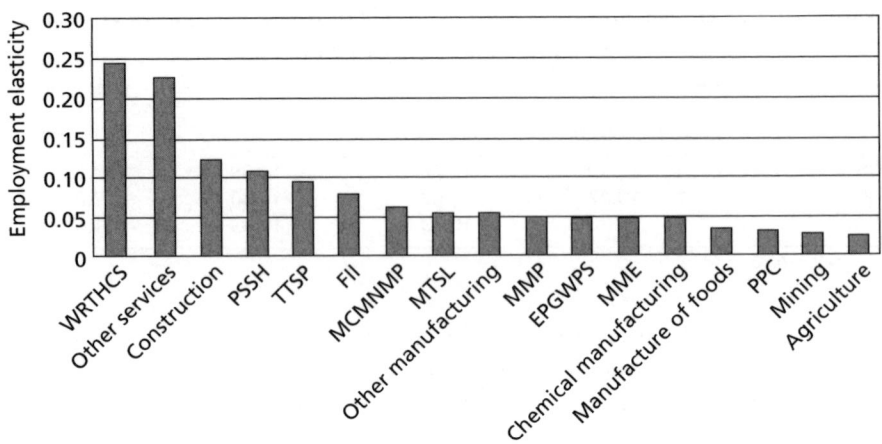

Note: WRTHCS refers to 'wholesale and retail trades and hotels and catering services', PSSH refers to 'public services and services of households', TTSP refers to 'traffic, transport, storage and post', FII refers to 'financial intermediation and insurance', MCMNMP refers to 'manufacture of construction material and non-metallic mineral products', MTSL refers to 'manufacture of textile, sewing and leather', MMP refers to 'manufacture of metal products', EPGWPS refers to 'electric power, gas and water production and supply', MME refers to 'manufacture of machinery and equipment' and PPC refers to 'processing of petroleum and coking'.
Source: Based on data in Zhu (2006)

Figure 9.5 *Employment elasticity by sector*

in both production consumption and resident consumption. In recent years, the share of resident energy consumption has been relatively stable, accounting for about 10 per cent of the total, while production dominates energy consumption with 3.1 per cent in the primary sector, 73.1 per cent in the secondary sector, and 13.7 per cent in the tertiary sector. Thus, one can conclude that the industrial sector acts as overwhelming consumer of energy and thus can be potentially the biggest energy saver and emissions reducer.

How do energy-consuming sectors perform in absorbing employment? From Figure 9.6 we can observe energy consumption, value added and employment by sector. Among non-agricultural sectors, manufacturing is the major energy consumer, utilizing 67.8 per cent of total energy. While manufacturing generates a higher proportion of value added and employment than other sectors – that is, it produced 37.9 per cent of value added in 2006 and 32.4 per cent of employment[2] – such contributions do not sufficiently cancel out its heavy energy use. In contrast, the service sector has a different profile. For example, sectors of wholesale, retail, hotel and catering consume only 2.6 per cent of total energy but generate 10.8 per cent of value added and create 21.2 per cent of employment. This implies that sectoral structure adjustment is potentially a highly valuable approach to reaching a desirable combination of low energy consumption, fast growth and full employment.

While China has become increasingly integrated into the global economy, changes in its industrial structure reveal some unique features. First, the Chinese experience of structural evolution follows the common pattern of

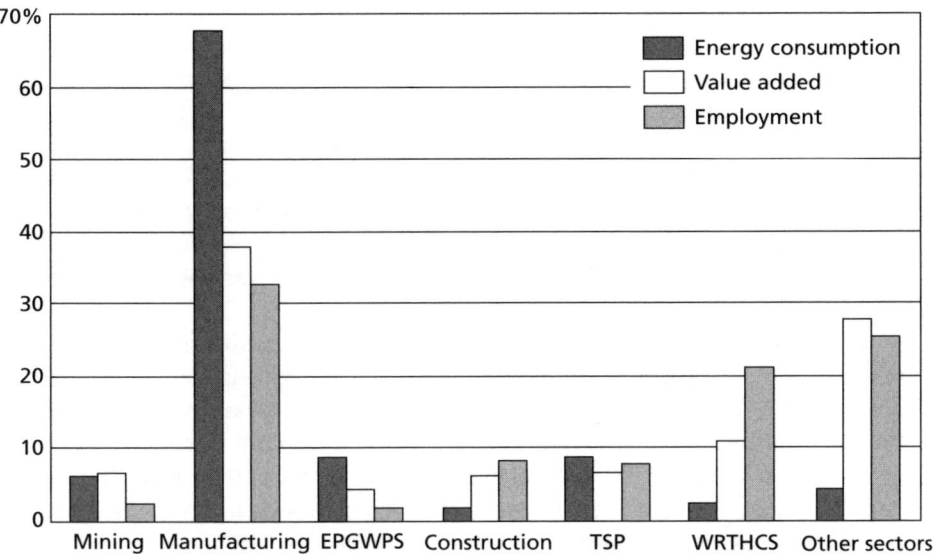

Note: (1) Energy consumption data are from 2007, value added data from 2006 and employment data from 2005; (2) EPGWPS refers to 'electric power, gas and water production and supply', TSP refers to 'transport, storage and post', WRTHCS refers to 'wholesale and retail trades, hotels and catering services'.
Source: Energy consumption and value added data are from NBS (2008); employment data are calculated from a 2005 1 per cent population sample survey.

Figure 9.6 *Energy consumption, value added and employment in non-agricultural sectors*

continuous growth of non-agricultural sectors. In 2007, the share of agricultural value added declined to 11.3 per cent of GDP. Second, economic growth is driven by manufacturing intensification, as a result of economic globalization. Thanks to the simultaneous expansion of trade and foreign direct investment, which provides an increasingly expanding international market for China's manufacturing, the basic pattern of non-agricultural economy has not fundamentally changed. The shares of the secondary and tertiary sectors in total GDP, for example, were 48 per cent and 44 per cent respectively in 1978, and they remain almost unchanged at 49 per cent and 43 per cent, respectively.

The uniqueness of China's sectoral structure change can be validated by international comparison. According to the World Bank (2008a), 20 countries with similar development levels to China averaged a primary sector share of 10 per cent of value added, a secondary sector share of 37 per cent of value added, and a tertiary sector share of 53 per cent of value added (see Table 9.15 in the appendix to this chapter). While non-agricultural share is identical to those countries, China's secondary sector share is 12 percentage point higher.

As already mentioned, industrial structure cannot be changed in a short period of time. In addition, because a country's sectoral structure is determined by its endowment of production factors and thus comparative advantage as revealed in international trade, overuse of industrial policy tends to deviate a country's industrial structure away from its comparative advantage and therefore jeopardize its competitiveness. Therefore, industrial structure change will not help gain energy savings and emissions reduction in the short term. However, implementation of compulsory policies involves consideration of policy costs, of which employment loss is one. It is worth noting that while heavy energy-consuming sectors often perform poorly in terms of energy efficiency, energy intensity and energy efficiency can be separately targeted in order to reach a goal, particularly when involving consideration of the employment costs of energy use-pattern revision.

A strategy focused on energy consumption requires different implementation mechanisms from a strategy focused on energy efficiency. While the former can have a direct impact by restraining the emissions of enterprises through administrative intervention, policy measures tend to generate a shock for enterprises and sectors. Moreover, this kind of strategy has a relatively narrow target. As discussed above, resolving emissions problems by using economic shocks leaves no room for sectors to improve their energy efficiency since they are subjected to shutdown under such a policy. However, the latter of the two strategies tends to gradually improve energy use efficiency of enterprises and sectors. Therefore, employment loss as a result of such policy measures is only marginal.

In recent years, administrative measures have dominated the campaign for energy saving and emissions reduction (see below). As is shown in Table 9.2, the top ten energy users (sectors) consumed more than 80 per cent of total industrial energy used, while they created about 42.3 per cent of industrial jobs in 2003. As a result of policy intervention targeting energy consumption, these sectors lost jobs – namely a 4.9 percentage points decline in their share of total industrial employment.

Ignoring efforts of improving energy efficiency can lead to increases in energy consumption. As is shown in Table 9.2, given the rigidity and distortion of energy prices, energy use in the low energy-efficient sectors increased significantly. The energy consumption share of the top ten low-energy efficient sectors increased from 65 per cent in 2003 to 76 per cent in 2007. The implica-

Table 9.2 *Employment performances of top ten energy-consuming sectors and top ten low energy-efficiency sectors*

| | By energy consumption (%) | | | By energy efficiency (%) | | |
	2003	2007	Change	2003	2007	Change
Share of industrial energy consumption	81.4	83.1	1.7	64.9	75.7	10.8
Share of industrial employment	42.3	37.4	−4.9	26.3	23.7	−2.6
Share of industrial value added	40.6	47.3	6.7	28.3	34.9	6.6

Source: National Bureau of Statistics (various years) *China Statistical Yearbook*, Beijing: China Statistics Press

tions are clear – that is, the marginal effects generated by an energy consumption strategy have reached their limit and implementation of energy efficiency-focused policy measures should become the prioritized strategy in the campaign for energy savings and emissions reduction.

China's Environmental Kuznets Curve

In his paper on the relationship between economic growth and income inequality, Kuznets (1955) argues that in the economic growth process, income inequality first increases then declines after reaching a turning point. This inverted U-shaped curve was later applied by environmental economists to depict a similar relationship between economic growth and environmental quality (Grossman and Krueger, 1995). The Environmental Kuznets Curve (EKC) can also be a useful framework to examine whether the Chinese economy possesses an inherent momentum to transform its growth pattern into a more environment-friendly pattern and in particular to understand the political economy behind the transformation. More concretely, we investigate the ways that changes in development stage impact on policy decisions on the environment from two aspects.

From the perspective of the region, we can deduce which regions should be targeted when implementing emissions reduction with lower costs according to their development stage (and the positions on the EKC).

The impact of emissions reduction on employment in the regions will vary. Due to differences in industrial structure and natural resources between regions, energy consumption per unit of value added will also different. Furthermore, due to the different industrial structures of the regions, emissions reduction will cause more shocks to employment in the regions with more high energy-consumption industries.

Due to a lack of data on CO_2 emission, it is hard to examine whether CO_2 emissions follow the pattern of the EKC. However, we can deduce this by looking at SO_2 emissions. In our recent study (Cai et al, 2008) it was found that there are significant differences in SO_2 emission among regions. As shown in Figure 9.7, SO_2 emissions in the coastal provinces follow the pattern of the EKC. According to our estimation, the turning point of the EKC appears when per capita GDP reaches ¥18,963 (at 1990 constant price). Beyond this point, emissions should decrease. Provinces with this pattern and that passed the turning point in 2007 include Beijing, Tianjin, Shanghai and Zhejing, while Guangdong and Jiangsu are very close to the point in terms of per capita GDP. In other words, under current circumstance, many provinces in eastern China already have the capacity and incentives to reduce their SO_2 emissions and follow a low-carbon growth path.

However, with accelerating economic growth, the central and western provinces continue the pattern of increasing emissions. The empirical results show that per capita income has a significant and positive effect on emissions, which implies that the central and western regions remain in a phase of increas-

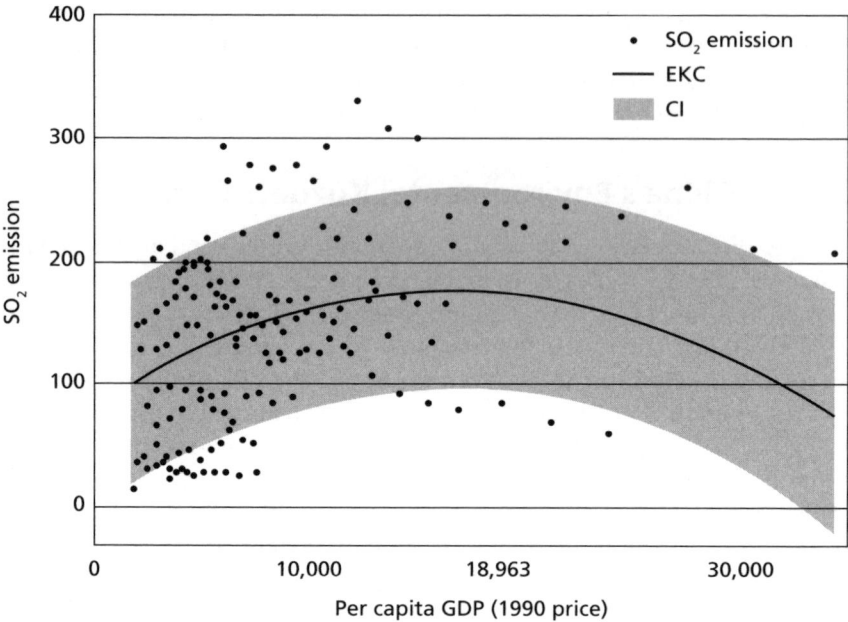

Note: CI = confidence interval.
Source: Cai et al (2008)

Figure 9.7 *Environmental Kuznets Curve: coastal provinces*

ing emissions. Figure 9.8 also shows that, with some variations, most of the provinces in the central and western regions adhere to the trend of steadily increasing SO_2 emissions. In comparison with the predicted EKC of the eastern provinces, the picture does not show any sign of the EKC pattern – the proportion of total emissions coming from the central and western regions has been increasing. For example, the emission of SO_2 in the central and western regions is twice as large as that of the eastern region. Considering the relationship between SO_2 emission and CO_2 emission, we are confident that CO_2 emission in the central and western regions has also been increasing. This also indicates that central and western regions should be the main focus of policies to reduce emissions.

While one can expect a future turning point from increase to decline in emissions for China as a whole, most Chinese provinces are still a long way from that turning point. The central and western provinces especially are still pursuing economic growth at the cost of the environment in order to catch up with their eastern counterparts. If this pattern continues, China will suffer a further worsening of the environment before reaching its spontaneous turning point. The experiences of spatial transfer of industries show (as also implied by the EKC) that the latecomers to economic growth tend to receive industries transferring from the more advanced regions in accordance not only with their comparative advantage but also with their acceptance of environment degrada-

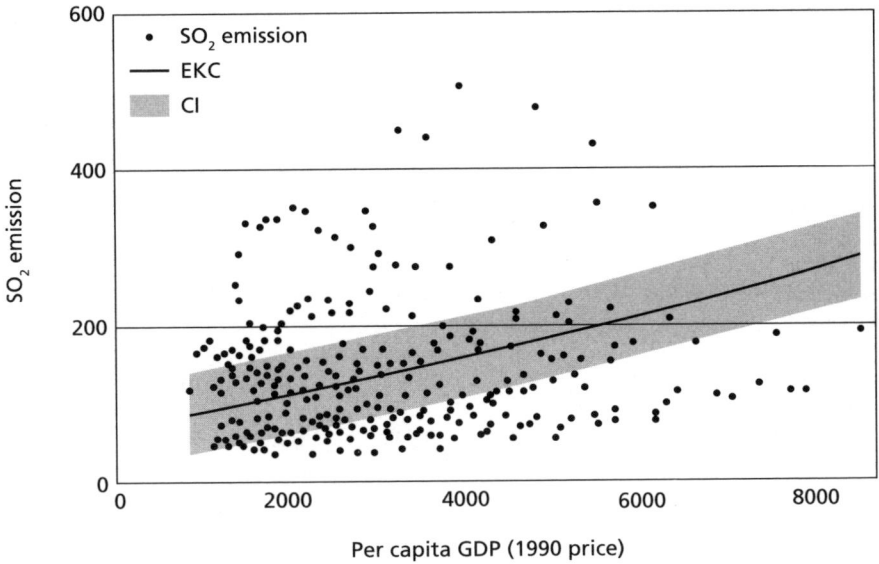

Note: CI = confidence interval.
Source: Cai et al (2008)

Figure 9.8 *Environmental Kuznets Curve: central and western China*

tion. Thus, the central and western regions may welcome polluting industries that are leaving the eastern regions. Given the strong desire for growth in the less-developed provinces and the large income gap between Chinese provinces, single incentives such as per capita income are not sufficient to lead those regions to the EKC turning point. Genuine changes must rely on the introduction of incentives and regulations derived from the need for the transformation of the growth pattern.

From the empirical results, one can see a great heterogeneity of SO_2 emissions among regions, which suggests distinct policy packages are needed for different regions in terms of emissions reduction. For most coastal provinces, which have either passed through or are moving towards the EKC turning point, the path of inertia and intrinsic forces will lead them to reduce emissions spontaneously. As far as the central and western provinces are concerned, it is hard to predict when they will reach the EKC turning point since emissions in these areas are accelerating. In this regard, it is vital to implement regulations to limit their emissions behaviour since China's total emissions are already among the largest in the world.

Table 9.3 shows the regional distributions of employment and value added of top ten energy-consuming industrial sectors in 2006. The proportion of total employment of ten industrial sectors in the eastern region is 56 per cent. That is, the high energy-consuming sectors bring abundant employment opportunities. Thus, if the eastern region is the main target for emissions reduction, it will lose many jobs.

Table 9.3 *Regional distribution of employment of top ten energy-consuming industrial sectors in 2006*

Sector	Eastern region		Central region		Western region	
	Number (tens of thousands)	Proportion (%)	Number	Proportion (%)	Number	Proportion (%)
Smelting and pressing of ferrous metals	421.2	5.72	73.76	1.00	59.59	0.81
Raw chemical materials and chemical products	505.5	6.87	85.19	1.16	62.01	0.84
Non-metallic mineral products	645.5	8.77	108.75	1.48	76.33	1.04
Production and supply of electric power and heat	354.9	4.82	90.48	1.23	64.78	0.88
Processing of petroleum, coking, processing of nuclear fuel	95.4	1.30	28.53	0.39	12.01	0.16
Smelting and pressing of non-ferrous metals	175.2	2.38	38.96	0.53	40.47	0.55
Mining and washing of coal	531.9	7.23	222.38	3.02	85.54	1.16
Textile industry	1034.5	14.06	94.47	1.28	46.82	0.64
Extraction of petroleum and natural gas	137.7	1.87	28.02	0.38	26.5	0.36
Papermaking and paper products	218.1	2.96	26.38	0.36	16.43	0.22
Total		55.99		10.83		6.67

Source: NBS (2007)

Table 9.4 gives the regional distribution of value added of the top ten energy-consuming industrial sectors. The proportion of the total value added of the ten industrial sectors is 30 per cent. The proportion of total value added of the ten industrial sectors of the central and western regions is only 6 per cent. If CO_2 emissions follow the EKC, emissions reduction measures that focus on the central and western regions result in less value added losses. We can say that either from the perspective of employment or value added, emission reductions that focus on the central and western regions will result in fewer losses. Considering that the eastern region already follows the EKC shape and emissions in the central and western region will continue to increase, emission reduction measures that focus on the central and western regions will be more efficient.[3]

Table 9.4 *Regional distribution of value added of top ten energy-consuming industrial sectors in 2006*

Sector	Eastern region		Central region		Western region	
	Number (hundreds of million yuan)	Proportion (%)	Number	Proportion (%)	Number	Proportion (%)
Smelting and pressing of ferrous metals	6779.4	3.87	897.4	0.51	643.8	0.37
Raw chemical materials and chemical products	5269.8	3.01	553.8	0.32	422.7	0.24
Non-metallic mineral products	3166.3	1.81	423.4	0.24	280.8	0.16
Production and supply of electric power and heat	7871.3	4.49	1216.5	0.69	1054.4	0.60
Processing of petroleum, coking, processing of nuclear fuel	2616.9	1.49	473.2	0.27	216.7	0.12
Smelting and pressing of non-ferrous metals	3625.8	2.07	173.6	0.10	102.2	0.06
Mining and washing of coal	3266.1	1.86	1145.2	0.65	435.9	0.25
Textile industry	1296.2	0.74	115.4	0.07	60.4	0.03
Extraction of petroleum and natural gas	1669.3	0.95	378.0	0.22	355.7	0.20
Papermaking and paper products	4990.0	2.85	1356.3	0.77	1031.0	0.59
Total		23.15		3.84		2.63

Source: NBS (2007)

Emission reduction strategies compatible with increasing employment

Emissions reduction will inevitably affect employment and economic growth. In the short term, the relationship between emission reduction and employment needs to be dealt with appropriately in order to achieve low-energy consumption and high employment. First, sectors with high energy consumption and low employment should assume the largest share of responsibility for emissions reduction, thereby causing the fewest shocks to employment. Second, though energy consumption for non-production purpose is only a small part of total energy consumption, slowing its growth will contribute to emissions reduction without having direct adverse impacts on employment and economic growth.

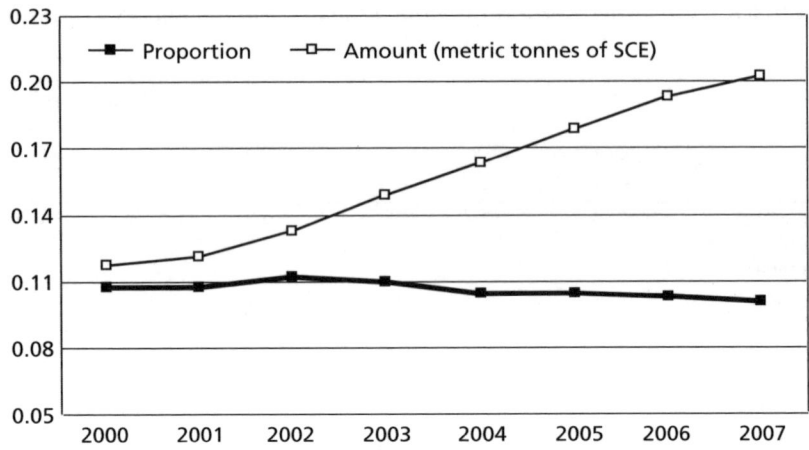

Note: (1) Proportion refers to the proportion of energy consumption for non-production purposes in total energy consumption; (2) Amount refers to annual average per capita energy consumption for non-production purposes.
Source: National Bureau of Statistics (various years) *China Statistical Yearbook*, Beijing: China Statistics Press

Figure 9.9 *Annual average per capita energy consumption for non-production purposes and as a proportion of total energy consumption*

Slowing the growth of energy consumption for non-production purposes

With economic development and improvement of people's living standards, energy consumption for non-production purposes increases. As shown in Figure 9.9, the percentage of energy consumption for non-production purposes in total energy consumption has remained at 10 per cent. However, per capita energy consumption for non-production purposes has been increasing. In 2000, it was about 0.118t SCE and in 2007 it increased by 72 per cent and reached 0.203t SCE. The annual average increase rate was 8.1 per cent.

Since energy consumption for non-production purposes is final consumption, slowing its growth through improving energy utilization efficiency will not cause employment loss directly. At the same time, under a set goal of emissions reduction, an increase of energy consumption for non-production purposes means that energy consumption for production will have to be reduced (which can potentially impact on employment). The impact of containing energy consumption for non-production purposes on emissions reduction and employment can be investigated by examining its potential to achieve energy savings and emissions reduction and its impacts on employment under different situations. Our analysis is based on three assumptions: first, under a set goal of emissions reduction, if energy consumption for non-production purposes cannot be reduced, energy consumption for production will have to be reduced. In other words, if energy consumption for non-production purposes is reduced, the pressure on energy consumption for production will

Table 9.5 *Increase in energy consumption for non-production purposes and its impacts on employment*

Energy consumption for non-production purposes (10,000t SCE)	Increase of energy consumption for non-production purposes (10,000t SCE)	Proportion of employment (%)	Number of employed persons (10,000)
Total number of population × per capita energy consumption for non-production purposes	Energy consumption for non-production purposes of the projected year – energy consumption for non-production purpose of 2007	Growth rate of energy consumption × employment elasticity with respect to energy consumption (0.1)	Proportion × number of employed persons in industrial sectors (108,290,000)
Annual average per capita energy consumption for non-production purposes is kept at the level of 2007			
2010 27,267.4	477.7	0.025	2.72
2015 27,962.6	1172.9	0.062	6.68
2020 28,354.4	1564.7	0.083	8.91
Annual growth rate of per capita energy consumption for non-production purposes is kept at the level from 2000 to 2007			
2010 34,444.5	7654.8	0.40	43.59
2015 52,141.5	25,351.8	1.33	144.37
2020 78,046.8	51,257.1	2.70	291.89
Annual growth rate of per capita energy consumption for non-production purposes is kept at 5 per cent			
2010 31,565.4	4775.7	0.25	27.20
2015 41,313.6	14,523.8	0.76	82.71
2020 53,466.5	26,676.8	1.40	151.91
Annual growth rate of per capita energy consumption for non-production purposes is kept at 3 per cent			
2010 29,795.8	3006.1	0.16	17.12
2015 35,422.2	8632.5	0.45	49.16
2020 41,639.4	14,849.7	0.78	84.56

be less and more jobs will be kept. Second, slowing the growth of energy consumption for non-production purposes will not directly cause employment losses. We assume that emissions reduction from energy consumption for non-production purposes has been achieved through decreases in unnecessary energy waste and improvement of energy utilization efficiency, keeping the same levels of consumer satisfaction and thus not reducing consumer demand. Third, employment elasticity with respect to energy consumption is stable. If we take energy consumption as one of the determinants of labour demand, employment elasticity with respect to energy consumption will decrease gradually with improvements in energy-saving technology. However, the elasticity is assumed to be constant during the period we examine.

If we estimate the potential of slowing the growth of energy consumption for non-production purposes, based on per capita energy consumption, we need to know the following basic information: total number of population, number of employed persons (industrial sectors), and employment elasticity with respect to energy consumption. In Table 9.5, the increase in energy consumption for non-production purposes is projected, based on different scenarios of per capita energy consumption for non-production purposes. If an increase in energy consumption for non-production purposes needs to be offset by a decrease in energy consumption for production, we can then calculate the employment loss accordingly.

In Table 9.5, the number of population is the projected number according to the total fertility rate (TFR) in 2007. The increase of energy consumption for non-production purposes takes energy consumption for non-production purposes in 2007 as the benchmark. Employment elasticity with respect to energy consumption is from estimates in Table 9.6. Since industrial sectors are intensively energy consuming and have the strongest restrictions on emissions, we use the number of employed persons in the industrial sectors when we estimate employment loss. Since 2002, the National Bureau of Statistics has not issued data on the number of employed persons by sector, so instead we estimate the number of employed persons of industrial sectors according to a 2005 1 per cent population sample survey,[4] resulting in a figure of 108 million.

Table 9.5 shows that, if the above assumptions are reasonable, slowing the growth of energy consumption for non-production purposes will contribute substantially to emissions reduction and have significant impacts on employment. If annual average per capita energy consumption for non-production purposes is kept at the level of 2007, emission increases from energy consumption for non-production purposes will be able to be offset by emission decreases from energy consumption for production, leading to just a small loss in employment. If the annual growth rate of per capita energy consumption for non-production purposes is kept at the level from 2000 to 2007, 435,900 employment posts will be lost to offset the emission increases in energy consumption for non-production purposes.

Employment elasticity with respect to energy consumption and its implications

To investigate the impacts of emissions reduction on employment, we need to know the relationship between employment and energy consumption. Employment elasticity with respect to energy consumption is the indicator that is able to reflect this relationship. We can deduce this elasticity by estimating the labour demand function. We employ the following equation to estimate employment elasticity with respect to energy consumption of industrial sectors:

$$\ln L_{i,t} = \alpha_0 + \beta_1 \ln Y{i,t} + \beta_2 \ln K_{i,t} + \beta_3 \ln w_{i,t} + \beta_4 \ln e_{i,t} + v_t + u_i + \varepsilon_{i,t} \quad (9.1)$$

The left side of the equation is the number of employed persons of sector i in year t, the right side is value added of sector i in year t, annual average balance of net value of fixed assets of sector i in year t, annual average wage of sector i

Table 9.6 *Labour demand function of industrial sectors and manufacturing sectors: two-way fixed effect model*

	Industrial sectors		Manufacturing sectors	
Value added	0.12	2. 93	0.10	2.08
Annual average wage	−0.72	−6.31	−0.67	−4.76
Energy consumption	0.10	2.55	0.12	2.78
Annual average balance of net value of fixed assets	0.50	11.56	0.53	10.82
Fixed effect: years	Yes		Yes	
No. of observations	262		206	
R²				
Between		0.85		0.89
Within		0.71		0.72
Overall		0.70		0.69

in year t and energy consumption of sector i in year t. β_1, β_2, β_3 and β_4 are parameters to be estimated, which are employment elasticity with respect to output, employment elasticity with respect to capital stock, employment elasticity with respect to wage and employment elasticity with respect to energy consumption, respectively. The control u_i reflects persistent sector-specific differences, such as prices, regulatory differences and preferences, and the control v_t, year dummy picks up changes over time, such as changing prices or technologies. $\varepsilon_{i,t}$ is randomly disturbing factors besides time and sector.

According to the above equation, we estimate the labour demand function for industrial sectors and manufacturing sectors from 2001 to 2007. Data on energy consumption, value added and annual average net value of fixed assets are for state-owned industrial enterprises and non-state-owned industrial enterprises above a designated size and data on annual average wages are for employed persons in urban units by sector (with data coming from years 2001–2008 of the *China Statistical Yearbook*). Due to adjustments to sector classifications, data for some sectors in some years are missing. The results are shown in Table 9.6.

Using logs on both the left and right side of the equation, the coefficient of the variable is actually employment elasticity with respect to the variable. According to the results, employment elasticity with respect to output is 0.12 (it is 0.10 for manufacturing), employment elasticity with respect to wage is −0.72 and employment elasticity with respect to energy consumption is 0.10. According to our previous study based on survey data at enterprise level (Cai et al, 2006), employment elasticity with respect to output is very similar to this. Employment elasticity with respect to wage is a little higher, but still lower than the estimations based on data for developed countries (Hamermesh, 1993). Here we focus on employment elasticity with respect to energy consumption. In the labour market, wage is the most important factor deciding labour demand. However, acting as an intermediate input, energy consumption will have derived effects on labour demand.

The results show that under the current technology level, if the energy consumption of industrial sectors decreases by 10 per cent, the number of employed persons may decrease by 1 per cent. Keeping other conditions constant, a decrease in energy consumption will result in the loss of 1 million jobs (based on 108.29 million employed persons in industrial sectors in 2005).

Emission reduction strategies compatible with enlarging employment

Implementing emission reduction in a gradual way can make emissions reduction and employment growth compatible. Emissions reduction without damage to economic growth and employment should a priority of emission reduction strategies. Slowing the growth of energy consumption for non-production purposes should be one such strategy. Our analysis shows that energy consumption for non-production purposes has been increasing. Slowing the growth of energy consumption for non-production purposes will contribute much to emissions reduction and will cause only limited loss of employment.

It should be pointed out that the positive effects of slowing the growth of energy consumption for non-production purposes are not only reflected in the above aspects. Since energy consumption for non-production purposes impacts on end-use energy consumption and the direct consumption behaviour of consumers, slowing the growth of energy consumption for non-production purposes will lead to innovations in energy-saving technology brought about by changes in consumers behaviours. At the same time, changes in the energy consumption pattern of consumers will also lead to demand for energy-saving products and drive demand for new products and employment opportunities. Thus energy consumption for non-production purposes is relevant to every citizen. Slowing the growth of energy consumption for non-production purposes helps boost people's awareness of the importance of energy saving and emissions reduction and helps construct a conservation-oriented society. The main instrument to slow the growth of energy consumption for non-production purposes is correction of the current distorted energy price system. Thus the price tool is the mechanism with the lowest cost and the optimal impact.

Our analysis shows that employment elasticity with respect to energy consumption in industrial sectors is positive, while employment elasticity with respect to energy consumption in manufacturing sectors is greater. Emissions reduction will lead to job losses. Keeping other conditions constant, if energy consumption decreases by 10 per cent, 1 million employment posts will be lost. However, the relationship between energy consumption and employment in industrial sectors is not static. The relationship between energy consumption and employment should be re-evaluated when factor prices, including energy prices, change and enterprises adjust their production structures and technology types in a gradual manner. So the most important strategy to ensure compatibility between emissions reduction and employment growth is the implementation of emissions reduction in a gradual manner.

Practices of emissions reduction: Employment loss and employment creation

When reducing emissions by closing enterprises with high pollution and production capacities and out-of-date technologies, some jobs will be lost. At the same time, developing low-coal and renewable energy and improving the energy structure will create employment opportunities. Thus there is the scope for both employment loss and employment creation when implementing emission reductions.

Employment loss

On 14 March 2006, the fourth session of the 10th National People's Congress approved the 11th Five-Year Plan for national economic and social development. According to the plan, during the 11th Five-Year period, energy consumption per unit of GDP will decrease by 20 per cent and total emissions of major pollutants will decrease by 10 per cent. In 2005, total emission of chemical oxygen demand (COD) was 14.14Mt and the plan was to decrease this to 12.7Mt by 2010. In 2005, the total emission of SO_2 was 25.49Mt, planned to decrease to 22.95Mt by 2010.

In order to achieve the goal of emissions reduction, several measures have been implemented in China, including measures to improve sewage treatment capacity by constructing sewage treatment plants and reducing emission by constructing new equipment such as coal-fired desulphurization sets. A second strategy involves reducing emissions by closing enterprises with high pollution rates and closing production facilities that depend on outdated technology. A third is to strengthen the auditing of clean production facilities, improve statistics on pollution and emissions reduction, strengthen inspections and monitoring systems, and to implement price, financing and trade policies that are favourable to emissions reduction.

Notable achievements have been made since 2005. In 2006, emissions of COD and SO_2 were 14.282Mt and 25.888Mt, respectively, increases of 1 per cent and 1.6 per cent compared with 2005. In 2007, emissions of COD and SO_2 were 13.818Mt and 24.681Mt respectively, decreases of 3.2 per cent and 4.7 per cent compared to 2006. These were the first recorded decreases in COD and SO_2 emissions. In the first half of 2008, the emission of COD was 6.742Mt, down by 2.48 per cent compared to the first half of 2007, while that of SO_2 was 12.133Mt, down by 3.96 per cent compared to the first half of 2007.

The implementations of the first and third measures for emissions reduction will not have a significant effect on employment. However, the implementation of the second measure will lead to job losses. In different

Table 9.7 *Number of job losses caused by closing enterprises*

Sector	Number of enterprises in 2004	Number of employed persons in 2005	Number of closed enterprises in 2007	Number of job losses in 2007
Papermaking and paper products	39,721	2,139,368	2018	108,689
Raw chemical materials and chemical products	75,179	3,842,559	500	25,556
Textile industry	83,103	8,978,160	400	43,215
Total	—	—	—	177,460

Source: Number of enterprises in 2004 from NBS (2004); number of employees in 2005 is calculated according to 2005 1 per cent population sampling survey data; number of closed enterprises is from *Report on China's Environmental Status in 2007* (People's Republic of China Ministry of Environmental Protection, 2008); number of job losses is calculated by the authors.

provinces the second measure has involved the closure of paper mills, chemical plants and textile, printing and dyeing enterprises and the closure of production facilities such as small thermal power-generating units and cement, iron, steel and flat glass facilities with outdated technology.

In 2007, some enterprises and production facilities with outmoded technology were closed, including 2018 paper mills, 500 chemical plants, 400 textile, printing and dyeing enterprises, 14.38 million kilowatts of small thermal power-generating units, and plants with a production capacity of 52Mt of cement, 46.59Mt of iron, 37.47Mt of steel and 6.5 million shipping containers of flat glass. All the enterprises and production facilities that were closed belonged to sectors with energy consumption ranking in the top ten. Therefore, the closure of these facilities was designed to make a significant contribution to emissions reduction. Taking Henan Province as an example, in 2008, total SO_2 reduction was 226,000t. The reduction from closing enterprises and production dependent on outmoded technology was 97,000t, which was 43 per cent of total emissions.

We can estimate the number of jobs lost as a result of closing enterprises and outmoded production facilities. Looking firstly at the number of job losses from closing paper mills, chemical plants and textile, printing and dyeing enterprises, we can derive the proportion of closed enterprises by dividing the number of closed enterprises by total number of enterprises within the sector. This proportion, multiplied by the total number of employed persons within the sector, will be the number of job losses; however, due to the limitations of the data, we only use the number of enterprises by sector in 2004 and the number of employed persons by sector in 2005, which may cause a bias in our results. We assume that the increased number of enterprises and number of employed persons are the same, then we can estimate the size of employment loss in 2007, using the method described above. Table 9.7 gives the number of job losses caused by closing enterprises: 108,689 caused by closing paper mills, 25,556 by closing chemical plants, and 43,215 by closing textile, printing and

Table 9.8 *Number of job losses caused by closing production capacity*

Sector	Production capacity in 2007	Number of employed persons in 2000	Number of employed persons in 2007	Closed production capacity in 2007	Number of job losses in 2007
Electronic power production	3281.553 billion kilowatts	147,2632	1,572,837	14.38 million kilowatts	–
Cement manufacturing	1361.1725 million metric tonnes	1670526	1,784,196	52 million metric tonnes	68,161
Iron	476.5163 million metric tonnes	488421	521,655	46.59 million metric tonnes	51,003
Steel	489.288 million metric tonnes	832632	889,288	37.47 million metric tonnes	68,102
Glass and glass products	539.1807 million shipping containers	698947	746,507	6.5 million shipping containers	8999
Total	—	—	—	—	196,265

Source: Number of production capacities in 2007 from NBS (2008); number of employed persons in 2000 is calculated according to 2000 census data; number of employed persons by sector in 2007 is calculated by the authors; closed production capacities in 2007 is from *Report on China's Environmental Status in 2007* (People's Republic of China Ministry of Environmental Protection, 2008); number of job losses is calculated by the authors.

dyeing enterprises. The total number of job losses was thus 177,460.

We then looked at the number of job losses caused by closing production facilites with outmoded technology, including small thermal power-generating units, cement, iron, steel and flat glass in order to derive the proportion of closed production capacity by dividing the amount of closed production capacity by the total amount of production capacity within the sector. This proportion multiplied by the total number of employed persons within the sector is the number of job losses; however, due to the limitations of the data, we could only derive the number of employed persons by sector in 2000, which may cause bias in our results. We can calculate the average yearly increase rate of total employment. If we assume that the increase rates of employed persons of the above sectors are the same as that of total employment, we can estimate the number of employed persons of these sectors in 2007 according to the average yearly increase rate of total employment. Thus we can estimate the number of job losses in 2007. Table 9.8 gives the number of job losses caused by closing production capacity. The total number of job losses 196,265.

In total in 2007 the number of job losses caused by closing enterprises and production capacity characterized by outmoded technology was 374,000 or 0.05 per cent of total employment loss. The size of job losses caused by closing paper mills was 108,689 or was 29 per cent of total employment loss. The number of job losses caused by closing production capacity in the cement and steel sectors was 68,161 and 68,102 respectively, or 36 per cent of total employment loss (though it should be remembered that these figures are only rough estimations).

Employment creation

The IPCC (2007) evaluation report on the causes of global climate change highlights that the energy consumption of human beings has increased the emission of GHGs. The activities of human beings are very likely to be the main cause of global warming (reliable at 90 per cent). An important mechanism to manage climate change is the reduction of GHG emissions. China is trying to reduce emission of GHGs by adjusting the industrial structure, transforming the economic growth pattern, saving energy, and developing renewables, among other measures, in order to cope with global climate change.

These measures have different impacts on employment. Some of them cause employment loss and some create employment opportunities. Among these measures, developing low-coal and renewable energies and improving the energy structure will not only reduce GHG emissions but also create employment opportunities. The measures include strengthening the development and utilization of water energy, nuclear energy, petroleum, natural gas and coal-bed gas, supporting the development and utilization of renewables such as biofuels, solar energy, geothermal energy and wind energy, and improving the proportion of clean high-quality energy in rural and remote areas and areas with appropriate conditions, all through the stimulus of national policies and financing. China has abundant renewable energy reserves that have already undergone development and deployment. The renewables with the greatest potential and a research foundation include water energy, bioenergy, wind energy and solar energy.

According to China's medium- and long-term development scheme for renewable energy, in 2005, the total amount of renewables developed and utilized (not including biofuels developed and utilized in traditional ways) was about 166Mt SCE, or 7.5 per cent of primary energy consumption. In 2010 and 2020, this figure it will reach 300Mt and 600Mt SCE respectively. The development and utilization of renewables will cause a significant decrease in coal consumption and supplement the shortage of natural gas and petroleum. It will also help improve the environment. When the development goal of 2010 is reached, the amount of utilized renewable energy resources will be equivalent to a decrease of 4Mt of SO_2 emission, a decrease of 1.5Mt of nitrogen oxides emissions, a decrease of 2Mt of soot, 600Mt of CO_2 emission, 1500 million cubic metres of water use, and the preservation of 150 million mu (10 million hectares) of forest land. The development and utilization of renewable energy resources, equipment manufacturing and its affiliated industries will generate many jobs and will drive the development of its relevant industries such as machine, electric, chemical and materials industries. The number of employed persons in the industries related to renewables is forecast to reach 2 million in 2020.

Table 9.9 *Emissions reduction strategy compatible with employment*

	Short run	Long run
Goals		
Energy consumption	20% decrease in energy consumption per unit of GDP	——
Energy efficiency	20% decrease in energy consumption per unit of GDP	——
Economic growth	7.5% of annual growth rate	——
Employment	Total employment increase of 45 million; transfer of rural surplus labour of 45 million	——
Targeting		
Industries	Manufacturing with high energy consumption and low employment	Industries with high energy consumption, household energy consumption
Regions	Central and western China	Nationwide
Measures		
Industrial policies	Shutdown of firms with high energy consumption and low employment	Support the development of industries with low energy consumption and high employment
Pricing	Remove the subsidy on energy consumption	Form pricing system at basis of market mechanism
Economic compensation	Compensation and transfer programmes across regions	Labour mobility and trade in emissions across regions
Technology substitution	——	Substitute high energy-consumption technology with new technologies with low energy consumption

Implementation mechanisms for emissions reduction policies and their coordination with employment policies

The implementation of emissions reduction measures brings costs, of which employment loss is one of the most significant. The impact of previous industrial policies on employment has often been ignored when restructuring an economy. Radical restructuring, in particular, can lead to a shock reduction in employment; avoiding this should be a concern for policy-makers. When considering employment loss as the price potentially paid for emissions reduction, the following aspects are worth noting in an attempt to harmonize employment and emissions reductions.

First, policy-makers should avoid negative labour market shocks when enforcing the strategy of emissions reduction. They should be wary of pursuing the goal through the closure of polluting firms. The role of market mechanisms should be valued: price signals can induce enterprises to employ energy-saving technologies, thus helping reduce any negative employment shocks. When using market mechanisms, the need for the inspection of enterprises is also avoided, which reduces implementation costs. In addition, when using energy price as a leverage, all enterprises are fairly challenged by a uniform price and therefore all of them have the incentive to reduce emissions. In contrast, administrative measures only achieve reductions from those enterprises that are closed. Policy-makers therefore need to carfully consider the safeguarding of jobs. As shown in Table 9.9, a considered plan should harmonize the goals, targeting industries and regions, and applying measures for short, medium and long terms.

Second, the economic structure that is mainly determined by the development stage needs to be a concern when implementing emissions reduction measures, since the economic structure makes a difference to the employment structure. Many developing countries, in the process of industrialization, rely on secondary industry that creates most job vacancies and serves as a means to absorb surplus labour from rural areas. For developed economies that have already completed the process of industrialization, employment is dominated by service industries characterized by low emissions. Thus, the same strategy of emissions reduction will cause more employment loss in a developing country. In particular, the economic and employment structure in China is unique. Compared to economies with similar per capita GDP, China's manufacturing sector is 10 per cent larger (see Table 9.15 in the appendix to this chapter), which means that China would pay a higher price in terms of unemployment if committing to the the same reductions.

Third, policy-makers can take advantage of the development of medium- to small-sized enterprises so as to realize the compatibility of employment and emissions reduction. Among those firms closed in China in order to reduce emissions, medium-scale enterprises account for a large share. In addition to having weak bargaining power, these have been repeatedly discriminated against because of their outmoded technology and poor energy consumption efficiency. In most economies medium-sized enterprises provide the greatest share of employment opportunities and their role in employment creation should be more highly valued in China. For this reason, in addition to applying a price leverage that induces medium-scale enterprises to innovate with technology, industrial policies that facilitate labour-intensive technology use in these enterprises will help reduce conflicts between employment creation and emissions reductions.

Fourth, policy-makers have to balance the varying interests of different regions by establishing transfer programmes and labour mobility mechanisms, reducing the incentive to pursue economic growth at the expense of increased emissions in the least-developed regions. The pursuit of high GDP growth in less-developed areas has resulted in the highest emissions increases in recent years.

On the one hand, there is no agreement over payments for pollution or compensation for reductions across regions, which means that the least-developed regions will not be compensated if they sacrifice economic growth and reduce emissions. Accordingly, they do not pay the price of high economic growth with high emissions. Hence, with the premise of respecting equal development rights among regions and imposing the same obligation to reduce emissions, agreement on the trade (or transfer) of emissions reductions across regions will help control emissions in the least-developed regions of China. Also, the reductions achieved in these regions will impact less on employment and economic growth.

On the other hand, a free labour market across regions also works to decrease the incentives of the poorest regions to pursue high economic growth and maintain employment growth. The least-developed regions pursue rapid economic growth because local governments are responsible for employment and benefits to local residents. With an integrated labour market, people in the least-developed regions can move to developed areas at lower costs. Therefore, the developed regions help solve the employment dilemma for least-developed regions. In addition, with the development of an integrated welfare system across regions, local governments will be less responsible for welfare provision and this will weaken their incentive to pursue rapid economic growth.

Finally, the strategy of emission reductions should include a plan to provide training services for enterprises so as to facilitate the extension of energy-saving technology. Too some extent, the extension of such technology is a win–win strategy. The new technologies, services and products reduce emissions while also creating job opportunities. Given that limited human resource capacity constitutes a barrier to the development of enterprises, the government can play an active role by supporting research, training and other public services for enterprises, thus perhaps also facilitating the balance between emissions reductions and employment.

Appendix

Table 9.10 *Basic indicators of top ten energy-consuming industrial sectors in 2007*

Sector	Energy efficiency (metric tonnes SCE/ ¥10,000)	Energy consumption (10,000 metric tonnes SCE)	Number of employed persons (10,000)	Value added (¥100 million)	Proportion of energy consumption	Proportion of employ-ment	Proportion of value added
Papermaking and paper products	2.29	3342.68	138.30	1459.13	1.76	1.76	1.49
Extraction of petroleum and natural gas	0.68	3677.49	90.67	5400.08	1.93	1.15	5.51
Textile industry	1.51	6207.57	626.26	4113.51	3.26	7.95	4.20
Mining and washing of coal	1.82	7170.75	463.69	3931.36	3.77	5.89	4.01
Smelting and pressing of non-ferrous metals	2.85	10,686.37	156.27	3748.27	5.62	1.98	3.83
Processing of petroleum, coking, processing of nuclear fuel	5.08	13,176.51	80.64	2592.52	6.93	1.02	2.65
Production and supply of electric power and heat	2.50	18,474.59	256.96	7390.78	9.71	3.26	7.54
Non-metallic mineral products	5.01	20,354.84	448.41	4059.32	10.70	5.69	4.14
Raw chemical materials and chemical products	4.43	27,245.27	380.28	6144.77	14.33	4.83	6.27
Smelting and pressing of ferrous metals	6.34	47,774.37	304.43	7540.00	25.12	3.87	7.70
Total					83.1	37.4	47.3

Source: NBS (2008)

Table 9.11 *Basic indicators of top ten energy-consuming industrial sectors in 2006*

Sector	Energy consumption (10,000 metric tonnes SCE)	Number of employed persons (10,000)	Value added (¥100 million)	Proportion of energy consumption	Proportion of employ-ment	Proportion of value added
Smelting and pressing of ferrous metals	42,812.3	296.1	7004.5	24.4	4.0	7.7
Raw chemical materials and chemical products	24,779	357.8	5398.8	14.1	4.9	5.9
Non-metallic mineral products	19,948.4	426.4	3656.2	11.4	5.8	4.0
Production and supply of electric power and heat	17,416.9	259.1	6912.5	9.9	3.5	7.6
Processing of petroleum, coking, processing of nuclear fuel	12,360.1	76.8	2314.23	7.1	1.0	2.5
Smelting and pressing of non-ferrous metals	8633.3	136.8	3198	4.9	1.9	3.5
Mining and washing of coal	6786.5	463.7	3587.2	3.9	6.3	3.9
Textile industry	5756.5	615.4	3963.0	3.3	8.4	4.4
Extraction of petroleum and natural gas	3626.2	93.3	5986.7	2.1	1.3	6.6
Papermaking and paper products	3443.7	134.8	1386.4	2.0	1.8	1.5
Total				83.1	38.9	47.7

Source: NBS (2008)

Table 9.12 *Basic indicators of top ten low-energy efficiency industrial sectors in 2006*

Sector	Energy efficiency (metric tonnes SCE/¥10,000)	Energy consumption (10,000 metric tonnes of SCE)	Number of employed persons (10,000)	Value added (¥100 million)	Proportion of energy consumption	Proportion of employ-ment	Proportion of value added
Smelting and pressing of ferrous metals	6.11	42,812.3	296.1	7004.5	24.4	4.0	7.7
Non-metallic mineral products	5.46	19,948.4	426.4	3656.2	11.4	5.8	4.0
Processing of petroleum, coking, processing of nuclear fuel	5.34	12,360.1	76.8	2314.2	7.1	1.0	2.5
Raw chemical materials and chemical products	4.59	24,779.0	357.8	5398.8	14.1	4.9	5.9
Production and supply of gas	3.37	646.3	14.5	191.7	0.4	0.2	0.2
Smelting and pressing of non-ferrous metals	2.70	8633.3	136.8	3198.0	4.9	1.9	3.5
Production and supply of electric power and heat	2.52	17,416.9	259.1	6912.5	9.9	3.5	7.6
Papermaking and paper products	2.48	3443.7	134.8	1386.4	2.0	1.8	1.5
Production and supply of water	2.38	748.8	46.1	315.1	0.4	0.6	0.3
Mining and processing of non-metal ores	2.37	896.7	44.2	378.1	0.5	0.6	0.4
Total					75.2	24.4	33.8

Source: NBS (2008)

Table 9.13 *Basic indicators of top ten energy-consuming industrial sectors in 2003*

Sector	Energy consumption (10,000 metric tonnes SCE)	Number of employed persons (10,000)	Value added (¥100 million)	Proportion of energy consumption	Proportion of employment	Proportion of value added
Smelting and pressing of ferrous metals	24,069.7	255.9	2894.5	20.1	6.7	4.5
Raw chemical materials and chemical products	17,108.2	311.3	2526.4	14.3	5.9	5.4
Production and supply of electric power and heat	13,276.9	238.4	3696.1	11.1	8.6	4.1
Non-metallic mineral products	12,656.1	396.2	1792.7	10.6	4.2	6.9
Processing of petroleum, coking, processing of nuclear fuel	8991.3	59.7	1319.6	7.5	3.1	1.0
Smelting and pressing of non-ferrous metals	5408.8	106.6	924.6	4.5	2.1	1.9
Mining and washing of coal	5395.8	376.6	1180.8	4.5	2.7	6.6
Extraction of petroleum and natural gas	4618.4	72.7	2447.8	3.9	5.7	1.3
Textile industry	3469.0	499.2	1954.3	2.9	4.5	8.7
Papermaking and paper products	2371.5	114.0	698.4	2.0	1.6	2.0
Total				81.4	40.6	42.3

Source: NBS (2008)

Table 9.14 *Basic indicators of top ten low energy-efficiency industrial sectors in 2003*

Sector	Energy efficiency (metric tonnes SCE/¥10,000)	Energy consumption (10,000 metric tonnes SCE)	Number of employed persons (10,000)	Value added (¥100 million)	Proportion of energy consumption	Proportion of employ-ment	Proportion of value added
Smelting and pressing of ferrous metals	8.3	24,069.7	255.9	2894.5	20.1	6.7	4.5
Manufacture of chemical fibres	7.3	2199.9	34.2	302.6	1.8	0.7	0.6
Non-metallic mineral products	7.1	12,656.1	396.2	1792.7	10.6	4.2	6.9
Processing of petroleum, coking, processing of nuclear fuel	6.8	8991.3	59.7	1319.6	7.5	3.1	1.0
Raw chemical materials and chemical products	6.8	17,108.2	311.3	2526.4	14.3	5.9	5.4
Production and supply of gas	6.6	513.3	14.7	77.2	0.4	0.2	0.3
Smelting and pressing of non-ferrous metals	5.8	5408.8	106.6	924.6	4.5	2.1	1.9
Mining and processing of non-metal ores	4.7	777.6	45.6	166.9	0.6	0.4	0.8
Mining and washing of coal	4.6	5395.8	376.6	1180.8	4.5	2.7	6.6
Mining and processing of ferrous metal ores	3.7	554.2	27.4	149.8	0.5	0.3	0.5
Total					64.9	26.3	28.3

Source: NBS (2008)

Table 9.15 *Industrial structure comparison between China and countries with equivalent per capita income*

	Per capita gross national income (PPP)	Proportion of primary industry (%)	Proportion of secondary industry (%)	Proportion of tertiary industry (%)	Population (million)	CO_2 emission per capita (metric tonnes)
Armenia	5890	19	47	34	3	1.1
Azerbaijan	5960	9	67	24	8	3.5
Philippine	5980	14	33	53	85	1.0
Peru	6080	7	34	60	28	1.0
Jordan	6210	3	32	66	6	3.3
Algeria	6900	8	61	30	33	5.1
Ukraine	7520	10	33	57	47	6.6
F.Y.R.O.M	7610	13	29	58	2	5.2
Columbia	7620	12	34	54	46	1.3
Panama	7680	7	16	76	3	1.9
China	7740	12	47	41	1312	3.2
Kazakstan	7780	7	39	54	15	10.7
Namibia	8110	11	31	58	2	1.2
Dominica	8290	12	26	61	10	2.3
Tunis	8490	11	28	60	10	2.1
Iran	8490	10	45	45	69	5.7
Brazil	8800	5	31	64	189	1.6
White Russia	8810	9	43	47	10	6.3
Turkey	9060	13	22	65	73	3.1
Thailand	9140	10	46	44	65	3.9
Average (excluding China)	7601	10	37	53	37	3.5

Note: CO_2 emissions per capita data from 2003.
Source: World Bank (2008b)

Notes

1   The figures on China's CO_2 emissions are disputable. Apart from China's official denial of the figures estimated and published by Western scholars, there is also a problem of transfer emissions – that is, over the past two decades, an increasing amount of polluting, energy-consuming and high-emissions products have been produced in China but mainly consumed outside China.

2   Because the National Bureau of Statistics no longer published sectoral employment data after 2003, the figures used here are calculated by the authors based on a 1 per cent population sampling survey conducted in 2005.

3   Due to lack of data on CO_2 emissions by region, we cannot examine its regional distribution by sector. Since CO_2 emissions are linked to SO_2 emissions and both the amount and proportion of SO_2 emissions in the central and western regions have been increasing (Cai et al, 2008), we are confident that it is reasonable to deduce trends in CO_2 emissions from SO_2 trends.

4   Employment elasticity with respect to energy consumption is calculated from data on 'stated-owned industrial enterprises and non-state-owned industrial enterprises above designated size by sector'. We assume employment elasticity with respect to energy consumption of non-state-owned industrial enterprises below the designated size is similar.

References

Bhagwati, J. N. (1996) 'The miracle that did happen: Understanding East Asia in comparative perspective', Keynote speech at the Conference on Government and Market: The Relevance of the Taiwanese Performance to Development Theory and Policy, Cornell University, 3 May

Cai, F. (2008) 'Approaching a triumphal span: How far is China towards its Lewisian turning point?', *UNU-WIDER Research Paper*, No. 2008/09, Finland: World Institute for Development Economic Research

Cai, F. and Wang, D. (2005) 'China's demographic transition: Implications for growth', *The China Boom and Its Discontents*, Canberra: Asia Pacific Press, no 6, pp4–11

Cai, F., Du, Y. and Wang, M. (2006) 'Employment in China's fast growing region: What can we learn from it to make employment expansion keeping up with economic growth?', Report for AAA Project of the World Bank, mimeo

Cai, F., Du, Y. and Wang, M. (2008) 'The political economy of emission in China: Will a low carbon growth be incentive compatible in next decade and beyond?', *Economic Research Journal*, vol 6

Grossman, G. and Krueger, A. (1995) 'Economic growth and environment', *Quarterly Journal of Economics*, vol 110, no 2, pp353–377

Hamermesh, D. (1993) *Labour Demand*, Princeton, NJ: Princeton University Press

Hayward, S. (2005) 'The China syndrome and the environmental Kuznets Curve', *Environmental Policy Outlook*, December, Washington, DC: American Enterprise Institute for Public Policy Research

IPCC (Intergovernmental Panel on Climate Change) (2007) *Climate Change 2007: Synthesis Report. Contributions of Working Groups I, II, III and IV to the Fourth Assessment Report of the Intergovernmental Panel on Climate Change*, Geneva, Switzerland

Kaneko, S. and Managi, S. (2004) 'Environmental productivity in China', *Economics Bulletin*, vol 17, no 2, pp1–10

Kim, S. and Kuijs, L. (2007) *Raw Material Prices, Wages, and Profitability in China's Industry – How Was Profitability Maintained When Input Prices and Wages Increase So Fast?*, World Bank China Research Paper No 8, Washington, DC: World Bank

Krugman, P. (1994) 'The myth of Asia's Miracle', *Foreign Affairs*, November/ December, vol 73, pp62–79

Kuiijs, L. and Wang, T. (2005) *China's Pattern of Growth: Moving to Sustainability and Reducing Inequality*, World Bank China Office Research Working Paper No 2, Washington, DC: World Bank

Kuznets, S. (1955) 'Economic growth and income equality', *American Economic Review*, vol 65 (March), pp1–28

NBS (National Bureau of Statistics) (2004) *China Economic Census Yearbook*, Beijing: China Statistics Press

NBS (2007) *China Industrial Statistics Yearbook 2007*, Beijing: China Statistics Press

NBS (2008) *China Statistical Yearbook 2008*, Beijing: China Statistics Press

People's Republic of China Ministry of Environmental Protection (2008) *Report on the State of the Environment in China 2007*, Beijing: Ministry of Environmental Protection

Stern, N. (2007) *The Economics of Climate Change: The Stern Review*, Cambridge and New York: Cambridge University Press

Thomas, M. (2007) 'Climate change and the Stern Review: An overview and comment from Future in Our Hands Network',
www.climatecooperation.org/index.php?title=Stern_Review/Mike_Thomas_2
World Bank (1997) *Clear Water, Blue Skies: China's Environment in the New Century*, Washington, DC: World Bank
World Bank (2008a) *World Development Indicators*, Washington, DC: World Bank
World Bank (2008b) *World Development Report 2008: Agriculture for Development*, Washington, DC: World Bank
Zheng, J. and Hu, A. (2004) 'An empirical analysis on inter-provincial productivity growth during the reform period', *Center for China Situation Studies Working Paper*, no 1, pp2–26
Zhu, J. (2006) 'The employment effect of heavy industry', *China Labour Economics*, vol 3, no 2

Part IV

Climate Change Mitigation: A Fair, Effective and Efficient Global Deal

10

International Mechanisms for Greenhouse Gas Mitigation Finance and Investment

Michael Lazarus and Clifford Polycarp

Introduction

Efforts to avoid dangerous climate change will require a profound transition in the way the world goes about its business. Such a transition will require deep changes in how energy and materials are produced and consumed, in how agricultural is practised, and in how lands are managed. It will require a large-scale deployment of climate-friendly technologies, the ability to finance technology development and investments of hundreds of billions of dollars annually, and a policy framework that can deliver this finance and technology to developing countries.

Domestic policies, economic signals and, most importantly, international financing instruments have so far been inadequate to shift current investment patterns, raise additional funding and direct funds to key investment gaps at the pace and scale needed for such a transition to take place. This shortfall has long been recognized. Benefiting from years of experience from the CDM, Global Environment Facility (GEF), multi- and bilateral programmes, policy analysts and policy-makers are addressing the question of finance with renewed vigour. Indeed, government negotiators made the reform of existing – and the creation of new – financing mechanisms a core element of their agenda for Copenhagen.

This chapter examines the scale of investment needed both globally and in China. We survey the sources of funding currently available and propose to ascertain the adequacy of the funds available or likely to be available in the coming years. We then examine some of the new financing mechanisms proposed, including reforms to the CDM, and the potential conflicts and complementarities in using them simultaneously. Finally, we propose some ways in which climate funds, carbon markets and domestic policies can work together in an adequate and efficient manner.

Mitigating climate change: Projected investment needs

In this section, we examine the premise that current international financing instruments are inadequate to deliver finance at the speed and scale needed to achieve a low-carbon growth trajectory. We review the findings of several recent assessments of investment and financial flows, and then distil key implications for various sectors of China's economy. As we will show, estimates of future investment and financing needs, globally and for China, vary by a factor of five or greater. Such differences are not surprising, given the uncertainties inherent in projecting future technology and resource availability and costs, as well as differences in socioeconomic assumptions and emission reduction ambitions. They are also due to disparities in what is considered or reported as investment or financial requirements and in how baselines are defined. Box 10.1 lays out some of the key components of investment and financing cost estimates.

By defining the scale and nature of future production and consumption patterns, the baseline scenario is a critical determinant of total investment and financial requirements for GHG mitigation. By gauging the extent to which policy or technology development might lead to BAU investment in renewable energy or other technologies that reduce GHG emissions, the baseline scenario is also a critical determinant of incremental investment and financial requirements. Climate policy, by definition, focuses on incremental needs. The extent to which investment in energy efficiency, renewable energy or other technologies occurs in a BAU scenario due to high fossil-fuel prices, local environmental benefits or other reasons, estimates of incremental investment needed to achieve a climate objective will be lower. Higher assumptions regarding BAU economic and demographic growth will tend to increase both total and incremental investments required. Not surprisingly, variations in BAU scenarios explain some of the differences among estimates of investment and financial flow requirements.

The scale of investment needs

Several recent studies estimate the scale of investment needed to achieve deep reductions in GHG emissions. The most detailed and widely cited of these

BOX 10.1 DEFINING TERMS:
MITIGATION COSTS AND INVESTMENT REQUIREMENTS

Though terms such as investment cost, financing requirements and mitigation cost have distinct meanings and applications, they are not always used consistently across the policy literature. For instance, a recent EU climate communication suggested that 'JRC and other independent institutes estimate the net global incremental investments [to reduce global emissions] in the order of €175 billion by 2020' (SEC, 2009). The €175 billion figure cited here (from Russ et al, 2009) represents the incremental mitigation cost rather the incremental capital investment, which the communication would seem to imply. Mitigation (or emissions abatement) cost estimates, such as those provided by Russ et al (2009) or McKinsey & Company (2009a, 2009b) include energy, resource and operational expenditures (or savings), often represent the financing of capital investments (the cost of capital) and apply discount rates. Thus, mitigation costs represent the full incremental costs of abatement and are often presented in 'cost curves' representing the abatement cost per tonne of CO_2. Incremental investment estimates, as reported in UNFCCC (2007), IEA (2008b) and McKinsey & Company (2009a, 2009b), represent only the capital investment associated with physical assets at the time of construction (not amortized). As a result, mitigation costs and investments requirements represent fundamentally different quantities: for energy efficiency investments in fact, the investment requirements can be high while mitigation costs are negative (resulting from higher net economic savings).

Term	Represents	Studies reporting	Application
Investment requirement	(Initial) capital investment associated with physical assets	UNFCCC (2007, 2008),[a] IEA (2008b)[b]	Capital requirements needed (for example in the form of debt, equity, or grants)
Financial flows	Recurrent costs, such as operation and maintenance, fuel and other resource costs, and other payments that do not add to physical assets; may also include costs of capital	UNFCCC (agriculture/forestry only); incorporated in mitigation cost studies below	Funding needed to cover ongoing annual costs
Incremental	The difference between costs in the mitigation scenario and the baseline scenario	All	Determines the additional investment, financial flow or mitigation costs required to achieve mitigation scenario over the baseline scenario
Mitigation or abatement cost	Full incremental cost including both investment and financial flows, discounted (generally at social discount rates of 2–5%) to return a net present value (NPV); often expressed in cost per CO_2t equivalent abated	Russ et al (2009), McKinsey (2009),[c] IEA's WEO and Energy Technology Perspectives (ETP) (2008a, 2008b)	Costs to society (bottom-up); can suggest the carbon price (revenue) needed to make a mitigation option profitable; used in 'cost curves'; always incremental to baseline, baseline assumptions critical

a These studies define investment flow as the initial capital expenditure for a physical asset and a financial flow as the ongoing expenditure that does not involve investments in physical assets.
b The IEA's ETP does not explicitly define capital investments but is comparable to the UNFCCC's definition, as both studies rely on the IEA's WEO, which defines investment as capital spending by businesses and by households and individuals, on cars, equipment, appliances and other energy-related expenditures.
c McKinsey & Company (2009a) defines investment somewhat differently as annualized capital expenditure repayments and the annual financial flows needed as the operational and maintenance expenditure including the savings from reduced energy consumption.

BOX 10.2 PROJECTED SCALE OF GLOBAL INVESTMENT AND ECONOMIC ACTIVITY IN 2030

- GDP: US$79,600 billion
- Total investment: $22,300 billion
  - Manufacturing: $15,500 billion
  - Electricity, gas, water supply: $1650 billion
  - China energy infrastructure: $100 billion (Sinton et al, 2005)
- Incremental mitigation investment
  - Based on 450–550ppm CO_2 equivalent target: $210 billion
  - China: $36 billion
  - Based on 450ppm CO_2 equivalent target (IEA ETP Blue Map Scenario): $1100 billion (IEA, 2008a)

Source: UNFCCC (2007) unless otherwise indicated

---

studies, *Investment and Financial Flows to Address Climate Change* (UNFCCC, 2007), was requested by the Conference of Parties to inform international deliberations on long-term cooperative action (LCA) and the review of existing and new financial mechanisms. The 2007 report and a subsequent update in 2008[1] have since become a benchmark for ongoing discussions regarding investment needs and financing mechanisms.

The UNFCCC (2007) study found that slightly over $200 billion per year in incremental investment and financial flows[2] would be required under a scenario that returns global emissions to 25 per cent below 2000 levels by 2030. This amount corresponds to roughly 0.3 per cent of projected GDP, or 1 per cent of projected overall global investment in 2030. In its 2008 update, the UNFCCC suggests that the incremental investment needed for reducing energy-related CO_2 emissions may be 170 per cent higher than its prior estimate, due largely to a rapid escalation in projected capital costs for energy supply facilities, an increase spurred by then-record high oil and commodity prices (see UNFCCC, 2007).[3]

In contrast, McKinsey & Company (2009a) estimates the upfront investment costs, incremental to BAU investments, to be more than four times higher, at approximately $1000 (€810) billion in 2030 (see Table 10.2).[4] As noted below, the McKinsey & Company analysis is more ambitious in terms of emissions trajectory and implementation of new technologies. Over 60 per cent of the incremental investment in the McKinsey mitigation scenario occurs in the transportation sector, largely due to a transition to hybrid and electric vehicles, a much stronger technology push than posited in the UNFCCC (2007) and underlying IEA (2006) analyses.

As shown in Table 10.1, the net, or incremental, annual investment and financial flow for mitigation is $210 billion globally by 2030, roughly half the level of gross[5] additional annual investment in mitigation technologies ($425 billion per year). The difference represents the savings in fossil fuel and other

Table 10.1 Annual emission reductions and investment needs: UNFCCC mitigation scenario (in 2030) and McKinsey abatement scenario (average 2026–2030 for global, 2010–2030 for China estimates)[6]

Sectors	UNFCCC GHG Emission Reductions (Gt CO$_2$-eq) Global	Non-annex I	UNFCCC Investment and Financial Flows (USD billion) Global	Non-annex I	China	McKinsey GHG Emission Reductions (Gt CO$_2$-eq) Global[c]	China	McKinsey Incremental Investment (USD billion) Global	China
Mitigation investment									
Power generation, of which	9.4	5.0	$149	$73	$36	14.4/10	2.8	$185	$60
Renewables, hydro, nuclear[a]			$85	$47	$19				
CCS			$63	$27	$17				
Industry, of which	3.8	2.3	$36	$19	$12	5.0/7.0	1.6	$140	$20
Energy-related			$20	$7	$3				
CCS			$14	$11	$9				
Other			$2	$1	$0				
Transport, of which	2.1	0.9	$88	$36	$11	3.2	0.6	$375	$90
Efficiency (incl. hybrids)			$79	$32	$11				
Biofuels			$9	$4	$1				
Buildings (an appliances)	0.6	0.3	$51	$14	$4	1.3/3.5	1.1	$250	$60
Waste	0.7	0.5	$0.90	$0.60	$0.10	1.5		$10	
Agriculture	2.7	0.4	$35	$13	$4	4.6		$0	
Forestry	12.5	12.4	$21	$21		7.8		$55	
Technology RD&D			$45						
Other									
Total mitigation investment			$425	$176	$68				
Avoided fossil fuel and energy infrastructure investment									
Transmission and distribution[b]			–$101	–$48	–$18				
Fossil fuel power generation			–$55	–$31	–$11				
Fossil fuel supply			–$59	–$32	–$3				
Investment savings			–$215	–$111	–$32				
Net mitigation	31.7	21.7	$210	$65	$36	37.9	6.7	$1000	$190–250

Note: a Combined additional investment needed in renewables, nuclear and hydropower; b Does not consider increased transmission and distribution needed to provide electricity access to unserved populations in developing countries; c Where two numbers are shown for emission reductions, the first number assigns all electricity emission reduction to the power generation sector (similar to UNFCCC), while the second allocates emission reductions due to decreased electricity demand to industry and buildings sectors.

Source: UNFCCC (2007); McKinsey (Global) (2009a); McKinsey (China) (2009b)

energy infrastructure that could be avoided if investments are shifted to lower-GHG alternatives. It is important to bear in mind these investment levels are over and above those of a BAU scenario that includes most of today's climate and energy policies, such as China's Renewable Energy Law and the EU emissions trading system, and thus represent very ambitious targets.

UNFCCC (2007) based its energy-related GHG mitigation investment estimates on the IEA's 'Beyond Alternative Policy Scenario' (BAPS) in its 2006 WEO (IEA, 2006). This scenario returns energy CO_2 emissions to 2004 levels by 2030 and total GHG emissions to below current levels as the result of major efforts to reduce deforestation and increase forest cover. While ambitious, this scenario is consistent with a 450 to 550ppm CO_2 equivalent concentration stabilization pathway and still has significant probability of exceeding a 2°C climate stabilization goal (Meinshausen et al, 2006).

Notwithstanding the limitations in comparability between the two studies, the higher estimates in the McKinsey & Company (2009a) report *Pathways to a Low-Carbon Economy* compared to the UNFCCC report can be largely explained by two factors. First, McKinsey's estimate of 2030 global baseline emissions is slightly over 10 per cent higher, 70 versus 62Gt CO_2 equivalent, as shown in Table 10.2. Second, and more importantly, the McKinsey analysis embodies a more ambitious pursuit of, and optimism regarding, low-GHG energy and industrial options. This is especially the case for options with higher marginal costs, such as electric vehicles, CCS, early retirement of coal-fired power plants, and a higher penetration rate for renewables, among others. As a result, global energy and industrial emission reductions are over 50 per cent higher in McKinsey (26Gt CO_2 equivalent) than in UNFCCC (16Gt CO_2 equivalent).

Table 10.2 compares the emission goals, emission trajectories and mitigation cost estimates of four recent and prominent mitigation studies. The IEA's ETP report represents the other major global assessment of investment needs associated with deep reductions in GHG emissions (IEA, 2008a).[7] The ETP's deepest mitigation scenario, 'BLUE Map', is also more ambitious than the UNFCCC and IEA's WEO scenarios (2008b) out to 2030 and, much like the McKinsey study, begins implementation of several yet-to-be fully commercialized technologies in order to achieve much deeper emission reductions in the longer run: plug-in hybrid, fuel cell and electric vehicles, solar photovoltaics (PV) and concentrating plants, and CCS in industry as well as power generation. The ETP report also accounts for significant investments in technology learning: research, development and diffusion activities to commercialize and drive down costs. As a result, the incremental investment costs in the ETP Blue Map scenario are also considerably higher than those in UNFCCC's analysis, averaging $1100 billion per year from 2010 to 2050, similar in scale to McKinsey (2009a) estimates for 2026–2030.

Table 10.2 *Comparison of selected ambitious global climate mitigation scenarios that estimate investment implications*

	UNFCCC IFF 2007 (based on WEO 2006 BAPS)	ETP 2008 BLUE MAP	McKinsey 2009 (Global)	WEO 2008 450 Policy Scenario
GHG stabilization target	450–550ppm CO_2 equivalent	445–490ppm (IPCC Category 1)	510ppm (peak) 450ppm (long-term stabilization)	450ppm
2030 baseline GHG emissions	62Gt CO_2 equivalent total (includes policies adopted by mid-2006)	42Gt CO_2 equivalent energy	70Gt CO_2 equivalent total	60Gt CO_2 equivalent total
2030 mitigation scenario GHG emissions	29Gt CO_2 equivalent total 26Gt CO_2 energy	23Gt CO_2 energy	32Gt CO_2 equivalent total	39Gt CO_2 equivalent total 26Gt CO_2 energy
2030 GHG emission reductions	19Gt CO_2 equivalent non-forestry; 12.5Gt CO_2 equivalent forestry		14Gt CO_2 equivalent energy efficiency; 12Gt CO_2 equivalent low-carbon energy; 12Gt CO_2 equivalent terrestrial carbon	
2030 incremental investment needs	$210 billion (0.3% of GDP)	$1100 billion (avg. of 2010–2050), 1.1% of GDP; 60% in transport sector	$1000 billion (1% of GDP)	$190 billion (net avg. of 2010–2030)
Average/marginal abatement cost (2030)		Measures up to a marginal cost of $200/t CO_2 Average cost of $38–117/t CO_2	Measures up to the average cost of $75/t CO_2 equivalent (€60/t CO_2 equivalent)	$180/t CO_2 equivalent
Energy CO_2 emissions peak	Not indicated	30Gt CO_2 in 2012	Not estimated	33Gt CO_2 equivalent in 2020

Implications for investment needs in China

Under a global low-carbon future, a significant fraction of mitigation investment will need to occur in China, in the order of 20 to 40 per cent of incremental investment by 2030, based on the studies reviewed here.[8] In fact, if the relative finding of the UNFCCC and McKinsey studies are a good indication, the more ambitious the reduction goal, the greater the share of investment occurs in China.

UNFCCC estimates the gross incremental mitigation investment needed in China in 2030 to be $68 billion, over half in new generation and CCS in the power sector, as shown in Table 10.1. Accounting for savings in conventional fossil fuel and energy infrastructure investments, net incremental mitigation investment needs are $36 billion. In contrast, McKinsey & Company (2009b) pegs China's net incremental mitigation investment required at $190–250 (€150–200) billion annually, averaged over the 2010 to 2030 period. McKinsey estimates that a third of these investments will have positive economic returns; a third will have a slight to moderate economic costs, while the rest will have substantial economic costs. These investments also face barriers, such as employment dislocation, limitations in technical and administrative capacity or competing consumer preferences, which could limit their full realization or increase their implementation costs relative to study estimates. Efforts to overcome these barriers could entail additional costs, for example, to compensate affected communities (lost jobs in more carbon-intensive activities) or train workers in the use of new technologies. However, the cost of overcoming these barriers is typically difficult to estimate. Some represent transfer payments (for example compensation) rather than direct societal costs. Furthermore, low-GHG technologies may also entail social and environmental benefits (improved health from reduced particulate emissions) that are also excluded from the estimates above.

As with the use of models to guide policy more generally, the value of these mitigation investment studies lies more in their insights on relative outcomes and response to policies than in their precise numerical estimates. This is especially true for long-term mitigation scenarios, which embed explicit assumptions regarding future economic drivers and technology costs and availability, as well as implicit assumptions regarding the response of human institutions and behaviour. With those caveats in mind, our analysis suggests that under an ambitious GHG reduction scenario:

- **Power generation is likely to be the predominant sector for mitigation investment in China over the next decade.** The UNFCCC (2007) analysis suggests that a significant fraction of the $36 billion investment needed could be largely financed by shifting investment from conventional sources, as suggested in Table 10.1 (see avoided infrastructure investment at the bottom of the table).[9] McKinsey & Company (2009b), however, estimate that China's incremental power generation sector investments to be quite significant, averaging approximately $60 (€50) billion each year. This figure is reflective of a far more aggressive pursuit of (more costly) low-carbon technologies, and in particular wind and solar, than in the UNFCCC analysis or in current policy (Renewable Energy Law) in China.[10] For reference, in 2007, China invested $12 billion in renewable energy (excluding large hydro), second only to Germany, which invested $14 billion (Zhang, 2008). This investment was made to implement the targets under China's Renewable Energy Law, and is a part of the baseline in both the UNFCCC and the McKinsey studies: in other words, their estimates of $36 and $60 billion per year, respectively, are over and above the significant investment in renewable energy already expected to take place.

- **Transportation may prove to be the most investment-demanding sector in the longer run.** At a global level, the IEA (2008b) projects that 70 per cent of the added energy investment cost through to 2050 will be in the transport sector – fuel cell, hybrid and/or electric vehicle infrastructure and equipment – in its ETP Blue Map scenario (IEA, 2008a). McKinsey (2009b) estimates that in China, the transportation sector will account for over 60 per cent of its incremental mitigation investment needs through to 2030. This projection is based on an optimistic outlook for electric vehicle potential, their wide adoption starting in 2016, and rising quickly to account for 100 per cent of new fleet by 2020.[11] McKinsey (2009b) envisages that China could emerge as a global leader in electric vehicle technology, thus catering not just to domestic but also global demand. However, major breakthroughs in electric vehicle technology (for example batteries) will be needed. 'Technology learning' investments will be particularly vital over the coming two decades to develop, diffuse and drive down costs of these and other yet-to-be-commercialized technologies.

- **Alternatively, carbon capture and storage could represent the predominant investment need in China from 2020 onward.** In UNFCCC's mitigation scenario nearly three-quarters of the projected net incremental investment needs in China are for CCS in 2030.[12] Whereas other mitigation technology investments in other sectors avoid investments in conventional technologies and fossil fuels infrastructure, CCS is purely an end-of-pipe cost that can, in fact, increase fossil fuel consumption and infrastructure needs.[13] Much like electric vehicle technologies, there are large uncertainties regarding the ultimate costs and feasibility of CCS and they depend on learning rates, the location and reliability of storage sites (in China most coal plants are located far from suitable storage locations). Similar to electric vehicles, China could emerge as the leading global supplier of CCS equipment (McKinsey & Company, 2009b). Should CCS not prove viable at the needed scale, the pace of investment in other low-carbon power generation technologies would need to accelerate at a much brisker pace to meet emission reduction objectives.

- **China's energy intensity target, if achieved and sustained, could require investment flows in energy efficiency on a scale of those envisioned in more ambitious mitigation scenarios.** Energy efficiency in buildings and appliances presents the highest mitigation potential in the short term, with mostly positive economic results, but nonetheless significant investment requirements. The 11th Plan's target of improving energy intensity by 20 per cent (discussed below) is estimated to require an annual investment of $35 to $40 billion through to 2010 (Energy Foundation, 2007). This estimate compares with up to $80 billion per year[14] in incremental efficiency investment through to 2030 in McKinsey (2009b). Motivating and sustaining private investment at this level will be challenging – public investment in energy efficiency was $6 billion in 2008[15] – as efficiency investments in buildings and industry face significant market barriers.[16]

- **Investment in changing transportation modes, reducing vehicle miles travelled (VMT), traffic management systems, and designing urban areas to reduce energy-intensive transportation needs could be significant,** but it is generally not a significant component of modelled scenario exercises. Take, for example, rail investment which is a major component of China's recent stimulus package. China is expected to build 3000km of high-speed railways within the next 15 years. Considerable potential may exist to increase the share of goods transported by rail instead of truck, which can reduce fuel use by a factor of three to ten; however, such assessments tend to be highly site specific and generally not included in mitigation studies.

- **The recovery of waste and by-products from emissions-intensive industries[17] in China presents the second highest abatement opportunity after efficiency improvements in buildings and appliances, also with some of the lowest costs of abatement.[18]** Because investment requirements tend to be fairly small as well, as noted below, non-CO_2 emissions tend to be highly amenable to carbon finance, as well as fund-based approaches.

- **The financial flows and opportunities to retain and increase carbon stocks in forest, agricultural, and other lands could be quite significant globally,** contributing to a third of emission reductions in the McKinsey analysis and nearly half the share in the UNFCCC.[19] Agriculture and forestry mitigation opportunities are less significant in China than globally (for example less than 10 per cent). New international financing mechanisms to direct financial flows to retaining forests and increasing carbon stocks are a distinct and separate focus in the ongoing climate negotiations. Therefore, we do not address these sectors further in this chapter.

Although the specific numerical estimates of the investments needs are uncertain and should be viewed with caution, they do place the scale of investment needed within a specified range: $200 to $1000 billion globally and $35 to $250 billion for China over the next two decades. While the overall estimates are a useful indicator of the scale of funds that need to be mobilized, the insights they provide on sectoral and technological needs are also useful for the design of policy and financing mechanisms, as discussed below. Moreover, there is greater convergence in the estimates for the power generation sector (which vary by a factor of two) where the abatement costs are more certain, than in the transportation sector (which vary by a factor of four or five). The divergences observed can be largely attributed to the extent of pursuit of higher-cost technologies and technologies whose prospects remain uncertain, such as large-scale CCS and electric vehicles.

A significant amount of reductions can be achieved at very low or negative costs, once the savings from lower energy use and other efficiencies are taken into account. In China, building and appliance energy efficiency and industrial waste recovery present such opportunities. A significant amount of reductions could thus be achieved by adopting and accelerating policies that help to overcome market barriers, such as information and incentive programmes and energy codes and standards.

Current financing

Finance for mitigation in developing countries is currently driven largely by domestic policies and regulations, predominantly focusing on measures to promote energy efficiency and renewable energy. A major share of the investment in mitigation activities are financed using domestic capital and are often led by domestic state-owned enterprises. To some extent, domestic investment is complemented with public and private foreign capital and carbon finance.

Among mitigation technologies, the concentration of investment in energy efficiency and renewable energy activities is not surprising, given that these activities contribute to other public policy goals, namely energy security, air quality, cost savings and competitiveness. Carbon finance, however, has been most instrumental in driving investment towards low-cost mitigation opportunities with more limited public benefits, leading to significant emission reductions in non-CO_2 gases in the industrial and waste management sectors. The investments generated for mitigation activities in recent years indicate that investors respond quickly to changes in policies and incentives (UNFCCC, 2007).

In this section, we review China's energy policies that contribute to reducing GHG emissions, investment trends in renewable energy and energy efficiency, and the role of international public funds and the carbon market (CDM) to date.

Domestic policies

Appropriate policies and incentives at the domestic level enable shifts in both public and private investments into more climate-friendly alternatives (UNFCCC, 2007). Broadly speaking, policies that reduce the time to market for renewable energy projects – for example, by speeding up the planning process – and provide investors with greater certainty about future support, will foster investment (Boyle et al, 2008). Domestic measures also help optimize the use of available funds by spreading risk across private and public investors (UNFCCC, 2007). Domestic policies sometimes provide project developers a critical source of revenue (for example via feed-in tariffs and mandatory purchase requirements). More generally, they provide legitimacy and a higher level of public recognition helpful in capital markets (Miller, 2008).

China's Renewable Energy Promotion Law, which aims to increase the share of renewable electricity generation to 10 per cent by 2010 and 16 per cent by 2020, provides a variety of incentives, including preferential grid

Table 10.3 *Current and planned renewable energy capacity in China*

	Hydro	Wind	Solar PV	Biomass power	Tidal	Bio-ethanol	Bio-diesel	Solar water heating	Biogas billion	Geo-thermal
	GW	GW	GW	GW	GW	billion litres		million m²	m³	GW
2007 capacity	145	6	0.1	3	–	1.6	0.1	130	10	0.032
2020 NDRC target	300	30	1.8	30	0.1	12.7	2.4	300	44	12 Mtce

Note: NDRC is the National Development and Reform Commission.
Source: Boyle et al (2008)

access, subsidies via feed-in tariffs and tax breaks. As shown in Table 10.3, by 2020, China aims to increase hydroelectric capacity to 300GW (twice 2007 levels), wind and biomass power generation to 30GW each, biomass and ethanol and biodiesel production to 15 billion litres combined (Boyle et al, 2008; Zhang, 2008).

China's efforts have rapidly increased investment in renewable energy from less than a $0.2 billion in 2004 to over $12 billion in 2007 (see Table 10.5). China is now second only to Germany in renewable energy investment (Zhang, 2008). Most of this investment has been in wind energy, as China more than doubled its wind capacity from 2.6GW to 6GW in 2007 (Boyle et al, 2008). China's Renewable Energy Law and targets are projected to require an average investment level of $17 billion per year through to 2020, an increase of about 35 per cent over 2007 levels (Boyle et al, 2008; Energy Foundation, 2007). This is similar to the level of incremental (beyond existing policy) renewable energy investment required in the UNFCCC (2007) mitigation scenario, though still well short of the levels suggested in McKinsey (2009b).

China has established even more ambitious targets for improving energy efficiency and transitioning economic growth to less energy-intensive sectors. In its 11th Five-Year Plan, China set a target of reducing energy intensity – energy used per unit of GDP – by 20 per cent from 2005 to 2010 (NDRC, 2007). Among chief efforts to achieve this goal is the 1000 Enterprise Energy Savings Program, which commits large state-owned enterprises in key energy-supplying and energy-consuming industrial subsectors to the goal of saving 100 million tonnes of coal equivalent (tce) energy by 2010. In 2006 alone, 20 million tce were saved (Zhang, 2008). China has also launched a $0.5 billion (¥7 billion) fund to promote energy-saving projects (ADB, 2008).

In recent years, the government has reversed a two-decade-long decline in energy efficiency funding. While energy investment devoted to efficiency reached 13 per cent in the mid-1980s, it dropped to 5 per cent from the late 1990s through to the early 2000s (LBNL, 2008). Though the budget for energy-saving measures increased from $3 billion (¥23.5 billion) in 2007 to $6 billion (¥41.8 billion) in 2008 (Zhang, 2008), reaching the 20 per cent energy intensity goal may require significant additional resources (up to $30–40 billion a year) and policies. Energy intensity in China decreased by 1.8 per cent

Table 10.4 *Selected China policies and measures, emissions and investment implications*

	Selected policies and measures	Annual emissions savings by 2010 (Mt CO_2-eq)	Associated investment needs (estimated)	Fraction of plan savings/investment in China CDM pipeline (See Table 10.7)
Renewable energy	Renewable Energy Law	500Mt CO_2: Hydro	≈$17 billion per year ($251 billion from 2005 to 2020)	15% (≈75Mt CO_2 equivalent/year): hydro[20]
	Wind Power Concessions Programme	60Mt CO_2: Wind, solar, geo, tidal		50% (≈30Mt CO_2 equivalent/year): wind, solar, geothermal, tidal
	Brightness Programme	30Mt CO_2: Bioenergy		40% (≈10Mt CO_2 equivalent/year): bioenergy
Energy efficiency	11th Five-Year Plan 20% intensity improvement target; standards and regulations; 1000 highest energy-consuming enterprises; fuel economy standards; vehicle excise taxes based on engine size	550Mt CO_2 (Plan) 1500Mt CO_2 (Target)	$30–40 billion per year to achieve target	<1% (≈1Mt CO_2 equivalent/year)
Thermal power efficiency	Energy Law; Coal Industry and Electric Power Law; phase out 'backward units'	110Mt CO_2	No estimate	70% (≈75Mt CO_2 equivalent/year) including fuel switch, new coal, industry own generation
Nuclear energy	Target new power stations in coastal regions; R&D on fast reactor and fusion technologies	50Mt CO_2	No estimate	Not eligible
Coal bed and coal mine methane	Exemptions from mining fees; tax preferences; tariff support; use of CDM	200Mt CO_2 equivalent	≈$1.4 billion/year (based on CDM project costs)	15% (≈30Mt CO_2 equivalent/year)

Source: NDRC (2007); UNFCCC (2007); Boyle et al (2008); Lin et al (2008); author calculations

in 2006 and a further 3.6 per cent in 2007, and will need to decrease by an average of 5.3 per cent per year through to 2010 to reach the 20 per cent target (Zhang, 2008). Indeed, from 1980 onward, China sustained energy intensity reductions at this level until the take-off of 'global workshop' energy-intensive industries such as steel and cement in the late 1990s.

China's recent policies to shift economic growth to less energy-intensive service and high-tech industry sectors and to shut down 'backward' industrial facilities are important factors in achieving the 20 per cent target (State Council Information Office, 2008). The government has been clamping down on energy-inefficient facilities, pursuing the decommissioning of more than 130GW of inefficient coal-fired power plants (ADB, 2008). Other measures to reduce energy intensity include new buildings efficiency targets, progressive taxation biased against gas-guzzling cars, and fuel economy standards for China's passenger fleet that are more stringent than those in Australia, Canada or the US (Zhang, 2008).

China's efforts to promote energy efficiency and renewable energy through domestic policies are highly ambitious yet often underappreciated. They lead to significant reductions in CO_2 emissions even though they are driven by other policy goals of addressing energy security, rural electrification and public health (Boyle et al, 2008). Some estimate that the energy intensity goal, if achieved, could reduce CO_2 emissions by over 1.5 billion tons annually by 2010 and would stand as one of the most significant carbon mitigation efforts in the world today (Lin et al, 2008).

As Table 10.4 illustrates, China has assembled an impressive suite of policies and measures that, according to some estimates, could yield emission savings totalling 1–2 billion Mt CO_2 by 2010. In addition to the energy efficiency, renewable energy and older coal plant phase-out measures noted above, China has plans to increase nuclear power investments and promote coal bed and coal mine methane capture. As indicated in the last column of the table, a significant fraction of planned renewable energy investment by 2010 is already in the CDM pipeline, as well as in investments in improving thermal power efficiencies. The fraction of energy efficiency investment supported by the CDM is negligible.

Investment trends in energy efficiency and renewable energy

This section looks at recent trends in renewable energy (excluding large hydro) and energy efficiency investment. Investment in other technologies, such as CCS, nuclear and/or transportation systems, could be equally important in the long run. However, energy efficiency and renewable energy (EE/RE) account for the bulk of current investment in 'mitigation technologies' and recent analyses of EE/RE investment trends provide some useful insights into the ability to mobilize capital at a rapid pace.

According to the United Nations Environment Programme's (UNEP) Sustainable Energy Finance Initiative (SEFI), new global investments in renew-

Table 10.5 *New investment in EE/RE by region, 2004–2007 ($ billion)*

Region	2004	2005	2006	2007
OECD				
EU Europe	$7.2	$16.7	$26.0	$55.8
US	$3.2	$9.2	$23.7	$26.5
Other OECD	$2.6	$3.6	$5.6	$9.5
OECD total	$12.9	$29.5	$55.3	$91.7
Non-OECD				
China	$0.2	$1.9	$5.4	$12.5
Brazil	$0.1	$0.5	$4.0	$5.7
India	$0.6	$0.9	$0.9	$3.1
Other non-OECD	$0.9	$1.7	$2.6	$4.7
Non-OECD total	$1.8	$5.1	$12.9	$26.0
Total	$14.7	$34.6	$68.1	$117.7

Note: New investment includes venture capital/ private equity, public markets and asset finance only. It does not include R&D or what UNEP terms 'small scale projects', which explains the difference between the $117.7 billion shown here for 2007 and the $148 billion figure cited in the text.
Source: Boyle et al (2008); Global Trends Dataset

able energy (excluding large hydro) and energy efficiency stood at $148 billion in 2007, up 60 per cent over 2006 levels and a nearly fourfold increase since 2004 (Boyle et al, 2008).[21] Of this, energy efficiency investment (as defined by Boyle et al, 2008) accounted for only $2 billion of this $148 billion in 2007, a figure that may seem surprisingly low, given that it is less than China's budget for energy saving ($3 billion) in 2007, and that energy efficiency investment represents the dominant source of GHG emission reductions globally as well as in China. Indeed, this figure largely represents early-stage investment by energy efficiency technology companies, such as lighting or smart meter manufacturers, rather than the much more significant investment by government, state enterprises and energy users themselves.[22] Energy efficiency investment, given its widespread nature and the challenge of defining what is an 'efficiency investment', is exceedingly hard to track comprehensively. What UNEP's analysis does show, however, is that early-stage investment (venture capital and private equity) in efficiency technologies doubled from 2006 to 2007, indicating significantly greater investor interest.

For renewable energy, international tracking of investment data provides a much fuller picture of relevant market activity. In particular, UNEP found that asset finance, namely investment in new renewable energy capacity, has been the main driver for the surge in global investment, rising 68 per cent to reach $84 billion in 2007 (Boyle et al, 2008). Growth across regions was strong, but growth in Europe and China was particularly impressive, as shown in Table 10.5.

As Figure 10.1 shows, asset finance, or direct investment in renewable energy projects, is clearly the dominant form of investment, though investors have also begun to move towards developing China's domestic renewable energy industry, especially on the manufacturing side. An increasing number of

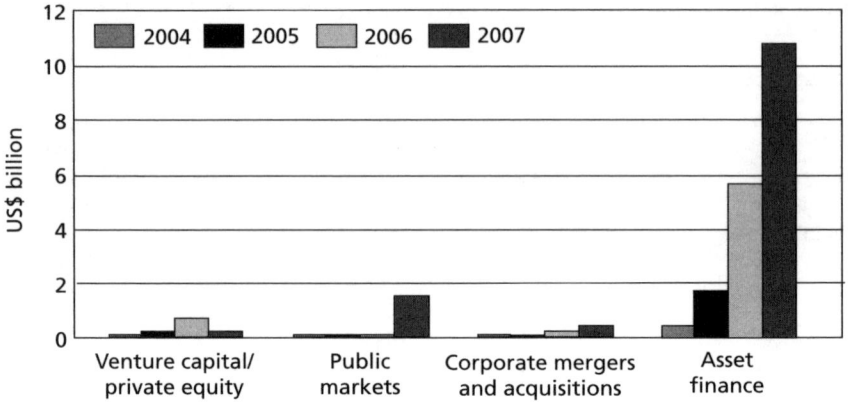

Source: SEFI, New Energy Finance, as cited in Boyle et al (2008)

Figure 10.1 *Sustainable energy transactions in China, 2004–2007*

foreign investors have been, at least until recently, attracted by the heady combination of strong economic growth and the government's commitment to clean energy. The successful initial public offering (IPO) of Chinese solar company Suntech at the end of 2005 sparked investor interest. Since then, a steady stream of solar IPOs followed into early 2007, with many more in the pipeline (Boyle et al, 2008). Overall, there is still relatively little private venture capital in China, though the China Environment Fund and others are actively seeking to scale up 'clean tech' investment.

With the government's renewable energy targets acting as a driver in China, domestic sources have dominated renewable energy investment. State-owned utilities have financed much of this activity directly, providing equity and raising debt, often as high as 80 per cent of total finance, largely from state-owned banks (personal communication, Sebastian Meyer, Ecofys/Azure, November 2008). Unlike the US and Europe, where most wind projects are developed through the project finance route off the balance sheets of the promoting companies, finance for most wind and biomass projects in China has been in the form of asset finance on the balance sheets of the companies (personal communication, Sebastian Meyer, Ecofys/Azure, November 2008; Boyle et al, 2008). The use of corporate rather than project finance may limit the uptake of riskier projects. Major scale-up of wind and biomass capacity may ultimately require a greater diversity in financing instruments.

Outside of solar energy, international private investment continues to play a marginal role in China. Low interest rates and easy access to domestic capital have generally tended to make foreign private capital less competitive for project investments (personal communication, Sebastian Meyer, Ecofys/Azure, November 2008; Boyle et al, 2008). Several factors lead to greater internal rates of return (IRR) for domestic finance, including bank dealings, attractive interest rates, higher debt leveraging tolerance and greater access to CDM revenues: as noted below, China requires at least 50 per cent domestic owner-ship of CDM projects (Ecofys/Azure, 2008). As a result, for a given hypothetical investment, return on equity can be considerably more attractive if renewable energy projects are financed domestically. Ecofys/Azure (2008) show that for a given hypothetical wind project investment, domestic finance might yield a 14 per cent equity return while international finance would yield only a 9 per cent return on equity. To the extent that domestic capital is abundant, as it has been in China, and domestic renewable energy policies themselves are the principal driver for investment, as they are in China, favour-ing domestic investment may not pose a constraint on overall renewable energy investment, at least for commercial renewable energy technologies.

While conditions will change in response to the global economic slowdown and to future developments in international climate change policy, recent trends in renewable energy finance are very encouraging. Globally, the level of investment in renewable energy generation facilities in 2007 ($84 billion in asset finance) already exceeds the total renewable energy investment in the UNFCCC mitigation scenario for 2030 ($79 billion) (see UNFCCC, 2007). Though, as noted above, other scenarios project significantly greater invest-ment needs for 2030, recent rapid growth in renewable energy investment suggests that, given the right policy and economic signals, capital can be rapidly mobilized at scale.

At the same time, clean energy infrastructure projects face significant challenges in accessing relatively small amounts of capital in early stages of development. The clean energy sector tends to be populated by individual project developers or small companies without the balance sheets to raise significant capital resources (Carmody and Ritchie, 2007). Often these actors also lack the technical, commercial, financial and legal skill sets required to transform a project concept into a robust, investable and bankable transaction.

Role of international public funds

International bilateral and multilateral funds have helped increase the uptake of renewable energy and energy efficiency and build capacity in developing countries. Much of this international investment has been made under the rubric of official development assistance (ODA).[23] ODA has the explicit goal of supporting economic and social development in developing countries, distinct from climate change or other goals that a donor country may pursue. In contrast, the investments that flow through the GEF, and through the carbon market (discussed in the next section) explicitly support climate change mitiga-tion and adaptation goals. Funding flows for climate change goals are intended to be a compensatory payment from developed countries to developing countries governed by the 'polluter pays' and the 'common but differentiated responsibilities' principles enshrined in the 1992 Rio Declaration on Environment and Development. As such, they are often expected to be distinct and additional to ODA.

Table 10.6 *Multilateral and bilateral funding for energy, renewable energy and energy efficiency, annual average, 1997–2005 ($ billion)*

	All energy	*Renewable energy*	*Energy efficiency*
Bilateral	2.23	0.71	n.a.
Multilateral, of which	4.97	0.68	0.38
World Bank	2.88	0.39	0.17
Inter-American Development Bank	0.78	0.17	0.08
Asian Development Bank	0.73	0.04	0.00
European Bank for Reconstruction and Development	0.58	0.00	0.09
GEF (non-ODA)	0.12	0.07	0.04
Total bilateral and multilateral	7.20	1.38	0.38

Source: Tirpak and Adams (2008)

Investments in conventional fossil energy resources, driven by electrification (or development) goals in developing countries, continue to dominate the energy portfolios of bilateral and multilateral agencies. As shown in Table 10.6, only 20 per cent of international public finance from 1997 to 2005 ($1.8 billion per year) was directed towards energy efficiency ($0.4 billion/year) and renewable energy ($1.4 billion per year), the remaining 80 per cent went largely towards conventional energy projects and infrastructure.

In recent years, strong economic growth has reduced the multilateral and bilateral ODA – which we refer to here simply as 'aid' – directed to large- and medium-sized Asian countries. Net aid to China fell by two-thirds from the early 1990s to the early 2000s. International aid is increasingly concentrated in sub-Saharan Africa and the Middle East (Tirpak and Adams, 2008).

Although international funds, in pursuit of development goals, also support activities that lead to GHG mitigation in sectors other than energy, it is not always easy to identify these investments. Often clean energy investments themselves are embedded in funds provided to other sectors. Some funders now calculate the share of their portfolio that can be attributed to specific themes. For example, Asian Development Bank projects that clean energy will comprise 18 per cent of its 2008–2010 portfolio (ADB, 2007).

Under the Climate Convention

CDM and the GEF are the principal financial mechanisms of the Climate Convention.[24] The GEF is the largest dedicated financing source for the commercialization of new low-carbon energy technologies in developing countries (Miller, 2008). Since 1991, it has received pledges worth over $3.3 billion, of which $1 billion is for the 2007–2010 period. Further, it leveraged co-financing of another $14 billion through the first half of 2007 (Haites, 2008).

While seemingly small, based on its 2 per cent share of total multilateral and bilateral funding (see Table 10.6), GEF's resources have been strategically allocated to promising near-term and long-term mitigation technologies, policies, projects and capacity-building activities. For example, GEF's 2003–2006 budget included nearly $300 million each for energy efficiency and renewable energy, over $100 million for the development of long-term mitigation technologies,[25] and over $80 million in sustainable transport (see UNFCCC, 2007).

Several GEF projects have met with considerable success and have played important roles in leveraging policy change. A $33 million GEF grant spurred transfer of coal-fired industrial boiler designs to China. The China Efficient Refrigerators Project laid the groundwork for enactment of new national efficiency refrigerator standards and transformation of the domestic manufacturing industry. The GEF, together with the World Bank, also invested more than $200 million in support of development and implementation of the Renewable Energy Promotion Law (Birner and Martinot, 2005; Miller, 2008).[26]

In sum, the GEF's strategic and leveraging impact with respect to GHG mitigation investment may be considerably greater than the low overall investment figures in Table 10.6 might suggest.

Bilateral energy assistance

Bilateral energy development assistance represents approximately 31 per cent of the funding of all international ODA, or aid, for the energy sector, averaging over US$2.2 billion during the period 1997 to 2005. Japan, Germany and France were the leading donors, while India, Indonesia and China were the leading recipients of bilateral energy assistance.[27] In China, Japan is the lead bilateral donor and 28 per cent of China's energy aid was spent on renewable energy projects (ADB, 2008; Tirpak and Adams, 2008).[28]

Bilateral assistance provides an important mechanism for promoting the transfer of low GHG-emitting technologies to developing countries. However, bilateral assistance for energy appears to have remained static while overall ODA has increased.[29] Renewable energy accounted for about 30 per cent of bilateral funding (aid and climate funds through GEF), 1997 to 2005, as shown in Table 10.6, with most of this directed to hydropower projects.[30] There is an apparent shift to funding for less GHG-intensive energy projects, but the variability and low levels of funding suggest that this trend is fragile. It is not well established as an existing policy among donors or supported strongly among recipient countries (Tirpak and Adams, 2008).

Multilateral finance outside the Convention

Multilateral financial institutions such as the World Bank, its private financing arm the International Finance Corporation (IFC), its risk-guarantee arm the Multilateral Investment Guarantee Agency (MIGA) and the regional development banks play an increasing role in financing low-carbon energy projects. However, they do so primarily as part of their obligation to support economic and social development in developing countries. Together, they supported about 6 per cent of global efficiency and renewable energy investment in 2007. Collectively, their investments in efficiency and renewable energy projects have grown dramatically to over $8 billion in 2007, from an average of $1 billion annually from 1997 to 2005.[31]

The World Bank is the single largest funder (among bilateral and multilateral donor agencies) of renewable energy and energy efficiency projects in developing countries, although as recently as 1997 to 2005, conventional power, coal, and oil and gas projects accounted for 75 per cent of its portfolio (Tirpak and Adams, 2008). Clean-energy-related lending by the World Bank has been increasing in recent years, as discussed below, and was almost $1.5 billion in the financial year 2007, or 40 per cent of total energy commitments (World Bank, 2007a in Miller, 2008). In China, the World Bank has been largest multilateral funder in the energy sector with a total lending of $6 billion.

Although the Asian Development Bank's exposure to energy efficiency and renewable energy has been fairly modest in the past (see Table 10.6), focusing mostly on the expansion and upgrade of transmission lines (Tirpak and Adams, 2008), it reached its $1 billion target for 2008 only a few months into the year (Boyle et al, 2008). By the end of 2006, the Asian Development Bank had lent China more than $3.5 billion in the energy sector; 45 per cent was for power projects and the remainder was for industrial energy efficiency and environment improvement (ADB, 2008). Since 2000, there has been a focus on renewable energy, energy efficiency, capture of fugitive methane from coal mining, and improvement of urban air quality (ADB, 2008).[32]

Financing programmes with targeted technical assistance and partial risk guarantees have had recent success in engaging local banks in lending for clean energy investments, primarily efficiency upgrades.[33] The IFC's China Utility-Based Energy Efficiency Finance Program (CHUEE) works with several Chinese banks,[34] offering a facility under which the IFC shares the risk for all loans within the energy efficiency financing portfolio. By December 2007, 37 loans were approved with a total loan value of $0.1 billion (¥0.645 billion). Ultimately, the project aims to mobilize $0.7 to $1.4 billion (¥5 to 10 billion) in finance for energy efficiency projects and reduce 20Mt CO_2 and other GHGs by 2010 (IFC, 2008). For context, this amount represents 2 to 4 per cent of the investment estimated to be needed to meet China's 20 per cent target for reducing energy intensity (¥250 to 300 billion).[35] The project is considered a success as it has been able to mobilize risk-averse domestic capital into smaller projects that may not have been possible without the IFC project (personal communication, Sebastian Meyer, Ecofys/Azure, November 2008).

The European Investment Bank (EIB), which does not share an international development obligation as other regional development banks, has established a multi-annual $4 billion (€3 billion) facility to support projects that contribute to energy sustainability and security of EU energy supply. The facility will be used to fund projects in several developing countries, including China, until the end of 2013. As a part of this facility, the EIB has provided a $0.6 billion (€0.5 billion) loan to China for climate change mitigation as part of the EU–China Partnership on Climate Change.[36]

Recent trends witnessed at the international finance institutions indicate a willingness to take on expanded mandates to fund low-carbon energy programmes in partnership with their largest clients. The World Bank Group, in cooperation with other leading banks, has been preparing approaches for significantly expanding the scale of their support for clean energy finance in large developing countries – particularly the 'Plus 5' countries: China, India, Brazil, Mexico and South Africa (Miller, 2008). This is leading to the creation of new funds and the exploration of innovative financing instruments, as discussed below.

Role of the carbon market

Spawned by the binding emission limits of the Kyoto Protocol, global carbon trading had grown to an over $60 billion market by 2007, doubling in size from 2006 (Capoor and Ambrosi, 2008). Most industrialized countries plan to use emissions trading systems to regulate domestic emissions. While the EU's ETS is today's largest carbon scheme in terms of trading activity, ETS are in design or operation in Australia, New Zealand, and the states and provinces in the US and Canada. These signs, together with ongoing international negotiations, suggest that the carbon market should continue to grow rapidly for years to come.

The Kyoto Protocol enables industrialized countries to meet their emissions obligations by purchasing certified emission reductions (CERs) from project-based activities in developing countries through the CDM. Trading in CDM project-based 'emissions offsets' ($13 billion in 2007) and in EU ETS emissions allowances ($50 billion) comprise 98 per cent of today's carbon market transactions. Of the $13 billion CDM market, $7.4 billion represents primary transactions, new emission reduction credits entering the market (Capoor and Ambrosi, 2008).

The CDM plays an increasing role in supporting, and specifically in providing revenue to, projects that reduce or remove GHG emissions in developing countries. Figure 10.2 illustrates the rapid emergence of the CER market, since CDM rules and infrastructure were effectively established around 2004. As depicted in the left hand panel of Figure 10.2, industrial gas projects – hydrofluorocarbons (HFCs) and N_2O – dominated the early CDM market. These 'low-hanging fruit', with their attendant low costs and high profit margins, have been largely harvested, and renewable energy projects – particularly hydro, wind and biomass projects in China, India and Brazil – have now become the principal source of new CERs. As depicted in the right-hand panel of Figure 10.2, China has been the leading host to CDM projects and source of CERs, capturing over half the market from the time China established its CDM infrastructure in 2005.

The role that the CDM has played in leveraging new investment for GHG emission reduction projects is somewhat more controversial. The extent to which the CDM itself has served to mobilize new project investment – or to put it in carbon market terms, the extent to which such investment is 'additional'

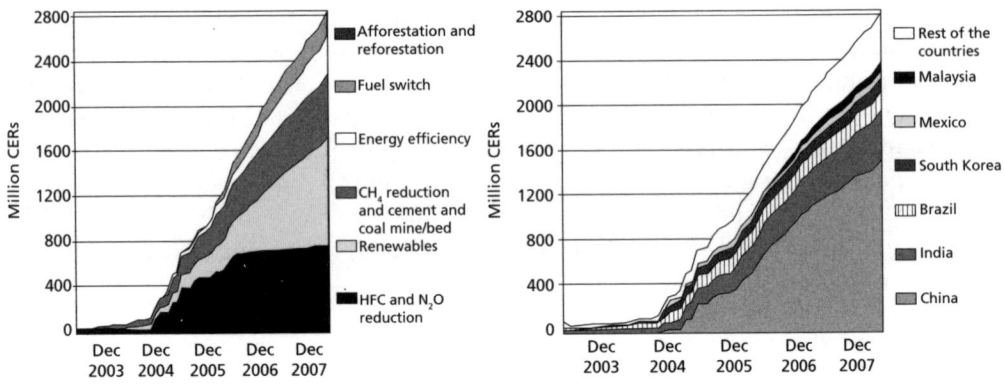

Source: UNEP-Risø (2008)

Figure 10.2 *Cumulative expected CERs by 2012, CDM pipeline*
(as of 1 November 2008)

to what would have otherwise occurred – has been widely debated, as discussed below. Notwithstanding the question of causality, the estimated capital invested in CDM projects is significant and has continued to grow, though the recent economic downturn and collapse in carbon prices threatens new CDM investment in the near term.

As of November 2008, projects in the CDM pipeline – those officially applying or approved for CERs – represented over $125 billion in new project investment, about $100 billion of which occurred in 2007 and 2008.[37] Of this investment, approximately two-thirds ($83 billion) is associated with renewable electricity projects (hydro, wind, biomass and other), and about a quarter ($30 billion) with lower-emitting fossil fuel power, fuel switch and industrial energy efficiency projects. Less than 10 per cent of total investment is associated with all other project types; this outcome is due to the low investment costs of industrial gas and methane projects, and to the low CDM activity for other project types (for example transportation, forestry or building efficiency), which has been disappointing to many.

Future growth of the carbon market and its ability to leverage capital for mitigation activity will be a function of the ambition and scope of future emissions limits (demand for CERs) and the future design of carbon finance mechanisms, as discussed below. As an indicator of the uncertainty, Haites (2007) projects that CDM could be as small as a $5 billion market (smaller than today) or as large as a $125 billion market (over ten times larger than today) by 2030, depending on, among other factors, whether large developing countries such as China adopt binding limits. In terms of investment and emissions reductions (CERs), this range implies from less than $100 billion to over $1 trillion in annual capital investment in CDM projects, and $0.4 billion to $6.0 billion annual CERs.[38] While the high end represents significant capital flows, the level of emissions reductions from the CDM alone is likely to remain short of those implied by various climate stabilization scenarios for non-Annex 1 countries (22Gt CO_2 equivalent in UNFCCC (2007) and 25Gt CO_2 equivalent in McKinsey & Company (2009a)).

The CDM in China

China's unique role in the CDM is not restricted to its market dominance: its internal CDM regulations are also distinct in several respects. First, they require Chinese enterprises to own at least 51 per cent of a CDM project, which helps to direct CDM to support or complement government policies. Second, they create a minimum (floor) price for selling primary CERs (slightly above $10/t CO_2 equivalent). The floor price and the ownership requirements help to retain CER rents within China. They also discourage foreign CDM investment, as well as investment in riskier projects that might fetch a lower price than the floor level (Ecofys/Azure, 2008).[39] Third, China taxes CDM projects, collecting 65 per cent of the revenue from HFC projects, 30 per cent from N_2O, and 2 per cent from all other projects. The national CDM Fund, which supposedly now exceeds $1 billion in value, will be directed to climate change purposes.

If the projects in China's CDM pipeline as of late 2008 deliver their full expected CERs, this could amount to annual GHG emission reductions of more than 300Mt CO_2 equivalent, of which over 100Mt CO_2 equivalent are from projects already registered. As illustrated in Table 10.4 above, CDM projects account for a significant fraction of China's targeted emission reductions for several sectors and technologies, the notable exception being energy efficiency.

Table 10.7 shows that the capital investment associated with China's CDM projects is currently around $70 billion. This represents about 60 per cent of global CDM project investment.

Figure 10.3 illustrates the investment trend over time, showing that investment associated with projects entering China's CDM pipeline in 2007 reached $30 billion in 2007. Although data shown in this chart are through to mid-2008 only, our preliminary calculations suggest that the 2008 pipeline is likely to exceed $30 billion in associated investment. Economic slowdown and uncertainty regarding the CDM post-2012 may slow CDM project activity and investment in 2009.

As shown in both Table 10.7 and Figure 10.3, hydropower projects currently constitute the greatest share of both CER generation and project investment, with wind investment a close second. Investment in new natural gas combined-cycle (NGCC) power plants has been the third leading project type in terms of capital investment. According to Wara (2008), all NGCC facilities built between 2005 and 2010 have applied for CDM credit. Similar assertions have been made regarding wind and hydro facilities (below 500MW in size).[40]

Other domestic policies support the construction of natural gas and hydro facilities in China, and yet a large majority has applied as CDM activities with

Table 10.7 *Emissions reductions and investment associated with Chinese projects in the CDM pipeline (as of 1 November 2008)*

Project type	Annual CERs (million tCO$_2$e/yr)	Total MW (electric)	Average investment per expected CER [a] ($US per tCO$_2$e/yr)	Total capital cost (million US$)[a]
Power generation	193.7	73.2	$344	$66,627
Biomass energy	11.0	1.6	$234	$2568
Hydro	75.5	22.5	$290	$21,912
Solar	0.1		$5349	$765
Wind	31.8	13.8	$553	$17,598
Natural gas (and coal)	30.9	27.9	$534	$16,467
Industrial own gen[b]	44.5	7.3	$165	$7316
Industry – process and non-CO$_2$	115.8	0.9	$20	$2340
Cement additives	1.3		$97	$130
Coal bed/mine methane	26.6	0.9	$35	$921
Oil/gas methane recovery	0.4		$63	$25
HFC destruction	65.7		$17	$1116
N$_2$O abatement	21.8		$7	$147
Ag/For/Waste/Other	9.1	0.3	$83	$756
EE service	0.1			
Agriculture	0.1		$12	$1
Biogas	1.5	0.0	$94	$143
Landfill gas	7.3	0.3	$73	$533
Reforestation	0.2		$484	$79
Total	318.6	74.4	$219	$69,722

Note: [a] Based on analysis of 1191 Chinese CDM projects through to mid-2008 courtesy of Erik Haites and Stephen Seres.
[b] Includes small fraction of industrial energy efficiency.
Sources: UNEP-Risø CDM Pipeline Database, updated 1 November 2008 (rejected projects removed) and authors' calculations

the attendant claim that carbon finance was critical to their construction (Wara, 2008). This situation has engendered major critiques of the role of the CDM, and calls for significant reform or abandonment (Haya, 2007; Schneider, 2007; Wara, 2008). Since non-additional CERs can effectively lead to increased global emissions unless accounted for in terms of deeper targets or by other means, China as well as other countries' project portfolios have raised questions regarding whether the CDM is fulfilling its environmental as well as its sustainability objectives, and has inspired many of the proposed reforms discussed below. As one renewable energy and CDM consulting firm notes, while 'it is clear that although CDM does not represent a significant factor in shifting China's overall energy mix, its influence on a project-level can be sufficient to impact the viability of individual technology types. ... The additional revenue provided by CDM has also enabled project owners to finance more efficient and higher quality projects in the renewable sector' (Ecofys/Azure, 2008).

When considering total capital accumulation and its relationship to increased power capacity, there is no question that CDM has brought billions of dollars of added annual revenue into certain sectors within the Chinese energy market. While China is providing over half of all expected CERs

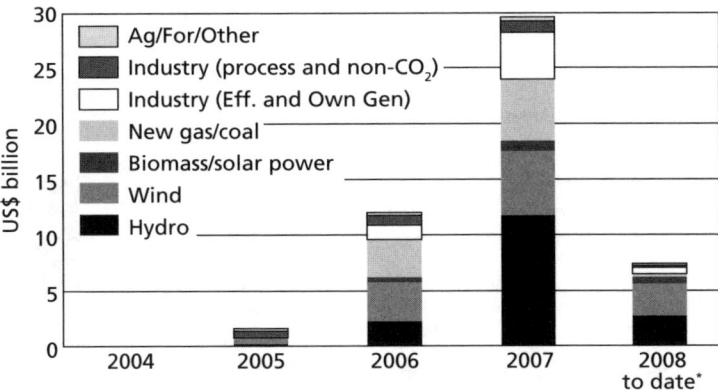

Note: In contrast to Table 10.7, this chart does not reflect new projects in pipeline in latter half of 2008.
Source: Authors' analysis based on data collected by Erik Haites and Stephen Seres from Project Design Documents through to mid-2008.

Figure 10.3 *Evolution of investment associated with China's CDM pipeline (to mid-2008)*

globally, the CDM has yet to have a major impact on China's energy system, and foreign investment plays a very limited role. As a long-time China energy expert has remarked, 'CDM does not make much of a difference to China, but China is definitely making a difference to CDM' (Zhang, 2008).

According to a survey of sustainable energy market actors in China, while local commercial institutions are open to financing renewable energy projects, the carbon value of these projects cannot be taken into account (Ecofys/Azure, 2008). This impacts greatly on those technologies that rely heavily on CDM investment, particularly biomass, biogas and coal mine methane. Further, the lack of access to financing for smaller private projects, or projects in sectors of overcapacity such as steel and cement, is negating the uptake of marginal emission reduction projects and ensuring that only opportunities with significant IRR and profitability are pursued. The lack of adequate institutional capacity at all levels in the CDM project approval stage is also affecting market efficiency, leading to long delays in the registration process and the reduced economic benefit endowed on project owners.

Notwithstanding the global economic slowdown and questions around the actual 'leveraging' of investment due to the CDM, recent trends in mitigation investment are encouraging. As illustrated in Figure 10.4, the trajectory of investment growth in renewable energy investment, public finance of energy efficiency and other lower-carbon activities flowing through the CDM pipeline, suggest that the lower end of projected incremental investment needs (UNFCCC, 2007) is quite achievable (approximately \$40 billion by 2030 net of avoided investment). However, large gaps remain for energy efficiency, early stage project finance, R&D and technology transfer, and more ambitious scenarios imply that investment in mitigation technologies must grow by an

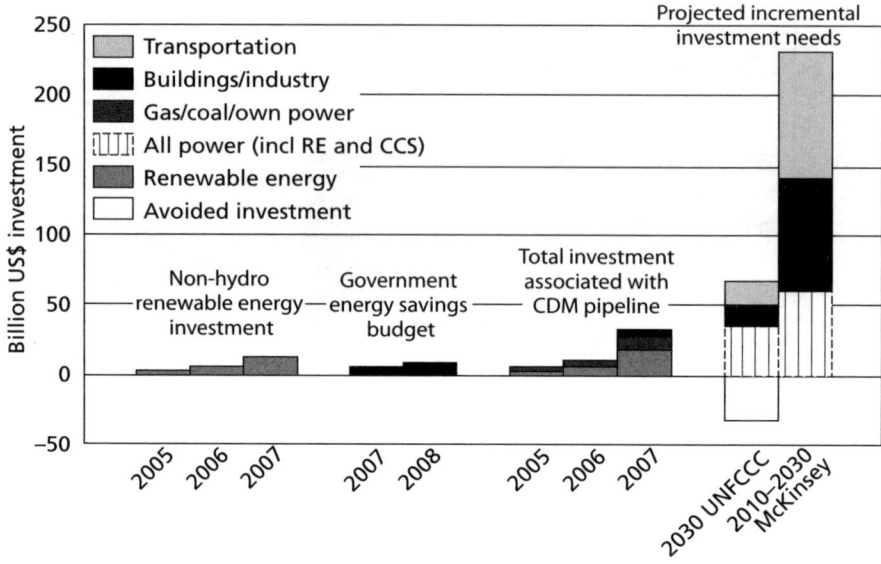

Source: UNFCCC (2007)

Figure 10.4 *Comparison of current finance levels and projected finance needs in China*

order of magnitude to about $200 billion on average over the next two decades. It will be difficult to cover the mitigation financing gap until an adequate policy and financial architecture provides predictable and sufficient financial flows, and a long-term price signal is created (World Bank, 2008).

Closing the gap: International climate funds and the carbon market

While estimates of the scale of additional funds needed to meet mitigation goals vary among different studies, there is a consistent conclusion that the scale is likely to be an order of magnitude greater than the funds currently available. From the preceding discussion, it is also evident that domestic policies and regulations have a pivotal role to play in China's future mitigation efforts, as they create the enabling frameworks and incentives required to attract investment. International funds and carbon markets will be most effective if they can complement and incentivize domestic policy action, share the investment risk of new technologies, and accelerate investments in deployment and diffusion of mitigation technologies.

Figure 10.5 presents a taxonomy of leading proposals for new and enhanced climate funds and carbon market structures. We briefly describe these options and then compare them in terms of their ability to overcome upfront investment barriers, to engage the private sector and to deliver the

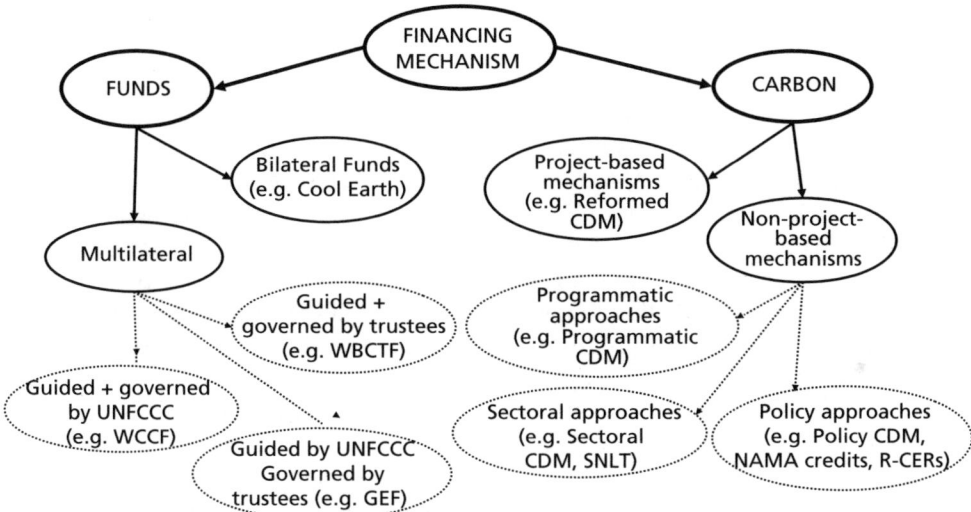

Figure 10.5 *Taxonomy of proposed international climate financing mechanisms*

scale of investment needed across different regions, sectors and technologies. Ultimately, the challenge will be to find the mix of mechanisms that can work in an efficient and sustainable manner. Therefore, we close this section by discussing options to coordinate or integrate a full suite of investment mechanisms.

International climate funds for mitigation

Proposals for international climate funds encompass a broad set of issues, including how to raise, manage and deploy funds in developing countries, how to address and balance needs for adaptation and forestry (REDD) and emission mitigation, and how to accelerate technology innovation and development. They also differ in terms of the means for raising new funds, the uses of funds, the form in which the funds should be made available (for example grants, debt or equity), and in the fund management administration. We touch on only some of the key elements of leading proposals, most of which still remain at the conceptual level.

New funds guided and/or governed by the Climate Convention

Since the adoption of the Bali Action Plan by UNFCCC parties in December 2007, the Ad-hoc Working Group for Long-Term Cooperative Action (AWG-LCA) has provided the principal international forum for exploring new and enhanced multilateral financing options under the Convention.

Overarching financing mechanism and the Climate Technology Fund,
under the Conference of Parties
Under the AWG-LCA, the Group of 77 and China have proposed a broad
conceptual framework that would place policies, programme priorities and
eligibility criteria of a multifaceted mitigation, adaptation and technology
financing system under the authority and governance of the UNFCCC's
Conference of Parties (FCCC/AWGLCA, 2008a). It would be funded largely
from government budgets of industrialized (Annex 1) countries, over and
above existing resources, including ODA, at a suggested scale of 0.5 per cent to
1 per cent of their GNP. Funding would be provided essentially in grant form,
particularly for adaptation, and to some extent in the form of appropriate
concessional loans.

A Conference of Parties-appointed Board with equitable and balanced
representation of all parties would establish specialized funds (for example a
Multilateral Climate Technology Fund for technology-related financing),
funding windows (for mitigation, deployment and diffusion of low-carbon
technologies, R&D, patents, etc.) and a mechanism to link various funds under
its governance. The funds would be administered by a trustee or trustees
selected through a process of open bidding, be advised by an expert group or
committee, and be supported by a technical panel or panels addressing specific
issues.

World Climate Change Fund (Green Fund)
Mexico has also proposed a multilateral and multifaceted fund under the
authority of the COP to target mitigation, adaptation and technology, and to
complement existing financing mechanisms (FCCC/AWGLCA, 2008b). The
World Climate Change Fund (WCCF) would aim to mobilize $10 billion per
year initially and increasing amounts in subsequent years. In contrast to the
G77/China proposal, under the WCCF, all countries would be expected to
contribute to the fund; developed countries would be net contributors and
developing countries would be net beneficiaries. The WCCF proposal suggests
a cap on withdrawals by a single country equal to 15 per cent of the total fund
amount. Such a limit would largely impact China and India, if CDM activity
and estimated mitigation potentials – both well in excess of 15 per cent of non-
Annex 1 totals – are good indicators of the overall future demand for
mitigation and adaptation funds. Further, Mexico proposes that countries
choosing not to join the fund should also be excluded from its benefits.

The size of each country's contributions would be differentiated based on
their responsibility and capability, using GHGs, population and GDP as indicators for differentiation. An objective formula would be negotiated based on the
criteria of 'polluter pays', equity, efficiency and payment capacity, and subject
to periodic review. Mexico proposes that the fund also be capitalized through
some of the innovative mechanisms discussed below, such as auctioning
permits or taxing air travel, in addition to public financing.

Technology funds

Various proposals have been put forward on financing technology research, development, deployment and diffusion. China has proposed creating the Multilateral Technology Acquisition Fund (MTAF), which would be aimed at supporting the development, transfer and deployment of environmentally sound technologies through a public–private partnership that links public finance with the carbon market capital market and technology market (such as loan guarantees or subsidies for demonstration projects).[41] The goal is to leverage a larger amount of private finance by a smaller amount of public finance. This is significant because this is a break from China's, and indeed the G-77's, stated position that public finance should be transferred to developing countries to cover the full incremental costs. The fund would be capitalized mainly through public finance from industrialized (Annex 1) countries, and could be sourced from fiscal budgets for R&D, from tax revenues on carbon transactions, from the auction of emission permits in the carbon market, or from energy or environmental taxes.

India proposes creating several funding 'verticals', or global thematic funds, such as for mitigation, adaptation and technology R&D capacity building, under the overarching financing mechanism proposed by the G-77/China.[42] For example, there could be a Technology Acquisition and Technology Transfer Fund for 'available' climate-friendly technologies, a Venture Capital Fund for emerging climate technologies, and a Collaborative Climate Research Fund. The Alliance of Small Island States (AOSIS) too has proposed the creation of The Technology Fund for the acceleration of the development of renewable energy technologies.[43]

Scaling up existing funds

With its established infrastructure, the GEF could be scaled up to meet some of the additional investment needs for mitigation, with significantly increased contributions and a broader mandate. Similarly, the Special Climate Change Fund (SCCF) could be used. As part of its fourth replenishment for the period 2007–2010, only $990 million was allocated to the GEF Trust Fund and, at the end of March 2008, pledges to the SCCF totalled only $90 million, most of which will be used for adaptation activities. Thus, these mechanisms are likely to be inadequate, unless vastly modified, to meet the scale of investment needed.

Multilateral funds outside the Convention

Climate investment funds

In mid-2008, the multilateral development banks, including the World Bank and the regional development banks, established two climate investment funds – the Clean Technology Fund (CTF) and the Strategic Climate Fund (SCF).[44] The funds are intended as an interim financing mechanism, providing developing countries with grants and concessional loans to address climate change. The CTF is designed to promote scaled-up demonstration, deployment and transfer of low-carbon technologies in the power sector, transportation and energy efficiency in buildings, industry and agriculture. Access to the funds will be demand driven for countries eligible for ODA and will build on the existing country programmes of the multilateral development banks (MDBs). The SCF will provide financing to pilot new development approaches and scale up activities. Such programmes could include mitigation or adaptation activities.

In September 2008, donors pledged $6.1 billion to the funds, including $2 billion from the US,[45] $1.5 billion from the UK and $1.2 billion from Japan. Each fund will have a Trust Fund Committee with equal representation from donor and recipient countries. The Committee will approve programming and project priorities and the financing of programmes, among other activities. The CIFs are expected to complement existing bilateral and multilateral financing mechanisms, and build on GEF activities.

Global Climate Financing Mechanism

The European Commission and the World Bank have suggested issuing bonds and using the funds generated to finance initiatives aimed at helping the poorest developing countries deal with climate change (Haites, 2008; European Commission, 2009). The concept is to raise money in the capital market to fund critical investments immediately and to repay the bonds from future ODA commitments, carbon-linked revenue (such as auctioned allowances for national ETS) or from other innovative sources. The focus of the Global Climate Financing Mechanism (GCFM) would be on grants for urgent adaptation needs, such as disaster risk management, and possibly mitigation actions that contribute to domestic poverty reduction strategies in least-developed countries (LDCs) and small island developing states (SIDS). The general concept of bond issuance for rapid fundraising could also apply more broadly to meet mitigation investment needs.

Bilateral initiatives

There is significant scope for bilateral initiatives to fund climate investments in China, especially, given the numerous energy technology collaborations already under way, such as the US–China Ten Year Framework on Energy and Environment Cooperation established in 2008[46] and the UK–China near-Zero Emissions Coal with CCS initiative, and many others. Two notable, broad climate-focused bilateral initiatives are discussed here.

Cool Earth Initiative[47]

Japan announced the establishment of a $10 billion fund to support efforts in developing countries to combat climate change over a five-year period. Roughly, up to $8 billion will be used for mitigation activities, half in the form of low-interest loans. The rest will be in the form of capital contributions, guarantees, trade and investment insurance, and other support by Japanese government institutions and private funds. The funds will be used to support clean energy access and to improve power plant efficiencies in developing countries.

International Climate Protection Initiative[48]
Germany is using 30 per cent of the revenue raised in 2008 from auctioning EU ETS allowances to address climate change internationally. The Initiative aims to support both investment and capacity building in energy efficiency, renewable energy and fluorocarbons emission reduction in developing and transition economies. In 2008, the total budget available for the international initiative was $150 million, with smaller amounts allocated for subsequent years. The Initiative includes support for a feasibility study of CCS in China.

Innovative mechanisms for raising funding
Some proposals are notable for their innovation in methods to raise climate funding.

Global carbon tax[49]
Switzerland has proposed a $2/t CO_2 tax on fossil fuel emissions applicable to all countries with per capita emissions greater than 1.5t CO_2 emissions. The revenues raised would be split between mitigation, technology transfer and adaptation activities. Although the tax has been conceived primarily for financing adaptation, it could also be used to finance mitigation activities in developing countries. The tax at this level could be expected to generate roughly $50 billion in additional revenues annually.

Levy on international air travel
Müller and Hepburn (2006) suggest that the rapidly rising emissions associated with international air transport be addressed through an international air travel adaptation levy (IATAL) or through an ETS where allowances are auctioned generating revenues for adaptation. They recommend a levy set at an average €5 ($6.5) per passenger per flight, which would generate €10 billion ($13 billion) annually. (See also UNFCCC, 2007.)

Using foreign exchange reserves
A small portion of the foreign exchange reserves of developing countries could be transferred into funds that invest in energy efficiency, renewable energy and other mitigation measures. An appropriate mix of investments would make it possible to maintain the value of the reserves contributed and earn a small return. However, since such investments would be less liquid than treasury bills, and liquidity is important for foreign exchange reserves, it would be prudent to use less than 5 per cent of the reserves for such funds. With global foreign exchange reserves at close to $4 trillion at the end of 2004, just 5 per cent of the reserves could provide additional capital of about $200 billion. (See also UNFCCC, 2007.)

Access to renewables programmes in developed countries
Renewable energy programmes in developing countries, such as feed-in tariffs, renewables obligations and targets with renewable energy certificates, could

allow a share of the renewable energy supply to be met by sources in developing countries that meet the programme requirements. If, for example, entities with renewable energy obligations in Annex 1 countries were to purchase certificates from developing countries equal to 5 per cent of their compliance obligations, this could generate up to $500 million. (See also UNFCCC, 2007.)

Efficiency penny
The UN Foundation has proposed that G8 countries impose a small surcharge on end-use energy consumption of electricity, natural gas and transportation fuels, and devote at least 25 per cent of the revenue to energy efficiency policies, programmes and projects in developing and transition economies. Such a surcharge could raise roughly $20 billion per year from G8 countries – approximately $8 billion from electricity markets, $6 billion from natural gas markets and $6 billion from oil markets (see Haites, 2008).

Taxing or auctioning allowance units
Norway has proposed that an appropriate international body auction a share of the assigned amount units of all parties or impose a tax at the time of issuance of the allowances. It believes that this measure will generate adequate, predictable and sustainable resources for meeting the financing needs under the UNFCCC. Auctioning 2 per cent of allowances could generate an annual income of $15 to 25 billion, depending on the stringency of the cap (FCCC/AWGLCA, 2008b).

Taxing or auctioning allowances for international aviation and marine emissions
Emissions from the international air and marine transport sectors could be regulated in conjunction with the International Civil Aviation Organization and the International Maritime Organization. This could be done through a sector-wide ETS or emissions levy.[50] The UNFCCC estimates that auctioning allowances equal to the projected international marine and aviation emissions could generate annual revenue of nearly $30 billion in 2020. It is expected that airlines and shipping companies would increase the prices of their services and that the higher costs would be borne mainly by residents of industrialized countries.

Donated Special Drawing Rights
In 2002, George Soros of Soros Fund Management and Joseph Stiglitz of Columbia University proposed that the International Monetary Fund (IMF) allocate a new form of special drawing rights[51] (SDRs) to its members to meet a share of the estimated $50 billion needed to achieve the UN Millennium Development Goals (MDGs). Under the proposal, developed countries would be expected to donate their new SDRs to meet specific MDGs. A modification of the proposal could incorporate climate mitigation and adaptation as a goal. The UNFCCC estimates that up to $45 billion could be generated in two tranches (UNFCCC, 2007).

Debt-for-clean-energy swap

Under debt swap programmes, creditors such as the multilateral development banks or bilateral donors negotiate agreements with their debtors (i.e. developing country governments) whereby they cancel a portion of the debt owed to them in exchange for a commitment by the debtor government to convert the cancelled amount into local currency for investment in clean energy projects. The positive impact of debt reduction at low cost, combined with increased investment in priority sectors such as EE/RE, makes such debt conversion programmes attractive. Other sources of financing can be used to pay for imported clean energy technologies, while the proceeds from debt swap programmes could be used to finance recurring local costs, such as salaries, project operation and maintenance, or costs associated with locally produced hardware. The proceeds could also be used as collateral to secure domestic bank financing for clean energy projects (see UNFCCC, 2007).

A wide landscape of climate funds

As is clear from the preceding discussion, the landscape of new climate funds is very wide, if not terribly deep. While there are many ambitious ideas potentially capable of increasing the scale of climate funding by an order of magnitude or more, progress – in terms of spending commitments and investment plans – is currently occurring largely outside the direct UNFCCC process, in the form of bilateral initiatives and the World Bank's climate investment funds. This situation could change after the Cancún meeting in late 2010, a situation further complicated by the ongoing global economic financial troubles.

Carbon market options

Carbon markets can be effective in mobilizing private capital for GHG mitigation activities, as they simultaneously send a price signal to developers of emission reduction projects in developing countries (through the CDM) and create demand for these reductions among private entities subject to emissions compliance obligations. As noted above, the carbon market has grown rapidly, especially in China, although the extent to which it has thus far promoted new investment, especially from China's private sector, appears more limited.

If the current Kyoto architecture is preserved, with the CDM intact and industrialized countries taking on deeper commitments, emission reduction purchases could generate over $100 billion a year in carbon finance by 2030.[52] As estimated above, the capital investment associated with a carbon market of this size could amount to $1 trillion annually. Such optimistic scenarios suggest that carbon finance, assuming that markets can be designed to avoid past shortcomings in administrative complexity, transaction costs and environmental integrity could eventually yield incremental investment flows at scale.

Numerous proposals seek to overcome these shortcomings, from reform and expansion of the current project-based mechanisms under the Kyoto

Protocol (Clean Development Mechanism and Joint Implementation) to new policy-based and sectoral approaches.[53] Below, we briefly describe some of the leading concepts for making the post-2012 carbon market more efficient (maximizing GHG emissions abated per dollar), balanced, equitable and sustainable.

Reform and expansion of project-based carbon finance (CDM and JI)

Including new eligible project types in the CDM

Several new project types including some Land Use, Land Use Change and Forestry (LULUCF) activities,[54] CCS and nuclear projects have been proposed for inclusion under the CDM. The inclusion of CCS and nuclear activities are particularly relevant in the case of China as they play an important role in China's mitigation scenarios. With CCS, issues related to leakage, potential environmental impacts, cost issues and the potential for perverse outcomes (i.e. incentivizing high carbon-emitting projects) need to be address among other issues. Environmental, safety and cost issues are also relevant for nuclear activities, in addition to the security issue associated with the proliferation of nuclear material.

Improving environmental integrity and additionality assessment

STANDARDIZED, MULTI-PROJECT BASELINES AND PERFORMANCE-BASED ADDITIONALITY TESTS

The CDM Executive Board could increase investor certainty and reduce transaction costs by defining standardized baselines (determining the amount of credits generated) and performance-based additionality tests that would eliminate the need to assess additionality on a project-by-project basis. Such an approach could be useful in China's wind sector. Maosheng (2008) notes that a conservative baseline ensures environmental integrity, while the elimination of additionality tests provides certainty that promotes investment in the sector. Such an approach is currently used in some voluntary carbon market protocols, such as those developed by the Climate Action Reserve and by the US Environmental Protection Agency Climate Leaders programme. However, such performance-based additionality tests have largely been applied to project types where the level of BAU activity has generally been limited, such as landfill and manure methane capture and combustion. Applying such an approach to project types with considerable BAU activity, such as natural gas, hydro or wind power projects, is more challenging as the potential for significant free riders could be high, risking credibility and environmental integrity (Kartha et al, 2005).

POSITIVE OR NEGATIVE LIST OF PROJECT ACTIVITY TYPES

Positive and negative lists would either allow or disallow emission reductions achieved by certain project types or technologies, based upon agreed criteria. One of the most significant issues that would need to be considered here is the possible differences arising out of different national circumstances. The process of coming up with such lists, the criteria for deciding which projects are put on the respective lists, and the consequences for existing project activities if an activity is added to a negative list or removed from a positive list could prove controversial.

MULTIPLICATION FACTORS TO INCREASE OR DECREASE CERS ISSUED FOR SPECIFIC PROJECT ACTIVITIES

The rate at which quantified emission reductions are credited with CERs could be discounted in order to achieve net mitigation benefits and/or improve environmental integrity of the CDM by accounting for the uncertainty of additionality assessment (Schneider, 2008). Multipliers could also be applied on a differentiated basis to promote particular project types, such as those with high sustainable development, innovation or technology transfer benefits, or those from less-developed countries. As with positive and negative lists, the process and criteria for setting multipliers could be challenging, especially owing to different national circumstances. Such challenges are in fact shared to some extent by all proposals to reform or modify carbon finance instruments, in particular options such as sectoral no-lose targets or policy CDM.

Improve specified host parties' access to CDM activities

GHG abatement projects in the least-developed countries, small island developing states and other similar groups could gain preferential access to the carbon markets by, for example, exempting small-scale activities in these countries from proving additionality, or provision of funding for validation, verification or certification activities.

Co-benefits as criteria for registration of project activities

Host countries are already required to consider sustainable development criteria in their project approval processes. Specific criteria could be applied across all countries at the registration process to address concerns that development benefits, technology transfer and poverty alleviation are not adequately addressed. Projects deemed to have significant co-benefits could also receive preferential treatment in the form of reduced fees, accelerated registration, etc.

Carbon finance beyond the project level

Programmatic CDM

In recent years, CDM has been expanded to allow 'a programme of activities (PoA)' to register to generate CERs. A PoA involves a voluntary action by a private or public entity, which coordinates and implements any policy or measure, such as an incentive scheme (see Baker & McKenzie's *CDM Rulebook*, 2008). No PoA has yet been registered, however. The approval of programmatic projects currently appears to be as complicated as the project-based approach, and additionality continues to be a barrier (Sanchez, 2008). However, the success of small-scale CDM projects, despite high transaction costs, holds promise for the success of programmatic CDM. Further, programmatic CDM could provide a testing ground for policy-based CDM (Leguet and Elabed, 2008). Opportunities in China that might be enhanced through programmatic or policy-based CDM include boiler rehabilitation and retrofitting, energy saving for buildings, small biogas utilization systems, energy efficiency motors, compact fluorescent light bulbs (CFLs) and energy-efficient household appliances (CASS in Ecofys/Azure, 2008). Donor financing through a PoA also provides the option of policy co-financing and enables strong MRV (Hampton, 2008).

Policy-based approaches
POLICY CDM

Policy CDM would allow activities that reduce emissions as a result of government policy to claim credits (Ward, 2008). Policy CDM would require the development of methods to assess whether a government policy is actually additional, a significant challenge given that governments are motivated by numerous political and strategic reasons. Policy CDM also carries the potential to reduce the role of the private sector, since governments could capture emission reduction opportunities as well as economic rents that would otherwise flow to project developers and carbon market intermediaries.

NAMA CREDITS

The Republic of Korea has proposed that Nationally Appropriate Mitigation Actions (NAMAs), implemented by developing countries, be verified and earn credits that developed countries could use to comply with their emission reduction commitments. Demand for NAMA credits would come from developed countries taking on more stringent targets. NAMA crediting would require an institutional architecture administering and issuing credits, and the form of the credits – whether fungible and equivalent to certified emissions reductions (CERs) or assigned amount units (AAUs) – would need to be determined. In other ways, NAMA crediting would face the same issues as policy CDM, including uncertain additionality and the potential for perverse incentives (i.e. reluctance to legislate or regulate without credits).

R-CERS

Müller and Ghosh (2008) propose adapting CDM such that developed countries take on obligations to obtain a certain number of CDM credits to be retired. These retirement-CER (R-CER) obligations would serve as a link between measurable, reportable and verifiable (MRV) mitigation activities in developing countries and MRV finance from developed countries distinct from the CDM's traditional offsetting role.

Sector-based approaches
In recent years, several parties have shown increased interest in migrating from project-based to sectoral crediting mechanisms. The EU's recent climate

communication states that project-based CDM 'has demonstrated its limits both in scale and environmental integrity' and that 'sector-wide mechanisms ... should be developed to provide broad incentives, and credit only actions that are additional and go beyond low cost options' (SEC, 2009). Sectoral approaches have been promoted to increase the scale of mitigation activity in key emitting sectors, to avoid the high transaction costs of project-based CDM, and to address concerns related to international emissions leakage and competitiveness (if sectoral approaches are applied globally), among other reasons (Hohne et al, 2008). Candidate sectors for sectoral approaches might include electricity generation, cement, steel or aluminium production, or upstream emissions from oil and gas production (Ward, 2008), sectors with relatively homogeneous products and high emissions intensities.

SECTORAL CDM

Under sectoral CDM, an emission baseline would be established at sectoral level, for example emissions per tonne of steel or cement produced and activities producing similar output at a lower emissions rate would be credited, subject to the modalities and procedures defined by the CDM Executive Board. Issues that would need to be address include the selection of sectors, the definition of the sector boundaries and baselines, and provisions to ensure the participation of all activities within the boundary.

SECTORAL TARGETS

This approach involves setting a target for a specific sector, expressed in terms of total sectoral emissions or sectoral emissions intensity (for example t CO_2 equivalent/t product). Sectoral targets can be binding or 'no-lose'; both variants would yield fungible credits for emission reductions below the target level. In the case of a binding target, compliance would no longer be voluntary; exceeding the target would require the purchase of emissions allowances or credits or trigger non-compliance penalties. A sectoral no-lose target (SNLT) would be purely voluntary, with no penalty for non-compliance (Ward, 2008). SNLTs could be set so that credit investments that reduce sectoral emissions below a level might be reached with domestically and internationally funded and supported policies and measures. Such an approach will require a process of vetting and establishing targets, and establishing the institutional structure for the administration and issuance of credits.

DOMESTIC SECTORAL REGULATION WITH TECHNOLOGY TRANSFER AGREEMENTS

Another variant of sectoral approaches involves domestic regulation, coupled with mandatory technology transfer requirements, and complemented by government-level bilateral or multilateral technology transfer agreements (Lütken, 2008). Such an approach, Lütken argues would significantly increase bilaterally or multilaterally financed technology transfer as well emissions reductions, while the CERs generated as stand-alone private sector CDM initiatives are crowded out.

Comparison of mechanisms

Clearly, there is no lack of options on the table for developing and enhancing international market-based and fund-based mechanisms for financing mitigation in China and other developing countries. The two approaches possess both complementary and overlapping attributes, as illustrated in Table 10.8. Climate funds can provide upfront financing for costly long-term investments, for which carbon finance has thus far been of limited effectiveness. In contrast, since it pays only upon delivery of monitored and verified activities, carbon finance can provide a more effective incentive for ongoing project, programme or policy performance. With carbon finance, the metrics used to measure performance of mitigation actions are inherently two-dimensional: carbon markets are driven by their units of commerce, i.e. tonnes of CERs and price per tonne delivered, and far less by the sustainable development and technology transfer objectives that are enshrined in the Kyoto Protocol (Olsen, 2007; Sutter and Parreno, 2007). Climate funds, by contrast, provide the flexibility to apply multiple criteria in the selection of mitigation actions. For instance, the World Bank's Climate Technology Fund's funding criteria include the potential of an action to transform sectors or markets, scale up to significant levels of emission reduction, or contribute to MDGs.

Carbon finance is ultimately sourced from the entities with emission reduction obligations in countries with binding emission limits. Since those entities are generally in the private sector, carbon finance has generally been perceived as more palatable than climate funds. Even though funds may originate from the private sector as taxes or allowance auctions, they typically pass through national budgets. The resistance of governments to commit money that is perceived as a domestic resource by domestic constituencies, i.e. the 'domestic revenue problem', is a major hurdle that proposals for climate funds must overcome (Doornbosch and Knight, 2008). This hurdle may be steepest, not surprisingly, for funds directed to China, given China's growing role in global economic activity and GHG emissions.

While various climate funds possess a number of common attributes, there are important distinctions among carbon market proposals. Table 10.9 outlines some of the key differences among general options receiving considerable attention: reforms to project-based CDM, policy-based crediting approaches, and sectoral approaches, most notably SNLTs.

Recent papers and proposals have focused greater attention on policy-based and sectoral approaches to carbon market crediting driven by widespread dissatisfaction with elements of project-based CDM to date among Annex 1 countries, and more widely by limitations in delivering large-scale additional reductions and raising finance (Hohne et al, 2008; Ward, 2008). The EU has proposed that for 'advanced developing countries and highly competitive economic sectors, the project based CDM should be phased out in favour of moving to a sectoral carbon market crediting mechanism'. Such a mechanism, they argue, can also 'pave the way for the development of cap and

Table 10.8 *Key attributes of fund versus market-based finance mechanisms*

	Carbon finance	Climate funds
Basic approach	Provides payment for mitigation delivered Follows pay-for-performance model; provides ongoing incentives	Can be grants, (concessional) loans, and/or equity investments Provide upfront incentive (pay for action/investment/policy)
Commodity and performance metrics	Creates fungible credit for 'MRVed' reductions Evaluated predominantly by a single performance metric: tonnes of reductions delivered	Do not (directly) create fungible credits Allow for broader set of performance metrics (technology development, market transformation, cost reduction)
Fund sources and use of market forces	Mobilizes private capital and market intermediaries Engages market forces to find and deliver reductions	Typically flow through national budgets Reduce role of market forces; fund management directs investment to regions, sectors ('picks winners')
Upfront investment barriers	Has limited ability to raise incremental capital (collateral value of CERs insufficient)	Could deliver large amounts of upfront investment capital
Integrity and fairness concerns	Presents varying levels of concern depending on mechanism design; concerns are endemic to all offsets due to basis on counterfactual baseline Influences level of reductions in capped regions; cap setting should take into account expected performance of finance instruments	Do not directly affect level of effort in capped regions, thus avoids environmental integrity concerns Raise concerns related to financial additionality (relative to ODA), predictability, sustainability and equity in raising and distributing funds

trade systems' (SEC, 2009). China by contrast has argued that 'project-based mechanisms operate generally well under the current rules and thus the relevant overall structures shall be maintained' (FCCC/KP/AWG, 2009).

Not surprisingly, sectoral mechanisms have been viewed with scepticism in developing countries that have expressed reluctance to adopt binding emission limits. Furthermore, sectoral mechanisms are motivated in part by Annex 1 countries as means to address concerns regarding the relative competitiveness of their emissions-intensive industries. The prospects for sectoral mechanisms are thus intertwined with some of the more difficult issues in international climate negotiations relating to commitments.

A common thread among policy-based crediting (for example NAMA crediting, R-CERs, policy CDM) and sectoral proposals is the ability to support the types of domestic policies that have already proven effective at delivering significant emission reductions, such as China's energy efficiency and renewable energy programmes outlined above. As Fan et al (2009) point out, project-based mechanisms are not suited to supporting the costs and challenges of further extension of China's policies, such as continuing the ambition of China's energy savings target to future five-year periods, which could be quite significant.

However, significant implementation challenges remain. SNLTs would be challenging to implement, with significant requirements for data development, institutional capacity building and technical assistance to negotiate meaningful sectoral targets (Ward, 2008). Policy-based crediting faces additionality issues similar to the current CDM, and will require methods capable of reasonably measuring policy impacts. For example, how could the amount of new wind power resulting from implementation of a preferential feed-in tariff be verifiably estimated? Furthermore, if offsets are awarded to government, policy-based crediting may crowd out private sector involvement and innovation in carbon markets, unless domestic measures are taken to pass through the economic value of the offsets to investors. (For example, if a UNFCCC body were to award offset credit to governments for implementing a renewable energy feed-in tariff, the governments could directly transmit the price signal through the tariff incentive itself. Standards-based policies such as a low-carbon fuel standard or enhanced building code would require an indirect mechanism to incentivize investment, such as tax incentives or funds for technology development.)

The intent of international climate policy may well be to extend binding emissions limits to countries or sectors as they reach a certain level in their capacity (for example as reflected in per capita income) and responsibility (for example as reflected in recent or cumulative historical GHG emissions per capita) to reduce emissions. If so, then the ability to transition from 'no-lose' baseline-and-credit systems (all of the carbon finance options shown in Table 10.9), should be considered. Project- and policy-based crediting mechanisms may create incumbent interests (carbon market players, project developers, and host and government institutions) who are vested in continuing to be remunerated for emission-reducing activities. Such a barrier to later adoption of binding limits can be lessened by explicitly designing mechanisms to be transitional, for example with articulated end dates, so as to encourage mitigation actions in early years. As noted above, sectoral approaches such as SNLT provide the clearest pathway forward to binding limits.

Given the complementary strengths of various climate mitigation funds and carbon market approaches, it is reasonable to expect that some combination of financing mechanisms will be pursued. Efficiency and effectiveness will necessitate that these approaches be coordinated if not integrated, and combined together with different countries' own mitigation efforts, to ensure that they can deliver the scale of additional resources and mitigation needed.

Integrating carbon finance, climate funds and domestic action

As noted above, current estimates of China's incremental investment needs consistent with a 450–550ppm CO_2 equivalent global emissions pathway range from under $50 billion to over $250 billion annually within the next two

Table 10.9 *Attributes of specific carbon finance and fund proposals*

	Carbon market finance			Climate funds
	Revised project-based CDM	Policy-based crediting	Sectoral no-lose targets	
Related proposals	Positive lists, CER discounting	South Korea NAMA; credited SD-PAMs, R-CERs		Various bilateral and multilateral technology and investment funds, e.g. GEF, CTF, WCCF, MTAF
Form of mitigation finance	Direct payment for 'MRVed' emission reductions			Equity, debt or grant finance
Source of funds/support	Countries and/or entities with emissions obligations ...		... plus domestic commitments	Various proposals: emitters via allowance auction, aviation/maritime fuel tax, national budget contributions, financial sector with government guarantees
Interaction with own mitigation efforts	Potential for perverse incentive to delay investments or policies in order to access carbon market finance ...		...may be somewhat reduced, as own mitigation is prerequisite to crediting	Can be designed specifically to support
Governance/ decision-making	UN governs system; market/ government direct investments ...		... and host country negotiates target, may agree to specific actions	A variety of proposals: UN, multilateral development banks, bilateral institutions possible
Sectors, projects, locations with significant promise	Less-developed regions with limited policy infrastructure, non-CO_2 abatement	Where policy impacts significant, MRV workable, (efficiency standards, building codes, renewable standards)	Large, relatively homogeneous sectors: electricity, aluminium, cement, steel	Technologies (early stage), programmes and policies difficult to finance through carbon market (e.g. efficiency, mass transit, urban infrastructure)

decades (UNFCCC, 2007; McKinsey & Company, 2009a). Table 10.10 provides a sectoral breakdown of these estimates. It also shows specific mitigation options by incremental mitigation costs (inclusive of fuel and other recurrent costs), and highlights (in bold) those estimated to have significant emission reductions across various studies (Tsinghua University of China, 2006; Wetzelaer et al, 2007; McKinsey & Company, 2009a). Together with the lessons from existing financing mechanisms (discussed above), and the complementary attributes, these results can inform the design of a comprehensive financing framework.

Table 10.10 *Sectoral breakdown of mitigation potential, investment needs and mitigation costs*

	2030 mitigation potential (Gt CO_2 equivalent)	Annual incremental investment needed ($ billion)	Mitigation options by rough incremental mitigation costs (high abatement potential in bold)		
			High (≥$50/t CO_2 equivalent)	Medium ($5–50/t CO_2 equivalent)	Low/net benefit (≤$50/t CO_2 equivalent)
Power generation	≈ 4	≈ 5 to 60	CCS, some solar and wind	Nuclear, wind/solar, hydro, plant efficiency, coal–gas switch	
Industry	≈ 2	≈ 10 to 20	CCS, coal-to-gas, advanced technologies		Energy efficiency, waste and by-product recovery
Buildings (and appliances)	≈ 2	≈ 5 to 60		Some design, control systems	Design/ appliances/ lighting
Transportation	≈ 1	≈ 10 to 90	Electric vehicles	Biofuels	Vehicle efficiency

Notes: Regarding relative costs and potentials, while several studies provide specific mitigation cost-per-tonne metrics, the uncertainty and variability in the underlying assumptions suggest that while these figures may be used as rough 'guideposts', the cost metrics should not be interpreted as definitive. This is especially the case (1) where technology costs could decline significantly over time due to learning-by-doing and learning-by-researching effects, and (2) where options involve the displacement of fossil fuels of highly uncertain future cost (for example oil and gas), since mitigation cost by definition is estimated relative to avoided fuel costs. This is the case for most options considered. For high abatement potential mitigation options, some studies suggest potential yields of 0.4Gt CO_2 equivalent or more in emission reductions by 2020 or 2030.
Source: Highest estimate shown among China mitigation studies reviewed: McKinsey & Company (2009a); Tsinghua University of China (2006); IEA (2007b). Low values from UNFCCC (2007) for 2030; high values from McKinsey (2009a) average for 2010–2030. Relative mitigation costs and potentials are drawn from McKinsey (2009a); Tsinghua University of China (2006) and Wetzelaer et al (2007).

Several factors can help to match specific financing mechanisms to sectors, technologies and regions. These include the scale of mitigation potential, incremental mitigation costs, incremental investment needs, responsiveness to price signals (i.e. lack of market barriers), generation and allocation of economic rents, ability to address carbon market challenges (permanence, leakage, additionality, MRV), presence of significant non-GHG social benefits, and the effectiveness of government policies and programmes.

Low-cost and cost-effective options in **vehicle, industrial and building energy efficiency, waste and by-product recovery,** and urban design opportunities are likely to be most effectively and efficiently addressed through a combination of (support for) domestic policies and international funds providing debt finance, capacity-building grants and access to advanced technologies.

Analysis by McKinsey & Company (2009b) suggests that approximately one-third of the incremental mitigation investments needed in China will have positive economic returns. Significant market barriers limit the price responsiveness of these options, as reflected in their limited uptake by the CDM to date. Furthermore, the significant difference between carbon prices and the cost of mitigation could create large economic rents if traditional carbon finance instruments are used (see Box 10.3). Domestic efficiency policies have proven to be quite effective in China, as noted above. Nonetheless, the capital requirements for deeper efficiency improvements may be quite high, as shown in Table 10.10. International funds could be critical in providing the credit needed, though less so in China than other large developing countries, given its relatively high domestic savings. They can nonetheless play an important role in guaranteeing private capital for small-scale activities, as has been successfully demonstrated in the IFC's CHUEE programme in China, and for financing early-stage technologies. NAMA-type funds, policy CDM crediting and technology agreements can support acceleration or deepening of efficiency targets and standards, and the commercialization of more costly and innovative energy efficiency technologies such as hybrid vehicles or smart grid devices (advanced meters).

Mitigation opportunities where the abatement costs are expected to be in the range of anticipated carbon prices – such as many **renewable electricity technologies, biofuels and some advanced efficiency technologies** – represent the 'sweet spot' for carbon finance. Support for renewable energy standards or incentives, low-carbon fuel standards or product-based emissions intensity goals could be provided through policy CDM or sectoral target approaches. Project-based CDM could be enhanced through positive and negative lists to target technologies and fuels that exceed common practice or high performance thresholds.

For industrial sectors amenable to sectoral approaches due to **high emissions intensity and homogeneous products** (steel, cement, aluminium), SNLTs could support both low-cost abatement options (through policies and measures) and higher-cost technologies (through carbon finance).

Where the marginal abatement costs are significantly higher than carbon prices in the near term, and/or carbon market challenges are significant (for example permanence of sequestration), instrument- and technology-focused international funds, such as the Multilateral Climate Technology Fund proposed by the G77 and China, could support R&D to bring down the costs and support demonstration and deployment. As suggested by available investment assessments, realizing significant gains from **CCS, electric and other vehicle technologies, and higher-cost renewable energy technologies** will require significant grant-based financing and R&D support, in tens of billions of dollars annually, based on the estimates in Table 10.10 and discussed above.

Should new financial mechanisms (for example technology funds) support sectors and activities that are also eligible for generating carbon credits? This is a question for which perspectives differ.[55] The desire to bring all available

BOX 10.3 ADDRESSING RENT CAPTURE
IN CARBON MARKETS

The potential to generate large economic rents from the carbon market presents both a risk in terms of political viability and an opportunity to leverage greater emissions reductions and ancillary benefits. A significant fraction of the early CDM pipeline to date was dominated by such projects: HFC and N_2O avoidance or destruction projects with abatement costs often near $1/t CO_2 equivalent compared with market prices often exceeding $10/t CO_2 equivalent. This situation can be addressed by:

- Domestic programmes such as the China Environment Fund. By establishing a minimum CER sale price and collecting 65 per cent of value of CER sales for HFC and N_2O projects, China is effectively capturing a significant fraction of the producer surplus.
- Separate reduction or set-aside programmes for targeted project types, financed with allowance or other revenues that seek to pay for mitigation closer to cost.
- A single-buyer carbon bank or fund that charges compliance entities for offsets at the price of allowances, and uses these revenues to purchase a larger quantity of emission reductions closer to the actual costs. This could be accomplished through direct incentives where abatement costs are generally known, or through reverse auctions to enable price discovery. The outcome would be similar to that of the CER discounting approach as described by Schneider (2008).

instruments to bear upon the scale-up of mitigation investment and technology development and diffusion must be balanced with the efficient use of funds and fair competition in carbon markets, which might be compromised.

If a number of international carbon market and fund-based financing mechanisms are pursued, as the preceding discussion suggests is both likely and necessary, their coordination and integration will help to ensure that inefficiencies are minimized, and that together they can achieve the needed global scale of mitigation investment.

One option for moving forward on climate finance is reliance on negotiated bilateral or regional agreements (US–China, EU–China). Such an approach avoids the challenges of reaching international consensus and creating new institutions. Under such an approach, fund-based support for technology, policies or other NAMAs might be agreed among two or more parties, and then monitored, reported and verified to the UNFCCC. It has the advantages of simplicity and flexibility; however, the overall adequacy of global finance could be difficult to ensure, unless there were an overarching system of international finance or emission reduction obligations to ensure global goals were met.

The ICP suggested by Fan et al (2009) in one of the companion papers to this book – in essence a government-to-government negotiated plan or NAMA crediting approach – also represents a bilateral approach. If implemented in concert with emission reduction obligations based on responsibility and capacity, such as those embodied in the GHG development rights framework that Fan et al (2009) also adopt, it would also help to ensure that global emission goals can be met. However, the ICP would present the same additionality concerns noted above for policy CDM or NAMA crediting. Government-to-government agreements might also crowd out investment by the private sector if such bilateral agreements were an alternative to traditional carbon finance. Like other bilateral or major emitter approaches, it also raises larger equity questions about whether the global climate change problem can be left to the largest polluters to resolve between themselves. Will it result in an outcome that is optimal for the global community unless such plans are approved multilaterally?

Another option for integrating funds and carbon markets is to have one finance the other. Some have suggested placing added taxes on carbon market transactions, however, as noted, the scale of revenues is likely to be rather limited, and most have raised this concept in the context of adaptation finance. Another approach would be to establish a single carbon market aggregator or bank that purchases emission reductions from non-Annex 1 countries at their actual incremental cost plus profit margin and sells them at expected or actual emission allowance prices. This approach would effectively capture the potentially very large economic rents (producer surplus) that have thus far been a source of controversy in the CDM, and could direct these resources to fund-based climate finance mechanisms. The carbon bank approach would provide an international institution that participates in, and effectively distinguishes activities between, the carbon market and carbon funds. However, there are several challenges to such an approach, including: how to pay for emission reductions at their marginal incremental cost (for example reverse auctions at the regional and project-type level), how to maintain incentives for project development in an already high-risk carbon finance market, and how to protect carbon funds from future carbon price uncertainty. Another approach that could accomplish a similar objective, with similar drawbacks, would be to place a limit on allowable emission reduction credits (for example CERs) issued during a given period and auction issuance rights.[56]

International agreement on a full integration of fund- and market-based approaches may be overly ambitious, given the many competing interests and sensitive issues related to finance generation, use and governance. A more modest approach would be to establish a single umbrella body to coordinate financial mechanisms and review and adjust them as necessary.

Conclusions

Renewed efforts are needed to enhance public support in four areas: (1) improving the efficiency and integrity of carbon finance; (2) using carbon finance and other mechanisms to support and strengthen domestic policies and measures; (3) generating the additional public and private finance needed to develop, deploy and transfer advanced technologies; and (4) overcoming financing barriers where they exist such as in early-stage project finance and in least-developed countries.

Experience with carbon finance to date, and the CDM specifically, has been mixed. CDM is bringing billions of dollars of added annual revenue into certain sectors of the Chinese energy market. The annual investment associated with Chinese projects in the CDM pipeline averaged about $30 billion in 2007 and 2008.[57] Yet, the extent to which carbon finance has actually 'leveraged' these investments or brought them to fruition is unclear.

With some reforms likely, carbon finance will probably be retained in its project-based mode for some time. Project-based crediting has proven particularly effective for project types that pose the fewest challenges in terms of the scale of investment needed, additionality assessment, and interactions with domestic policies, such as moderate-cost methane reduction projects in agriculture and waste management. These project types are also among those most likely to have claimed technology transfer benefits (Seres and Haites, 2008). Project-based CDM is also the most feasible carbon market mechanism for less-developed countries lacking the infrastructure and capacity to undertake and 'MRV' policy and sectoral crediting approaches.

Domestic policies and measures in China have been, and are likely to continue to be, the dominant drivers of emission reducing activity and finance, as witnessed with China's renewable energy policies. Mechanisms to recognize, support and deepen domestic policies – as implied by the Bali Action Plan – will be central to future international climate agreements. Proposals to monitor, report and verify 'nationally appropriate mitigation actions' as called for in the Bali Action Plan, as well as to provide some form of tradable or non-tradable credit, may hold particular promise in strengthening regulatory, fiscal and other measures that have proven effective at mobilizing investment in energy efficiency, renewable energy and other emission-reducing technologies. Similarly, SNLTs could provide another means to combine support for domestic policies with carbon finance, while providing the clearest path forward to binding emission limits.

The scale and target of needed financing could increase and change rapidly under an ambitious climate stabilization trajectory as more expensive technologies are forced earlier. As noted above, annual incremental investment needs in China could grow to $250 billion or higher by 2030, much of this for emerging technologies such as electric vehicles or CCS for which there is limited domestic and international support today. Current financing is also well short of what might be required to maintain or exceed the 11th Plan's target of improving energy intensity by 20 per cent every five years.

International bilateral and multilateral funds have helped increase the uptake of renewable energy and energy efficiency and build capacity in China and other developing countries, but far greater efforts will be needed in the future. Recent initiatives such as the World Bank's $6 billion Climate Investment Funds may provide a promising laboratory for mechanisms that can be funded at much larger scales.

Significant, new or expanded investment and technology funds will also be essential to overcome the many financing gaps and technology barriers that the carbon market cannot effectively address such as R&D, early stage finance, small-scale energy efficiency funding or acquisition of intellectual property rights (IPRs). Whether there is sufficient political will to create a large new climate fund as proposed by Mexico, G77/China, India and others – or simply to scale up existing mechanisms such as the GEF – remains to be seen. This could be a political challenge for some industrialized countries, should funds be seen as going directly from national budgets to large emerging economies, such as China's. Ultimately, an integrated approach, such as the ICP or an international coordination body, will be essential to ensure that a mix of carbon market- and fund-based mechanisms will function together in an efficient and adequate manner.

Notes

1 The 2007 UNFCCC report on investment and financial flows was updated in November 2008 to include new information on investment and financial flows needed, and an assessment of options, tools and mechanisms to enhance financing.

2 The UNFCCC reports only investment (and not financial flows) for the energy, building, transport and industrial sectors, while the estimates for the agriculture and forestry sectors, and technology R&D largely represent only financial flows.

3 Baseline emissions are only marginally higher for 2030 (up 1.2 per cent) than in the prior study.

4 For consistency, all monetary figures are presented in US dollars. The following default currency conversions are used throughout this chapter: 1€ = $1.25; $1 = ¥7.

5 This estimate reflects only the *gross* mitigation investments for the power sector; some of the other costs are incremental to avoided BAU investments.

6 There are differences in the abatement potential and corresponding abatement costs McKinsey estimates for China in its global (McKinsey & Company, 2009a) and China-specific (McKinsey & Company, 2009b) studies. For example, the abatement potential for China in the global study is estimated to be 8.4Gt CO_2 equivalent and in the China study it is 10.4Gt CO_2 equivalent. We present estimates for China only from McKinsey's China study, since it is more recent and provides greater detail.

7 The ETP is based on a different modelling framework and assumptions than IEA's WEO, upon which much of the UNFCCC energy CO_2 analysis is based. It has an energy-only focus, greater technological detail and a longer-term perspective to 2050.

8 UNFCCC (2007) estimates shown in Table 10.1 suggest a value of 20 per cent. McKinsey & Company (2009a) provides an estimate for China of $400 (€300) billion for 2026–2030, which is roughly 40 per cent of their global estimate.

9 For example, for China, the UNFCCC analysis suggests that by 2030, $29 billion per year could be saved in new fossil-fuelled power plants ($11 billion/year) and in transmission and distribution ($18 billion/year, largely in distribution). These savings are also enabled by reductions in conventional energy needs due to a projected $7 billion per year investment in buildings and industry.

10 As noted below, the Renewable Energy Law targets 30GW of wind and 1.5GW of solar power by 2020. The McKinsey analysis suggests a potential for 380GW of

wind and 300GW of solar by 2030; their assumption for the baseline scenario is that growth from 2020 to 2030 would have led to 108GW and 13GW respectively.

11 In contrast, the UNFCCC mitigation scenario assumes more gradual change and a focus on efficiency improvements, with the roll-out of hybrid electric vehicles reaching 60 per cent of the vehicle fleet in 2030.

12 The UNFCCC study projects $26 billion investment in CCS, $17 billion in power plants and $9 billion in industrial processes, compared with the overall net $36 billion in added investment in 2030.

13 CCS effectively increases coal consumption by 120 million tonnes in the McKinsey mitigation scenario due to lower generation efficiencies.

14 This represents the sum of buildings and industry investment, which is largely but not wholly in efficiency; it also includes costly non-efficiency investments in industrial CCS, for example, but not power sector efficiency.

15 Government support is only a fraction of this amount. The government increased funding for energy saving from $3.5 billion (¥23.5 billion) in 2007 to $6 billion (¥41.8 billion) in 2008 (State Council as cited in Zhang, 2008).

16 For a good discussion of the various market barriers to energy efficiency, see Section 1 of IEA (2007a).

17 Steel making, chemicals production, cement manufacture, coal mining and waste management.

18 McKinsey estimates 0.8Gt CO_2 equivalent in reduction by 2030 at costs of $-4 to 18 (€-3 to 14) per tonne CO_2 equivalent from municipal solid waste and wastewater management, clinker substitution and co-firing of agricultural waste in the cement manufacturing, coal mine methane capture and fluorocarbon destruction.

19 Nearly half of the global financial flows shown under agriculture in Table 10.1 are associated with agroforestry.

20 No projects in the CDM pipeline exceed 500MW.

21 UNEP/NEF (New Economics Foundation) analysis of energy efficiency and renewable energy is limited to categories that it defines as sustainable energy projects and for which data could be collected: all biomass, geothermal and wind generation projects of more than 1MW, all hydro projects of between 0.5 and 50MW, all solar projects of more than 0.3MW, all marine energy projects, all biofuels projects with a capacity of 1 million litres or more per year, and all energy efficiency projects that involve financial investors (Boyle et al, 2008).

22 Based on data collected by New Energy Finance.

23 See *The Story of Official Development Assistance* for a detailed discussion on the history and the rationale of development assistance, available at: http://www.oecd.org/dataoecd/3/39/1896816.pdf

24 Other mechanisms under the Convention include the Special Climate Change Fund (SCCF) and the Adaptation Fund. As its name suggest, the latter is focused on adaptation activities. The SCCF also funds mostly adaptation activities, but it also funds technology transfer activities under its Programme for Transfer of Technology. Of the $74 million received by March 2008, $14 million was allocated to technology transfer. A key goal of the SCCF is to complement the activities funded by the GEF and other bilateral and multilateral agencies.

25 GEF technology commercialization projects have supported a wide range of technologies; the largest recipients have been solar thermal power plants, fuel cell-powered buses and power plants using biomass more efficiently. These efforts

achieved modest results with many of the projects either being dropped or cancelled. The GEF has assigned this strategy lower priority going forward (Miller, 2008).

26 GEF assessments of international models and experience, for example, assisted in the choice of a feed-in tariff approach to renewable energy support.

27 Japan provided two-thirds of the bilateral energy aid during the 1997–2005 period followed by Germany and France, which account for 12 per cent and 3.4 per cent respectively. India and Indonesia accounted for over 17.5 per cent of the total bilateral energy aid to developing countries, while China accounted for 15.5 per cent of the total share, each receiving over $2 billion.

28 Other bilateral agencies (from Australia, the EU, France, Germany, Japan, the UK and US) are also active in China (ADB, 2007).

29 This is possibly due to the provision of multi-sector funds and general programme assistance, which may incorporate or obscure support for energy activities (Tirpak and Adams, 2008).

30 Non-hydropower renewable energy averaged about $0.2 billion per year, or 10 per cent of the total bilateral energy funding, largely in wind energy projects. Coal-fired plants received 15 per cent of bilateral funding from 1997 to 2005, most of it coming in the late 1990s. Funding for gas power plants has increased in recent years, overtaking coal plant investments.

31 The EIB led the group with $3.2 billion followed by the World Bank at $1.4 billion and the European Bank for Reconstruction and Development (EBRD) at $1.2 billion (Boyle et al, 2008).

32 In the 1990s, the energy portfolio focused on overcoming critical capacity constraints by expanding the capacity for electricity generation and transmission, and improving industrial processes to reduce pollution.

33 Such projects require some concessional resources but have very high leverage. As banks become comfortable with the different lending criteria needed to evaluate investments justified by energy saving, the projects become self-sustaining and even self-replicating as other banks duplicate product offerings. In 2008 it was reported that the IFC had such projects in eight countries including Russia and China, two of the largest and least energy-efficient countries. The IFC was proposing to scale up such activities to support $500 million worth of lending annually, without the need for donor funds (Miller, 2008).

34 Industrial Bank, the Bank of Beijing and Shanghai Pudong Development Bank.

35 This assumes that efficiency investments (rather than economic and sectoral changes) are needed to meet the 20 per cent target.

36 The EIB's China Climate Change Framework Loan is a large-scale, multi-investment scheme designed to support investment projects in a broad range of categories, including renewable energy, energy efficiency, capture and use or storage, and afforestation projects. This is the first loan to be provided by the EIB under the € billion Energy Sustainability and Security of Supply Facility (www.eib.org/projects/press/2007/2007-123-eur-500-million-to-support-climate-change-mitigation-in-china.htm; www.eib.org/attachments/general/briefing_bg2007_clean_energy_en.pdf)

37 Authors' estimate based on UNEP/Risø database (CERs by project category) and Haites and Seres' estimates of $ invested per tonne CER/year.

38 Haites (2007) estimates a market size ranging from 400 (low end of low estimate) to 6000 (high end of high estimate) million CERs per year in 2030. We apply an estimated average capital cost of $200 per CER/year, based on current trends.

Haites (2007) reports an average capital cost of $137 per CER/year for the CDM pipeline to mid-2007, which since increased to over $200 per CER/year (average of full CDM pipeline) due to the decline in industrial gas projects. As this illustrates, estimates of capital cost per CER/year are also highly variable, thus projections should be viewed with the usual caution.

39 The carbon price paid by a buyer for a CER is often discounted for the level of risk borne by the buyer. For riskier projects, a buyer may be unwilling to pay $10/t CO_2 equivalent if the value of credits, if and when generated, is much higher.

40 According to Haya (2007), in China, 'the majority of large hydro projects nearing completion are now applying for CDM credits. Yet there has been no substantial increase in the number of hydros under construction compared to recent years when hydros did not receive any credits. Most credits that may be generated by these projects should therefore be considered to be "hot air" – fake credits which will increase global GHG emissions.'

41 http://unfccc.int/files/kyoto_protocol/application/pdf/china_bap_280908.pdf

42 http://unfccc.int/files/kyoto_protocol/application/pdf/indiafinancialarchitecture171008.pdf

43 http://unfccc.int/files/meetings/ad_hoc_working_groups/lca/application/pdf/barba-dos_on_behalf_of_aosis.pdf

44 http://go.worldbank.org/58OVAGT860

45 In March 2009, the US Congress removed the Bush administration's pledge to the Clean Technology Fund from the 2009–2010 budget due to the Bank's proposed inclusion of coal projects. See www.climatefundsupdate.org/news/whatsnew-march2009.

46 www.treas.gov/press/releases/hp1311.htm

47 www.ma.emb-japan.go.jp/pdf/resume_Financial_Mechanism_for_Cool_Earth_Partnership_.pdf

48 www.bmu-klimaschutzinitiative.de/en/home_i

49 http://unfccc.int/resource/docs/2008/awglca3/eng/misc02a01.pdf

50 The EU plans to auction 15 per cent of the EU-ETS allowances for aviation from 2012 onwards. It plans to use the revenues to finance mitigation actions in the EU and in developing countries.

51 SDRs are a form of intergovernmental currency that supplements financial liquidity in its member countries.

52 This would represent about 6 billion t CO_2-eq in primary CER sales per year, at a price of $17 per t CO_2-eq. This amount corresponds to a 450–500ppm CO_2 equivalent (350–400 CO_2 only) stabilization pathway scenario wherein Annex 1 CER purchases are similar in volume to their own domestic emission reductions (see UNFCCC, 2007).

53 See FCCC/KP/AWG/2008/3 at http://unfccc.int/resource/docs/2008/awg5/eng/03.pdf

54 New LULUCF activities proposed include REDD activities, the restoration of wetlands and other sustainable forest and land management activities, provided non-permanence and methodological issues can be resolved. Caps have also been proposed on the use of CERs generated by LULUCF activities for compliance.

55 For the first time, GEF finance and CDM support were combined, together with Montreal Protocol support, in an industrial chillers project in India. The World Bank's Climate Investment Funds are considering the option of coupling CDM finance with fund investments where needed. It has been suggested, for example,

that a large multifaceted sustainability project, like the development of a new Chinese eco-city, could be supported jointly by carbon finance and direct finance mechanisms (personal communications, World Bank Carbon Finance Unit).

56 This concept was recently floated by California agency staff with respect to issuance offsets for compliance with its state emissions limit. See meetings of 23 March 2009 at: www.arb.ca.gov/cc/capandtrade/meetings/meetings.htm# publicmeetings.

57 The contribution of carbon finance itself is $1–3 billion per year.

References

ADB (Asian Development Bank) (2007) *Clean Energy Financing Partnership Facility: Establishment of the Clean Energy Fund and Clean Energy Trust Funds*, Manila: Asian Development Bank

ADB (2008) ADB Country Partnership Strategy: People's Republic of China 2008–2010, Asian Development Bank, February, www.adb.org/Documents/CPSs/ PRC/2008/CPS-PRC-2008-2010.pdf

Baker & McKenzie (2008) 'What is a programme of activities?', in *CDM Rulebook*, www.cdmrulebook.org/pageid/452

Birner, S. and Martinot, E. (2005) 'Promoting energy-efficient products: GEF experience and lessons for market transformation in developing countries', *Energy Policy*, vol 33, pp1765–1779

Boyle, R., Greenwood, C., Hohler, A., Liebreich, M., Sonntag-O'Brien, V., Tyne, A. and Usher, E. (2008) *Global Trends in Sustainable Energy Investment 2008: Analysis of Trends and Issues in the Financing of Renewable Energy and Energy Efficiency*, Kenya: UN Environment Programme, Sustainable Energy Finance Initiative and New Energy Finance Limited

Capoor, K. and Ambrosi, P. (2008) *State and Trends of the Carbon Market 2008*, Washington, DC: World Bank

Carmody, J. and Ritchie, D. (2007) *Investing in Clean Energy and Low Carbon Alternatives in Asia*, Asian Development Bank, http://aequero.com/docs/ Publications_Presentations/Investing_in_Clean_Energy_and_Low_Carbon_ Alternatives_in_Asia_-_November_2007.pdf

Doornbosch, R. and Knight, E. (2008) 'Discussion paper: What role for public finance in international climate change mitigation?', Round Table on Sustainable Development, SG/SD/RT(2008)3, Paris: OECD

Ecofys/Azure (2008) *The Value of Carbon in China: Carbon Finance and China's Sustainable Energy Transition*, Beijing: WWF Hong Kong

Energy Foundation (2007) *Energy in China: The Myths, Reality, and Challenges*, Energy Foundation Annual Report 2007, San Francisco: Energy Foundation

European Commission (2009) Towards a Comprehensive Climate Change Agreement in Copenhagen, Commission Staff Working Document, SEC (2009) 102, Brussels

Fan, G., Cao, J., Yang, H., Li, L. and Su, M. (2009) 'Toward a low carbon economy: China and the world', available at http://gdrights.org/wp-content/uploads/2009/06/ ce50midtermreport.pdf

FCCC/AWGLCA (Framework Convention on Climate Change/Ad Hoc Working Group on Long-Term Cooperative Action under the Convention) (2008a) 'MISC.2/Add.1', http://unfccc.int/resource/docs/2008/awglca3/eng/misc02a01.pdf

FCCC/AWGLCA (2008b) 'MISC.2', http://unfccc.int/resource/docs/2008/awglca3/ eng/misc02.pdf

FCCC/KP/AWG (2009) 'MISC.3', Submission by China on Emissions Trading and Project Based Mechanisms under AWG-KP, http://unfccc.int/resource/docs/2009/ awg7/eng/misc03.pdf

Haites, E. (2007) 'The carbon market', UNFCCC, http://unfccc.int/files/cooperation_ and_support/financial_mechanism/application/pdf/haites.pdf

Haites, E. (2008) *Negotiations on Additional Investment and Financial Flows to Address Climate Change in Developing Countries*, New York: UNDP

Hampton, K. (2008) 'Programmatic CDM in a Post-2012 Context', Powerpoint Presentation, 9 September, Copenhagen: Climate Change Capital

Haya, B. (2007) *Failed Mechanism: How the CDM is Subsidizing Hydro Developers and Harming the Kyoto Protocol*, Berkeley, CA: International Rivers, http://internationalrivers.org/files/Failed_Mechanism_3.pdf

Hohne, N., Worrell, E., Ellerman, C. Vieweg, M. and Hagemann, M. (2008) 'Sectoral approach and development', Ecofys

IEA (International Energy Agency) (2006) *World Energy Outlook 2006*, Paris: IEA

IEA (2007a) *Mind the Gap: Quantifying Principal-Agent Problems in Energy Efficiency*, Paris: IEA, www.iea.org/textbase/nppdf/free/2007/mind_the_gap.pdf

IEA (2007b) *Sectoral Approaches to Greenhouse Gas Mitigation*, Paris: IEA

IEA (2008a) *Energy Technology Perspectives 2008*, Paris: IEA

IEA (2008b) *World Energy Outlook 2008*, Paris: IEA

IFC (International Finance Corporation) (2008) *China Utility-based Energy Efficiency Programme (CHUEE)*, Introduction, Washington, DC: IFC

Kartha, S., Lazarus, M. and LeFranc, M. (2005) 'Market penetration metrics: Tools for additionality assessment?', *Climate Policy*, vol 5, no 2, pp147–165

LBNL (Lawrence Berkeley National Laboratory) (2008) *China Energy Databook*, http://china.lbl.gov/databook

Leguet, B. and Elabed, G. (2008) 'A reformed CDM to increase supply: Room for action' in K. Holm Olsen and J. Fenham (eds) *A Reformed CDM: Including New Mechanisms for Sustainable Development*, Roskilde: UNEP Risø Centre, www.cd4cdm.org/Publications/Perspectives/ReformedCDM.pdf

Lin, J., Zhou, N., Levine, M. and Fridley, D. (2008) 'Taking out one billion tons of CO_2: The magic of China's 11th Five Year Plan?', *Energy Policy*, vol 36, pp954–970

Lütken, S. E. (2008) 'Developing country financing for developed country commitments?', in K. Holm Olsen and J. Fenham (eds) *A Reformed CDM: Including New Mechanisms for Sustainable Development*, Roskilde: UNEP Risø Centre, www.cd4cdm.org/Publications/Perspectives/ReformedCDM.pdf

Maosheng, D. (2008) 'Clean Development Mechanism: Assessment of experience and expectations for the future', in K. Holm Olsen and J. Fenham (eds) *A Reformed CDM: Including New Mechanisms for Sustainable Development*, Roskilde: UNEP Risø Centre, www.cd4cdm.org/Publications/Perspectives/ReformedCDM.pdf

McKinsey & Company (2009a) *Pathways to a Low-Carbon Economy: Version 2 of the Global Greenhouse Gas Abatement Cost Curve*, http://globalghgcostcurve.bymckinsey.com/

McKinsey & Company (2009b) *China's Green Revolution: Prioritizing Technologies to Achieve Energy and Environmental Sustainability*, www.mckinsey.com/ locations/greaterchina/mckonchina/reports/china_green_revolution.aspx

Meinshausen, M., Hare, B. Wigley, T., Van Vuuren, D., Den Elzen, M. and Swart, R. (2006) 'Multi-gas emissions pathways to meet climate targets', *Climatic Change*, vol 75, pp151–194

Miller, A. (2008) 'Financing the integration of climate change mitigation into development', *Climate Policy*, vol 8, no 2, pp152–169

Müller, B. and Ghosh, P. (2008) *Implementing the Bali Action Plan: What Role for the CDM?*, Oxford Institute for Energy Studies, Energy and Environment Comment, www.oxfordenergy.org/pdfs/comment_1008-2.pdf

Müller, B. and Hepburn, C. (2006) *IATAL – An Outline Proposal for an International Air Travel Adaptation Levy*, Oxford: Oxford Institute of Energy Studies

NDRC (National Development and Reform Commission) (2007) *China's National Climate Change Programme*, www.ccchina.gov.cn/WebSite/CCChina/UpFile/File188.pdf

Olsen, K. (2007) 'The clean development mechanism's contribution to sustainable development: A review of the literature', *Climatic Change*, vol 84, pp59–73

Russ, P., Ciscar, J-C., Saveyn, B., Soria, A., Szábó, L., Van Ierland, T., Van Regemorter, D. and Virdis, R. (2009) *Economic Assessment of Post-2012 Global Climate Policies – Analysis of Gas Greenhouse Gas Emission Reduction Scenarios with the POLES and GEM-E3 Models*, European Commission, Joint Research Centre, Institute for Prospective Technological Studies , http://ftp.jrc.es/EURdoc/JRC50307.pdf

Sanchez, S. (2008) 'Reforming CDM and scaling-up finance for sustainable urban transport', in *A Reformed CDM – Including New Mechanisms for Sustainable Development*, Perspective Series, UNEP-Risø Centre, Kenya

Schneider, L. (2007) 'Is the CDM fulfilling its environmental and sustainable development objectives? An evaluation of the CDM and options for improvement', Oeko Institute, http://assets.panda.org/downloads/oeko_institut__2007___is_the_cdm_fulfilling_its_environmental_and_sustainable_developme.pdf

Schneider, L. (2008) 'A CDM with atmospheric benefits for a post-2012 climate regime', Oeko Institute, http://www.oeko.de/oekodoc/622/2007-162-en.pdf

SEC (2009) 'Towards a comprehensive climate change agreement in Copenhagen', 101/102, Commission of the European Communities, http://ec.europa.eu/environment/climat/future_action.htm

Seres, S. and Haites, E. (2008) 'Analysis of technology transfer in CDM projects', prepared for the UNFCCC Registration & Issuance Unit, http://cdm.unfccc.int/Reference/Reports/TTreport/TTrep08.pdf

Sinton, J. E., Stern, R., Aden, N. and Levine, M. (2005) 'Evaluation of China's energy strategy options', China Energy Group, Berkeley: Lawrence Berkeley National Laboratory, www.efchina.org/csepupfiles/report/2006102695218686.4335721472912.pdf/NEP_LBNL_EN.pdf

State Council Information Office (2008) 'White paper: China's policies and actions on climate change', www.china.org.cn/government/news/2008-10/29/content_16681689.htm

Sutter, C. and Parreno, J. C. (2007) 'Does the current Clean Development Mechanism (CDM) deliver its sustainable development claim? An analysis of officially registered CDM projects', *Climatic Change*, vol 84: pp75–90

Tirpak, D. and Adams, A. (2008) 'Bilateral and multilateral financial assistance for the energy sector of developing countries', *Climate Policy*, vol 8, no 2, pp135–151

Tsinghua University of China (2006) *GHG Mitigation in China: Scenarios and Opportunities through 2030*, Beijing: Center for Clean Air Policy

UNEP-Risø (United Nations Environment Programme – Risø) (2008) 'The Pipeline', by Jørgen Fenhann, Roskilde UNEP Risø Centre, www.cd4cdm.org/pipeline/index.htm

UNFCCC (United Nations Framework Convention on Climate Change) (2007) *Investment and Financial Flows to Address Climate Change*, Bonn: UNFCC

Wara, M. (2008) 'Measuring the Clean Development Mechanism's performance and potential', *55 UCLA Law Review*, no 1759, pp1759–1803

Ward, M. (2008) *The Role of Sector No-lose Targets in Scaling up Finance for Climate Change Mitigation Activities in Developing Countries*, Prepared for the International Climate Division, London: Department for Environment, Food and Rural Affairs (DEFRA)

Wetzelaer, B. J. H. W., van der Linden, N. H., Groenenberg, G. and de Coninck, H. C. (2007) *GHG Marginal Abatement Cost Curves for the Non-Annex I Region*, ECN-E—06-060, Petten: Energy Research Centre of the Netherlands

World Bank (2008) *Development and Climate Change: A Strategic Framework for the World Bank Group, Consultation Draft*, 21 August, http://go.worldbank.org/WWT4W1LH60

Zhang, Z. (2008) 'Is it fair to treat China as a Christmas tree to hang everybody's complaints? Putting its own energy saving into perspective', 10 October, http://ssrn.com/abstract=1285618

11

Emissions Trading and the Global Deal

*Christian Flachsland, Gunnar Luderer, Jan Steckel,
Brigitte Knopf and Ottmar Edenhofer*

Introduction

Avoiding dangerous climate change will require a Global Deal that introduces a set of international institutions addressing the market failures giving rise to climate change (Edenhofer et al, 2008a, 2008b; Stern, 2007, 2008). A global carbon market could be a central element of such a Global Deal, addressing simultaneously the key issues of environmental effectiveness via the global cap, cost effectiveness by trading of permits and equity through fair allocation of emission permits. We aim at providing an overview of the rationale and elements of a Global Deal on climate change and the role of international emissions trading therein, with a particular focus on the options for developing country participation in international emissions trading, in particular China.

Our contribution proceeds in four steps. First, we summarize key elements of a Global Deal on climate change that is in our view required to ensure an environmentally effective, cost-effective and equitable response to global warming. Then we discuss options for building a global carbon market by outlining and comparing top-down and bottom-up approaches. We examine developing countries' options for joining an international carbon market after 2012 and in the long term, and then present preliminary modelling results on mitigation costs and the first-mover advantage in climate policy.

A Global Deal for climate change[1]

Climate change represents an unprecedented challenge to global society. Unmitigated climate change will introduce large-scale risks to ecosystems and human societies, while its mitigation represents a major task for the world economic system. In this section we briefly discuss the key challenges of climate change and propose the outline of an institutional architecture – a 'Global Deal' – to tackle the issue of climate change in an effective, efficient and equitable manner. It is important to note that in our view such a Global Deal should provide a guiding vision that can be implemented via a set of policy roadmaps, eventually merging into an integrated climate policy architecture.

As shown by the IPCC (2007a), climate change is already under way and could lead to an increase of global mean temperature of up to 5°C or more relative to pre-industrial levels. This implies large-scale shifts in global and regional climates, ecosystem patterns and human activities. Global warming could push components of the climate system ('tipping elements') past critical thresholds so that they switch into qualitatively different modes of operation, resulting in considerable consequences for human and ecological systems (Lenton et al, 2008). Examples of such tipping elements include, among others, the Arctic Sea Ice, where summer minima have been decreasing at alarming rates in recent years, and the Greenland Ice Sheet, which stores ice masses equivalent to a sea-level rise of 7m.

Key impacts of climate change include flooding of coastal areas and river deltas, more intense droughts and desertification, increased occurrence of weather extreme events, and water scarcity due to melting glaciers and changing precipitation patterns (IPCC, 2007b). In general, developing countries are more vulnerable to climate change. In addition to small islands, Africa and the Arctic, many regions in Southern and Southeast Asia are particularly vulnerable. The seven Asian mega-deltas account for a collective population of about 200 million and will be strongly affected by increasing sea levels. With increasing warming, the Himalayan glaciers could decay at very rapid rates, giving rise to increased flooding in the next decades and a decrease in water availability later on. Possible perturbations of the monsoon circulation would have a dramatic effect on water resources and agriculture. While no level of climate change is inherently 'safe', stabilization of global warming at 2°C above pre-industrial level is expected to prevent the most severe impacts (see also Smith et al, 2009).

Climate change raises serious questions regarding global equity. Most of the carbon emissions historically occurred in the industrialized countries, and in the past there has been a strong link between capital accumulation and historic emissions (see Figure 11.1). At the same time, developing countries are particularly vulnerable to climate change. Due to high exposure to climate risks and limited adaptive capacity, they are projected to feel the bulk of impacts. Unmitigated climate change will further increase global inequalities.

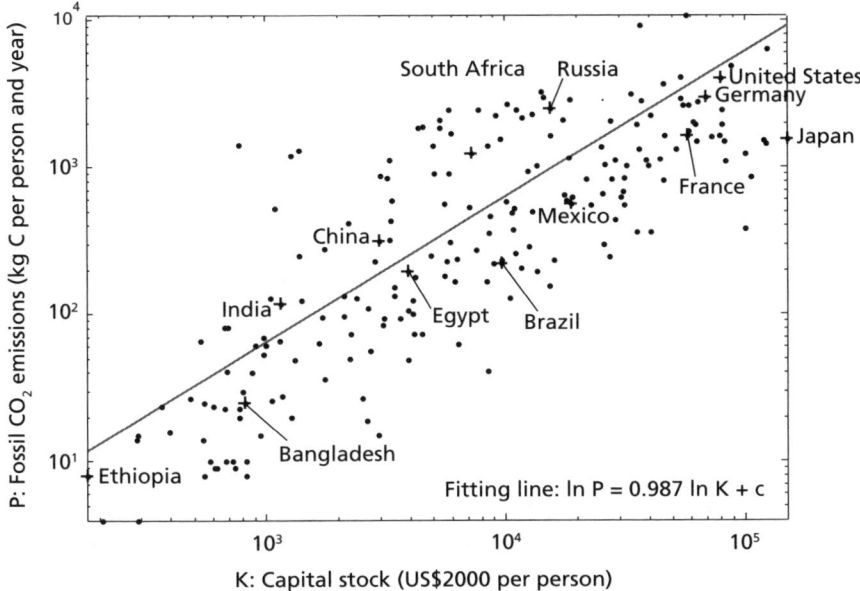

Note: Capital stock is defined as the investment over the past 20 years, with 5 per cent depreciation.
Source: Based on data from Füssel (2010)

Figure 11.1 *Correlation of capital stock and cumulated per capita emissions*

In recent years several studies have shown that the cost of limiting global warming to around 2°C above pre-industrial levels will be relatively low at around 1–3 per cent of global GDP (Edenhofer et al, 2006; IPCC, 2007b; Stern, 2007). However, this requires that action is taken quickly and effective technologies and institutions are put into place on a global scale (see below).

Given that the costs of limiting the rise of global mean temperature to around 2°C are relatively moderate, and that major impacts of climate change regarding tipping elements and ecosystem changes may be avoided when limiting global warming to 2°C, we suggest that this is a reasonable target for international climate policy, given our current knowledge base. This target has also been adopted by the EU and a total of more than 100 countries worldwide including major players such as China, the US and India (EU, 2007; Major Economies Forum, 2009; Meinshausen et al, 2009).

However, achieving this target will require an institutional framework that can deliver on the criteria of environmental effectiveness (reducing emissions in accordance with the 2°C limit), economic efficiency (doing so at least costs) and equity (taking into account different responsibilities and capabilities in mitigating and responding to climate change).

A *global carbon market* builds on existing efforts to implement international emissions trading and achieves environmental effectiveness by setting a global cap for global emissions; realizes efficiency through trading of permits and the emergence of a single carbon price; and allows addressing equity considerations through international allocation rules. The emerging price for emissions should stretch across all sectors[2] and countries. Clearly, a global carbon market implies considerable challenges for international economic policy coordination, and it will be difficult to achieve global consensus on the related distributional issues. Thus, a global carbon market might possibly emerge step by step, for example by continuing the current approach of the Kyoto Protocol beyond 2012, or by linking regional ETS (see the following sections for a more detailed treatment).

A large-scale transformation of the global energy systems will be needed to achieve the deep emission reductions required to avoid dangerous climate change. In this context, experience learning has a large potential to reduce the costs of the transition towards low-carbon technologies. However, market failures where, for example, innovators cannot fully capture the benefits from developing a new technology because others will imitate them will lead to an undersupply of R&D as well as deployment of innovative *low-carbon technologies*. Other market failures relate to principal-agent problems, the provision of information and differences between privately and socially optimal discount rates. Thus, policy intervention is required in addition to carbon pricing. Policy instruments on the national level include enhanced R&D funding for low-carbon technologies, publicly supported demonstration projects for complex technologies such as CCS, and market introduction programmes for renewable energies. In addition, sustainable energy provision for developing countries is of key importance for a long-term and global solution of the climate problem and comes with numerous ancillary local and regional benefits. Mainstreaming low-carbon development into development policy, promoting technology sharing and setting up a low-carbon fund for least-developed countries and regions are important policy options to foster leapfrogging of developing countries into a low-carbon future.

Deforestation and forest degradation accounts for roughly 20 per cent of global anthropogenic GHG emissions (Stern, 2007). According to most estimates, these emissions can be reduced at low costs. Also, reducing defor-estation comes with significant ancillary benefits due to the preservation of ecosystems and their services. Important challenges in establishing an environ-mentally effective REDD regime lie in ensuring permanence of forest conservation and limiting leakage. Options for providing incentives for REDD are full-scale integration into the global carbon market, fund-based schemes or hybrid approaches.

Finally, it is clear that in order to solve the climate problem, mitigation and *adaptation* must go hand in hand to meet the principle of 'avoiding the unman-ageable and managing the unavoidable'. The funding required to finance adaptation to climate change in the developing world is significant. As the adaptation fund set up under the Kyoto Protocol is inadequate in meeting these needs, a broadened funding mechanism should provide a sufficient and reliable financial basis for adaptation activities in developing countries.

In view of the scale of the challenge, the historic responsibility of industrialized countries, the vulnerability of the developing world and the rapidly increasing energy demand in emerging economies, it is evident that the global community will not be able to solve the climate problem in an effective, efficient and equitable manner unless it seeks close international cooperation along the policy roadmaps outlined above.

Architectures for international emissions trading

While the previous section illustrated that a global carbon market could form a central pillar of a Global Deal on climate change, this section analyses options for its construction. It is essential to bear in mind that other policy instruments are required: (1) as substitutes where ETS are not (yet) implemented, for example if there are problems related to accurate monitoring of emissions or the substantial institutional requirements for setting up efficient trading systems represent an obstacle to their implementation, and (2) as complements where trading schemes do not address all relevant market failures.

We proceed by first summarizing recent developments in international emissions trading. As it turns out that there is large uncertainty regarding the evolution of carbon markets beyond 2012, we proceed by discussing options for international emissions trading post-2012. Then we assess these options with respect to their performance regarding environmental effectiveness, cost effectiveness and political feasibility.

Status quo

Recent years have witnessed a considerable amount of political activity directed at the establishment of ETSs.[3] Four major developments can be distinguished, characterizing the status quo.

First, the government-level cap-and-trade system for Annex 1 countries was established by the 1997 Kyoto Protocol and technically specified in the Marrakesh Accords of 2001. It covers GHG emissions of 37 states that represent 29 per cent of global CO_2 emissions (CAIT, 2008).[4] The US does not participate in this system. The Kyoto compliance period runs from 2008 until the end of 2012. So far, hardly any transactions have taken place between governments (Point Carbon, 2008).

Second, the CDM was also established under the Kyoto Protocol. It enables non-Annex 1 countries to participate in international emissions trading without adopting caps. So far, credits representing 200Mt CO_2 equivalent have been issued by the UNFCCC CDM Board, with an expected total of credits representing ~1.5Gt CO_2 equivalent reductions in 2008–2012. About 40 per cent of credits expected for the 2008–2012 period are generated in China, and another 25 per cent in India, indicating a heavy regional concentration (partly reflecting the economic size of these countries) (UNEP, 2008). In 2007, the overall CDM market value[5] was estimated at $13 billion (World Bank, 2008).

The JI mechanisms established by the Kyoto Protocol only achieved a market value of $0.5 billion. The future of both mechanisms after 2012 is unclear. However, various options to modify the current mechanism are discussed (see below).

Third, the EU launched its facility-level EU ETS in 2005. With 2Gt CO_2 it caps ~40 per cent of European GHG emissions from about 10,000 facilities, and EU policy-makers have emphasized that the EU ETS will continue to operate beyond 2012 (EU Council, 2007). The EU ETS market value in 2007 was estimated to be $50 billion. With a global carbon market value of $64 billion, the EU ETS and CDM combined represented the bulk of global emissions trading activity in 2007 (World Bank, 2008).

Finally, a number of regional cap-and-trade systems are now being planned or emerging in several OECD countries. On the national level these include Australia, Canada, Japan, New Zealand, Switzerland and the US. Sub-national initiatives for emissions trading also exist in the US (the RGGI, California, the Western Climate Initiative (WCI), and the Midwestern Greenhouse Gas Accord), Canada (some provinces are members to WCI) and Japan (Tokyo and Kyoto) (see Figure 11.2 for a timeline of these systems as of mid-2008; see also Tuerk et al, 2009 for more details). Also, as the first non-Annex 1 (NA1) country, South Korea, considers the implementation of a domestic ETS (Businessgreen, 2008). These regional activities are flanked by the recent establishment of the International Carbon Action Partnership (ICAP), a forum that was created with the explicit intention of exploring the '... potential linkage of regional carbon markets' (ICAP, 2007).

These developments can be understood as manifestations of two different approaches towards the establishment of ETS. First, there is the top-down approach, characterized by a centralized multilateral decision-making process and embodied in the UNFCCC negotiations. Second, there is the bottom-up approach, associated with decentralized decision-making of individual nations or sub-national entities that implement ETS unilaterally, bilaterally or plurilaterally (Zapfel and Vainio, 2002).

These processes yield two different types of institutional architectures for international emissions trading. The backbone of 'top-down' architectures is emissions trading between governments while 'bottom-up' architectures rest upon the implementation and possible linkage of regional systems, based on facility-level emissions trading.

To sum up, the status quo of international emissions trading is characterized by considerable heterogeneity: there are several approaches to emissions trading that are not implemented in a coordinated manner. Also, the further evolution of the international emissions trading regime after 2012 is highly uncertain, as there is no international agreement and consensus on how to proceed with the development and potential integration of both international government-level and domestic entity-level trading systems. In the next section we discuss various approaches for future international emissions trading.

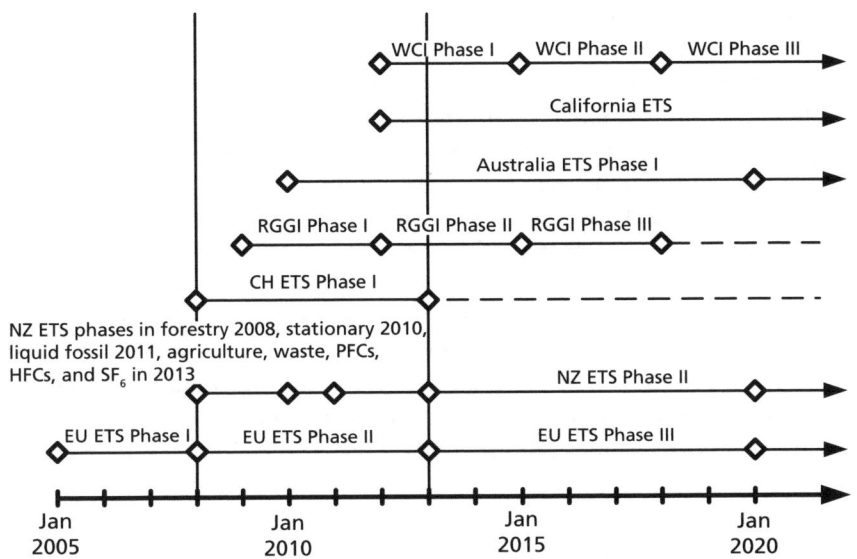

Note: The two vertical lines indicate the timespan of the Kyoto Protocol's first commitment period.
Source: Flachsland et al (2009a)

Figure 11.2 *Timeline for emerging ETS as of mid-2008*

Options[6]

From a theoretical point of view, a global entity-level ETS for all sectors[7] and regions may be regarded as the 'ideal' theoretical benchmark for an environmentally and cost-effective global emissions trading architecture.[8] It would ensure broad coverage, rule out leakage effects and minimize concerns for market power. All GHG-emitting entities would face the global marginal costs of emissions abatement. However, it will be politically difficult to implement such a system at least in the short to medium term due to the unprecedented level of international economic policy coordination required, distributional questions, transaction costs of facility-level emissions trading (for example related to monitoring emissions), and questions of legal enforcement of compliance (Victor, 2007). Thus, alternative approaches for international emissions trading after 2012 are currently being discussed. Subsequently we attempt a summary of this debate by distinguishing two 'top-down' and three 'bottom-up' architectures for international emissions trading.

Top-down

A *global government-level cap-and-trade architecture* implies that every country adopts a well-defined and limited GHG emissions budget for its entire economy, and that emission allowances can be traded between governments (see for example Vattenfall, 2006). This kind of architecture would emerge if all countries assumed Annex B status in the Kyoto Protocol.

By contrast, a *Kyoto-II architecture* would continue the current Kyoto Protocol approach beyond 2012. Only a limited group of countries, for example Annex 1 countries and major emitting economies including for example China, implements a cap-and-trade system on the level of governments. All other countries, for example remaining developing countries, could participate for example by means of 'trade without cap' as represented by the current CDM approach or another mechanism. Given the critique of the CDM in its current form, for example with regard to additionality and transaction costs (see Michaelowa et al, 2003; Schneider, 2007; Fan et al, 2009), various reform proposals are currently discussed (see below).

For any government trading scheme, additional domestic policy instruments need to be implemented to translate the international price signal to domestic actors, for example domestic cap-and-trade, taxes, standards and technology support schemes (Hahn and Stavins, 1999). We abstract from such domestic policies here to focus the analysis on the structure of international emissions trading.

Bottom-up

In the presence of two or more independent ETS that are installed at the national, supra- or sub-national level, and that do not have any intentional linkages between them, one can speak of *fragmented markets*. Even though international trade in goods already induces a certain tendency towards permit price convergence across different ETS,[9] prices will in general vary and thus prevent a cost-effective outcome. The degree of inefficiency increases – *ceteris paribus* – in proportion to the price differential between carbon markets.

If at least two regional cap-and-trade systems accept credits from the same baseline-and-credit scheme, an *indirect link* between them is established (Egenhofer, 2007). Depending on the supply curve for credits, cap levels, MAC curves and quantity limits on the import of credits, indirect linking will lead to a complete or incomplete convergence of the allowance price in indirectly linked cap-and-trade markets (see Flachsland et al, 2009a for details).

Therefore, a carbon market architecture with indirect linkages between regional trading systems will improve cost effectiveness vis-à-vis the fragmented market case. By how much depends on the level of price convergence across systems, which largely depends on the flatness of the credit supply curve. Assume, for example, unlimited supply of CDM credits at a fixed price, and a link of several cap-and-trade markets to the CDM. If the pre-link price in the capped regions was above the CDM price, the prices in all markets would converge at the CDM price level. Therefore, developing country baseline-and-credit schemes that clear the way for large-scale investment opportunities into abatement, for example in the power sector of developing countries, would be conducive to cost effectiveness in a market architecture, characterized by indirect linkages between regional trading schemes.

Finally, *direct linking* occurs whenever two (or more) regional ETS mutually recognize each others' allowances, i.e. they accept emission certificates issued in other systems as valid for compliance within their own system. A direct linking architecture is thus established through a concerted linking-decision of different regional trading systems (Tangen and Hasselknippe, 2005; Edenhofer et al, 2007; Victor, 2007; Jaffe and Stavins, 2008).[10] Evidently, an immediate consequence of linking is the formation of a common emissions price.[11]

The benefit of enhanced cost effectiveness comes, however, at the cost of contagiousness: once two ETS are linked, changes in the design or regulatory features in one system that influence the price formation automatically diffuse into all other systems.[12] For instance, if only one country decides to adopt a price ceiling in the form of a price cap, the entire linked market is in effect capped at the same price. Arbitrage trading will ensure that prices in each system do not exceed the price ceiling. Thus, there is a partial loss of control for domestic regulators over their own system, necessitating a high degree of coordination – and mutual trust – in the management of the joint carbon market. Relevant design issues with implications for the whole linked carbon market include, *inter alia*:[13]

- the setting and modification of emission caps;
- upper and lower bounds for permit prices;
- links to baseline-and-credit schemes, for example the CDM;
- banking and borrowing provisions;
- compatible registries;
- rules for MRV of emissions;
- penalties and enforcement of compliance.

To address these issues, institutional provisions in the form of, for example; linking agreements and joint regulatory bodies are required, both before and during the linking operation (Flachsland et al, 2008; Mehling, 2009; Tuerk et al, 2009). In fact, as a first step in that direction, several countries and regions with existing or emerging regional cap-and-trade systems and with an openly expressed interest in linking have already joined forces and established the ICAP in 2007. Part of its mandate is to assess barriers to linking and work out solutions where such impediments may exist (ICAP, 2007).

Discussion

In this section we analyse the five carbon market architectures outlined above with respect to the three criteria – environmental effectiveness, cost effectiveness and political feasibility – in order to identify their relative strengths and weaknesses.

First, *environmental effectiveness* refers to the capability of an emissions trading architecture to bring about significant reductions in global emissions. Its potential for doing so depends, first of all, on the share of global emissions that are actually covered by the emissions trading regime. Also, leakage effects[14] and reduction targets need to be taken into account. Thus, the percentage reduction of global emissions can be expressed by the following equation:

$$\text{Global Reduction [per cent]} = \frac{\text{Regime Reduction [\% to BAU]} \times \text{Regime Emissions [t]}}{\text{Global Emissions [t]}) \times (1 - \text{Leakage Rate})} \qquad (1)$$

Because we abstract from specific reduction targets in our analysis of carbon market architectures (to be able to focus on their broader properties), we should rather speak of the *potential* environmental effectiveness of a trading architecture.

Against this backdrop, a top-down architecture with global cap-and-trade obviously offers the best possibility for significant cuts in global emissions, as it features the highest coverage and excludes carbon leakage. On the other end of the spectrum, a bottom-up architecture consisting of fragmented markets is unlikely to significantly curb global emissions, as it can be expected that only few countries will implement trading schemes with ambitious reduction targets that remain unconnected to major baseline-and-credit schemes such as the CDM. The situation is less definite for the other, 'intermediate' architectures: with a sufficient number of committed participants, the indirect linkages and especially the direct linking approach may come close to the environmental effectiveness of a Kyoto-II architecture. In fact, while a bottom-up approach may be more likely to start out with lower initial emissions coverage, it can expand step by step, thereby gradually increasing the share of global emissions that it covers. Eventually, it may resemble our first-best theoretical benchmark configuration. Both the Kyoto II and all of the bottom-up schemes will have to face the challenge of controlling emissions leakage. In fact, this will be the case for every architecture not covering all major economic sectors and regions. However, Kyoto II as well as the formal and indirect linking architectures can be extended to provide economic incentives for emission control to third countries in the form of appropriately designed trading mechanisms (see Chapter 4). In general, short-term concerns over leakage can be mitigated if most or all of those countries that are close trade competitors participate in a linked carbon market with a single permit price.[15]

Our second criterion, *cost effectiveness*, requires the minimization of the costs of achieving a given emissions reduction target. It is well known that cost effectiveness depends on the equalization of MAC across regions and sectors (see Tietenberg, 2003). In theory, an ETS realizes this by creating a uniform price signal reflecting marginal abatement costs across all covered entities. In practice, however, the emerging emissions price under a permit trading scheme may deviate from marginal abatement costs, in particular if: (1) one or more actors possess market power,[16] (2) regulators trade on behalf of firms but do not have full information on the abatement costs incurred by the latter (Kerr, 2000), and (3) not all economic sectors are included in the scheme.

Among participating countries, top-down architectures always ensure a complete equalization of the permit price. However, there are concerns over market power[17] and doubts about the proper revelation of domestic MAC, reducing the cost effectiveness of these architectures. Also, it can be generally questioned whether governments will act as cost minimizers on the international carbon market given, for example, geopolitical interests (Hahn and Stavins, 1999). By contrast, bottom-up approaches lead to permit price equalization only in the formal linking case or – under the condition that the credit supply curve is sufficiently flat and no restrictions are imposed on credit imports – in the indirect linking case. However, the price signal may be more robust since facility-level trading systems are better suited to resolve the information asymmetry between governments and companies and are less prone to market power distortions.

Finally, the question of *political feasibility* is related to participation levels and the requirements for coordination and burden sharing, as well as transaction costs of a trading architecture. Evidently, in order to establish an integrated[18] trading architecture, players need to agree on a common regulatory framework that will be more difficult to achieve with a larger number of players. For example, some players will favour higher emissions prices to foster technological development, while others may focus on cost containment by implementing price caps. With an increasing number of participants, such differences will become both more probable and more difficult to reconcile. Regarding burden sharing, regional caps have distributional implications, as allocations represent each player's cost-free endowment and thus largely determine the required effort. In consequence, bargaining over burden sharing becomes a strategic game where self-interested players have an incentive to free-ride on the mitigation efforts of others by implementing targets with low stringency (Helm, 2003; Rehdanz and Tol, 2005). We assume that a rising number of players that need to agree over caps in trading architectures negatively effects political feasibility. Finally, trading architectures can be compared in terms of the transaction costs that arise from creating the necessary institutional structures. We assume that high transaction costs reduce political feasibility.

Top-down approaches resemble 'all-or-nothing' options with respect to political feasibility: without international consensus on burden sharing, complete political standstill is imminent. This constitutes a very tangible threat, given that any kind of agreement can rather easily be blocked by countries with vested interests. Similarly, agreement on the design details of the trading and accounting system will be more difficult to achieve than for bottom-up approaches with fewer participants. In fact, the direct linking approach will always enable cooperating regions to jointly reduce emissions in a cost-effective manner, even in the absence of a global accord on burden sharing and regulatory design. Indirect linking provides less certainty regarding permit price harmonization, but implies even less need for political coordination (Jaffe and Stavins, 2008). Transaction costs of top-down architectures are at least

Table 11.1 *Comparison of six carbon market architectures*

| | *Integrated* ←--------------------------→ *Fragmented* | | | | | |
| | *Benchmark* | *Top-down* | | | *Bottom-up* | |
	Global entity cap & trade	*Global government cap & trade*	*Kyoto II*	*Direct linking*	*Indirect linkages*	*Fragmented systems*
Environmental effectiveness						
Coverage	++	++	+ depends on developing country participation	o (+) depends on participation	o (+) depends on participation	− depends on participation
Prevention of leakage	++	++	+	o	-	—
Cost effectiveness						
Price convergence	++	++	++	++	+	−
Overcoming MAC information asymmetry	++	o	o	++	++	++
Limiting market concentration	++	−	−	++	+	o
Political feasibility						
Ease of achieving agreement on cooperation	—	—	-	+	++	++
Low transaction costs	—	—	-	o	+	++

Note: The ratings, from very high (++) to neutral (o) to very low (—) represent a relative measure of differences between architectures[19]

Source: Adopted from Flachsland et al (2009a)

initially relatively high, because a larger number of players need to implement the institutional infrastructure required for participating in the common carbon market.

Table 11.1 summarizes the key characteristics of the five carbon market architectures under consideration, plus our benchmark case. It illustrates how the choice between integrated top-down and fragmented bottom-up architectures corresponds to a trade-off between high environmental effectiveness on the one hand, and political feasibility on the other. The picture is less clear-cut for cost effectiveness.

In the context of ambitious climate policy goals such as the EU's 2°C objective, global emissions need to stabilize and start declining within the next two decades. Also, in view of global BAU GHG projections, large amounts of emissions need to be mitigated. Thus, if carbon markets become the major tool for coordinating emissions reductions in a future international climate regime, top-down architectures appear quasi indispensable to meet this challenge as they enable the instant implementation of globally significant reduction targets. Moreover, their major weakness – low political feasibility due to the need to resolve the burden-sharing issue – can in a way be understood as strength: the very crux of the climate problem – the equity issue – is addressed at once, which keeps up the pressure on negotiators and prevents procrastination up to a point in time where low stabilization becomes unfeasible. Thus, within this long-term perspective bottom-up architectures appear as imperfect substitutes of top-down approaches, representing a mere fallback option if a global agreement cannot be achieved instantly. Consequently, they would mainly serve to bring new momentum to the currently stagnant efforts to establish a global, integrated system.

By contrast, the two approaches can be viewed as complementary in the sense that bottom-up architectures may serve as essential building blocks for more comprehensive top-down architectures. In this way, efficient regional facility-level carbon markets can already be put into place, while the delicate question of burden-sharing is deferred for some time. For example, it would be conceivable that after the Kyoto Protocol's expiry in 2012 a group of countries willing to adopt binding economy-wide caps proceeds with the Kyoto Protocol's intergovernmental cap-and-trade system, and formally link their emerging domestic trading systems within this overarching structure. By devolving intergovernmental permit trading to the entity level, the economic performance of the international carbon market would be improved.[20] Within participating countries, sectors remaining outside the domestic facility-level systems would be covered by different instruments such as standards, and governments could trade on behalf of these remaining sectors on the international carbon market.

But unlike the Kyoto scheme, this architecture can be designed as an open system, where countries can join by linking-up their domestic ETS whenever they feel ready, or whenever the political momentum in the country has reached a sufficient level.[21] Such an approach could be environmentally and economically more effective than pure bottom-up approaches, while being less prone to political deadlock than the pure top-down approach.

Options for developing countries

Developing countries of today will be responsible of a growing share of global emissions in the future (see for example Stern, 2008). Therefore, they will need to actively participate in a global climate regime if the global effort aimed at avoiding dangerous climate change is to be successful. If international emission trading emerges as a major instrument for reducing global emissions, there are a number of options for developing countries' participation in such a regime. Clearly, their choice of approaches and timing will be related to the emissions trading architecture adopted by industrialized countries post-2012, and to other elements of the future climate regime. Least-developed countries might

Table 11.2 *Overview of mechanisms for developing country participation in international emissions trading*

Cap-and-trade			Intensity baseline-and-credit		Project baseline-and-credit	
Government		Entities	Government		Government & Entities	
Economy-wide	Sectoral	Sectoral	Economy-wide	Sectoral	Sectoral, bundles of projects	Single projects

choose trading approaches different to those of major industrializing nations such as China, motivated by differing economic circumstances and institutional capacities, but also due to unequal future shares in global GHG emissions.

Mechanisms for participation of developing countries in international emissions trading can be distinguished by some principal features. First, the trading mechanism may adopt the cap-and-trade or baseline-and-credit rationale. Another important specification is whether targets are binding, i.e. additional allowances need to be purchased if prescribed emission levels are exceeded, or non-binding, where this is ruled out and developing countries can only become permit sellers. Also, a distinction needs to be made regarding the actors participating in international emissions trading, the options being governments or entities. Finally, the scope of a mechanism is an important feature. Here we distinguish economy-wide coverage, restriction to some sector(s) and project-level approaches.

Table 11.2 displays an overview of the options, focusing on those we consider particularly relevant and promising for the post-2012 debate. The remainder of this section deals with characterizing the main properties of these seven approaches.

Before we turn to specific approaches, we briefly outline five key criteria for analysing these options. The *environmental effectiveness* of a trading mechanism for a developing country relates to the amount of emissions it covers, and to the level of caps or baselines that are implemented and achieved. Regarding baselines there might be concerns about additionality. *Cost effectiveness* of emissions trading requires that marginal abatement costs are equalized among emissions sources. For example, high transaction costs can preclude harmonization of marginal abatement costs. Also, there is a wide range of techno-economic options for reducing emissions,[22] and different instruments will introduce different incentives to exploit these options. Also, there will be *distributional considerations*, where a major concern will be the overall economic impact of an emissions trading mechanism on developing countries' economies, the distribution of potential rents from trading for example among developed and developing countries, and the impact of carbon pricing on low-income groups. *Institutional feasibility* relates to questions of financial and human resources required for implementation, for example for setting up emission registries. Also, how well emissions trading fits into the regulatory and legal context and traditions may vary from country to country. Finally, *co-benefits* may arise from implementing trading schemes, such as improved international sharing of technologies, improved local and regional air quality from emission reductions, and enhanced energy security (for example if less fossil fuels need to be imported). The presence of co-benefits could facilitate the adoption of trading mechanisms in developing countries, and may also be taken into account in the calculation of caps or baselines.[23]

Economy-wide cap

Similar to Annex B countries that ratified the Kyoto Protocol, developing countries could adopt economy-wide caps where governments trade on behalf of the economy. Importantly, in this case the country still has the choice to implement domestic policy instruments for reducing emissions, with options including emissions trading for companies, carbon taxes, combinations of these approaches, standards, and R&D support schemes for low-carbon technologies (including subsidies for low-carbon technologies). The government may use revenues from international emissions trading for implementing these domestic instruments, for example via tax cuts (see Hahn and Stavins, 1999).

It is important to note that the adoption of national caps does not necessarily mean that emissions reductions are required from developing countries: for example a national cap for China may be equal or slightly below expected BAU emissions. Also, if this is acceptable to industrialized countries, caps may initially be implemented on a no-lose basis, ensuring that developing countries do not need to buy permits if emissions rise beyond the caps. There would only be an incentive to reduce emissions below the cap via selling emissions. This implies that as long as a developing country's emissions are below the cap, expanding emissions has an opportunity cost because permits used for emitting cannot be sold on the international market. Thus, even no-lose trading can provide an incentive to avoid carbon leakage.

The question of specifying caps for developing countries relates to the discussion about alternative international allocation rules (for example contraction and convergence, equal per capita, GDRs, etc.). Different international allocation rules will yield different economic outcomes for developing countries (see below). Also, as discussed above, the cost effectiveness of government-level emissions trading can be generally questioned, for example due to concerns over market power.

Absolute sectoral caps – government trading

Instead of setting a cap for the entire economy, it is also conceivable to set caps for particularly emissions-intensive and suitable sectors only, where the government would trade internationally on behalf of these sectors. Again, domestic policies are required to translate the international carbon price signal to companies.

Setting a series of sectoral caps in major developing countries can be one way of implementing an international sectoral cap-and-trade regime, which would mitigate concerns over carbon leakage: if a developing country cannot sell a permit due to the expansion of domestic emissions, for example from the aluminium or steel sector, this represents an opportunity cost.[24] In fact, addressing carbon leakage in such a manner could enable more ambitious reduction targets in industrialized countries. If sectoral caps are implemented, it is important to ensure that the boundaries of the sector definition are clear so that facilities cannot 'leak' out of a sector (for example via a redefinition of affected facilities).

Sectoral entity-level cap-and-trade with international linking

Developing countries may also implement facility-level cap-and-trade systems for certain sectors (for example electricity, iron and steel), and enable covered entities to trade permits internationally (EU Commission, 2009). Such an approach may also be part of an international sectoral agreement. It would eliminate the need for government trading on behalf of companies. In fact, this approach is identical to the efforts by industrialized countries that set up domestic cap-and-trade systems to link them internationally. As with all other approaches discussed in this section, setting the cap will crucially determine environmental and distributional effects. Also, the covered sectors will directly face the international permit price, which they will pass on to consumers wherever possible in competitive markets. A key requirement for facility-level trading is the presence of a reliable MRV regime for facility emissions, which might be difficult to ensure in some developing countries with weak institutions.[25] In any case, a reliable MRV should be implemented before introducing a trading scheme, as was demonstrated by the European experience (see Chapter 8).

Economy-wide intensity target

An economy-wide intensity-based trading system is characterized by an intensity benchmark for the entire economy, for example in the form of emissions per unit GDP. Credits are generated ex post by taking into account actual GDP and emission levels in relation to the intensity benchmark. As the intensity target is applied to the whole economy, credits are issued to the government. Again, domestic policy instruments are required.

The major argument in favour of intensity over absolute targets is that with uncertainty about economic development, there is less risk that a booming economy will exceed the cap (in case of binding absolute targets necessitating permit purchases). Vice versa, in case of an economic recession, no excess permits become eligible for sale (as would be the case with absolute caps), which may be a concern for some industrialized countries.[26] The major argument against intensity targets is that they do not provide certainty about emissions reductions, as expanding GDP (or another benchmark indicator) enables more emissions. However, intensity targets may be an instrument for transition periods, where the major policy objective is to start decreasing emissions growth rates of developing countries.

Sectoral intensity targets

In analogy to absolute caps, intensity targets can also be applied for single sectors instead of the entire economy. The CCAP has worked out a widely discussed proposal featuring sectoral no-lose intensity targets (Schmidt et al, 2006; see also EU Commission, 2009). Developing countries would implement intensity targets for particularly emissions-intensive sectors such as electricity or iron and steel; when actual emission intensities are below the baseline, the corresponding emission reductions are certified and can be sold on the international carbon market. If emission intensities exceed the baseline, there is no need to purchase permits for developing countries.

Benchmarks need to reflect the technological potential of reducing intensities to ensure that there is an incentive to actually overachieve on the target. Also, benchmarks may reflect an own-contribution of developing countries to emissions mitigation by setting them lower. As long as sector emissions intensity is below the benchmark, there is a disincentive to operate facilities with high emission intensities. However, an incentive for operating additional facilities with intensities below the baseline is introduced, possibly giving rise to concerns over subsidies for leakage. As with all other government-based trading mechanisms, this proposal leaves open the question of which domestic policy instruments to choose.

Importantly, this mechanism fails to incentivize the abatement options of reducing the demand for a sector's products, which may be used more efficiently or substituted by other materials given an appropriate price signal.

Sectoral projects, bundles of projects

Sectoral projects[27] operate by implementing some kind of programme or policy, for example implementing a technological standard or a tax, and then calculating the amount of induced emissions reductions determining the amount of credits for sale (Samaniego and Figueres, 2002; Cosbey et al, 2005). In general, both governments and companies could initiate such projects. The major challenge is to calculate the emission reductions actually induced by the programme. Also, it needs to be determined whether a policy would not have been implemented in the absence of the emissions trading mechanism (additionality criterion).

It is also conceivable to bundle several individual projects (see next subsection) under one single baseline. This can broaden the scope of single projects, facilitate the process of baseline development and reduce transaction costs.

Single projects

Instead of designing policy instruments for the entire economy or sectors, GHG abatement may be rewarded on the project level, an approach employed by the current CDM. Methodologies for setting baselines and determining additionality need to be developed for each type of emission reduction activity. A number of methodologies have been developed to date, whose availability reduces the transaction costs of similar future projects. The CDM might be continued within sectors or countries not covered by other trading approaches (as otherwise issues of double counting would arise).

While the CDM market has experienced considerable growth in recent years (see above), there are a number of criticisms. First, project-based CDM is seen to be too narrow in its scope to trigger financing of large-scale decarbonization in developing countries (Fan et al, 2009). The need for developing methodologies on the project level introduces transaction costs (Michaelowa et al, 2003; Fan et al, 2009) and institutional barriers. Importantly, there have been concerns over the additionality of projects (Michaelowa and Purohit, 2007; Schneider, 2007).

Various CDM reform proposals have been made to address major criticisms of the CDM. Some aim to upscale the instrument in general, while others aim to address shortcomings within the current structure. We see two dimensions with respect to upscaling the CDM: first, to widen the scope of the instrument, for example in bundling projects or introducing sectoral targets, as discussed above. Second, it is conceivable to allow additional technologies to be included into the CDM, as it is currently discussed for CCS (de Coninck, 2008).

In regard to modifications within the current structure, one proposal is to discount credits generated by the CDM (CERs). In this case, emission reduction credits would only be issued for a certain percentage of the certified emission reductions. This approach might serve several targets: first, discounting can lead to a net emission reduction achieved by the CDM projects themselves (and not merely by the targets set in cap-and-trade systems) (Chung, 2007). Second, discounting CERs could also be one way to reflect uncertainties that are connected to the additionality of projects. Finally, discounting may be used to reduce the rents produced in low-cost CDM abatement projects such as N_2O or HFC-23.

Discussion

While this section only provides a brief sketch of options for developing-country participation in international emissions trading, a number of themes emerge. First, determining caps or baselines for developing countries is a very sensitive issue, and there are different approaches for choosing them (for example with view to a global emission reduction schedule, distributional considerations, technical feasibility and abatement costs, uncertainty over future economic developments). Second, if governments engage in international trading on behalf of economic sectors, they need to implement additional domestic instruments, and they can distribute revenues from international permit sales to the economy. Third, for introducing market-based (taxes, trading) carbon pricing on the entity level, a reliable emissions monitoring infrastructure is required. Fourth, cap-and-trade mechanisms can introduce opportunity costs for expanding emissions in developing countries, thus mitigating concerns over carbon leakage in industrialized countries and enabling more ambitious reduction targets by the latter.

Stern (2008) proposes that major developing countries adopt sectoral no-lose intensity targets in an interim period post-2012, with a 'commitment to commit' to binding caps in 2020.[28] He proposes to implement intensity-based no-lose mechanisms as transitory instruments. This may trigger carbon finance flows to developing countries and a social-learning process regarding emissions trading in the latter, and can facilitate implementation of the emission monitoring infrastructure required for international trading without putting economic development at risk.

Towards a quantitative assessment of mitigation costs

Regional distribution of mitigation costs and the role of allocation rules

In the scientific literature, several studies have shown that the inter-temporally aggregated costs of ambitious mitigation are rather moderate at about 1–3 per cent of gross world product (Edenhofer et al, 2006; Stern, 2007; IPCC, 2007c). A central question determining the acceptability of a global ETS for developed and developing countries alike, however, is the distribution of these costs among world regions.

It is fundamental to note that in ETS there is a difference between the initial endowment (allocation) of emission rights and actual emissions after emissions trading has occurred. Countries that emit less than their endowment can sell permits, while regions with higher emissions can buy these. As an example, Figure 11.3 displays an illustrative projection for the allocation of emission rights and actual regional emissions after trading in a global ETS. Since cheap abatement opportunities are more abundant and per capita emissions are low in most developing countries, the developing world as a whole is projected to be a net seller of emission allowances to the industrialized countries.

Under the assumption of a well-functioning carbon market, integrated assessment models featuring an inter-temporal optimization calculus enable the separating of the welfare consumption effects induced by the allocation of emission allowances from those induced by energy trade effects and domestic abatement. For a given global stabilization target, and with equalization of marginal abatement costs due to efficient trading, the amount of emission reductions performed in a region is independent (separable) of the amount of

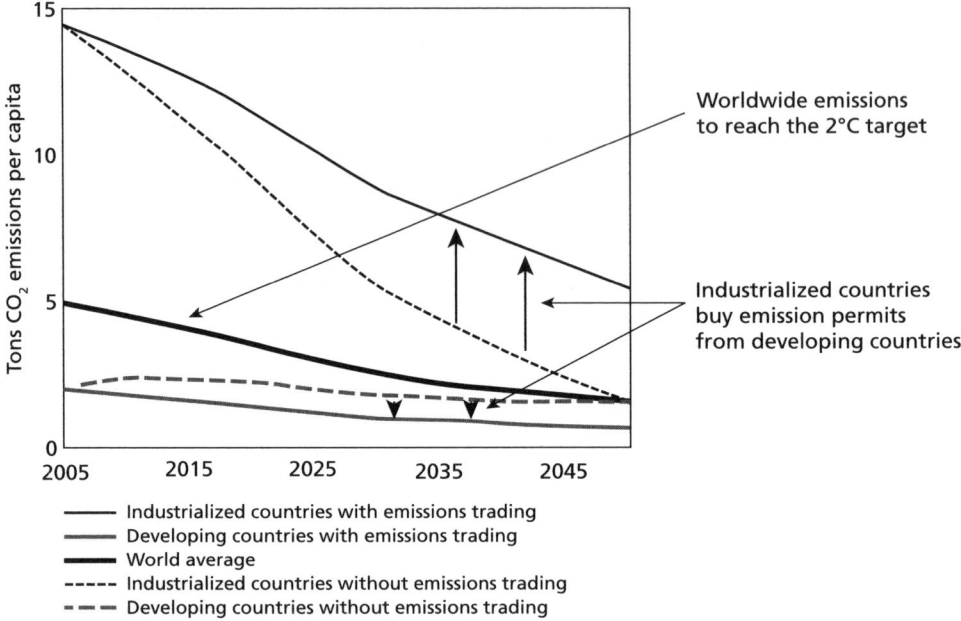

Note: Annex 1 and NA1 regions are allocated emission rights according to the contraction and convergence rule (see below). Actual emissions deviate from this distribution as Annex 1 countries buy emission permits from NA1 countries.
Source: Leimbach et al (2009)

Figure 11.3 *Difference between allocated and actual emissions after emissions trading in a global ETS*

emission allowances allocated. An increase in emission endowments thus merely results in an increase of the region's revenue from selling emission rights to other regions (or not needing to buy allowances). By the same token, the global carbon price is also independent of the allocation rule. Thus, in the presence of a global ETS, efficiency and equity can be addressed separately by means of adjusting regional permit allocation.

We can therefore decompose the regional mitigation costs, measured in terms of consumption losses, into the following three terms:

1   Domestic abatement costs – the cost of reducing emissions from BAU level to the level consistent with the global carbon price. As described above, in the presence of an efficient global carbon market, for a given global reduction target this term is independent of the regional cap.
2   Trade in energy and other goods – global mitigation efforts will strongly affect international trade as they will reduce demand for carbon-intensive energy carriers such as coal and oil as well as energy-intensive goods while increasing demand for low-carbon energy carriers and technologies. Similar to domestic abatement costs, the welfare effect incurred by trade in goods is independent of the international allocation of emission rights.

3   Trade of emission allowances – depending on the international allocation rule, a region will be a net buyer or seller of emission permits. For a given number of permits to be sold or bought, the expenses or revenues from emissions trading (i.e. the value of permit endowment) are proportional to the level of the global carbon price. The level of the permit price strongly depends on the efficiency of global mitigation efforts (for example broad coverage of sectors and regions in a global carbon market, technology policy for fostering low-carbon innovation) and the portfolio of available mitigation technologies.

In the context of international burden-sharing for mitigation of climate change, the discussion usually focuses on the last term, since rules for allocation of emission rights can be directly influenced in the negotiation process. It is, however, fundamental to note that the first two terms are equally important contributors to the distributional outcome of a global mitigation effort, which can, under certain circumstances, exceed the allocation effect.

We use the hybrid energy-economy ReMIND-R (Leimbach et al, 2009) to assess the regional distribution of mitigation costs for various allocation rules. Further in-depth model-based analyses on the distributional effect of climate policy are currently under way (Knopf et al, 2011). A discussion of the decomposition of mitigation costs by the above-mentioned three categories is given in Luderer et al (2009) and Lüken et al (2009). First results on welfare effects, both in relative and absolute terms, are shown in Figure 11.4. The allocation rules considered include:

• Allocation of emission rights in proportion to population.
• Allocation of emission rights in proportion to a region's share in global GDP.
• Contraction and convergence (Meyer, 2000), where the regional shares of global emissions rights converge linearly from status quo (2005 emissions) to equal-per-capita in 2050.
• Default GDR (static). While the previous three schemes are based on a rule for allocating the global *allowable* emissions under a climate stabilization constraint, the GDR framework, as introduced by Baer et al (2007), defines a rule for distributing the *reductions* required relative to the baseline level. For this purpose, a composite index based on historic responsibility and economic capacity is defined. The historic responsibility is quantified in terms of the cumulative emissions from 1990 to 2005. The capacity is determined by the amount of individual income in excess of a pre-defined per-capita development threshold. Thus it depends on the income distribution within nations. The higher historic responsibility and the higher the capacity, the higher is the aggregated RCI. Here we applied the GDR framework in a static form: for each time step, the global mitigation burden in terms of the difference between the baseline emissions and the emissions in the climate stabilization scenario is distributed in propor-

(a)   (b)

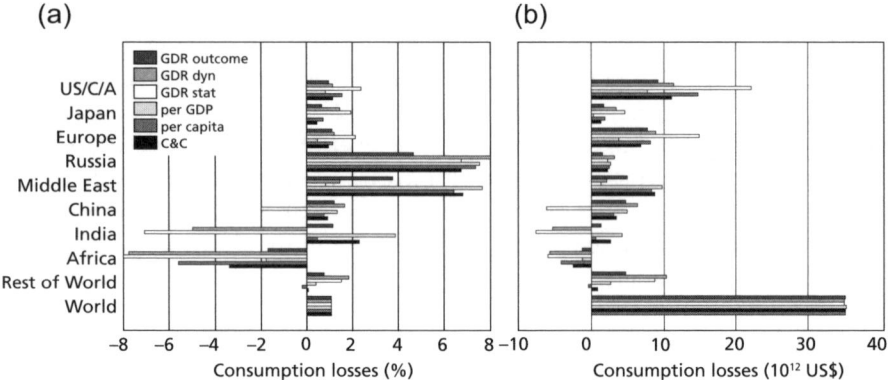

Note: Mitigation costs are calculated as consumption losses relative to baseline discounted at 3 per cent and given (a) as percentage changes and (b) in absolute numbers. A limit on global mean temperature increase of 2°C relative to pre-industrial levels was assumed as a climate policy target.[29]
Source: author's calculation using ReMIND-R

Figure 11.4 *Distributional effects for the three generic allocation schemes (contraction and convergence, equal per capita emissions, allocations in proportion to GDP) as well as three stylized variants of the GDR framework (static allocation-based, dynamic allocation-based, static outcome-based)*

tion to the 2005 value of the RCI. The allocation of a region thus equals its baseline emissions minus the region's share of the global mitigation gap.

- GDR (dynamic) – distribution of the global mitigation gap similar to the last scheme, albeit with dynamic adjustment of the capacity component according to regional GDP growth rates to account for the fact that GDP and therefore capacity in the regions change over time.

- Outcome-based GDR (static) – distribution of the global mitigation costs induced by climate policy in proportion to the 2005 static value of the RCI. Instead of considering allocation rules and then calculating the distribution of costs in terms of regional GDP losses from mitigation, one may approach the equity issue the other way around by fixing some desired outcome regarding the relative regional distribution of mitigation costs and calculating corresponding allowance allocations. Here we adopt the distributional suggestions from the GDR framework to calculate the allowance allocations that would lead to a distribution of mitigation costs in line with GDR. Allocations are adjusted such that each region's aggregated consumption losses until 2100 (including domestic abatement costs, effects on energy trade and emissions trading) correspond to its global share of the RCI.

All simulations were conducted under the assumption of an immediate start of global mitigation action with a set-up of an international carbon market from 2010. The time horizon considered for all simulations presented here is 2005–2100. Welfare effects are measured in terms of consumption losses relative to baseline aggregated over time at a discount rate of 3 per cent.

The ReMIND-R model projects that among the first group of allocations (C&C, per capita, per GDP), the allocation effect is small compared to the effects related to trade and domestic abatement – variations for a given region between allocation schemes are smaller than variations for a given allocation scheme between regions. In general, exporters of fossil resources (Russia and Middle East) suffer highest consumptions losses, while Africa benefits, largely due to the sale of emission rights. For the other industrialized countries and China, the consumption losses are between 1 and 2 per cent relative to baseline, close to the world average. Africa is the only region exhibiting appreciable welfare gains from the global mitigation policy. Consumption gains are highest for the per capita allocation scheme.

The GDR framework is distinguished by the fact that it distributes the global mitigation effort (i.e. the emission reduction obligations) in terms of historic responsibility and economic capacity. Our analysis shows that this approach can substantially alter the distributional effect of climate policy.

In the default GDR scenario, which corresponds to the approach suggested in the original Baer et al (2007) paper, the difference between the global baseline emissions trajectory and the climate stabilization trajectory is determined and distributed among nations in proportion to the RCI. In the presence of a global carbon market, net sellers of emission rights benefit substantially from an allocation-based scheme – their average domestic abatement costs will be substantially lower than the global carbon price, which is determined by the global marginal costs of an additional unit of emission reduction. As demonstrated by the model results, this effect gives rise to substantial welfare gains for Africa, India and China if the static, allocation-based GDR rule is applied. The high-income industrialized countries (US/Canada/Australia, EU, Japan) are characterized by high historic emissions and per capita GDP, thus their allocations are substantially smaller than in the C&C, common but differential convergence (CDC) and GDP shares scenarios. In fact, as demonstrated by the calculations in Baer et al (2007), the static and allocation-based application of the GDR-framework results in negative emissions for the US, UK and Germany as early as 2020–2025, i.e. these countries would be obliged to purchase emission rights even in the hypothetical case of a complete elimination of domestic GHG emissions. In absolute terms, the mitigation costs borne by US/Canada/Australia and the EU more than exceed the global total average of mitigation costs (see Figure 11.4b). Not surprisingly, this allocation scheme would result in the highest mitigation costs of all six schemes considered here for the high-income industrialized countries.

For the static implementations of the GDR framework, the RCI was based on 2005 data for national incomes and historic emissions from 1990 to 2005 and applied throughout the entire simulation period. For long-term simulations until 2100, as presented here, this is a significant shortcoming. In the second edition of the GDR framework (Baer et al, 2008) this concern is addressed by using projections of the controlling parameters to estimate a time-dependent version of the RCI. In order to demonstrate the effect of dynamical adjustments, we calculated a time-dependant RCI by scaling the RCI with the regional GDP growth rates.[30] For each time step, the reduction relative to the baseline was then distributed in proportion to the time-dependant RCI. This dynamic allocation-based scenario results in a considerable increase of mitigation costs compared to the static case for fast-growing economies such as China, India and Russia. For high-income industrialized countries, by contrast, the aggregated relative welfare losses decrease to a level that is only moderately higher than the global average.

In the default version of the GDR framework, the global mitigation effort is distributed in terms of the global emissions reduction requirement. Thus, the higher a country's projections of future emissions, the higher is its baseline, and thus the larger its permit endowment. Since mitigation costs are the decisive metric for welfare effects of mitigation policy, an alternative application of the GDR framework is to distribute the global mitigation effort outcome-based, i.e. in terms of costs, according to the RCI. The regional pattern of costs changes markedly for the outcome-based approach compared to the default GDR. In this case, only Africa, which benefits from negative emissions induced from biomass use, has small negative mitigation costs. All other regions have positive mitigation costs. In terms of consumption losses relative to baseline, costs in the Middle East and Russia are projected to exceed the global average by more than a factor of two but the relative losses are considerably smaller than for the three generic allocation schemes. For the other regions, welfare losses are comparable to the global average costs. Overall, the variance of the regional mitigation costs in relative terms is smallest for this allocation scheme.

In general it can be concluded that the distributional effect of the GDR framework hinges critically on its implementation. For a static calculation of the RCI and a distribution of the mitigation burden in terms of emission reductions, it tends to strongly favour developing countries by endowing them with a high emissions baseline and the right to sell any emission reductions at a price that is much higher than their average abatement costs. However, the static allocation-based scheme envisages rapid emission cuts to high-income industrialized countries that eventually result in negative allocations of emission rights. These are required to maintain ambitious overall global emission reductions. If the approach is altered by either (1) dynamically accounting for future changes of nations shares of capacity or (2) relying on an outcome-based principle of distributing the mitigation costs, the regional distribution of costs changes markedly. In the dynamic case, the RCI of emerging economies increases rapidly, thus assigning them a larger share of the overall reduction effort compared to the static regime. In the case of an outcome-based application, the global net mitigation cost rather than the gross mitigation effort are distributed according to responsibility and capability. While the static, allocation-based scheme implies substantial welfare gains for China, an outcome-based or dynamic formulation of the GDR framework would result in costs that even exceed those projected for the C&C scenario.

It is very important to note that the sensitivity of regional welfare effects depends strongly on the carbon price: the higher the carbon price, the higher potential gains from the sale of excess emission rights. Due to assumptions of a well-functioning carbon market, flexibility introduced by international and inter-temporal trade, and the availability of a broad portfolio of mitigation options, ReMIND-R projects rather moderate carbon prices. Carbon prices are highly sensitive to restrictions or impediments, for example incomplete regional or sectoral coverage, unavailability of certain mitigation technologies, as well as changes in the model design and assumptions, resulting in a higher importance of allocation relative to other effects.

The case for early action

The previous section, as well as most assessments of mitigation costs, is based on the assumption of an immediate and global mechanism for pricing emissions. Obviously, delays and imperfections in the set-up of a global carbon market, as discussed in the third section of this chapter, will affect global mitigation costs as well as its regional distribution.

What happens if a global carbon market is not established at once, but with a delay of one or two decades? What are the economic consequences of a fragmented climate regime in which some regions adopt caps and establish a carbon market while others do not participate to begin with? Figure 11.5 shows ReMIND-R simulations conducted in the context of the RECIPE project that assessed economic effects of climate policy aiming at a stabilization of atmospheric CO_2 concentrations at 450ppm, corresponding to roughly 500–550ppm CO_2 equivalent (Luderer et al, 2009). The timescale considered is 2005–2100.

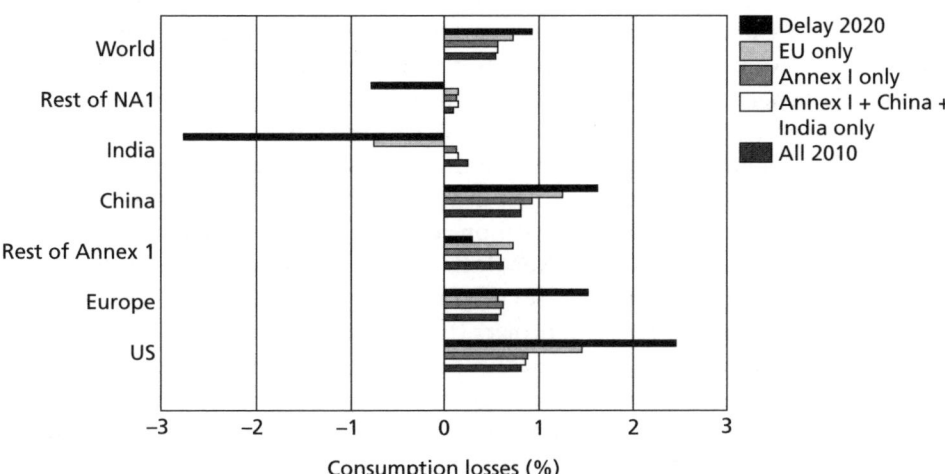

Note: A stabilization of atmospheric CO_2 concentration at 450ppm was assumed as the climate policy target, and allowances are allocated according to the C&C rule.
Source: author's calculation using ReMIND-R

Figure 11.5 *Global and regional consumption losses for scenarios on a delayed or initially fragmented global carbon market*

An important first result is that the stabilization target becomes unfeasible if global action is delayed beyond 2025. If all countries continue emitting according to BAU until then, it is no longer possible to stabilize atmospheric CO_2 concentrations at 450ppm. This is largely due to the build-up of carbon-intensive energy infrastructure in the absence of climate policy that, along with the extensive economic lifetime of investments, results in a lock-in to a high-emission development trajectory even beyond 2025.

Given a delay of global action until 2020, stabilization at 450ppm CO_2 is still feasible, albeit at significantly higher global economic costs. As part of RECIPE, the following scenarios of fragmentation were analysed:

- delay of action by all countries with emissions according to BAU until 2020 ('delay 2020');
- immediate action by the EU, BAU emissions by others until 2020 ('EU only');
- immediate action by all Annex 1 countries, BAU emissions by others until 2020 ('Annex 1 only');
- immediate action by all Annex 1 countries, China and India, BAU emissions by rest of NA1 countries until 2020 ('Annex 1 + China + India');
- immediate action by all countries ('all 2010').

Preliminary results concerning global and regional mitigation costs are depicted in Figure 11.5. An important aspect is the incentive system for a region to take early action on climate change. In all models, the EU has lower mitigation costs in the 'EU-only' than in the 'delay 2020' scenario – thus the result suggests that there is an incentive for the EU to take action even if the other regions do not participate. Similarly, the mitigation costs in the US decrease in 'Annex 1 only' compared to the 'EU only', thus suggesting that the US would benefit if it joined, along with the other Annex 1 countries, the EU in its mitigation efforts. We can spin this even further: if Annex 1 countries are committed to climate policy, China will increase its welfare by participating in a global carbon market. The effect of the participation of China and India in a global carbon market is, by contrast, projected to be almost neutral for India. The results presented here for ReMIND-R are confirmed by the other two hybrid energy-economy models WITCH (Bosetti et al, 2007) and IMACLIM that participated in the RECIPE project (Luderer et al, 2009).

It is important to note that these results rely critically on the assumption that all countries will have to join a global carbon market with a global cap leading to a 450ppm CO_2 stabilization, i.e. the set-up of a credible expectation of long-term equalization of per-capita emissions by 2050 irrespective of early action or delay. According to the RECIPE results, most countries fare better in adopting climate policy targets immediately rather then continuing according to BAU until 2020. There are two opposing forces affecting the welfare of early movers: regions adopting early mitigation targets accept a higher overall reduction burden. In most cases, however, this is more than overcompensated by the second effect – benefits of early action include avoiding stranded investments and enabling the sale of emission permits to late movers once the global carbon market is established.

Conclusions

A Global Deal on climate change requires a global long-term policy objective that balances the costs and risks of climate change impacts with the economic costs and risks of emissions mitigation. It appears that given our current knowledge, the 2°C objective endorsed by China and other major economies strikes a sensible balance between climate impacts, mitigation and adaptation. Institutions and technologies will be the key to a successful effort to curb global warming. We propose a set of international institutions for an environmentally effective, economically efficient and equitable response to global warming. This includes a global carbon market, policies for fostering low-carbon technology development, a mechanism to reduce emissions from deforestation and forest degradation (REDD), and an international adaptation framework.

Regarding the implementation of a global carbon market, the future of the international emissions trading regime beyond 2012 is highly uncertain, and several scenarios are conceivable. UNFCCC-driven top-down approaches promise higher instantaneous coverage of global emissions and thus higher environmental effectiveness than bottom-up approaches resting on unilateral, bilateral and plurilateral initiatives. However, in the case of top-down approaches, the need to achieve global agreement on emissions allocation and burden sharing raise the danger of political stalemate. Also, government-level emissions trading is prone to market power effects. By contrast, bottom-up approaches are constantly challenged by the question of whether they can significantly contribute to global emissions reductions at all. Both approaches can complement each other, if governments adopt caps and devolve international trading to companies covered by internationally linked domestic trading schemes. In effect, governments would only trade on behalf of sectors not covered by domestic trading. This would combine comprehensiveness of top-down approaches with cost effectiveness of facility-level trading.

There are a number of options for developing-country participation in a future carbon market. They can be distinguished according to their principal approach (cap-and-trade versus baseline-and-credit), their scale (economy-wide versus sectoral versus project-based), the carbon market players involved (government versus companies) and whether they are binding or non-binding. First, the current CDM approach may be continued. Discussions are under way to upscale the CDM to the sectoral level to enable bundling of projects, or to discount credits in order to address some of the criticisms that have been put forward. Beyond the current CDM and its reform, intensity-based approaches with economy-wide or sectoral coverage on the government level are conceivable. Finally, developing countries such as China can adopt economy-wide and sectoral caps where governments trade on international markets or link domestic entity-level cap-and-trade systems internationally. Setting caps or baselines will be a sensitive issue in any approach, as this crucially determines environmental and distributional outcomes. For any government-level trading approach, domestic instruments such as taxes, trading or standards need to be implemented, in addition to possibly utilizing funds received via international trading to foster low-carbon development. In a transition period, developing countries might adopt trading mechanisms, for example on the sectoral scale, to begin decarbonizing their economic growth, mitigate carbon leakage concerns of industrialized countries, trigger domestic social-learning processes regarding market-based climate policy instruments, and build the infrastructure required for a more comprehensive future approach.

Results from the integrated energy–economy–climate model, ReMIND-R, developed at Potsdam Institute for Climate Impact Research (PIK) suggest that in the context of a global cap-and-trade system the influence of allocations on regional mitigation costs are rather small, unless historical responsibility for emissions is taken into account. If the emission reduction burden is distributed according to the GDR framework, which accounts for historic responsibility, economic capacity and disparities within nations, welfare effects change significantly in favour of developing countries. For China, which is characterized by strong economic growth, results for a dynamic approach, i.e. with time-dependant adjustment of the burden-sharing index, are less favourable than for the static case. Moreover, our model analysis shows that regional effects hinge decisively on the question of if the mitigation burden is distributed in terms of the emissions reduction requirement or the net mitigation costs. In general, it is important to distinguish three economic effects of global climate policy. First, the incidence of domestic abatement costs; second, trade effects resulting from revaluation of energy resources and energy-intensive industrial products due to carbon pricing; and third, the initial endowment of emission permits (the allocation), the value of which will change with the carbon price. International allocation rules will only affect the magnitude of the last of these effects.

Results from a regionally disaggregated integrated energy–economy–climate modelling study indicate that climate stabilization at 450ppm CO_2 is feasible if global action is taken swiftly, but becomes unfeasible if global action is delayed beyond 2025. Assuming global cooperative action for a 450ppm target to begin in 2020, most countries benefit from adopting cap-and-trade earlier because they avoid a lock-in to emissions-intensive infrastructure. China also benefits from early participation along with Annex 1 countries compared to a scenario in which only Annex 1 countries are the first movers in implementing a cap-and-trade system.

To sum up, recent international developments suggest that emissions trading could become a central pillar of the human response to climate change, alongside other policy instruments addressing specific market failures. When establishing a global ETS, allocation of emission permits represents a distributional challenge for the international community. If a comprehensive agreement is difficult to achieve in the short term, linking regional trading initiatives directly or indirectly with the intention of creating a comprehensive system in the mid to long term may be a feasible fallback option. However, countries may be concerned over carbon leakage, rendering the adoption of ambitious reduction targets difficult in a system that forms from the bottom-up.

At Copenhagen and Cancún, countries have pledged their emission reduction objectives for the coming years. To achieve the 2°C goal, not only more ambitious and long-term emission cuts will be required, but also a set of policy instruments to deliver reductions in a cost-effective manner. In principle, there is infinite option space for international climate policy instrument configurations. International emissions trading could be a policy instrument that enables internationally diverse expectations to converge.

Notes

1 For a more detailed treatment of these issues, see Edenhofer et al (2008a, 2008b).
2 In some sectors, emissions trading may not be feasible due to uncertainties in monitoring emissions (see note 7).
3 We use the terms carbon market and ETS interchangeably to refer to both cap-and-trade and baseline-and-credit systems. Also, we use 'carbon' interchangeably to refer to carbon dioxide and other GHGs. The more general term emissions trading architecture is used to denote the overarching structure of relations between ETS that are implemented all over the world.
4 Throughout our report, data from CAIT (2008) refers to CO_2 emissions of the year 2004, excluding emissions from LULUCF.
5 Number of permit transactions times prices.
6 See Flachsland et al (2009a) for a more detailed treatment of the issues discussed in this section.
7 It is difficult to include some sectors in emissions trading due to uncertainties in the monitoring of emissions, for example in the agriculture sector, or for N_2O emissions from stationary sources (Herold, 2008; ICAP, 2009). However, the Lieberman-Warner proposal for an US ETS, for example, envisages coverage of 84 per cent of all US emissions (World Bank, 2008). In particular, emissions from small sources such as in the transport or building sector can be integrated upstream (Hargrave, 2000).
8 Babiker et al (2004) question whether international emissions trading is always beneficial for *all* participating countries. Evidently, in a first-best setting this would be the case due to a simple gains-from-trade argument. However, in some second-best constellations, such as in the presence of tax interaction and terms-of-trade effects, international emissions trading among companies can lead to welfare reductions for permit-selling countries. This is mainly because the associated increase in carbon prices pronounces pre-existing tax distortions (such as high energy taxes), the effect of which outweighs the primary gains from trading. However, reducing the pre-existing distortions would considerably reduce these welfare losses. Also, careful case-specific analysis is required.

9   In a trade-theoretic analysis, Copeland and Taylor (2005) argue that a cost-effective outcome (i.e. equalization of permit prices) can be reached even in the absence of permit trade between carbon markets, due to the effects of trade in goods on the prices of non-traded inputs. However, their results are derived within a stylized theoretical model and based on strong assumptions, for example identical technologies and tastes across all countries, which are – at best – idealizations of the real world. This means that their results should for practical purposes be interpreted in the sense that 'international trade in goods induces a *certain tendency* towards equalization of the permit price'.

10 We only consider bilateral linkages. A unilateral link is established if cap-and-trade system A accepts allowances from another system B for compliance, but not vice versa. In such a system, the allowance price in A would remain at or below the price level of B.

11 The permit price might differ by a constant factor if systems use different units of measurement, for example metric and short tonnes. The latter unit is in fact envisaged for RGGI.

12 Depending on the level of price convergence, this will also be the case in the indirect linking case.

13 These issues are treated in depth by Flachsland et al (2009b).

14 Leakage occurs if the regulation of emission-intensive industries in one country leads to an expansion of those industries' output in other less-regulated countries due to a shift in comparative advantage. The impact of this effect will depend on a number of factors, including the size of the carbon price differential, the trade exposure of affected sectors, and the relative importance of the expected persistence of the cost gap for investment decisions. International sectoral agreements, border tax adjustments and the free allocation of allowances (Neuhoff, 2008) have been proposed to address leakage concerns. In general, the available evidence suggests that this effect would not be a serious problem in most sectors, at least in the short to mid term (for example Houser et al, 2007).

15 We neglect the options of compensation schemes and border tax adjustments (see Neuhoff, 2008).

16 A case in point would be Russia's bargaining power with its large amounts of 'hot air' within the Kyoto trading framework. See also Böhringer and Löschel (2003).

17 In 2004, the biggest five emitters, i.e. the US, China, Russia, Japan and India accounted for 51 per cent of global CO_2 emissions, which would equip them with considerable market power in a government-level system (CAIT, 2008).

18 We speak of integration when we refer to a situation with either only one global trading system or sufficient linkages between different carbon markets to lead to an equalization of permit prices across sectors and regions. Conversely, fragmentation occurs if permit prices differ across regions and sectors, either because there are several trading systems with different prices for permits, or because there are no trading systems at all.

19 Note that the ratings for environmental effectiveness of the three scenarios 'Kyoto II', 'direct' and 'indirect linking' crucially depend on the level of participation (number and size of systems) and the design of schemes for countries not participating via cap-and-trade. Ratings should thus be interpreted as sort of 'average' assessments.

20 This approach is currently pioneered by the EU ETS, where transactions of EU allowances across country borders are mirrored by transfers of assigned amount units (AAUs) between national Kyoto registries.

21 As was the case with Australia and the Kyoto Protocol (Keohane and Raustiala, 2008).

22 In the energy sector, for example, we can distinguish three major options for reducing emissions: (1) improving energy conversion efficiency (less primary energy consumption per unit energy services), for example by building more efficient power plants, or introducing more efficient appliances; (2) decarbonizing energy production by substituting carbon-intensive primary-energy carriers by less carbon-intensive ones (for example from coal to gas; from gas to renewable energies), and CCS; and (3) shifting economic activity from energy-intensive sectors to less energy-intensive ones.

23 For example, developing countries may be willing to adopt more stringent caps or baselines to reflect the domestic co-benefits from a trading mechanism.

24 This will also be the case for some of the other approaches discussed in this section, in particular all cap-and-trade approaches, depending on the specific mechanism design and assumption about the developing country emissions trajectory.

25 A MRV regime is required for entity-level emission taxes as well.

26 In fact, if absolute caps are set in accordance with a globally shared view on the international emission reduction schedule, the latter concern should be obsolete. It does not matter how emissions reductions are ultimately achieved. Against the argument that the example of 'hot air' under the Kyoto Protocol should be avoided in the future, one may argue that it is completely different to over-allocate ex post – as was the case in Kyoto – or ex ante, where this might occur due to unforeseeable recessions.

27 In the current debate on whether and how the CDM could be modified post 2012, 'programmatic' and 'policy CDM' can also be classified as sectoral projects. NAMAs have recently also received increased attention in this context (Sterk et al, 2009).

28 This is a similar approach to the 'multistage' concept where developing countries 'graduate' to a different status (implying different mechanism and targets) depending for example on per capita income (for example Michaelowa, 2007).

29 The regions represented in the model are USA/Canada/Australia, Japan, Europe, Russia, Middle East, China, India, Africa and Rest of World. World refers to the global aggregate.

30 With a second parameter we scaled the regional RCI to the number given in Baer et al (2008) for the values of the RCI in 2020 and 2030.

Acknowledgements

We thank Hans-Martin Füssel for providing Figure 11.1.

References

Babiker, M., Reilly, J. and Viguier, L. (2004) 'Is international emissions trading always beneficial?', *The Energy Journal*, vol 25, no 2, pp33–56

Baer, P., Athanasiou, T. and Kartha, S. (2007) *The Right to Development in a Climate Constrained World: The Greenhouse Development Rights Framework*, Heinrich Böll Foundation, www.ecoequity.org/docs/TheGDRsFramework.pdf

Baer, P., Athanasiou, T., Kartha, S. and Kemp-Benedict, E. (2008) *The Greenhouse Development Rights Framework*, revised second edition, Heinrich Böll Foundation, www.ecoequity.org/docs/TheGDRsFramework.pdf

Böhringer, C. and Löschel, A. (2003) 'Market power and hot air in international emissions trading: The impacts of US withdrawal from the Kyoto Protocol', *Applied Economics*, vol 35, no 6, pp651–663

Bosetti, V., Massetti, E. and Tavoni, M. (2007) *The WITCH Model: Structure, Baseline, Solutions*, Nota di Lavoro 10.2007, Venice: Fondazione Eni Enrico Mattei

Businessgreen (2008) 'South Korean emissions trading moves closer', 2 September, www.businessgreen.com/business-green/news/2225120/south-korea -introduce-emissions

CAIT (2008) *CAIT (Climate Analysis Indicators Tool)*, World Resource Institute, http://cait.wri.org/

Chung, R. K. (2007) 'A CER discounting scheme could save climate change regime after 2012', *Climate Policy*, vol 7, pp171–176

Copeland, B. R. and Taylor, M. S. (2005) *Trade and the Environment: Theory and Evidence*, Princeton: Princeton University Press

Cosbey, A., Parry, J.-E., Browne, J., Babu, Y. D., Bhandari, P., Drexhage, J. and Murphy, D. (2005) *Realizing the Development Dividend: Making the CDM Work for Developing Countries*, Winnipeg: International Institute for Sustainable Development (IISD), www.iisd.org/pdf/2005/climate_realizing_dividend.pdf

de Coninck, H. (2008) 'Trojan horse or horn of plenty? Reflections on allowing CCS in the CDM', *Energy Policy*, vol 36, pp929–936

Edenhofer, O., Lessmann, K., Kemfert, C., Grubb, M. and Köhler, J. (2006) 'Induced technological change: Exploring its implications for the economics of atmospheric stabilization: Synthesis report from Innovation Modeling Comparison Project', *The Energy Journal*, Special Issue, pp57–107

Edenhofer O., Flachsland, C. and Marschinski, R. (2007) *Towards a global CO$_2$ market: Expertise for the Policy Planning Staff in the Federal Foreign Office*, Potsdam Institute for Climate Impact Research, May, www.pik-potsdam.de/ members/flachs/publikationen/towards-a-global-CO$_2$-market

Edenhofer, O., Luderer, G., Flachsland, C., Füssel, H. M., Popp, A., Knopf, B., Feulner, G. and Held, H. (2008a) 'A Global Contract on Climate Change', background paper for the conference A Global Contract Based on Climate Justice: The Need for a New Approach Concerning International Relations, Brussels, 11 November, www.pik-potsdam.de/members/flachs/publikationen-2/global-contract-full-text

Edenhofer, O., Luderer, G., Flachsland, C., Füssel, H. M., Popp, A., Knopf, B., Feulner, G. and Held, H. (2008b) 'A Global Contract on Climate Change. Summary', www.pik-potsdam.de/members/flachs/publikationen-2/global-contract-summary

Egenhofer, C. (2007) 'The making of the EU Emissions Trading Scheme: Status, prospect and implications for business', *European Management Journal*, vol 25, no 6, pp453–463

EU Council (2007) 'Presidency conclusions', Brussels European Council 8/9 March, 7224/1/07 REV 1, Brussels: EU Council

EU Commission (2009) 'Communication from the Commission to the European Parliament, the Council, the European Economic and Social Committee and the Committee of the Regions. Towards a comprehensive climate change agreement in Copenhagen', COM(2009)39 final, Brussels: European Commission

Fan, G., Cao, J., Yang, H., Li, L. and Su, M (2009) Draft paper prepared for the midterm review of the project 'China Economics of Climate Change', Beijing

Flachsland, C., Edenhofer, O., Jakob, M. and Steckel, J. (2008) *Developing the*

International Carbon Market: Linking Options for the EU ETS, Report to the
Policy Planning Staff in the Federal Foreign Office, www.pik-
potsdam.de/members/edenh/publications-1/carbon-market-08

Flachsland, C., Marschinski, R. and Edenhofer, O. (2009a) 'Global trading versus
linking: Architectures for international emissions trading', *Energy Policy*, vol 37,
pp1637–1647

Flachsland, C., Marschinski, R. and Edenhofer, O. (2009b) 'To link or not to link:
Benefits and disadvantages of linking cap-and-trade systems',*Climate Policy*, Special
Issue, vol 9, no 1, pp358–372

Füssel, H. M. (2010) 'How inequitable is the global distribution of responsibility,
capability, and vulnerability to climate change: A comprehensive indicator-based
assessment', *Global Environmental Change*, vol 20, pp597–611

Hahn, R. W. and Stavins, R. N. (1999) *What Has the Kyoto Protocol Wrought? The
Real Architecture of International Tradable Permit Markets*, Washington, DC: The
AEI Press

Hargrave, T. (2000) *An Upstream/Downstream Hybrid Approach to Greenhouse Gas
Emissions Trading*, Washington, DC: Centre for Clean Air Policy

Helm, C. (2003) 'International emissions trading with endogenous allowance choices',
Journal of Public Economics, vol 87, pp2737–2747

Herold, A. (2008) 'The significance of monitoring capabilities to decide about the
scope of a carbon market', presentation at 1st Global Carbon Market Forum on
Monitoring, Reporting, Verification, Compliance and Enforcement Session II, 19
May, www.icapcarbonaction.com/docs/mrvce_material/session2/
Session_II_AnkeHerold_OekoInst.pdf

Houser, T., Bradley, R., Childs, B., Werksman, J. and Heimayr, R. (2007) *Leveling the
Carbon Playing Field*, Washington, DC: Peterson Institute for International
Economics and World Resources Institute

ICAP (International Carbon Action Partnership) (2007) 'Political declaration', Lisbon,
29 October, www.icapcarbonaction.com/index.php?option=com_content&view=
article&id=12&Itemid=4&lang=en

ICAP (2009) 'Summary Report on the first Global Carbon Market Forum on
Monitoring, Reporting, Verification, Compliance and Enforcement 'Backbone of a
Robust Carbon Market', 19–20 May 2008, Brussels: ICAP,
www.icapcarbonaction.com

IPCC (Intergovernmental Panel on Climate Change) (2007a) *Climate Change 2007:
The Physical Science Basis, Contribution of Working Group I to the Fourth
Assessment Report of the Intergovernmental Panel on Climate Change*, S. Solomon,
D. Qin, M. Manning, Z. Chen, M. Marquis, K. B. Averyt, M. Tignor and H. L.
Miller (eds), Cambridge and New York: Cambridge University Press

IPCC (2007b) *Climate Change 2007: Impacts, Adaptation and Vulnerability,
Contribution of Working Group II to the Fourth Assessment Report of the
Intergovernmental Panel on Climate Change*, M. L. Parry, O. F. Canziani, J. P.
Palutikof, P. J. van der Linden and C. E. Hanson, (eds), Cambridge and New York:
Cambridge University Press

IPCC (2007c) *Climate Change 2007: Mitigation, Contribution of Working Group III
to the Fourth Assessment Report of the IPCC*, B. Metz, O. R. Davidson, P. R.
Bosch, R. Dave and L. A. Meyer (eds), Cambridge and New York: Cambridge
University Press

Jaffe, J. and Stavins, R. N. (2008) 'Linkage of tradable permit systems in international climate policy architecture', Discussion Paper 2008-07, Cambridge, MA: Harvard Project on International Climate Agreements

Keohane, R. and Raustiala, K. (2008) 'Toward a post-Kyoto climate change architecture: A political analysis', Discussion Paper 08-01, Cambridge, MA: The Harvard Project on International Climate Agreements, http://belfercenter.ksg.harvard.edu/files/Keohane%20%20Raustiala%20HPICA1.pdf

Kerr, S. (2000) 'Domestic greenhouse gas regulation and international emissions trading', in S. Kerr (ed) *Global Emissions Trading: Key Issues for Industrialized Countries*, Cheltenham: Edward Elgar Publishing

Knopf, B., Kowarsch, M., Lüken, M., Brunner, S. and Edenhofer, O. (2011) 'Emission trading and the distribution of emission allowances', in Edenhofer, O., Wallacher, J., Lotze-Campen, H. and Reder, M. (eds) *Climate Change: Overcoming Injustice: Linking Climate and Development Policy*, forthcoming

Lenton, T. M., Held, H., Kriegler, E., Hall, J. W., Lucht, W., Rahmstorf, S. and Schellnhuber, H. J. (2008) 'Tipping elements in the Earth's climate system', *Proceedings of the National Academy of Sciences of the United States of America*, vol 105, pp1786–1793

Leimbach, M., Bauer, N., Baumstark, L. and Edenhofer, O. (2009) 'Mitigation costs in a globalized world: Climate policy analysis with ReMIND-R', *Environmental Modeling and Assessment*, vol 15, no 3, pp155–173

Luderer, G., Bosetti, V., Steckel, J., Waisman, H., Bauer, N., Decian, E., Leimbach, M., Sassi, O. and Tavoni, M. (2009) 'The economics of decarbonization: Regionally and sectorally explicit results from the RECIPE project', RECIPE Working Paper, www.pik-potsdam.de/recipe

Lüken, M., Bauer, N., Knopf, B., Leimbach, M., Luderer, G. and Edenhofer, O. (2009) 'The role of technological flexibility for the distributive impacts of climate change mitigation policy', paper presented at the International Energy Workshop, 28 May, Venice, www.iccgov.org/iew2009/speakersdocs/Lueken-et-al_TheRoleofTechnological.pdf

Major Economies Forum (2009) 'Declaration of the Leaders of the Major Economies Forum on Energy and Climate', Major Economics Forum, www.g8italia2009.it/static/G8_Allegato/MEF_Declarationl,0.pdf

Mehling, M. (2009) *Global Carbon Market Institutions: An Assessment of Governance Challenges and Functions in the Carbon Market*, Climate Strategies Working Paper, 1 July

Meinshausen, M., Meinshausen, N., Hare, W., Raper, S. C. B., Frieler, K., Knutti, R., Frame, D. J. and Allen, M. R. (2009) 'Greenhouse-gas emission targets for limiting global warming to 2°C', *Nature*, vol 458, no 7242, p1158

Meyer, A., (2000) *Contraction and Convergence: The Global Solution to Climate Change*, Totnes, Devon: Green Books

Michaelowa, A. (2007) 'Graduation and deepening', in J. E. Aldy and R. N. Stavins (eds) *Architectures for Agreement: Addressing Global Climate Change in the Post-Kyoto World*, Cambridge and New York: Cambridge University Press

Michaelowa, A. and Purohit, P. (2007) *Additionality Determination of Indian CDM Projects*, Zurich: University of Zurich, Institute for Political Science

Michaelowa, A., Stronzik, M., Eckermann, F. and Hunt, A. (2003) 'Transaction costs of the Kyoto mechanisms', *Climate Policy*, vol 3, pp261–278

Neuhoff, K. (2008) 'Tackling carbon: How to price carbon for climate policy', University of Cambridge Electricity Policy Research Group, www.climate-strategies.org/uploads/Tackling_Carbon_final_230508.pdf

Point Carbon (2008) 'AAU deals delayed due to carbon price slump', *Carbon Market Europe*, vol 7, no 42, p4

Rehdanz, K. and Tol, R. (2005) 'Unilateral regulation of bilateral trade in greenhouse gas emission permits', *Ecological Economics*, vol 54, pp397–416

Samaniego, J. and Figueres, C. (2002) 'Evolving to a sector-based Clean Development Mechanism', in K. A., Baumert, O., Blanchard, S. Llosa and J. F. Perkaus, (eds) *Building on the Kyoto Protocol. Options for Protecting the Climate*, Washington, DC: World Resource Institute, pp89–108

Schmidt, J., Helme, N., Lee, J. and Houdashelt, M. (2006) *Sector-based Approach to the Post-2012 Climate Change Policy Architecture*, Washington, DC: Center for Clean Air Policy

Schneider, L. (2007) 'Is the CDM fulfilling its environmental and sustainable development objectives? An evaluation of the CDM and options for improvements', report prepared by Öko-Institut e.V. for WWF, 5 November, www.oeko.de/oekodoc/622/2007-162-en.pdf

Smith, J. B., Schneider, S. H., Oppenheimer, M., Yohe, G. W., Hare, W., Mastrandrea, M. D., Patwardhan, A., Burton, I., Corfee-Morlot, J., Magadza, C. H. D., Füssel, H. M., Pittock, A. B., Rahman, A., Suarez, A. and van Ypserle, J. P. (2009) 'Assessing dangerous climate change through an update of the Intergovernmental Panel on Climate Change (IPCC) "reasons for concern"', *Proceedings of the National Academy of Sciences*, vol 106, pp4133–4137

Sterk, W., Arens, C., Beuermann, C., Bongardt, D., Borbonus, S., Dienst, C., Eichhorst, U., Kiyar, D., Luhmann, H-J., Ott, H. E., Rudolph, F., Santarius, T., Schüle, R., Spitzner, M., Thomas, S. and Watanabe R. (2009) 'Towards an effective and equitable climate change agreement. A Wuppertal proposal for post-2012', www.wupperinst.org/uploads/tx_wibeitrag/Wuppertal_Proposal_Post2012.pdf

Stern, N. (2007) *The Economics of Climate Change: The Stern Review*, New York: Cambridge University Press

Stern, N. (2008) 'Key elements of a global deal on climate change', http://eprints.lse.ac.uk/19617

Tangen, K. and Hasselknippe, H. (2005) 'Converging markets: International environmental agreements', *Politics, Law and Economics*, vol 5, no 1, pp47–64

Tietenberg, T. (2003) *Environmental and Natural Resource Economics*, sixth edition, Boston, MA: Addison Wesley

Tuerk, A., Mehling, M., Flachsland, C. and Sterk, W. (2009) 'Linking carbon markets. Concepts, case studies and pathways', *Climate Policy*, Special Issue, vol 9, no 4, pp341–357

UNEP (United Nations Environment Programme) (2008) 'UNEP Risoe CDM/JI Pipeline Analysis and Database', http://cdmpipeline.org/cers.htm

Vattenfall (2006) *Curbing Climate Change: An Outline of a Framework Leading to a Low-carbon Emitting Society*, Stockholm, Vattenfall

Victor, D. (2007) 'Fragmented carbon markets and reluctant nations: Implications for the design of effective architectures', in J. E. Aldy and R. N. Stavins (2007) *Architectures for an International Global Climate Change Agreement. Addressing Global Climate Change in the Post-Kyoto World*, Cambridge and New York: Cambridge University Press, pp350–367

World Bank (2008) *State and Trends of the Carbon Market 2008*, Washington, DC: World Bank

Zapfel, P. and Vainio, M. (2002) *Pathways to European Greenhouse Gas Emissions Trading: History and Misconceptions*, FEEM Working Paper No. 85.2002, http://papers.ssrn.com/sol3/papers.cfm?abstract_id=342924

12

Meeting Global Targets through International Cooperation

Fan Gang, Li Lailai and Han Guoyi

Introduction

Since the time when the UNFCCC was conceived, the challenges of climate change have intensified in complexity. Global GHG emissions are at the top end of the IPCC projection range, and are accelerating. Developed countries are responsible for 70 per cent of all CO_2 accumulated in the atmosphere; the current emissions from developing countries are surpassing those of OECD countries by a margin and are expected to increase rapidly in the decades to come (IEA, 2008). To reduce poverty, developing countries need to speed up economic development; this is most likely to drive up energy consumption. Meanwhile, the impacts of climate change are being felt first and foremost by the poor. Technologies to reduce CO_2 emissions are available, but are not adequately applied, for reasons that have little to do with climate change.

It is clear that the efforts of developing countries are indispensable to achieving the target of stabilizng CO_2 concentration in the atmosphere at 450ppm. Therefore cooperation between developed and developing countries is a must. The question is how to design a mechanism for such cooperation so that there are adequate internal incentives for compliance and also sufficient external incentives for participation (Aldy and Stavins, 2008).

The rapidly emerging major economies are the major driving forces of this change, among which China has been the greatest centre of focus. In this chapter we focus on China, as a case of a large developing country, to design an effective international cooperation mechanism that can bridge China's national priorities and global climate security, basing on our analysis of the 'time window' as both an opportunity and cost for China in view of the limitations of the current international mechanism of the CDM. Our key arguments are the following:

- The next two decades or so are critical for both China and the global climate change pact. To meet the country's national development agenda and global mitigation targets, growth patterns must shift towards greater sustainability, i.e. towards building 'a resource-saving and environmentally friendly society' as the current Chinese government puts it. This is where international cooperation on climate change enters and plays a role.
- Under the UNFCCC principle of 'common but differentiated responsibility', international cooperation on climate change should aim to help developing countries, including China, to make the shift in the next two decades from an energy-intensive economy to a low-carbon one, facilitated by a new mitigation mechanism for emission reduction, technology transfer and supportive finance on a large scale and at the national level.
- Today the only mitigation mechanism that involves developing countries is the CDM, which operates at a project level on a limited scale. Developing countries largely are excluded from the global carbon market by the current 'cap-and-trade' regime. Given the urgency of the issue, an inclusive and participatory mechanism is badly needed for a post-2012 global climate deal, which puts the interests of sustainable development of developing countries in the centre and which operates at an effective scale.
- Based on an empirical analysis of CDM and its operation, we propose a new climate change mitigation regime – an ICP. The ICP requires three preconditions, based on the UNFCCC principle of 'common but differentiated responsibilities': (1) national, voluntary, intensity-based emission reduction targets are adopted by developing countries; (2) emission reductions, technology transfer and financial flows built into an ICP are subject to international standards of MRV; and (3) an international fund is established to finance the ICP. (The Montreal Protocol on the ozone layer is a good model and reference for this.)

Data used for our analysis include statistics published by China Climate Change Info-net[1] sponsored by the Department of Climate Change under the NDRC of China. We also used information from the Institute for Global Environmental Strategies' CDM Project Database, and from interviews with ten Chinese potential CDM project owners – companies in the energy-saving business. The secondary data collected and summarized by the Environment and Economics Policy Research Centre of the Chinese Ministry of Environment and other investigators are quoted and acknowledged as sources of information, in addition to other related data from the UN, World Bank, IMF, IEA, World Resources Institute, the US government's EIA and other international organizations.

'Time window' of no-cap as both opportunity and cost

'Time window'

Under the principles of UNFCCC and the Kyoto Protocol, the status of being a developing country provides China with a 'time window'. In the next decade, or over a longer period, China is not obligated to commit itself to quantified global GHG emission reduction targets, and the country is also entitled to receive international support to voluntarily reduce its emissions. Within this 'time window', China's actions contributing to the global climate change mitigation are centred on reducing its energy and pollution intensities while meeting national needs of development and poverty alleviation, as in many other developing countries. These actions are defined in the current regime of global mitigation mechanisms as 'no-cap'.

At the global level this time window is constrained by estimates of allowable emissions – a 'carbon budget' – based on the requirement for climate security (Wang and Watson, 2008). The 'carbon budget' so far reflects nothing but the increasing urgency for action to mitigate climate change. In its Fourth Assessment Report, the IPCC plotted an 'emergency path' that requires global emissions to peak around 2020 and fall sharply after that (Baer et al, 2007). The reality, however, is that GHG emissions are now at the top of the IPCC projection range and the rate of emissions is accelerating. Based on new evidence, climate science has shown that the available atmospheric 'space' to maintain global temperature increases at less than 2°C is running out quickly (Hansen et al, 2008).

In the Chinese context, the 'time window' is limited and will close soon. It is primarily affected by the following dynamics. First, it depends on when China's per capita CO_2 emission will reach or surpass the world average. Figure 12.1 compares per capita emissions of China to selected OECD countries under a 'no-regrets' trajectory (see Fan et al, 2008, for the detailed assumptions). China's per capita emissions rate was half of the world average a few years ago, is at the world level today, and will catch up with OCED countries in two decades. Other forecasts suggest that China could catch up with OECD countries sooner than this (Aldy, 2008; Auffhammer and Carson, 2008) because the low-carbon development being mainstreamed in the OECD countries will make a significant change. From 1990 to 2005, per capita emissions from the consumption of fossil fuels on mainland China grew at 7.25 per cent annually, while this figure was 1.01 per cent for OECD countries (EIA, 2007). Since the turn of the century, China's CO_2 emissions experienced alarmingly rapid growth, at an annual rate of around 10 per cent. China and the US, the two largest GHG-emitting countries, contribute 40 per cent of global total emissions.

Second, it depends on when China's per capita historical accumulated emissions will reach the world average. Figure 12.2 shows four potential

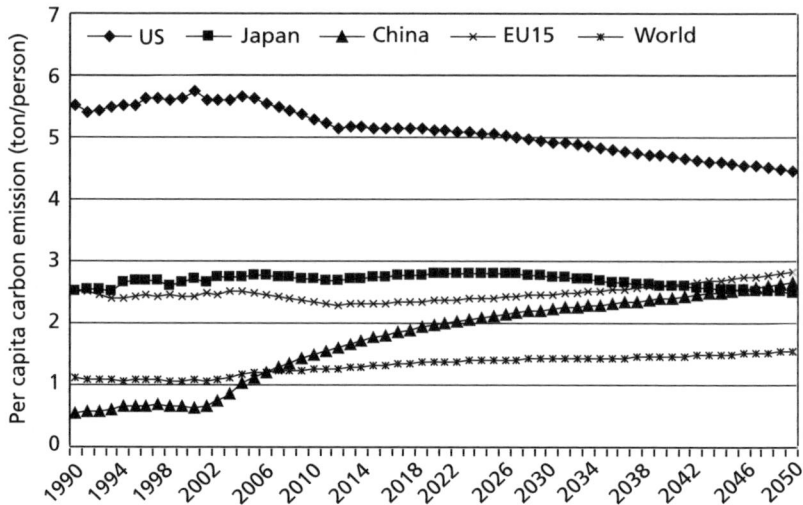

Source: Data from the World Resources Institute, IEA and the International Foundation for Science and World Development Indicators (World Bank)

Figure 12.1 *Forecast of per capita emissions (1990–2050) under a 'no-regrets' growth trajectory*

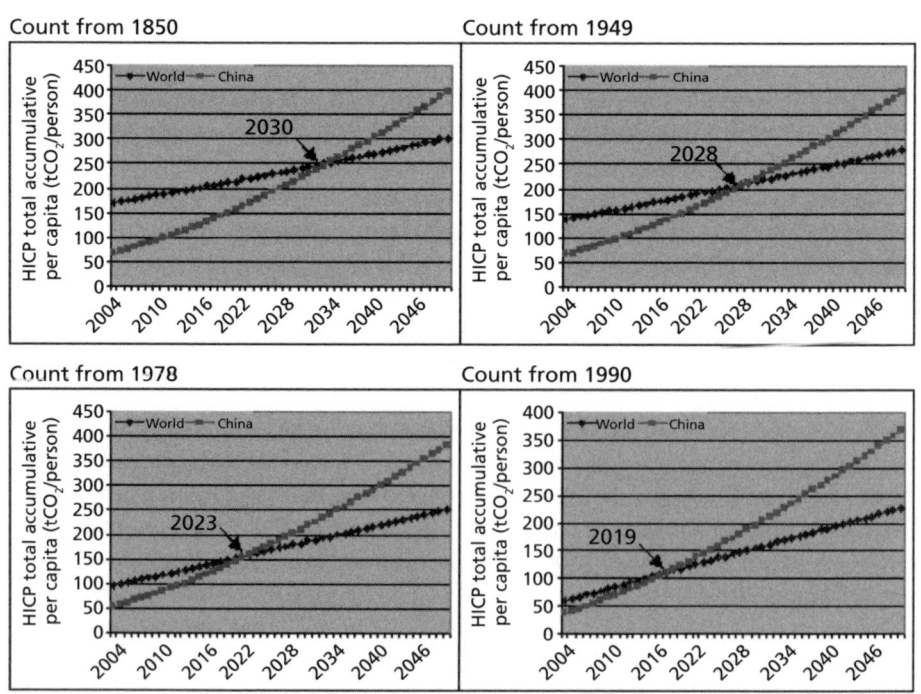

Note: HICP = Accumulated historical emisions per capita.
Source: Data from the World Resources Institute, IEA and the International Foundation for Science and World Development Indicators (World Bank)

Figure 12.2 *Forecast of per capita accumulated historical emissions under a 'no-regrets' scenario*

scenarios, based on a 'no-regrets' growth trajectory. While China's emissions are still considerably lower than the OECD average, and will be for some time, China's per capita accumulated historical emissions could reach the world average between 2020 and 2030, depending on the starting year from which historical emissions are calculated.

In view of several key variables – total emissions, per capita emissions and per capita historical accumulated emissions – China's responsibility for climate change will change and the time window for a no-cap international policy will close for China within the next two decades. Considering the global trend towards low-carbon economies, the timeline could be even shorter.

Opportunities of 'no-cap'

This time window offers China an important opportunity as China's industrialization and urbanization are at a critical point, and its need for energy and environment security is increasingly a barrier to realizing the country's development agenda. There is no loss for the country in seriously considering and taking actions to maximize the opportunity of the time window in pursuit of low-carbon growth, and in seeking an effective pathway to align the country's development with global climate security.

Within the time window, China has good bargaining power to access financing and technical support through international cooperation on climate change, to accomplish its national strategy of shifting the pattern of development towards sustainability by updating technologies and infrastructure. Without being 'capped', China can effectively link its national energy conservation and pollution reduction plans to national sustainable development goals as its contribution to global mitigation targets, in exchange for technology transfer and supportive finance in the form of grants and/or investments. Such a linkage could serve as a foundation for China's negotiations regarding an international climate change regime, and could maximize international support towards meeting mutually agreed targets, which would benefit both China and the global community.

This, however, will not automatically happen. It requires both China and the international community to steer the current negotiation towards flexible, innovative and practical ways to build cooperation for a global deal in which both developing and developed countries will jointly participate. In his 'key elements of a global deal on climate change', Nick Stern proposes 'a liquid international carbon market in order to allow for the most effective, efficient, and equitable emissions reductions', featuring 'cap-and-trade' measures (Stern, 2008). Without the participation of developing countries, this global carbon market will not be 'liquid'. The reality so far is that, like all developing countries, China is basically excluded from the 'global deal' by the current 'cap-and-trade' mitigation regime that largely ignores China's voluntary measures to reduce energy intensity as a basis for negotiation. Before analysing the current mitigation mechanism, we look first at the implications for China of not being part of a global deal.

Costs of 'no-cap'

Reducing carbon emissions requires investment and often this could conflict with the high demand to maintain short-term economic growth. Countries commonly prioritize economic growth over long-term global climate considerations. The urgent need to reduce poverty makes developing countries 'reluctant' to commit to emission reduction and participation in global climate efforts (Victor, 2008). Calculations that the costs of inaction in the long run will outstrip the cost of short-run investments are generally an insufficient incentive for developing countries. The challenges are: (1) how to produce economic growth through climate change mitigation; and (2) how to integrate mitigation measures into the sustainable development strategies of the country.

As the climate change challenge has started reshaping the future economy and carbon 'space' is becoming a scarce resource, there is a cost to the no-cap policy. In the context of China, this cost can be viewed qualitatively as a set of trade-offs:

- China has set ambitious development targets, but faces extraordinary challenges to successful implementation.
- The risk of 'lock-in' would result in unaffordable costs for China in the long run yet the short-term pressure for maintaining growth, employment and poverty reduction is daunting in the global recession.
- Energy efficiency has co-benefits for health and the environment, but it is difficult for China to deal with issues of additionality and to meet international standards of MRV.
- The 'additionality' question is pertinent and we need to look at it from a 'pace versus urgency' perspective.

Here we describe in detail China's national policies and programmes on energy intensity reduction and emission reduction, and the price they have paid and are still paying for these, as a result of not being part of the global deal.

It is well known that China has adopted quantified targets to reduce energy intensity by 20 per cent and total pollution discharge – SO_2 and chemical oxygen demand (COD) – by 10 per cent in its 11th Five-Year Plan period (2006–2010). Figure 12.3 illustrates the different CO_2 emission trajectories of China based on whether or not consistent reduction targets are adopted over the next ten years. Assuming that China continues to implement consistent reduction targets until 2020, and assuming that the GDP growth is steady at 10.04 per cent during the 11th Five-Year Plan, and at 7.67 per cent in both 12th and 13th Five-Year Plan periods (Jiang et al, 2008), the projection shows that by 2020 China would have avoided a cumulative amount of 58.1Gt CO_2 emissions, which would be the world's largest mitigation programme (see Fan et al (2008) for the detailed assumptions of this forecast).

Depending on the GDP growth rate, two or three more five-year plans (FYPs) with the same intensity reduction targets would allow China's carbon

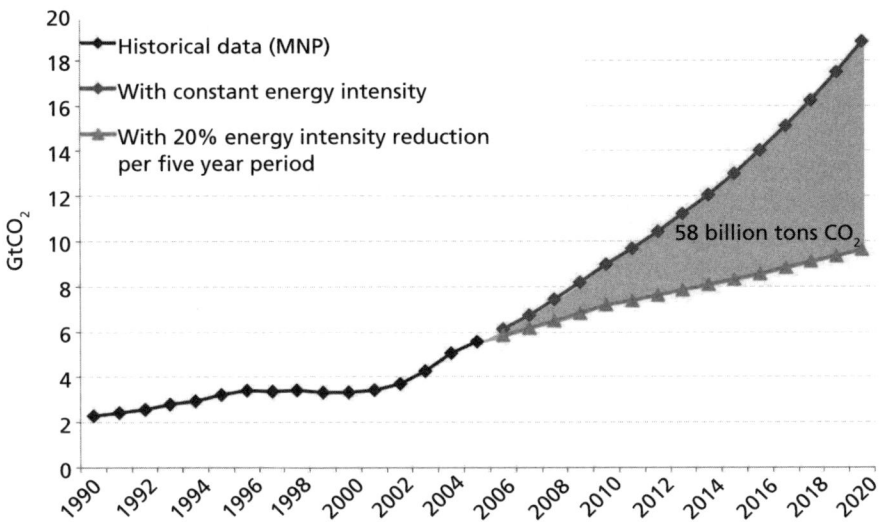

Source: Data from the World Resources Institute, IEA and the International Foundation for Science and World Development Indicators (World Bank)

Figure 12.3 *An illustrative exercise*

emissions to peak by 2020 or a bit later, as shown in Figure 12.4, where the growth rate is assumed to be 10 per cent during the 11th Five-Year Plan, 7.5 per cent during the 12th Five-Year Plan and 5 per cent during the 13th Five-Year Plan. In Figure 12.5, the growth rate is 10 per cent, 6 per cent and 5 per cent respectively.

This aggressive emission reduction initiative so far has been conducted by China entirely on its own. What is less known to the world is the increasing difficulty China is facing and the price that it has paid to meet the targets. According to the mid-term evaluation of energy conservation and emission reduction measures during the 11th Five-Year Plan period, conducted by the Development Research Center of the State Council, the anticipated targets have not been met although great progress has been achieved (see Table 12.1). To achieve the quantified targets, by June 2008 China had shut down or suspended small-scale thermal generating units producing 25.87 million kilowatts. China also reduced the production of cement by 100 million tonnes, steel by 50 million tonnes, coke by 30 million tonnes and paper by 5 million tonnes over a two and a half-year period. In these actions, the Chinese Central Government invested ¥21.3 billion in 2006, ¥23.5 billion in 2007 and ¥27 billion in 2008 (DRC, 2008). In addition, a number of policy instruments have been employed. Resource tax and consumption tax for high-emission vehicles have been increased, export rebates on energy-intensive and polluting products have been reduced and punitive duties have been levied. Electricity price policies have been adopted to encourage desulphurization by power plants and consumption of electricity produced through renewable energy, among other measures.

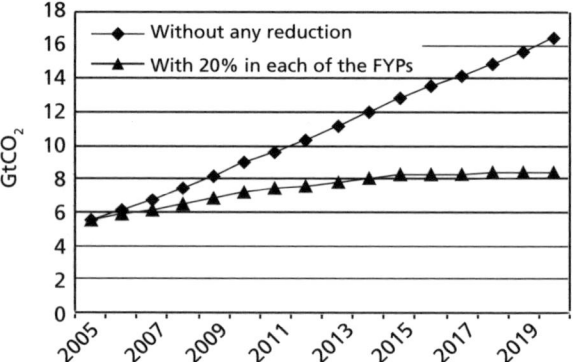

Source: Data from the World Resources Institute, IEA and the International Foundation for Science and World Development Indicators (World Bank)

Figure 12.4 *Forecast 1 of GDP growth and emissions*

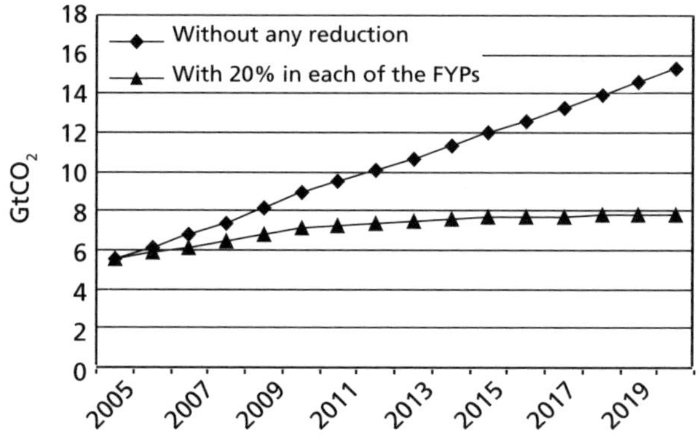

Source: Data from the World Resources Institute, IEA and the International Foundation for Science and World Development Indicators (World Bank)

Figure 12.5 *Forecast 2 of GDP growth and emissions*

A major programme launched to meet the targets is the 'Top 1000 Enterprises Energy Conservation Action Plan', which aims to reduce energy consumption by 100 million tonnes of carbon equivalent in five years. The chief executive officer of each participating company has signed an agreement with the Chinese government containing binding commitments. If targets are not met, strict penalties apply. By the time this plan is completed, the government will have invested ¥25 billion, in effect awarding ¥250 for each tonne of carbon equivalent of energy consumption reduced, not including implementation and administrative costs; ¥5.56 billion was spent in 2007. At the same time, 27 provinces have established specific funds for local implementation, and ¥3 billion has already been invested. Spending on technology updates and refurbishment already amounted to ¥50 billion in the same year (Yu, 2008; Price et al, 2008).

Table 12.1 *Progress of the 11th Five-Year Plan targets*

	The 10th Five-Year Plan period			The 11th Five-Year Plan period			
	2000	2005	Change (%)	2006 (%)	2007 (%)	January– June 2008 (%)	Targets met
Energy consumption per unit GDP (constant price in 2005)	1.19	1.22	2.5	−1.8	−3.7	−2.9	34% of the anticipated targets
SO₂ (10,000t)	1995	2549	27.78	−1.5	−4.7	−4.0	51% of the anticipated targets
COD (10,000t)	1445	1414.2	−2.1	10	−3.1	−2.5	35% of the anticipated targets

Source: DRC (2008)

Given the Chinese initiative and the price paid, two questions are worth asking: is China likely to support another one or two such five-year plans? Are the current mechanisms for international cooperation on climate change adequate or sufficient to help China continue its five-year plan targets over the next decade? We try to answer these two questions by examining the CDM, the only international mitigation mechanism that developing countries can access. We focus on the position and role of CDM in the global climate change mitigation regime and its use and implications in China.

Limitations of the current mitigation mechanism

The CDM is a project-based financing mechanism designed under the Kyoto Protocol to help Annex 1 countries meet their emission reduction targets cost-effectively, while supporting sustainable development in the NA1 countries. The logic of the CDM is simple. The *buyers* or *investors* from developed countries purchase CERs generated from CDM *project owners* in a project *host country*. CERs, which the buyer or investor acquires from the project owner, serve to offset the investor's or buyer's emission amounts so that they do not exceed the limit of their assigned emissions quota (Article 3.12 of the Kyoto Protocol).

The CDM project cycle is complex methodologically, technically and administratively, involving multiple stakeholders from public and private sectors (see Figure 12.6). Project planning starts with a Project Idea Note (PIN) and projects are completed when the CERs are issued after registration at the CDM Executive Board. In China many project owners (primary CER sellers), particularly small ones, have to rely on project developers to go through the procedures and make the deals because they lack the capacity to manage the project process. Investors or CER buyers are from Annex 1 countries and can be corporations, government bodies or NGOs.

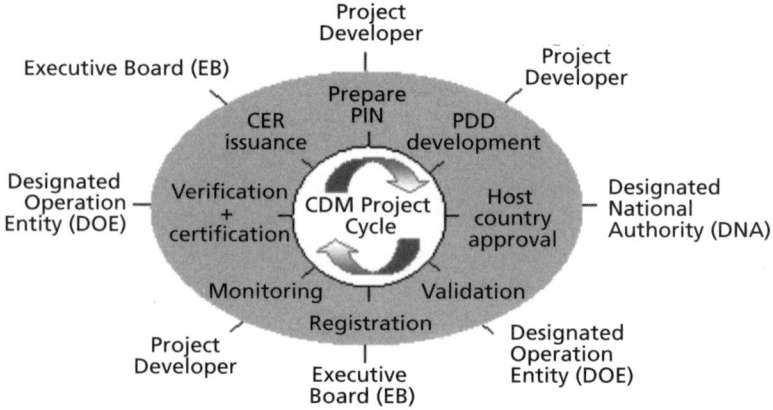

Figure 12.6 *CDM project cycle, tasks and responsible entities*

A CDM Executive Board supervises the authorizing and registering of projects and issuing of CERs. The Executive Board reports to the Conference of Parties to the UNFCCC and the Meeting of Parties to the Kyoto Protocol. A designated national authority (DNA) represents the project's host country, creates the criteria to determine whether a CDM project meets the country's sustainable development goals, and approves the CDM projects based on these identified criteria. Designated operational entities (DOEs) are domestic or international legal entities that have been accredited by the Executive Board. They validate, verify and certify CERs generated from CDM projects.

The limitations of the CDM have been widely discussed. Some have viewed the CDM as 'fundamentally flawed' (see for example, Victor 2008). Many weaknesses have been identified, such as the ubiquitous question of additionality, narrow coverage, high transaction costs and inadequate incentives for engaging developing country participation (Teng et al, 2008; Victor, 2008). We focus on China to explore what alternatives can achieve better results.

The CDM in the global carbon market

There are three types of global carbon market in operation. The EU ETS is by far the largest market, involving the sale and resale of EUAs. Project-based trading is primarily through the CDM and Joint Implementation. Access to this market by developing countries, including China, is strictly limited to supplying primary CDM credits or selling CERs. Before analysing CDM practice in China, we examine the position of the CDM in the global carbon market.

First, CERs traded through the CDM are low-priced in comparison to the EU ETS. International carbon prices in CDM transactions averaged $10.90/t CO_2 equivalent (€8.40) in 2006, while average prices ranged from $24.70 or €19 to US$22.10 or €17 in phase 1 of EU ETS (2005–2007) (Capoor and Ambrosi, 2007, 2008). In Phase II in 2008, the European Commission is expected to tighten the overall compliance caps allowed to each member state, and as a result carbon prices will go up. According to the analysis made by Deutsche Bank in June 2007 (Lewis, 2007), EU ETS Phase II carbon prices should be higher than €35 per tonne.

The price difference between CERs traded on the European market and CERs traded through the CDM would be about €24/t CO_2 equivalent. In effect, carbon prices established through the CDM are much lower than those traded on the European market. Assuming that this difference holds true till the year 2020, China's avoidance of 58Gt of CO_2 emissions through two more five-year plans with consistent energy saving and emission reduction targets (see Figure 12.3) would mean a financial turnover of €1.4 trillion, obtained from €24 multiplied by 58 billion. Simplified as it is, this exercise makes one point – without access to the global carbon market there is certainly a major cost to China in that regard. It is likely that the carbon price will be much higher when the limited carbon budget is taken into consideration. One indirect proof of this is the carbon tax rate placed by Sweden. Starting in 1991, Sweden added a CO_2 tax to its energy taxation policy. Since then it has been raised several times. The tax is currently €0.08/kg CO_2, the highest in the world (Swedish Ministry of the Environment, 2007), which means €80/t translated into the reduction cost in Sweden.

Second, the CDM plays a marginal role in the global carbon market, which has grown dramatically since 2005–2007 (see Table 12.2). The most active and largest carbon market is EU ETS, with a market share of over 78 per cent since 2005. In 2006, the total trading volume reached a value of $24.4 billion (€19 billion), up from $8 billion in 2005. This figure doubled within one year, going up to $50 billion in 2007 (Capoor and Ambrosi, 2007, 2008).

Developing countries supplied 537Mt CO_2 equivalent of primary CDM credits in 2006 to the global market, at a total market value of $5.8 billion, nearly 19 per cent of the total value of carbon credits traded globally. In 2007, developing countries' share of the primary CDM market fell to less than 12 per cent, and that of the secondary market went up to 8.5 per cent, where developing countries have no role to play any more.

Third, the 'cap-and-trade' regime prevents developing countries from accessing the global carbon market. According to Capoor and Ambrosi (2008), the demand for carbon credits by industrialized countries could exceed 2.4 billion metric tonnes over the 2008–2012 period, whereas supply is estimated at only 1.83 billion metric tonnes, suggesting a shortfall of close to 600Mt. This figure has increased from the 300Mt shortfall predicted by the World Bank in 2008.

Carbon cap-and-trade regimes allow for the import of credits from CDM transactions for compliance purposes. But the EU is likely to limit the import of credits from CDM because the process of import is perceived as containing risks, such as regulation, project development and performance, concern about environmental additionality, and the potential for significantly higher transaction costs. Even if CER imports are approved, the process does not engage

Table 12.2 *Carbon market at a glance: Volumes and values, 2005–2007*

	2005		2006		2007	
	Volume (Mt CO₂ equivalent)	Value ($ million)	Volume (Mt CO₂ equivalent)	Value ($ million)	Volume (Mt CO₂ equivalent)	Value ($ million)
Allowances						
EU ETS	321	7908	1104	24,436	2061	50,097
New South Wales	6	59	20	225	25	224
Chicago Climate Exchange	1	3	10	38	23	72
UK ETS	0	1	na	na		
Subtotal	328	7971	1134	24,699	2109	50,394
Project-based transactions						
Primary CDM	341	2,417	537	5804	551	7426
Secondary CDM	10	221	25	445	240	5451
Joint limplementation	11	68	16	141	41	499
Other compliance	20	187	33	146	42	265
Subtotal	382	2894	611	6536	874	13,641
Total	710	10,864	1,745	31,235	2983	64,035

Source: Capoor and Ambrosi (2007, 2008)

CDM host countries or project owners. A European Commission Draft Directive proposed on 23 January 2008 will lead to restrictions on the import of CDM credits generated after 2012, as long as an international agreement is not reached.[2] This will drive up the carbon price at EU ETS even higher, while developing countries are further limited to the lower end of the market. The political cause of this situation is quite simple: CDM project host countries' failure to agree to cap emissions.

CDM practice in China

Financial benefits, according to the Kyoto Protocol, are brought to the project owners of host countries by CDM transactions that would not happen otherwise. As of 10 September 2008, the Chinese DNA approved 1554 CDM projects. Of this number, 270 projects were registered with the Executive Board and the rest have been proposed to the Board for registration. These projects together represent an estimated annual emission reduction of 325Mt CO₂ equivalent, more than 50 per cent of the world total. Using €8 per tonne CO₂ equivalent as a guiding price[3] (Zhang, 2007) set up by the Chinese DNA, these projects would generate a financial turnover of €2.6 billion.

This financial turnover includes the transaction costs that are incurred at different phases of the project cycle. The two types of costs – upfront costs at the planning phase and costs at the operation phase – are described in the UNEP (2007) (see Table 12.3).

These costs represent two types of fees: (1) institutional fees charged by official CDM institutions such as the Executive Board and DOE, covering DNA approval, environmental impact assessment of CDM projects required in host countries and other compliance needs according to national regulations;

Table 12.3 *CDM project transaction costs*

Activity	Cost (large-scale, $)		Cost (small-scale, $)		Type of cost
	low	high	Low	high	
Planning phase					
Initial feasibility study, including PIN	5000	30,000	2000	7500	Consultancy or internal
Preparation of project design document	15,000	100,000	10000	25,000	Consultancy or internal
New methodology (if required)	20,000	100,000	20000	50,000	Consultancy or internal
Validation	8000	30,000	6500	10,000	DOE fee
Registration fee	10,500	350,000	0*	24,500	Executive Board fee
Total at planning phase	38,500	610,000	18500	117,000	
Operation phase					
UN adaptation fund fee	2% of CERs		2% of CERs		Executive Board fee
Initial verification (including system check)	5000	30,000	5000	15,000	DOE fee
Ongoing verification (periodically)	5000	25,000	5000	10,000	DOE fee
Share of proceeds to cover administration expense (SOP-Admin)	The fee paid at registration is effectively an advance that will be 'trued up' against actual CERs issued over the crediting period (if different to emission reductions projected at registration). SOP-Admin is not capped				Executive Board fee
Total at operation phase	Variable – minimum 2% of CERs plus 5000/year (if verification undertaken annually)				

Note: * No fee is payable, if total annual average emission reductions over the crediting period are below 15,000t CO_2-eq.

and (2) consulting fees charged by project developers based on market rates, which vary with the amount and the nature of the services provided.

Although there is no standard cost applied, the empirical data show that the transaction costs of a CDM project range from $150,000 to US$500,000 (€325,000) per project (Seres, 2007; Dechezleprêtre et al, 2009). In 2006 the World Bank estimated an average cost of $200,000 (World Bank, 2006). Meanwhile the costs are driven further up by the length of time spent on project development and by payment schedules that have other financial implications such as rates of discount and inflation.

Transaction costs relate to the scale of the project – the smaller a project, the higher its transaction costs. Table 12.4 tabulates and summarizes the Chinese CDM projects by size. At the low end, 29 per cent of the projects deliver less than 50,000t CO_2 equivalent of emission reductions annually. At the other end, larger projects delivering an annual reduction of more than 500,000t CO_2 equivalent make up 6.3 per cent of projects, while accounting for 50 per cent of emission reductions. In the middle range, over 60 per cent of projects deliver reductions between 50,000 and 500,000t CO_2 equivalent per project, making up 44 per cent of the total reduction in emissions. Taking into

Table 12.4 *Distribution of CDM projects by scale*

Projects by scale (annual estimated tonne CO_2 equivalent reduced)	No. of projects	%	Total estimated annual tonne CO_2 equivalent reduced	%	Average project scale (tonne CO_2 equivalent)
>10 million	3	0.19	31,358,570	9.64	10,453,000
1 million to 10 million	37	2.38	93,840,466	28.85	2,536,000
500,000 to 1 million	58	3.73	40,974,773	12.60	706,500
100,000–500,000	599	38.55	115,151,841	35.41	192,000
50,000–100,000	402	25.87	28,929,069	8.90	72,000
<50,000	455	29.28	14,970,865	4.60	32,900
Total	1554	100	325,225,583	100	209,000

Source: data on China climate change from www.ccchina.gov.cn

account transaction costs, the large projects are able to derive substantial income through the CDM, while medium ones are economically viable to various degrees, and small projects face financial uncertainties.

China's sale of CERs through the CDM represents the dominant share of all CDM transactions globally, in terms of project scales and financial turnover. Small projects offering emission reductions of less than 50,000t CO_2 equivalent account for 29 per cent of CDM transactions within China, but the percentage is 54 at the world level. Nearly 40 per cent of the Chinese CDM projects offer emission reductions of more than 100,000t CO_2 equivalent, while 21 per cent is the world level (Takahashi and IGES CDM Project Database, 2009). China has benefited financially from CDM transactions more than other countries in the world.

An important factor associated with the financial benefits for the project owners is the price. The Kyoto Protocol does not set a fixed price on CERs, and the price of 1t of CO_2 (= 1 CER) therefore depends on negotiation with the buyer, often through project developers. The Chinese DNA set up a 'floor' price of a CER to guide the price negotiations. This not only has helped the project owners achieve financial benefits but also contributed to the global CDM market in terms of stabilizing the price floor for the global supply of CERs. International Emissions Trading Association (IETA) statistics show that China's floor price was around $10.40–11.70 in 2006, while the vast majority of transactions worldwide were in the range of $8–$14.

It is important to note that while the financial benefits for China through the CDM are substantial relative to other countries, they are not significant compared with the amount of investment needed for China to upgrade its technology base and to achieve its full potential for climate mitigation actions. According to the recently released cost curve study of China by McKinsey & Company (2009), the average annual investment required between now and 2030 is around €180 billion to reach the technology abatement scenario.

Technology transfer

From the beginning, the CDM was expected to promote technology transfer as the only mechanism for global climate mitigation by promoting cooperation between countries with the technologies and those without them. By definition, technology transfer, more than hardware supply, involves 'sharing knowledge and adapting technology to local conditions' (UNDESA, 2008), and requires access to related technical and commercial information and the human skills needed to properly understand it and effectively use it. A critical aspect of the technology transfer process is the development of domestic capacities to absorb and master the received knowledge, to innovate using that knowledge, and to commercialize the results (International Center for Trade and Sustainable Development, 2008).

The main factors relevant to technology transfer include, and are not limited to, technology demand, technology supply and the institutional setting to facilitate or constrain the process of transfer. The types of CDM projects well represent the technologies built in and are used here to examine the degree of technology transfer and how different factors interact with each other, influencing the results.

Gap between the national demands and supplies

To cope with climate change, the Chinese Premier made five proposals. Among them is that 'Climate change must be tackled by relying on technological progress' (Wen, 2008). China's technology demands can be easily identified in both research publications and the government's announcements. The country's White Paper on climate change (Chinese State Council Information Office, 2008) has made clear that China has decided to focus R&D on the development of:

- technologies that save energy and enhance energy efficiency;
- technologies for renewable energy and new energy;
- technologies that can control, dispose of or utilize GHGs such as CO_2 and methane in industries;
- biological and engineering carbon-fixation technology;
- technologies for the clean and efficient exploitation and utilization of coal, petroleum and natural gas;
- technologies for manufacturing advanced equipment for coal- and nuclear-generated power;
- technologies for capturing, utilizing and storing CO_2;
- technologies that control GHG emissions in agriculture and other land uses.

At the top of this list are energy efficiency of energy supply and energy use in industries, transportation and building. Recovery and utilization of GHG emissions rank third on the list. According to publications of the Beijing High-level Conference on Climate Change Technology Development and Technology Transfer, organized by UNDESA (2008), priority needs for renewable energy technologies in the short, medium and long terms include lower-cost wind power and low-wind-speed turbines, PV building materials and large-scale solar systems, advanced bio-refineries and cellulosic biofuels, water photolysis and energy storage options. This list of technology needs is supported by academic research (Jiang et al, 2008).[4]

The demand for new technologies at the national level is clear. We proceed to examine whether and to what extent the CDM has contributed to meeting the demand for technology transfers to achieve emission reductions.

Table 12.5 categorizes CDM projects according to the technologies used in the projects, comparing them against the country's list of desired technologies. HFC-23 has been one of the most controversial issues since the beginning of the CDM, and the Chinese DNA has decided to reduce its approval of such projects; one reason is that HFC-23 projects contribute little to China's technological progress (Lu, 2007).

Putting HFC-23 and N_2O aside, renewable energy forms the largest category of CDM projects in China, making up 71 per cent of projects and contributing 37.3 per cent of emission reductions. Energy saving and efficiency improvement follow, making up 17 per cent of projects and 15 per cent of reductions. Together, energy efficiency and fuel substitutes make up 18.4 per cent of the projects and 19 per cent of the reduction. Methane recovery and utilization make up 6 per cent of projects and 10 per cent of annual reductions of tonnes CO_2 equivalent.

Further examination of renewable energy, the largest category, consisting of projects generating hydropower, wind power, biomass, solar power and power from solid wastes (see Table 12.6), shows that hydropower predominates; within this category, small hydro accounts for 40 per cent of projects and 13 per cent of emission reductions. Wind power makes up the next largest

Table 12.5 *Projects categorized by technologies used*

Projects categories	Projects Number	%	Annual estimated CO_2 reduced		
			Tonne CO_2 equivalent	%	Average tonne CO_2 equivalent
Renewable energy	1105	71.11	121,377,338	37.32	109,844
Energy saving and efficiency improving	271	17.44	50,994,932	15.68	188,173
Methane recovery and utilization	96	6.18	34,609,387	10.64	360,514
Fuel substitutes	31	1.99	25,776,928	7.93	831,514
N_2O decomposition	23	1.48	22,823,254	7.02	992,315
HFC-23	14	0.90	67,523,781	20.76	4,823,127
Forestation	3	0.19	47,124	0.01	15,708
Other	11	0.71	2,072,840	0.64	188,440
Total	1554	100	325,225,583	100	209,283

Source: China climate change data at www.ccchina.gov.cn

sector within this category, representing 22 per cent of both projects and CO_2 reduction. The proportion of projects using advanced technologies and providing major emission reductions, such as biomass and solid waste conversion to energy, is extremely low. There are no projects at all using advanced solar technology. There is a clear disparity between China's CDM practice and the kind of mitigation technologies that China requires.

The same disparity is identified in a project design document (PDD) survey of the 202 CDM projects registered with the Executive Board, conducted in 2008 by the Policy Research Center of Environment and Economics, Ministry of Environment Protection. Wind power projects are primarily being used as a source of CERs rather than to obtain key technologies (Tian and Li, 2008). The survey also shows that while a third of the project documents contain the technology transfer description, two-thirds of which are equipment import at commercial prices without significant 'international cooperation' defined by UNFCCC or 'Kyoto Protocol', the rest are about training for equipment operation. Because these project documents have no clear definition of 'technology transfer', in practice some technology owners continue holding the key technologies of equipment that has already been sold.

Low interest in technology transfer at the project level

The weak link between China's technology demands and actual CDM practice is related to the interests and nature of CER buyers and project developers. Given the complex procedures described earlier, over 50 per cent of the Chinese project owners rely on CER buyers or project developers to play a role in determining the origins of the technologies required (Seres, 2007) and facilitating technology transfer.

European companies are the predominant CER buyers, making up nearly 70 per cent of buyers in the Chinese CDM market, followed by Japanese companies (18 per cent). Smaller numbers of US, Canadian, Australian and Singapore companies make up the rest, in addition to a few multilateral organizations such as the World Bank.[5] The nature of CER buyers explains to an

Table 12.6 *Renewable energy projects*

Type of projects	Number of projects	Within category of renewable energy				Total CDM projects	
		%	Annual tonne CO_2 equivalent reduced	%	Average tonne CO_2 equivalent/ per project	% of projects	% of tonne CO_2 equivalent reduced
Hydro	367	33.21	67,465,549	55.58	183,830	23.62	20.74
Small hydro	437	39.55	15,961,808	13.15	36,526	28.12	4.91
Wind	241	21.81	27,739,476	22.85	115,102	15.51	8.53
Biomass	54	4.89	9,765,053	8.05	180,834	3.47	3.00
Wastes to power	3	0.27	337,992	0.28	112,664	0.19	0.10
Solar	3	0.27	107,461	0.09	35,820	0.19	0.03
Total	1105	100.00	121,377,339	100.00	109,844	71.11	37.32

Source: China climate change data www.ccchina.gov.cn

Table 12.7 *Who buys CERs from CDM projects*

CER buyers	Estimated annual tonne CO_2 equivalent reduced	%
Carbon funds, management or trader	218,170,545	67.08
Banks or financial institutions	58,344,639	17.94
Energy or power companies	44,744,010	13.76
One-side projects	2,902,295	0.89
Governments	1,064,094	0.33
Total	325,225,583	100

Source: China climate change data at www.ccchina.gov.cn

extent their business interests in the CDM. CER buyers and their transaction amounts are summarized in Table 12.7.

Carbon traders, including trading companies, carbon assets management or carbon funds, are responsible for 67 per cent of CDM transactions in China. They practice a common business model, purchasing CERs at a low price of $10 (€8) in China through primary transactions of the CDM, and selling them at higher prices of $14.30–19.50 (€11–15) in the European-dominated secondary CDM markets or the EU ETS when imports are allowed (Capoor and Ambrosi, 2007, 2008). It is not surprising that the price of CERs is their major interest rather than technology transfer. Among the ten Chinese companies interviewed that were in the energy-saving business, five were engaged in CDM efforts through potential CER buyers with the intention of transferring energy efficiency technologies from northern Europe to China. Over a period of eight months to two years, all had failed to achieve this. Dialogues with carbon trading companies showed they were only interested in buying CERs and had limited capacity to handle technology transfer.

Although energy companies often own advanced mitigation technologies, the carbon transactions they made with the Chinese CER sellers reflected their primary motivation to purchase CERs to offset their own emission reduction obligations. For example, oil and gas companies invested in CERs that were generated mainly from large or small hydropower projects.

On the demand side, project owners were also found to have little interest in technology transfer. Companies engage in CDM transactions based on their own business interests, which are not connected to China's need for technologies to mitigate climate change.

Institutional gaps

The lack of technology transfer in CDM practice reflects some major institutional gaps in promoting technology transfer. First, technology transfer is not in the mandate of the CDM, although the CDM is increasingly expected to be a mechanism of promoting such transfer. Nor is technology transfer a criterion for the Board to approve or register a CDM project (Teng et al, 2008).[6] Second, the Kyoto Protocol urges developed countries to formulate policies and programmes for the effective transfer to developing countries of environmentally sound technologies that are publicly owned or in the public domain, and to create an enabling environment for the private sector to promote and enhance the transfer of, and access to, environmentally sound technologies (Article 10). However, there is no specific proposal or mechanism defined for creating an 'enabling environment'. In practice, this gap is used to legitimize failure to facilitate technology transfer.

The CDM supports business transactions at the project level, and is not linked institutionally to the country's national strategy or targets, for example, to China's target of reducing energy intensity by 20 per cent in five years (2006–2010). Therefore, China's large-scale national emission reduction actions have benefited little from the CDM.

Summary

CDM projects have been small-scale, one-time transactions at the project and company level, and are traded primarily through dealers at low-end prices. Neither the DNA nor the Executive Board require these projects to enable technology transfer between countries. Procedures are complicated, transaction costs are high, and the process is long, requiring seven to ten years, which further raises transaction costs. National mitigation actions require updating or redevelopment of infrastructure on a large scale. Old technologies need to be phased out and new ones phased in to drive structural changes in the economy, as indicated by DRC's mid-term evaluation of China's implementation of its 11th Five-Year Plan. In other words, the project-based approach of CDM has not helped China to achieve its national energy saving and emission reduction targets while realizing its sustainable development agenda. As a result, China risks remaining 'locked-in' to old technologies.

The limitations of CDM, the urgency of the issue and the scale of action required to mitigate climate change require a new international mechanism. Such a mechanism must improve the level and scale of coverage and operation, respond better to national development strategies, and connect to mainstream global governance. We call this new mechanism the ICP. In the next section we lay out the basic principles and key components for ICP as an alternative mechanism to CDM.

Designing an inclusive mechanism for effective international cooperation – the ICP

In analysing a country's mitigation potential, both bottom-up and top-down models have been used. The former takes a sectoral approach in order to understand to what extent 'best practice' technology in various economic sectors can reduce GHG emissions. The use of 'best practice' technology in industry has received increasing attention particularly since the Bali Action Plan where sectoral agreements are proposed as an option for achieving MRVable actions to reduce emissions. Top-down models are macro models of the whole economy, which for example IPCC used to assess mitigation options and the macroeconomic cost of mitigation. The Stern Review is also based on a top-down model.

Our discussion of the ICP is firstly situated in an array of post-2012 climate policy architecture discussions, with particular attention to sectoral approaches (Michaelowa et al, 2005; Bodansky, 2007; Matsuhashi et al, 2007). At the same time the economy-wide effects interlinked with the sector-based mitigation actions are captured. In addition, a few innovative proposals made through the Harvard Project on International Climate Agreement (Aldy and Stavins, 2008) also provide us with useful references, including the multi-stage framework (Cao, 2008), the Climate Accession Deal (Victor, 2008) and the Technology CDM (Teng et al, 2008).

Sectoral approach to engaging developing countries

In general, sectoral approaches encourage emission reductions in developing countries through the deployment of low-carbon technologies in those sectors. 'Bottom-up' models are an analytical tool for understanding emission reduction capacity, crediting emission reductions to the sector and developing binding sectoral agreements or uniform, transnational standards.

In an earlier proposal by the CCAP in Washington, key developing countries would pledge to achieve a voluntary 'no-lose' GHG intensity target in key industrial sectors such as electricity, cement, steel, oil refining, pulp/paper, metals, etc. (Schmidt et al, 2006, 2008). Emission reductions achieved beyond the voluntary target would be eligible for sale as emissions reduction credits to developed countries, while failure to meet the target would not involve any penalties. Developed countries and international financial institutions would provide pledging countries with a 'Technology Finance and Assistance Package' in the form of rebates and incentives. The inclusion of the top ten largest GHG-emitting developing countries in each sector would ensure coverage of 80 to 90 per cent of GHG emissions from developing countries in each of the selected sectors (Schmidt et al, 2006).

Meckling and Chung (2009) have categorized three types of approaches to working with industry sectors: (1) government-led targets and timetables, (2) transnational technical cooperation, and (3) industry-led targets and timetables. A bottom-up model is used to design three steps or stages of implementation: (1) assessing technology and processes in selected sectors so as to determine policy measures, (2) devising incentive structures as well as benchmarks, targets and timetables, and (3) negotiating agreements.

The sectoral sustainable development policies and measures (SD-PAMs) (Höhne et al, 2008) have a common feature with the ICP proposal. In SD-PAMs, a developing country government incorporates the policies and measures of GHG emission reductions in a certain sector into its national sustainable development agenda. In return, the country receives financial or other support from developed countries according to a set of rules under the UNFCCC.

The Harvard Project on International Climate Agreements proposes 'climate accession deals' (CADs), arguing that any viable strategy of engaging the participation of developing countries must create stronger incentives for them to adjust their development patterns, and that enhancing the administrative capacity of developing countries is crucial (Victor, 2008). The proposal of a multistage framework allows developing nations to take on emission reduction responsibilities in stages, from 'no-regrets' mitigation options to voluntary targets to binding commitments (Cao, 2008). A technological CDM is proposed specifically to address technology transfer needs, based on the importance and urgency of avoiding major technology 'lock-in' in developing countries (Teng et al, 2008).

With the above brief review to set the context, we turn to the design of ICP, focusing on the Chinese context.

Principles

The ICP has been formulated based on the principles of the UN Convention and the Kyoto Protocol – equity, effectiveness and efficiency – and would be subject to the authority of the Meeting to the Convention (Seres, 2007; Makower et al, 2007, 2008, 2009; Stern, 2008; Dechezleprêtre et al, 2009). It emphasizes inclusiveness, transparency and participation, engaging with developing countries from the beginning and throughout the process. The ICP serves the same twofold goal identified for the CDM but aims specifically at large-scale emission reductions, putting the sustainable development interests of developing countries in the centre. In practice, the ICP is a mitigation initiative undertaken through sectoral approaches, which at the same time explores the feasibility of being applied at a regional level, for example, through a municipality or a province of China, capturing the economy-wide effects (Pizer et al, 2006).

The ICP requires three preconditions, based on the UN Convention's principle of 'common but differentiated responsibilities'. First, the national voluntary intensity-based energy saving and emission reduction targets adopted by developing countries should be recognized internationally. China's target of a 20 per cent reduction in energy intensity within five years (2006–2010) is an example of national voluntary action. Second, emission reductions, technology transfer and financial flows built into an ICP should reflect international MRV standards. Third, an international fund should be set up to fund the ICP. The Montreal Protocol on the ozone layer would be a good model and reference for this.

The ICP is formed on the basis of cooperation between an ICP host country (a developing country) where the emission reduction takes place, and one or more ICP partner countries (developed countries). The key players would include national governments of participating countries, private companies who hold or implement needed technologies and/or financial capital, an international body similar to the CDM Executive Board to support, facilitate and supervise the operation of the ICP, and a multilateral financial agency to manage the ICP fund. NGOs may also play an important role in the process.

In the Chinese context, the huge costs that China has been incurring to meet the national target of energy saving and emission reduction will result in large-scale avoidance of CO_2 emissions and could result in CO_2 emissions peaking by 2030 at the latest. This will occur if China's efforts remain consistent over the next two or three five-year plans, subject to GDP growth rates as described earlier. The purpose of the ICP is to offer both incentives and aid, so that major national policy actions as such will continue and be supported.

Description of an ICP

An ICP starts with a host country preparing a national emission reduction (energy-saving) plan based on voluntary and quantified targets for reducing energy intensity, derived from its national sustainable development strategy. In this plan, not only emission targets are established. The technologies, financial investment and institutional capacity needed to meet the targets are specified. The host country then invites an ICP partner country or countries to jointly implement the plan.

Partner countries review the proposal and negotiate a deal with the ICP host country. An agreement is then signed by the governments of all the participating countries, based on consensus. The partner countries have two to three basic commitments to this plan: (1) sharing the target of emission reduction, (2) transferring the needed technologies required for realizing the target, and/or (3) allocation of finance required for realizing the target.

A UN body is needed to join in the assessment of ICP proposals, to support and facilitate negotiations, and to supervise and evaluate ICP implementation at a later stage. Assessment of ICP proposals is conducted transparently, focusing on targets, joint actions of delivery, responsibilities and commitments, resources required, milestones and benchmarks for monitoring and evaluation.

The results of the ICP – emission reduction, technology transfer and financing – are measured, reported and verified according to the Bali Action Plan. The reduction in GHG emissions counts towards the emission reduction commitments of the partner countries as well as the host country, based on the rights and responsibilities agreed in the ICP. Regular assessment of implementation is conducted to ensure delivery of ICP targets.

For technological manageability and effectiveness, it is easier for an ICP to be operated on a sectoral basis. For example, in China's power generation sector, improving the energy efficiency of the national grid by 1 per cent using smarter real-time grid management systems, as practised in developed countries, would avoid 400Mt of CO_2 emissions annually (Victor, 2008). Transportation, including automobile manufacture and transport systems, is the fastest growing sector in China, accompanied by rapid growth in CO_2 emissions. The sector has great potential for energy saving and emissions reduction. For each of these major economic sectors there is a five-year national development programme, contributing to the overarching national five-year development plan. ICP can also be applied at a regional level, for example, in a municipality or a province of China, to achieve impacts on the whole economy. Both approaches require strong capacity for economic coordination (Victor, 2008). Building such capacity would be part of an ICP deal.

Policies and institutional settings in the host country form a national context for the implementation of an ICP. These contextual factors should be taken into consideration by partner countries when they assess the proposal. Open policies and market incentives in the host countries to attract needed technologies and investment from abroad would support an ICP, similar to those policies adopted by China in the 1980s and 1990s to attract foreign direct investment,[7] which have made major contributions to the country's fast economic growth for 30 years (Wu et al, 2009). Today a new round of open policies and open markets is needed to make progress towards China's sustainable development strategy, with the goal of moving the country towards a low-carbon economy.

Incentives

An ICP provides strong incentives for the host country. Essentially, the ICP defines and ensures long-term benefits for the host country because it starts with a proposal made by the host country that places national interests at the centre through its own national sustainable development strategy, from which emission reduction targets are drawn.

For convenience of description, a hypothetical case of an ICP proposal in the transportation sector is made here. China's 11th five-year programme for development of the automobile industry (2006–2010) highlights three strategic objectives: (1) enhancement of national innovation systems and R&D capacity, (2) creation of national automobile brands, and (3) development of vehicles powered by energy-saving or new energy technologies that can increase fuel efficiency by 15 per cent.[8] Technologies and approaches prioritized in the short, medium and long term for China's transport industry development include, among others, hybrid and plug-in hybrid electric vehicles, alternative and flex-fuel vehicles, cellulosic ethanol vehicles, intelligent transport systems, zero-emission vehicle systems, optimized multi-modal inter-city and freight transport, engineered urban designs and regional planning (Yang and Xing, 2008). The links between the country's transportation development programme, its technology priorities and national energy saving and emission reduction targets are quite obvious. These links need to be adequately identified to serve as a basis for negotiation of an ICP deal. At the completion of the ICP, China's transportation industry would have been upgraded, due to technology transfers and financial flows through this initiative of international cooperation on climate change mitigation.

In the ICP proposal, the objective of increasing fuel efficiency by 15 per cent by 2010 against the baseline of 2005 is broken down into specific targets, each one linked to concrete technical solutions. Technologies supporting each technical solution are specified, the finance required is calculated, and the needs for capacity building (individual or institutional) to manage such large-scale emission reductions are identified. With this proposal, China invites cooperation from partner countries with the technological and financial capacities and the institutional and technical experience. The agreement for joint implementation is signed by the governments and becomes effective after an assessment of the proposal and negotiations involving the participating countries and UN body.

Some technologies that are needed are available on the market but are too expensive for developing countries. They are held by private companies in the partner countries who have made substantial investments in R&D in expectation of commercial returns, partly through intellectual property right (IPR) fees. Some kinds of technology require further investments to adapt them to the Chinese context, including investments to update Chinese infrastructure. Some are not commercially viable yet, as further R&D is needed to lower the costs to the consumer. Others require major technological innovations. With regards to available technologies, IPR has been seen as the main obstacle to environmental technology transfer and deployment. Through an ICP, IPR could be negotiated through a package bridging the high demand for environmentally friendly technologies in China and the commercial interests of technology vendors in the partner countries. Such a package could benefit both parties because the volume of the transactions can reduce supply costs. A professional agency with expertise in IPR from the host country represents the technology demand to participate in the negotiation with the partner countries.

Through domestic open market policies, ICP host countries can link foreign investments with technology transfers to meet their own emission targets and also to upgrade key industry sectors. In this sense, the ICP has a long-term impact on sustainable development of the host country, phasing out polluting technologies and phasing in the new ones, avoiding 'lock-in' effects and supporting structural change in the economy.

For partner countries, they firstly gain through low-cost emission reductions in the host country. Second, they are helped to gain market share for their own climate change mitigation technologies. Today this market is largely located in developing countries, not only because they account for half of CO_2 emissions today with a tendency to fast growth in the short term, but also because their large populations and much lower consumption levels offer a big market potential. China's five-year programme to increase automobile production to 9 million a year to achieve a ratio of 40 vehicles per 1000 people, and to increase fuel efficiency by 15 per cent by 2010, is one example of the scale of this market. The ICP is a platform for partner countries to tap developing country markets for hybrid motor vehicle technologies and optimal inter-city transport management systems, to name just two examples. Compared to the CDM, which is limited to one-time transactions at the project level, the ICP offers a long-term vision for addressing the market as a whole.

Partner countries can also gain long-term leverage on CO_2 emissions with joint research projects to build new technology demonstration programmes targeting emission reductions and energy saving (Victor, 2008), and to tailor their technologies to the needs of host countries, making the laboratory-tested technologies commercially viable in host countries. Since the 1990s, along with foreign direct investment flows, there has been a steady trend of moving R&D work to China from overseas. By August 2005, 750 R&D centres had been set up by foreign companies in China to benefit from China's large domestic market and its comparative research advantages (Jiang, 2007). The percentage of foreign companies doing so increased from 6.7 per cent in 2000 to 17.1 per cent in 2007. According to research of the UN Conference on Trade and Development (UNCTAD), 61.8 per cent of transnational corporations selected China as their overseas R&D locations during the period 2005–2009. The purpose is to enhance their business competitiveness and ensure their market share in China. Today technologies are needed not only to fight climate change but to create major business. The countries or companies holding the technologies on China's list of desired mitigation technology are all potential beneficiaries. This is why an ICP can be attractive to private sector companies, which are always looking for market expansion in return for their investment.

Equally beneficial for both ICP host and partners will be the mutual trust built into the operating process. Monitored transparently against international standards, the results of emission reduction, technology transfers and financial flows are measured, reported and verified. Although the application of international monitoring and evaluation standards may raise implementation costs to the host country, the benefits brought in would surpass the costs by a sufficient margin, and the costs are shared among ICP participating countries. The same standards are applied to measure and verify the technology transfer and financial support committed by partner countries to ensure that commitments are fulfilled.

An ICP provides the host country with access to the global carbon market, making it truly 'liquid' (Stern, 2008). Being subject to the standards of international MRV systems and built on a common agreement, an ICP can contribute to preventing carbon leakage to some extent, which has been a major concern over the CDM. An ICP is viable and self-enforcing because it is a win–win solution (Victor, 2008). Current reviews of the 30-year experiences of economic reform in China support this conclusion.

Institutional capacity and funding for the ICP

Capacity

To operate an ICP, two other necessary conditions are needed. One is the institutional capacity to manage the ICP and the other is an international fund to support capacity building for ICP host countries. Managing an ICP is much more methodologically, technically and administratively complex than doing a CDM project. But it is more cost effective and has long-term impacts, compared to the CDM which is a one-time deal.

Managing a large-scale emission reduction programme to match the country's economic interests is new to most developing countries. It means economic coordination in the targeted sector across different administrative levels or, at the regional level, across sectors or administrative lines. If coordination is lacking, it will be difficult to monitor whether the policies of emission control are followed at province or company levels. Such coordination is built into a market economy operating within the legal and institutional frameworks of the country. A centralized, planned economy cannot offer this kind of coordination. The ICP provides adequate economic incentives for engagement but in order to make an ICP deal and to operate it, the government of the host country needs to create institutional capacity to manage or coordinate emissions reduction, technology transfer, investment and financing. This capacity will benefit China in the long run because it is needed for managing every ICP deal.

China introduces new technologies and attracts international investment through its ICP to move the country to a low-carbon economy. The capacity of economic coordination is much needed to meet both economic and environmental (emission reduction) objectives. In China's 11th Five-Year Plan, one of three criteria used to assess the value of new foreign direct investment is whether it brings in economic coordination capacity and good international experience of management. The other two criteria are whether new foreign direct investment is introducing advanced technologies and needed human resources. The ICP meets all three criteria. Capacity building through ICP in this respect delivers 'compound interest'.

Funding

An international fund needs to be set up and operated by a professional multi-lateral agency to support the ICP, from proposal preparation to capacity building. A large amount of financial resources will be needed to operate an ICP. Financing of such size can only be accessed on a public–private partnership basis. Public funds should take the lead to leverage and provide incentives for private investment. Financing can come from three sources. The first source would be a multilateral fund that is able to receive contributions from public sector agencies in developed countries and from large private sector or transnational corporations. Supervised by the UNFCCC authority body, the multilateral fund should be managed and operated by an international professional (financial) agency. It is similar to the three special funds under the UNFCCC operated by the GEF, including the SCCF, the Least Developed Country Fund (LDCF) and the Adaptation Fund (AF). Unfortunately, these funds are very small.

The multilateral fund plays a critical role in the ICP. It can effectively leverage public finances from both the host country and partner countries to attract investment from the private sector. In its international capacity and credibility, it leads the expansion of technology markets, shares and lowers the cost of technology transfer and deployment, and attracts financing based on public and private partnerships. It has also been shown that a multilateral fund can be very good at supporting innovation, pilot programmes and capacity building. In July 2008, a Climate Investment Fund (CIF) was approved by the Board of Directors of the World Bank, to contribute to demonstration, deployment and transfer of low-carbon technologies and to pilot new development approaches or to scale up activities aimed at a specific climate change challenge or sectoral response.[9] This fund would potentially meet ICP demands if the scale is adequate.

The second source is public-sector financing from ICP participating countries – both host and partners. The contribution is made according to agreed commitments and responsibilities. The purpose is to enable the fulfilment of these responsibilities and commitments. The Chinese government's investment in the 'Top 1000 Enterprises Energy Conservation Plan' is a form of such public-sector financing. Funds from partner countries could support the technology transfer and cooperation process, among other activities. Public financing has strong policy implications and plays an important role in leveraging and providing incentives for private investment. It leads to structural change in the economy by facilitating markets for low-carbon technologies and avoiding 'lock-in' impacts.

The third source is private companies. In partnership with the public sector, their financial contribution helps to build a solid international fund. Participation in an ICP at this stage also enables private companies to understand the programme, define their contribution and take a proactive role in seeking win–win opportunities. An important function of this fund is to prepare for effective operation of the ICP, which will bring long-term economic benefits to all participants while protecting the climate, a global public good.

Inputs from private companies are made in the form of investments at different stages of the ICP and will be the most important source of ICP finance. Holding the necessary technologies and capital, private companies play a critical role in the ICP. Multilateral funds and public finance combined with policy instruments provide them with powerful incentives to participate. Their investments will eventually scale up mitigation actions through market forces.

Conclusions

Meeting the global climate target requires a 'global climate emergency path' that is challenged by current uncertainty regarding China's emissions trajectory. Managing such uncertainty is a challenge to a large, 'non-capped' economy in the midst of rapid, energy-demanding industrialization and urbanization. It can only be addressed by international cooperation through innovative, inclusive and flexible mechanisms. The success of such cooperation will determine China's emissions trajectory. In such a context, we propose the ICP, by which deals can be negotiated, partnerships formed and – mostly importantly – actions can be taken quickly. We also argue there are strong incentives from all sides for an ICP deal to occur. What is needed is to craft the deal in such a way that it is closely aligned with national interests in creating sustainable development pathways, and at the same time helps to fundamentally alter emission patterns in the host economy.

Notes

1   www.ccchina.gov.cn/en/index.asp
2   www.CO$_2$-handel.de
3   'A guiding price of CER was set up by NDRC. This price was US$5 per CER before 2006, and now has gone up to €8' (Zhang, 2007).
4   For China's energy efficiency to reach the level of the developed countries by 2025, China should speed up the development of renewable energies, for example wind, hydro, solar including PV and concentrated solar power, as well as new clean power energy of the 3rd and 4th generations (Jiang et al, 2008).
5   http://cdm.ccchina.gov.cn/web/BuyList.asp
6   The Chinese 'measures for option and management of CDM projects in China' does have a sentence that 'CDM project activities should promote the transfer of environmentally sound technology to China.' This is not a mandatory requirement based on which DNA approves the projects.
7   During 1979–2007, use of foreign capital in China reached ¥957.4 billion, of which foreign direct investment accounted for more than ¥765 billion invested in 632,298 projects. The employment created by foreign companies in China increased from 60,000 in 1985 to 15.8 million in 2007. The foreign trade volume created by foreign companies in China increased rapidly from $831.6 billion in 2005 to over $1.25 trillion in 2007, making up nearly 60 per cent of China's total. (Wang and Chen, 2007).
8   'Research outline for the national 11th 5-year program for development of automobile related industries', 9 January 2007, www.autoinfo.gov.cn/autoinfo_cn/lbj/tbgz/syw/webinfo/1182388344493521.htm
9   Climate Investment Fund at www.worldbank.org/cif

References

Aldy, J. (2008) Testimony at US-China Economic and Security Review Commission's Hearing on China's Energy Policies and Environmental Impact, One Hundred Tenth Congress, Second Session, 13 August

Aldy, J. and Stavins, R. (2008) *Designing the Post-Kyoto Climate Regime: Lessons from the Harvard Project on International Climate Agreements*, Cambridge, MA, Harvard Kennedy School

Ambrosi, P. (2007) *Sustainable Development Operations, 2007*, Development Economics Research Group, Washington, DC: World Bank

Ambrosi, P. (2008) *Sustainable Development Operations, 2008*, Development Economics Research Group, Washington, DC: World Bank

Auffhammer, M. and Carson, R. T. (2008) 'Forecasting the path of China's CO_2 emissions using province level information', *Journal of Environmental Economics and Management*, vol 55, pp229–247

Baer, P., Athanasiou, T., Kartha, S. and Kemp-Benedict, E. (2007) *The Right to Development in a Climate Constrained World: The Greenhouse Development Rights Framework*, Berlin: Heinrich Böll Foundation

Bodansky, D. (2007) *International Sectoral Agreements in a Post-2012 Climate Framework*, Working Paper, Washington, DC: Pew Center on Global Climate Change

Cao, J. (2008) 'Reconciling human development and climate protection: Perspectives from developing countries on post-2012 international climate change policy', Discussion paper 08-25, Cambridge, MA: Belfer Centre for Science and International Affairs

Capoor, K. and Ambrosi, P. (2007) *State and Trends of the Carbon Market, 2007*, Washington, DC: World Bank

Capoor, K. and Ambrosi, P. (2008) *State and Trends of the Carbon Market, 2008*, Washington, DC: World Bank

Chinese State Council Information Office (2008) *China's Policies and Actions for Addressing Climate Change*, White Paper, Beijing: Chinese State Council Information Office

Dechezleprêtre, A., Glachant, M. and Ménière, Y. (2009) 'Technology transfer by CDM projects: A comparison of Brazil, China, India and Mexico', *Energy Policy*, vol 37, pp703–711

DRC (Development Research Center) (2008) 'Mid-term evaluation on implementation of energy conservation and emission reduction during the 11th 5-year plan period', Industrial Economics Research Department, Beijing: Development Research Center of the State Council

EIA (Energy Information Administration) (2007) *Annual Energy Outlook 2007*, Washington, DC: Office of Integrated Analysis and Forecasting, US Department of Energy, available on-line at http://tonto.eia.doe.gov/ftproot/forecasting/0383(2007).pdf

Fan, G., Cao, J., Yang, H., Li, L. and Su, M. (2008). Draft paper prepared for the midterm review of the project 'China Economics of Climate Change', Beijing

Hansen, J., Sato, M., Kharecha, P., Beerling, D., Berner, R., Masson-Delmotte, V., Pagani, M., Raymo, M., Royer, D. L. and Zachos, J. C. (2008) 'Target atmospheric CO_2: Where should we aim?', *Open Atmospheric Science Journal,* vol 2, pp217–231, doi:10.2174/1874282300802010217

Höhne, N., Worrell, E., Ellermann, C., Vieweg, M. and Hagemann, M. (2008) 'Sectoral approach and development', Input paper for the workshop: Where development meets climate – development related mitigation options for a global climate change agreement, Ecofys

International Center for Trade and Sustainable Development (2008) *Climate Change, Technology Transfer and Intellectual Property Rights*, Geneva: International Center for Trade and Sustainable Development

IEA (International Energy Agency) (2008) *World Energy Outlook 2008*, Paris: OECD/IEA

Jiang, K. Hu X., Zhuang, X., Liu, Q. and Zhu, S. (2008) 'China's energy demand and greenhouse gas emission scenarios in 2050', *Advances in Climate Change Research*, vol 4, no 5, pp296–303

Jiang, X. (2007) *Foreign Capital Economy in China: Contributions to the Growth, Structural Upgrading and Improved Competitiveness*, Beijing: China Renmin University Press

Lewis, M. (2007) 'Carbon emissions – banking on higher prices: We see EUAs at E35/t over 2008–20', Deutsche Bank Report. Frankfurt: Deutsche Bank

Makower, J., Pernick, R. and Wilder, C. (2007) *Clean Energy Trends 2007*, San Francisco: Clean Edge

Makower, J., Pernick, R. and Wilder, C. (2008) *Clean Energy Trends 2008*, San Francisco: Clean Edge

Makower, J., Pernick, R. and Wilder, C. (2009) *Clean Energy Trends 2009*, San Francisco: Clean Edge

Matsuhashi, R., Misumi, K. and Yoshida, Y. (2007) 'Comparative analyses of sector-based approaches and national numerical targets as Post-Kyoto frameworks', *Environmental Informatics Archives*, vol 5, pp36–41

McKinsey & Company (2009) *China's Green Revolution: Prioritizing Technologies to Achieve Energy and Environmental Sustainability*, Beijing: McKinsey & Company

Meckling, J. and Chung, G. Y. (2009) 'Sectoral approaches to international climate policy: A typology and political analysis', *Discussion Paper 2009-02*, Cambridge, MA: Energy Technology Innovation Policy, Belfer Center for Science and International Affairs, Harvard Kennedy School

Michaelowa, A., Tangen, K. and Hasselknippe, H. (2005) 'Issues and options for the post-2012 climate architecture – an overview', *International Environmental Agreements*, vol 5, pp5–24

Pizer, W., Burtraw, D., Harrington, W., Newell, R. and Sanchirico, J. (2006) 'Modeling economy-wide vs sectoral climate policies using combined aggregate-sectoral models', *Energy Journal*, vol 27, pp135–168

Price, L., Wang, X. and Hun, J. (2008) *China's Top-1000 Energy-Consuming Enterprises Program: Reducing Energy Consumption of the 1000 Largest Industrial Enterprises in China*, Berkeley, CA: Lawrence Berkeley National Laboratory

Schmidt, J., Helme, N., Lee, J. and Houdashelt, M. (2006) *Sector-based Approach to the Post-2012 Climate Change Policy Architecture*, Washington, DC: Center for Clean Air Policy

Schmidt, J., Helme, N., Lee, J. and Houdashelt, M. (2008) 'Sector-based approach to the post-2012 climate change policy architecture', *Climate Policy*, vol 8, pp494–515

Seres, S. (2007) 'Analysis of technology transfer in CDM projects', paper prepared for the UNFCCC, Registration and Issuance Unit

Stern, N. (2008) *Key Elements of a Global Deal on Climate Change*, London: London School of Economics and Political Science

Swedish Ministry of the Environment (2007) *A Swedish Strategy for Sustainable Economic, Social and Environmental Development*, Stockholm: Swedish Ministry of the Environment

Takahashi, K. and IGES Market Mechanism Project (2009) IGES CDM Project Database, IGES EnviroScope, available online at http://enviroscope.iges.or.jp/modules/envirolib/view.php?docid=968

Teng, F., Chen, W. and He, J. (2008) 'Possible development of a technology clean development mechanism in a post-2012 regime', *Discussion Paper 08-24*, The Harvard Project on International Climate Agreements, Cambridge, MA: Harvard Kennedy School, Harvard University

Tian, C. and Li, L. (2008) 'Policy study of technology transfer in CDM project implementation in China', *Environment Protection*, vol 392, pp62–64

UNDESA (2008) *DESA News*, vol 12, no 12, available online at www.un.org/esa/desa/desaNews/v12n12/techcoop.html

UNEP (United Nations Environment Programme) (2007) *Handbook of Financing a CDM Project*, Nairobi: UNEP

Victor, D. (2008) 'Climate accession deals: New strategies for taming growth of greenhouse gases in developing countries', *Discussion Paper 08-18*, The Harvard Project on International Climate Agreements, John Kennedy School of Government, Cambridge, MA: Harvard University

Wang, X. and Chen, L. (2007) 'Challenges and policy options to China to attract foreign capital', *Economic Affairs*, vol 6, pp75–80

Wang, T. and Watson, J. (2008) *Carbon Emissions Scenarios for China to 2100. SPRU*, Sussex Energy Group and Tyndall Centre, Brighton: University of Sussex

Wen, J. (2008) Beijing High-level Conference on Climate Change: Technology Development and Technology Transfer, Beijing, 7–8 November

World Bank (2006) 'CDM and World Bank project cycle', presentation at carbon finance workshop, Nairobi, Kenya, November

World Bank (2008) *State and Trends of the Carbon Market 2008*, Washington, DC: World Bank

Yu, C. (2008) 'Top-1000 Enterprises Energy Conservation Action Plan', presented at the 11th Senior Policy Advisory Council Meeting, China Sustainable Energy Program, November, Beijing

Zhang, W. (2007) 'The first CDM project in China', *Economics*, vol 7, available at http://finance.sina.com.cn/265/2007/0810/1545.html (in Chinese)

13

Policy Implications of Carbon Pricing for China's Trade

Frank Ackerman

China has a large positive balance of trade – both in monetary terms and in terms of the carbon embedded in trade (see Chapter 4). As the world moves towards policy measures and agreements that will place a price on carbon emissions, how will China's exports be affected? This chapter addresses two aspects of that question, first looking specifically at the border tax adjustment proposals that have circulated in some developed countries, and then more generally at the likely impacts of carbon prices on China's economy.

The conclusions are, in short, that border tax adjustments, as currently discussed, would do little good for developed countries and little harm to China. Only a handful of industries are heavily affected, and they are not, for the most part, ones in which China has a comparative advantage. More broadly speaking, carbon prices need not be globally harmonized in theory and may not be harmonized in practice for some time; when and if prices are harmonized, they will raise costs for China's carbon-intensive industries, but will also create a market opportunity for China to 'leapfrog' beyond the carbon-saving technologies developed in high-income countries.

Border tax adjustments: The debate

One of the obstacles to action on climate change is the fear that if some but not all countries introduce a price on carbon emissions, either through a tax or through permit trading, they will place their industries at a competitive disadvantage. Other countries with lower carbon prices, or none at all, will have lower costs of production and could win an increased share of world markets. Carbon-intensive industries could migrate to carbon-tax-free locations, so that some part of the expected reduction in emissions would be lost through 'leakage' out of the countries with carbon prices. A border tax adjustment is essentially a tariff on the carbon embedded in a country's imports, bringing the price of the embedded carbon up to the importing country's standard. This is intended to eliminate any unfair advantage from low carbon prices, at least within the importing country's own economy.

There are numerous practical problems with border tax adjustments. They would have to be differentiated by country of origin, since carbon prices could vary around the world. The taxes would also depend on elaborate calculations of embedded carbon: complex manufactured goods often contain components from more than one country, with differing carbon intensities and, perhaps, differing carbon prices. The focus here, however, is on the economic principles and impacts of border tax adjustments, not on the difficult details of implementation.

According to Joseph Stiglitz (2006), in a world that is trying to reduce emissions, failure to tax or otherwise price carbon could be considered an unfair subsidy, calling for a complaint to the WTO and the imposition of retaliatory tariffs against the non-compliant nation. At the time when this article appeared, the US was increasingly isolated in its rejection of the Kyoto Protocol; Stiglitz presented his proposal as a justification for Kyoto-compliant countries to impose carbon tariffs on US exports. Other authors making a case for border tax adjustments on grounds of efficiency and/or preventing carbon leakage include Demailly and Quirion (2006), Kopp and Pizer (2007), Ismer and Neuhoff (2007) and Goh (2004).

The spectre of countries using differential carbon prices as a basis for applying tariffs on each other's goods raises fears that climate policy could become an excuse for a new protectionism. It is far from certain that carbon tariffs would be compliant with WTO rules; for analyses of this complex question see Hufbauer et al (2009) and Werksman and Houser (2008). To minimize the practical problems and political resistance to border tax adjustments, attention has focused on targeting policies specifically to the most affected industries, where international differences in carbon prices could conceivably cause leakage of production and carbon emissions to lower-priced regions.

Emissions leakage, in significant quantities, can only occur in industries that are both internationally competitive and highly carbon intensive. The list of such industries is surprisingly short and relatively consistent around the world; they are generally the energy-intensive, primary materials industries. In the US in 2002, just six industries accounted for 81 per cent of manufacturing energy demand (and also for most of the non-energy process emissions of GHGs from manufacturing): petroleum refining, chemicals and plastics, pulp and paper, non-metallic mineral products (including cement), ferrous metals, and non-ferrous metals.

These six industries represented only 2.5 per cent of US employment, and less than 4 per cent of US GDP.[1] The affected sectors may be a subset of these broad industry classifications (as suggested in Houser et al, 2008), implying that an even smaller fraction of the US economy is at risk.

Even for the most directly affected industries, the impacts of differential carbon prices may be modest. An analysis by economists at Resources for the Future modelled the effect on the US economy of unilateral US adoption of a carbon price of $10 per ton of CO_2 (Ho et al, 2008). The increase in costs in the very short run, prior to any substitution of inputs or adjustment of production processes, was 5.0 per cent for cement, 4.2 per cent for petrochemicals, between 1.0 and 2.6 per cent for many other branches of the most affected industries, and less than 1.0 per cent for the rest of the US economy. In the long run, after industries adjusted to the change in energy costs, substantial losses occurred only in energy-producing industries; in all non-energy manufacturing sectors, the loss of employment was less than 0.7 per cent, and the decrease in output was less than 1.3 per cent.[2] Reduced demand for fossil fuels – which is the intended outcome of a carbon price – led to long-run declines in petroleum refining, coal mining and oil and gas production, ranging from 4 to 10 per cent.

A study of the European economy modelled a much greater carbon price (Manders and Veenendaal, 2008). In its scenario with no trading with other countries, that study projected a permit price of €52 ($69) per tonne of CO_2 in the industries covered by the EU ETS, and a carbon tax for other sectors of €29 ($39) per tonne.[3] These prices, imposed unilaterally in Europe, led to a 3.2 per cent loss of employment and a 4.5 per cent loss of production in the ETS sectors – which represent 9 per cent of total employment. Use of the CDM or a similar system for purchase of carbon reductions outside the EU allowed sharp reductions in these losses: unilateral European carbon pricing combined with use of the CDM to obtain one-third of the reductions needed by ETS sectors (a limit based on current EU policy proposals) would lower the permit price to €27 ($36) per tonne, and would lead to employment losses of only 1.2 per cent and production losses of 1.7 per cent in the ETS sectors. Another scenario, modelling a border tax on carbon-intensive imports in the absence of CDM, projected larger losses to ETS sectors than the CDM scenario.[4]

In short, the energy-intensive industries, where there is a credible competitive threat from countries with lower carbon prices, account for a very small fraction of the US and European economies. Border tax adjustments targeted specifically to these industries would affect only a small fraction of world trade. If such policies are politically necessary to achieve a global agreement on carbon reduction, they should be carefully designed and limited to the affected industries, in order to avoid retaliatory tariffs and a retreat into protectionism. In his recent proposals for a 'global deal', Nicholas Stern suggests that rich countries could offer assistance to developing countries in introducing new, lower-carbon technologies in energy-intensive industries, combined with imposition of carbon tariffs on those industries ten years after the agreement is signed, if developing countries have not established carbon prices or lowered emissions (Stern, 2009). The delay would allow firms and developing countries time to adjust to the new market regime and the effects of carbon prices. This is, however, only a small part of what is needed to achieve global agreement on reduction, and it is probably not the most controversial part of the package.

Why so little effect on China?

A border tax adjustment, targeted to the most carbon-intensive, internationally competitive industries, would have surprisingly little effect on China. Of the six energy-intensive manufacturing sectors listed above, China has a revealed comparative advantage[5] in only one, non-metallic mineral products (which includes cement).

As shown in the top panel of Table 13.1, China's gross exports of $35.4 billion to high-income countries in the six energy-intensive materials industries represented 9 per cent of its global total of $379.5 billion of exports in all sectors. The materials exports were widely distributed among high-income countries in geographic terms; by sector, chemicals and plastics accounted for more than half, followed by minerals.

Energy-intensive materials are not an insignificant category of exports, but they are not central to China's export success. That success is based on a comparative advantage in labour-intensive, rather than energy-intensive, industries. The great majority of China's exports are not in the most energy-intensive materials industries; the great majority of high-income countries' imports of these materials are not from China. Exports from China are a very small fraction of world supply of materials such as steel, cement, aluminium and chemicals (Houser et al, 2008).

The six energy-intensive materials industries are large, diverse categories, such as chemicals and plastics. Within such categories, China exports some products and imports others. As the second panel of Table 13.1 shows, China is a net importer of the products of these six industries, from high-income countries and from the world as a whole; modest net exports in non-metallic minerals are outweighed by much larger net imports in the other five sectors.

Despite being a net importer of energy-intensive materials, China is a net exporter of carbon embedded in these materials, as shown in the bottom panel of Table 13.1. As explained in Chapter 4, China's exports are much more carbon-intensive than its imports, making it entirely possible to be both a net importer in dollars and a net exporter in carbon.

Trade with high-income countries represents most of China's trade with the world, as seen by comparing the last two columns of Table 13.1. The one major exception appears in the calculation of carbon embedded in net exports of ferrous metals, where China is a net exporter to high-income countries but a net importer from the world. This reflects large, carbon-intensive imports of ferrous metals from Russia and other parts of the former Soviet Union.

Table 13.1 China's exports of energy-intensive materials to high-income countries

	US	Europe	Japan	East Asia	Other high-income	High-income total	World total
Gross exports (millions of 2001 US$)							
Paper products, publishing	1205	527	250	726	199	2908	3308
Petroleum, coal products	247	338	167	755	141	1648	3323
Chemical, rubber, plastic products	5767	5480	2661	3503	1171	18,581	24,244
Mineral products	2688	1382	1354	1495	338	7257	8655
Ferrous metals	463	267	399	963	109	2202	3089
Other metals	340	421	505	1438	111	2815	3361
Six-industry total	10,710	8416	5338	8879	2069	35,412	45,981
All industries, total	107,033	78,530	57,258	66,444	14,903	324,168	379,468
Net exports (millions of 2001 US$)							
Paper products, publishing	332	(158)	(224)	(623)	(486)	(1159)	(2710)
Petroleum, coal products	117	224	85	(1433)	108	899	(858)
Chemical, rubber, plastic products	2300	1034	(3584)	(8698)	70	(8878)	(10,142)
Mineral products	2329	866	357	580	319	4450	5534
Ferrous metals	(104)	(668)	(2491)	(1287)	(162)	(4713)	(6814)
Other metals	(413)	(436)	(880)	339	(946)	(2335)	(4069)
Six-industry total	4560	861	(6736)	(11,122)	(1098)	(13,534)	(19,060)
All industries, total	78,033	26,623	8694	(9340)	4332	108,342	108,236
Carbon embedded in net exports (thousands of tons of CO$_2$-eq)							
Paper products, publishing	2801	1031	210	1144	(29)	5157	3501
Petroleum, coal products	1203	1833	678	1306	770	5790	8936
Chemical, rubber, plastic products	21,873	22,046	2080	1120	4308	51,427	54,684
Mineral products	17,298	8435	1036	7511	2197	36,476	49,879
Ferrous metals	3461	1042	(819)	5836	(153)	9367	(6386)
Other metals	984	1522	(5147)	8118	(4954)	523	2012
Six-industry total	47,620	35,909	(1963)	25,034	2140	108,740	112,627
All industries, total	271,850	198,247	142,788	174,877	21,844	809,606	850,344

Notes: Europe = EU-27, Switzerland and Norway; East Asia = Korea, Taiwan, Singapore and Hong Kong; Other high-income = Australia, New Zealand and Canada
Source: Author's calculation from Multi-Regional Input-Output (MRIO) model of carbon embedded in international trade flows. Thanks to Glen Peters for making the MRIO outputs available for analysis.

Does the world need harmonized carbon prices?

This section steps back from the detailed debate about energy-intensive industries and border tax adjustments to consider the effects of carbon prices in more general terms. It argues on the one hand, that a consistent worldwide price for carbon may not be necessary in theory or, in the short run, likely in practice; and on the other hand, a harmonized carbon price would create both obstacles and new opportunities for China and other developing countries. The complexity of these issues emphasizes the need for climate policy to include much more than setting a price on carbon emissions.

It has become commonplace to call for a single global price of carbon, applicable everywhere. Price harmonization is thought to ensure efficiency in the worldwide distribution of abatement effort: with appropriate market institutions, investment in emissions reduction should flow to the countries where the costs of reduction are lowest. Fears about the effects of unharmonized carbon charges have slowed climate policy initiatives in some high-income countries, and have prompted the discussion of border tax adjustments, as discussed above.

Why, in theory, should the same carbon price apply everywhere? A single harmonized price, in a world of smoothly functioning international markets, would mean that investment in abatement would occur in every country, up to the point where the marginal cost of abatement equals the price of carbon emissions. No country would invest in abatement that was more expensive than the price of the abated carbon; at the same time, no country could profit by investing less, since all investments up to that cost would save money. Thus a globally harmonized price means that every country gives up the same dollar amount of consumption per tonne of carbon avoided.

This theory could be criticized for assuming unrealistic perfection in the functioning of global markets for investment. More fundamentally, it also assumes that investment up to the same carbon price in every country is an equitable distribution of abatement costs. This would only be true if the same price for investment, such as $20 per tonne of CO_2 avoided, represented the same amount of human welfare in every country. In fact, $20 represents a much bigger change in welfare in low-income than in high-income countries. So the support for a harmonized carbon price rests on the unexamined assumption that the world income distribution is equitable – or equivalently, that increases in per capita consumption are equally urgent everywhere (Chichilnisky and Heal, 1994; Sheeran, 2006). In an inequitable world, equal sacrifice of human welfare per tonne of avoided carbon requires higher carbon prices in richer countries, making it profitable for them to carry out higher-cost abatement efforts.

There is a possibility that the politics of climate negotiation will lead to a two-tiered price system, at least in the short run. As carbon allowance trading becomes more widespread in high-income countries, there may be moves to limit the amount of abatement that can be done overseas. Such limits are supported both by environmental moralism – we *should* be reducing our own carbon emissions, not just paying to reduce someone else's – and by economically parochial or protectionist sentiments – we should keep some of our emission-reducing investment at home to create jobs and incomes here, rather than sending it all to other countries. In an international trading system where high-income countries require a certain percentage of domestic abatement, there may be a higher price for domestic carbon allowances than for international ones.[6] As a result, higher-cost abatements will be undertaken in rich countries.

It is possible, however, that the momentum for a consistent worldwide carbon price will prevail in negotiations, at least for the long run. If this happens, developing countries will face a global carbon price, while local prices for labour, land and other inputs remain far below the levels of higher-income countries. Carbon emissions, or the credits for avoiding them, will account for a much larger fraction of the value of production in lower-income countries. The potential dissonance between expensive carbon and cheaper local inputs creates both an obstacle and an opportunity.

The obstacle is that development may be distorted in the direction of activities that yield marketable carbon reductions. Even undesirable activities may be promoted in order to generate carbon credits. Safeguards are needed to prevent 'carbon-allowance-seeking' investments; in any global carbon market, it will be essential to verify that emissions are not newly created in order to profit by reducing them. The temptation to seek such perverse allowances, unfortunately, is a natural consequence of a global carbon price in a low-cost local economy.

The opportunity created by this same pattern of prices is that much deeper reductions in carbon emissions will be economical in developing countries. In the simplest terms, saving a tonne of carbon is 'worth' more hours of labour at a lower wage rate. So there may be a category of carbon-saving investments and technologies that are profitable only in developing countries, where the trade-off between carbon and other inputs is more favourable to emission reduction. With appropriate public initiatives and financing for these technologies, developing countries could 'leapfrog' beyond the patterns of energy use in higher-income countries, establishing a new frontier for carbon reduction.

The potential for leapfrogging beyond the current technology frontier has been much discussed, but is difficult to achieve. The classic example is in telephones, where it is now possible to skip the expensive development of land lines and go directly to cell phones. This is not, however, an example of jumping to an entirely new technology; it became possible only after cell phones were invented and commercialized in high-income countries (Unruh and Carillo-Hermosilla, 2006). Likewise, research on the Chinese auto industry has found that the leading firms have shown little tendency towards leapfrogging beyond international standards; in fact, American auto companies, left to themselves, have often allowed their Chinese plants to lag behind their home-country technologies (Gallagher, 2006). Strong government policies and initiatives are required to achieve the potential for newer, cleaner vehicle technologies. Even for a country with the extensive resources and market potential of China, there is much that needs to be done to reach this new technological frontier.

To realize the opportunity created by a global carbon price in low-cost economies, there will be a need for R&D in appropriate, cutting-edge technologies for carbon reduction. As with many of the new energy technologies that will be needed around the world, decades of public investment may be required before the developing-country technologies are successful in the marketplace. This is one more reason why carbon prices are necessary, but not sufficient, for an equitable solution to the climate crisis.

Notes

1 Calculated from data from US EIA, www.eia.doe.gov/emeu/mecs/mecs2002/data02/shelltables.html and Bureau of Economic Analysis www.bea.gov/industry/gdpbyind_data.htm; see also Houser et al (2008).
2 Non-energy mining is projected to lose 1.0 per cent of employment and 1.1 per cent of output in the long run.
3 Currencies converted at €1 = \$1.33. The ETS covers electricity plus the six energy-intensive sectors of manufacturing listed above. The carbon tax of €29 was projected for the non-ETS sectors of the EU-15, i.e. the 15 members of the EU prior to 2004 – countries that generally have higher incomes than the newer members.
4 This analysis is based on a general equilibrium model, in which total EU employment is arbitrarily assumed to be constant. As a result, it is difficult to obtain meaningful estimates of overall impacts on the European economy. On this and other limitations of general equilibrium modelling of international trade, see Ackerman and Gallagher (2008).
5 See Chapter 4 for data and definitions. This discussion is based on 2001 data from the MRIO model introduced in that chapter.
6 Price differentials can only occur if the domestic abatement requirement is a binding constraint, that is, if an unconstrained market would have sold more international abatements than is allowed.

Acknowledgements

I would like to thank Glen Peters for generously sharing data and results. He is not responsible for the content of this chapter, which is my analysis based on his work.

References

Ackerman, F. and Gallagher, K. (2008) 'Looks can be deceiving: Measuring the benefits of trade liberalization', *International Journal of Political Economy*, vol 37, no 1, pp50–77

Chichilnisky, G. and Heal, G. (1994) 'Who should abate carbon emissions? An international perspective', *Economic Letters*, vol 44, pp443–449

Demailly, D. and Quirion, P. (2006) 'CO$_2$ abatement, competitiveness and leakage in the European cement industry under the EU ETS: Grandfathering versus output-based Allocation', *Climate Policy*, vol 6, no 1, pp93–113

Gallagher, K. S. (2006) 'Limits to leapfrogging in energy technologies? Evidence from the Chinese automobile industry', *Energy Policy*, vol 34, pp383–394

Goh, G. (2004) 'The World Trade Organization, Kyoto and energy tax adjustments at the border', *Journal of World Trade*, vol 38, no 3, pp395–423

Ho, M. S., Morgenstern, R. and Shih, J.-S. (2008) *Impact of Carbon Price Policies on US Industry*, RFF Discussion Paper 08-37, Washington, DC: Resources for the Future

Houser T., Bradley, R., Childs, B., Werksman, J. and Heilmayr, R. (2008) *Leveling the Carbon Playing Field: International Competition and US Climate Policy Design*, Washington, DC: Peterson Institute for International Economics

Hufbauer, G. C., Charnovitz, S. and Kim, J. (2009) *Global Warming and the World Trading System*, Washington, DC: Peterson Institute for International Economics

Ismer, R. and Neuhoff, K. (2007) 'Border tax adjustment: A feasible way to support stringent emission trading', *European Journal of Law and Economics*, vol 24, pp137–164

Kopp, R. and Pizer, W. (2007) *Assessing US Climate Policy Options*, Washington, DC: Resources for the Future

Manders T. and Veenendaal, P. (2008) 'Border tax adjustments and the EU-ETS: A quantitative assessment', Document 171, The Hague: CPB Netherlands Bureau for Economic Policy Analysis

Sheeran, K. A. (2006) 'Who should abate carbon emissions? A note', *Environmental and Resource Economics*, vol 35, pp89–98

Stern, N. (2009) *A Blueprint for a Safer Planet*, New York: Random House

Stiglitz, J. (2006) 'A new agenda for global warming', *Economists Voice*, July, www.bepress.com/ev

Unruh, G. C. and Carrillo-Hermosilla, J. (2006) 'Globalizing carbon lock-in', *Energy Policy*, vol 34, pp185–1197

Werksman, J. D. and Houser, T. G. (2008) *Competitiveness, Leakage and Comparability: Disciplining the Use of Trade Measures under a post-2012 Climate Agreement*, Washington, DC: World Resources Institute

Author Biographies

Frank Ackerman is the director of the Climate Economics Group at SEI's US Centre. He has written extensively about the economics of climate change and other environmental problems, and has directed policy reports for numerous national and international agencies. His most recent book is *Can We Afford the Future? The Economics of a Warming World* (Zed Books, 2009). He is a founding member of Economists for Equity and Environment, and a member scholar of the Centre for Progressive Reform. Dr Ackerman received his PhD in economics from Harvard University.

Melinda Bohannon currently works in the Department for International Development (DFID) as a senior policy adviser on international financial institutions. Prior to this she led the Stern Review Team at the Department for Energy and Climate Change, working closely with Lord Stern and the Grantham Research Institute on Climate Change and the Environment. She has served in a number of posts across UK government, including head of emerging market economies in the Foreign Office's Global Economy Group and economic adviser to DFID's Middle East and North Africa Department and Russia/Former Soviet Union Department. She holds a first class MSc in development studies from the London School of Economics (LSE) and a BA in Politics, Philosophy and Economics from Oxford University. She is a visiting fellow at the Grantham Institute.

Steffen Brunner studied International Affairs (BA) at St Gallen University and Environmental Policy (MSc) at the LSE. He is a PhD student at the PIK, conducting research on the political economy of incentive-based climate policy instruments, in particular emissions trading.

Cai Fang is Director of the Institute of Population and Labour Economics, Chinese Academy of Social Science (CASS). He has focused his research on poverty, regional development, demographic mobility, and labour and employment. He is especially concerned with issues related to development and reforms in China. Dr Cai received his PhD in economics from CASS, after which he took visiting scholarships at Stanford University and the Australian National University. In 2007 Dr Cai was elected as a Delegate and Member of the Standing Committee of the 11th National People's Congress.

Cao Jing is an assistant professor at the Department of Economics, School of Economics and Management, Tsinghua University. Her research interests include economics of the environment, energy economics, integrated modelling of economic and environment systems, climate change economics and modelling, productivity and economic growth.

Her recent research focused on computable general equilibrium modelling on carbon tax in China, integrated top-down and bottom-up modelling on induced technology change and economy-wide policy effects, total factor productivity and green productivity, energy intensity decomposition and carbon emission forecasts in China, burden-sharing rules of GHG mitigation allocations, and post-2012 climate architecture design from a developing country perspective.

Du Yang is Deputy Director of the Research Center for Human Resources and a member of the academic committee of the Institute of Population and Labour Economics, CASS. His research is focused on labour economics, economic growth, population economics and human capital theory. He received his PhD agricultural economics at Zhejiang University in 1999.

Ottmar Edenhofer is Professor of Economics of Climate Change (appointment together with Michael-Otto-Stiftung) at the Technical University Berlin and Co-Chair of Working Group III of the IPCC, which won the Nobel Peace Price in 2007. He is deputy-director and chief economist at PIK and is currently leading Research Domain III – Sustainable Solutions at PIK, which focuses on research on the economics of atmospheric stabilization.

Klas Eklund is Senior Economist of SEB, one of the largest commercial banks in the Nordic region, and adjunct Professor of Economics at the University of Lund. He is a member of the Group of Economic Policy Advisers, set up by the European Commission. Previously, Mr Eklund has held many different posts, including Deputy Under-Secretary of State, Swedish Ministry of Finance, and policy adviser to the prime minister, as well as chairman of several government committees. He has published a number of books and articles, including a best-selling Swedish economics textbook and a book on the economics of climate change.

Fan Gang is Director of National Economics Research Institute, China Reform Foundation, professor at the Graduate School of CASS, and member and Vice Secretary-General of the Chinese Economists 50 Forum. He serves as an adviser to the Chinese government and as a consultant to a number of international organizations. He is the author of over 100 academic papers and eight books on macroeconomics and the economics of transition. Dr Fan earned his PhD in economics at the Graduate School of CASS. He also holds the directorship of the China Reform Foundation.

Christian Flachsland is a PhD student at PIK. He has a degree in sociology, economics and philosophy from Potsdam University. His PhD thesis focuses on linking regional cap-and-trade systems in the context of international climate policy.

Han Guoyi is a research fellow at SEI's Stockholm centre. He specializes in nature–society synthesis in the areas of water resources, human dimensions of environmental change, natural disasters in developing countries, and applications of geographic information systems. His research interests include environmental impact assessment, regional environmental planning, risk analysis and hazard management, the dynamic nature of vulnerability and the resilience of human–environmental systems. More recently, he has been focused on mainstreaming natural disaster risk reduction, energy security and climate change policy in China, as well as the broad environmental and social implications of China's transition.

Karl Hallding heads SEI's China Cluster and has extensive experience from international cooperation with China on environment and sustainable development since the mid-1980s. He was the main author of UNDP's China Human Development Report 2002 – *Making Green Development a Choice* – and participated in the expert team behind the 2007 OECD Environmental Performance Review of China, where he was responsible for drafting the chapter 'Environmental–Social Interface'. His latest publications (2009) include *A Balancing Act: China's Role in Climate Change* where China's ambitious actions to reduce carbon emissions are placed in the larger context of energy security, development and international relations, and *China's Climate and Energy Security Dilemma: Shaping a New Path of Economic Growth*.

Charlie Heaps is the Director of SEI's US centre and a senior scientist in its climate and energy research programme. He is the designer and manager of SEI's Long-range Energy Alternatives Planning System (LEAP), a scenario-based modelling system for integrated energy planning and climate change mitigation assessment, as well as having developed a range of other software tools and websites for energy and environmental planning. Dr Heaps is also the founder and manager of COMMEND, a major international initiative designed to foster a community among energy analysts working on energy for sustainable development. In 2009, Dr Heaps was the lead author for the major study *Europe's Share of the Climate Challenge: Domestic Actions and International Obligations to Protect the Planet*, which analyses how Europe can achieve GHG emissions reductions of 40 per cent by 2020 and close to 90 per cent by 2050 relative to 1990 levels through a combination of radical improvements in energy efficiency, the accelerated retirement of fossil fuels and a dramatic shift towards various types of renewable energy forms.

Matthias Kalkuhl studied applied systems science at the University of Osnabrueck (Germany) and mathematics at the University of Granada (Spain). Since 2006, he has been PhD student at the PIK working on formal climate policy instruments analysis.

Sivan Kartha is a senior scientist at SEI's US centre, whose research and publications for the past 15 years have focused on technological options and policy strategies for addressing climate change. He has concentrated most recently on equity and efficiency in the design of an international climate regime. His most recent work has involved the elaboration of a GDR approach to burden-sharing in the global climate regime – an approach that places the urgency of the climate crisis in the context of the equally dire development crisis afflicting the world's poor majority.

Brigitte Knopf is a senior researcher at PIK. Her scientific work focuses on low-concentration pathways of CO_2 emissions for mitigating climate change. Her main interest is the transformation towards a low-carbon economy and the economic consequences and technological requirements for mitigation. She scientifically coordinated a model comparison within the ADAM project. Brigitte Knopf obtained her PhD in physics in the field of climate modelling and in the assessment of uncertainties. She is involved in the ongoing project on climate change and global poverty.

Elmar Kriegler is a senior scientist at Research Domain III 'Sustainable Solutions' at PIK. His research focuses on the assessment of climate change mitigation policies under uncertainty about technological and climate systems properties. He also works on the coupling of climate, energy and economy models. Kriegler studied physics at the University of Freiburg, obtained his PhD in Physics at the University of Potsdam, and was a Marie Curie fellow at Carnegie Mellon University in Pittsburgh.

Michael Lazarus has over 20 years of professional experience in energy and climate policy analysis. His current research focuses on domestic and international carbon markets, GHG mitigation finance globally, and on state and local energy and climate change initiatives within the US. He has worked throughout North America, Africa, Asia, Latin America and Europe with support from government agencies, development banks, foundations, utilities and non-profit groups. From 2002 to 2007, Mr Lazarus was a member of the Methodology Panel of the CDM and the project-based emission reduction trading programme of the Kyoto Protocol.

Li Lailai is Deputy Director of SEI and Centre Director of SEI Asia. Her research has been focused on the poverty cycle driven and reinforced by environmental degradation. She has extensive experience studying and generating solutions to poverty–environment loops, focusing on policy and institutional change and empowerment of the poor. She joined SEI in 2006 and has shifted research onto climate change, focusing on climate change mitigation issues in China.

Gunnar Luderer is a senior researcher at PIK working on the analysis of cost-efficient mitigation strategies in the context of integrated assessment modelling. He acts as a joint head of the energy system modelling group at PIK. He studied physics, economics and atmospheric sciences, and obtained his PhD in the field of atmospheric chemistry and physics. Before joining PIK in October 2007, he worked in the Climate Policy Unit of the German Federal Environment Agency.

Marie Olsson is a research associate with SEI's China Cluster. She has a background in political and Chinese studies, and holds an MSc in Asian politics (2010) from the School of Oriental and African Studies, University of London. She currently devotes much attention to the broader implications of Chinese energy and climate change policy in the context of rapid globalization. In 2009, Marie co-authored *A Balancing Act: China's Role on Climate Change*, commissioned by Sweden's Prime Minister's Office.

Robert Pietzcker joined PIK's Research Domain 'Sustainable Solutions' as a doctoral student after working briefly as a consultant with McKinsey & Company. In his doctoral thesis he will analyse the decarbonization of the transport sector and the interactions between transport and electricity sector in a hybrid energy–economy model. Previously, Pietzcker studied physics at the University of Freiburg and at McGill University in Montreal before graduating from the University of Jena.

Clifford Polycarp has worked on international environmental policy for the last eight years. He is an associate with SEI's Climate and Energy Program and his current research focuses on international climate change and energy policy as it relates to developing countries. Before joining SEI, Clifford spent six years in India working on international environment, climate change and energy policies, and carbon market development. He is a graduate of the Fletcher School of Law and Diplomacy, Tufts University, and St Xavier's College, Mumbai University.

Elizabeth A. Stanton is a senior economist with SEI-US and a Research Fellow at the Global Development and Environment Institute (GDAE) of Tufts University. An economist with a special interest in environmental policy and in economic inequality, she has focused much of her work on the interplay between climate protection and development. At SEI-US, Stanton has led studies commissioned by the United Nations

Development Programme, Friends of the Earth-UK and Environmental Defense, and authored dozens of reports on topics including the cost of inaction on climate change, the economics of emissions-reduction targets, and the balance of science, policy and equity in global climate protection. Currently, she is leading SEI-US work on the Consumption-Based Emissions Inventory (CBEI) model, several studies of the cost of climate inaction and state-by-state analyses of the impacts of climate legislation. She also serves on the Climate Taskforce of Economics for Equity and the Environment (the E3 Network). She earned her PhD in economics at the University of Massachusetts-Amherst.

Jan Steckel studied economics and engineering at the University of Flensburg. Since 2008 he has worked on his doctoral thesis at PIK, focusing on the question of how to integrate developing countries into global climate protection efforts. In his work he is particularly interested in China, where he spent two months in 2009 at Tsinghua University, Beijing.

Nicholas Stern, Lord Stern of Brentford, Kt, FBA is IG Patel Professor of Economics and Government at the LSE, where he is also head of the India Observatory within LSE's Asia Research Centre, and Chairman of the Grantham Research Institute on Climate Change and the Environment. Previously, having held academic posts at the Universities of Oxford and Warwick and the LSE, he was Chief Economist for the EBRD and subsequently Chief Economist and Senior Vice-President at the World Bank. In 2005, he was appointed by the UK government to conduct the influential Stern Review, which analysed the economic costs of climate change.

Su Ming is a PhD candidate at the China Center for Economic Research in Peking University. He is engaged in research on climate change and issues such as burden-sharing approaches and efficient mitigation measures. He received his bachelor's degree in energy engineering at Xi'an Jiaotong University in 2003 and a master's in thermo-physics at Tsinghua University in 2006. He has taken part in research projects in the Development and Research Center on the pathways of industry development and the reform of the Chinese energy price mechanism and energy management system. His published papers in international scientific journals include in *Journal of Membrane Science and Applied Thermo Engineering, China Securities*, and *China Development Observation* as well as Chinese scientific journals.

Helen Thai is a research associate with SEI's China Cluster. She has previously held several positions in Australia's federal government, including as a political and security analyst in the Defence Department, and as an Assistant Director in Water Reform Division, in the Department for Environment, Water, Heritage and the Arts. Helen has a master's in strategic affairs from the Australian National University, and a BA from Monash University. Her research interests focus on China's political development and foreign relations, particularly its climate change policies and climate security issues.

Wang Meiyan is Associate Professor, Institute of Population and Labor Economics, CASS. Her research is focused on demographic economics and labour economics.

Xing Weinuo is a research fellow at the Energy Research Institute, the National Development and Reform Commission of China.

Xu Shanda is Vice Minister of China's State Administration of Taxation. Xu graduated from the Automatic Control Department of the Tsinghua University in March 1970. He received an MA in agricultural economic administration from the China Academy of Agricultural Sciences in 1984, and in 1990 was awarded an MA in public finance from the University of Bath, UK. He also holds the title of Senior Economist and is qualified as a certified public accountant. Prior to his current position, Xu held a range of high-level roles with SAT and the Ministry of Finance.

Yang Hongwei is Director of the CDM Project Management Center at the Energy Research Institute of National Development and Reform Commission. He has played a key role in supporting the government of China on compliance activities under the UNFCCC and the Kyoto Protocol. He is a lead author for the IPCC's *2006 Guidelines on National Greenhouse Gas Inventories* and a contributing author to the IPCC's *Fourth Assessment Report on Climate Change*. Since 2002, he has been a member of the Chinese Delegation to the UNFCCC/Kyoto Protocol negotiations, and since 2003 he has been a lead reviewer for the technical review of national GHG inventories from Annex 1 Parties to the UNFCCC. He is a council member of China Energy Research Society.

Index